AF477325

FLEAS

PROCEEDINGS OF THE INTERNATIONAL CONFERENCE ON FLEAS
ASHTON WOLD / PETERBOROUGH / UK / 21-25 JUNE 1977

FLEAS

Edited by

R. TRAUB & H. STARCKE

Department of Microbiology,
University of Maryland School of Medicine, Baltimore, USA

A.A. BALKEMA / ROTTERDAM / 1980

ISBN 90 6191 018 8

Distributed in USA & Canada by MBS, 99 Main Street, Salem, NH 03079

Printed in the Netherlands

CONTENTS

PHYSIOLOGY AND MORPHOLOGY

PREFACE

The basic foundation for the study of Siphonaptera was laid by Nathaniel Charles Rothschild, and the delegates and participants of the 1977 International Conference on Fleas are pleased and proud to dedicate this volume to the memory of N.C.Rothschild and his wife, Rozsika.

The Conference was held at Ashton, 21-25 June 1977, at the childhood home of Dr Miriam Rothschild, now the property of her son, Dr Charles Lane. Approximately 80 scientists and their families participated and enjoyed the splendid facilities kindly provided by Dr Rothschild and Dr Lane. In addition to the papers presented, the titles of which are shown in the table of contents, informal discussions were held on each of the days of the Conference. Excursions and tours were organized for the benefit of the guests and for the delegates when formal sessions were not in progress. Transportation was provided with the co-operation of the Quest 4 Company, Ashton.

The organising committee consisted of Dr T.Clay, Mr F.G.A.M.Smit, Dr M.Rothschild, Mrs G.Brinck-Lindroth and Prof R.Traub. Of these, the last-named three served on the editorial committee. Mr Smit also rendered exceptional assistance through the presentation of pertinent notices and essays in the pamphlet, Flea News, which he prepares and distributes through the British Museum (Natural History). The lists of N.C.Rothschild's articles on fleas, included in this volume, were obtained from Flea News. Mr Smit also kindly crafted a beautiful glass vase for presentation to Dr Rothschild on behalf of the participants in token acknowledgement of their gratitude for having made the Conference possible in the first place, and for ensuring its success both as a scientific endeavour and as a personal experience. He also prepared a Certificate of Appreciation for Miss J.Beer, the organising secretary, in honour of her outstanding services to the Conference. A rising vote of thanks was also extended to Dr Rothschild's staff, family and friends, all of whom worked prodigiously on behalf of the Conference.

Although taxonomy was not a formal part of the agenda of the Conference, it has been deemed necessary to include in the Proceedings, an article on the reclassification of the genus *Pygiopsylla.* This was done so that the names of the new taxa could be used in subsequent papers in the volume, and so that other workers would have access to the diagnoses, illustrations and nomenclatorial changes for use in papers in preparation.

We are grateful to A.A.Balkema Publishers for publishing these proceedings. Editorial costs were borne by the Grant (No. AI-04242) of the National Institutes of Health, Bethesda, Maryland, to the Department of Microbiology, University of Maryland School of Medicine, Baltimore, Maryland, thereby rendering possible the valuable assistance of Miss Helle Starcke of our department, as co-editor.

Those who were so fortunate as to attend the Conference regard this volume as a tribute to the entire family of Nathaniel Charles Rothschild for their many contributions to Science and the good of humanity, and as a memento of an unforgettably pleasing occasion.

Robert Traub

MIRIAM ROTHSCHILD
Peterborough, UK

NATHANIEL CHARLES ROTHSCHILD, 1877-1923

Charles Rothschild's first publication, *The Lepidoptera of Harrow,* received a good review in the Entomological Record. No one guessed the author was a schoolboy. Most youthful writers have parents at home responsible for their early interests who, good naturedly, monitor their contributions, but N.C.R. and his brother Walter were strange mutations, for no one in the Rothschild family had ever previously been remotely interested in entomology. Finance, politics, books, *objets d'art,* breeding animals, growing orchids, philately, dairy farming, building railroads, medicine, the creation of Jewish agricultural settlements in Palestine, and social reform were numbered among their activities and interests. Natural history, however, was lacking. Nonetheless, his mother and father must have taken a keen interest in his curious hobby, for a few letters to his parents which, accidentally, escaped N.C.R.'s testamentary instructions to burn all his correspondence, are sprinkled with the Latin names of plants and butterflies which he captured or saw on his travels.[1] He also shared with his parents certain well-known family traits – wide horizons, zestful curiosity and boundless energy. Essentially he was an all-round naturalist, a great lover of animals and plants, with an encyclopaedic knowledge of specialized habitats ranging from the virgin forests of the Austro-Hungarian Empire to the Egyptian desert, and the shingle beaches and mosses of the United Kingdom.

His greatest contribution to the scientific scene was his promotion of conservation at the turn of the century. It has been said with an element of truth that he *invented* conservation, since, years ahead of his day, he realised the importance of protecting and preserving the habitat and special biotopes rather than the individual rare species threatened with extinction. Most of his ideas have been so generally and widely accepted that their origins are now virtually forgotten.[2] One of his best contributions in this field was the report he prepared for the Board of Agriculture[3] in 1915 surveying the areas worthy of conservation in the British Isles.

Had Charles Rothschild been born 50 years later, he might well have adopted Natural Science as a career, but at the time he felt it impossible to break with tradition, and for the sake of his father, dutifully joined the family banking firm of N.M.Rothschild & Sons. In letters to his friend Hugh Birrell he complained bitterly of the vitiated atmosphere of a London office, and dreamed of bug hunting in the Burmese jungle.[4] But the fact was – after a difficult beginning – he rather enjoyed the city and was full of relevant, creative ideas. What thoroughly irked him was the over-cautious, conservative attitude of his uncles who, for instance, prevented him from financing a new invention, the gramophone disc, for which he foresaw a great future. He was similarly thwarted in his efforts to include copper among the metals processed and handled by the Rothschilds, and his uncles again turned a deaf ear when, in 1903, he proposed setting up a branch of the business in Japan. Time proved that his judgement was excellent. Since he now worked from 9 a.m. to 5 p.m. in the city, entomology was of necessity relegated to the status of a delectable hobby. Although at no time had his zoological interests been as wide as those of his brother, he had, apart from the Siphonaptera, made fine collections of both living and pressed Iris (presented to the Royal Botanic Gardens at Kew after his death), Hungarian, Swiss and British Lepidoptera, Aculeate Hymenoptera, Longicorn beetles, Nycteribiid and Hippoboscid flies, parasitic bugs and British birds' eggs. The curating of this material was impeccable and all was eventually engulfed in the British Museum. Charles Rothschild's special fondness for the Siphonaptera[6] originated with a specimen of a Helmet Flea accidentally included in a small collection of slides purchased from W.Warren. The curious specialization of the head especially caught his fancy. Karl Jordan, 15 years his senior, was already working at the Tring Museum when he was asked to examine this particular slide. He freely admitted that his own enthusiasm was fired by Charles rather than by the Helmet Flea. 'He was so nice,' said K.J., his face wreathed in smiles at the distant memory of long vacations in the 1890's. Thus began the felicitous collaboration and life-long friendship which produced 37 joint publications and the illustrations for the descriptions of the 500 new species[7] and the collection which still forms the basis for the systematic study of the Order. Charles Rothschild collected the plague flea *par excellence,* appropriately named *Xenopsylla cheopis,* in Egypt, on a joint expedition with A.F.R.Wollaston,[8] and described it in 1903. Karl Jordan, as always, provided the illustrations.

At the age of 40, N.C.R. contracted a mortal illness and died six years later. How did he cover so much ground in the time at his disposal – working on his collections and his papers[9] not to mention his voluminous correspondence, only in the evenings, at week-ends and during his six weeks annual vacation? Looking back some 60 years, the writer recalls that on Saturday afternoons N.C.R. took his children out collecting, and after nursery tea – during which he kept the party, including the nursemaid and governess, in fits of spluttering laughter[10] – bowled inswingers to them in the cricket net. He never seemed in a hurry nor particularly busy. One can only conclude that, just as some men are born sprinters, so others can work much faster[11] than the rest of us.

In his will, N.C.R. left his collection of Hungarian butterflies to his wife, Rozsika, in memory of their first romantic meeting in the Carpathian Mountains and the many happy holidays they spent together in Hungary. Although she had no feeling for Natural History and never attempted so much as the identification of a Cabbage White, and consequently had no means of assessing Charles Rothschild's scientific ability, she loved and revered him, and deeply admired the combination of idealism, integrity, industry and delight which she felt characterised his approach to natural science. When he died she was determined that at least the flea collection should not be allowed to stagnate or regress. She also pressed Karl Jordan to begin work forthwith on the monograph based on the collection, which had been planned as a joint N.C.R. and K.J. publication. In this she was only partly successful. Karl Jordan certainly maintained and added to the collection and published a number of first-class papers based, as he said, on the unwritten monograph. But the years went by and there was no sign of the magnum opus. In the past it has been rather fashionable to marvel at Walter Rothschild's youthful perspicacity, when in his early 20's he selected the then unknown Ernst Hartert and Karl Jordan as his ornithological and entomological curators, and thereafter by inference, to minimise his personal contribution to their joint publications and new ideas emanating from his museum. What has been generally overlooked is the fact that the large Tring projects, the classical monograph of the Sphingidae, and the Avifauna of Lysan, for example, only materialised with Walter Rothschild as an active participant. Similarly, Karl Jordan suddenly discovered that he was unable to tackle the monograph without the collaboration of N.C.R. Belatedly, a rather pedestrian but worthy substitute was produced by Harry Hopkins and Miriam Rothschild in the form of an illustrated catalogue of the collection.[12] Although the first volume was published in 1953, 13 years after her death, its production was due mainly to Rozsika Rothschild's melancholy but ardent desire to see one facet of N.C.R.'s life's work fulfilled. Similarly, the nature reserve at Ashton and the village[13] which he rebuilt in 1900, and which she maintained virtually unchanged for 17 years, is a token of her passionate and unflagging devotion to his memory and her deep admiration for his creative gifts, which she was convinced never received the recognition they deserved.

Scientists in general and entomologists in particular achieve distinction, or even a tinkle of immortality, by virtue of their publications, and to a lesser degree by the usefulness of their collections. No doubt both Charles Rothschild's systematics[14] and his fleas will stand the test of time, but his major contribution to the contemporary scene was something entirely different and as elusive as the end of the rainbow. What is the opposite of a wet blanket? What is the converse of a thrower of cold water? What is the secret of those rare and perceptive individuals, those *animateurs,* those practical supporters and promoters of talent, who raise your spirits, restore your confidence, admire your gifts – whether these concern the drawings for Dyke's *The Genus Iris* or Frohawk's illustrations for *British Butterflies* or Oliver Pyke's pioneer bird photography, or merely old Holland's[15] latest capture – and somehow by sheer expectation raise both your performance and your self-esteem? It is perhaps the most enviable and rewarding of all gifts, but one which is the despair of the biographer.

NOTES

1. Before he was 26 he went round the world twice (although a desperately bad sailor who almost always finished a long sea voyage vomiting blood), and enjoyed nothing better than butterfly hunting in wild country in Ceylon, Egypt, Africa and Japan. A pioneer in photography, he filled album after album with pictures of singularly empty desert landscapes. Delighted with the varied creeds and customs which he saw on his expeditions, he despaired of the missionary.
2. It is unlikely that any of the 300 workers at the Natural History Museum in South Kensington know that they owe the modest green lawns and flowering trees and bushes around their building to his stubborn determination to preserve the garden from development.
3. 'Provisional schedule of areas in the U.K. considered worthy of preservation'. Prepared for the Board of Agriculture and submitted to the Board via the Society for the Promotion of Nature Reserves, 1915, printed privately.
4. N.C.R. detested the rigid Victorian formalities which he had experienced in childhood, and he not infrequently voiced, even in his own comparatively informal household, a wistful hankering for life in a mud hut. His wife, whose brand of humour was, to him, a never ending source of delight, would retort with her incorrigible Hungarian accent: 'And where will the secretary live?'
5. Eventually catalogued by Theodor (1967), his collection of Nycteribiids contained 56 undescribed species when it was presented to the British Museum.
6. This attracted a certain amount of publicity, since it was considered hilariously funny by the average Edwardian that a noted banker

should collect fleas. Driving home from the city, N.C.R. noticed a Daily Express poster which announced the attempted assassination of Mr Kensit and that '10,000 fleas are Mr Rothschild's hobby'. Framed, this poster was on view at the Conference. As late as 1939 Ripley's feature, 'Believe It or Not', carried a highly inaccurate statement to the effect that Mr Rothschild had paid £10,000 for a flea from a grizzly bear.

7. N.C.R.'s bibliography included some 150 scientific papers. Those dealing with the Siphonaptera will be found on pages 5-9.
8. It was frequently asserted that N.C.R. was the only person with whom A.F.R.Wollaston could camp without engendering a row, for the simple reason that it takes two to make a quarrel. However, on this occasion, true to type, the irascible 'Bear' walked out in a fit of temper. 'I just waited,' said N.C.R. 'In the desert it is unreasonable to go too far. He came back.'
9. N.C.R. had one assistant, F.J.Cox, who mounted fleas and labelled slides and a secretary who typed his correspondence. Owing to the destruction of his papers, there are no letters to consult, but due to an oversight one card index remained, and this contained the addresses – with notes – of some 100 correspondents interested only in the genus Iris! His widow, Rozsika Rothschild, recalled with amused affection that on returning at 6 p.m. from the city, the only entertainment he afforded his bride of a few months was to 'watch while I set my butterflies'!
10. N.C.R. preferred jokes to wit and apparently possessed a limitless store of funny stories which could be suitably tailored for any audience, and to fit any situation. Children like puns and riddles, and they enjoy the same ones repeated over and over again . . . N.C.R. always obliged, generally with something faintly tinged with natural history and, once again, 'mutton was cheap and venison was dear' . . . and 'what of those perennial Mustelids – weasely distinguished and stoatly different?'
11. K.J. could not throw any light on this subject. He merely remarked it took *him* 16 hours to describe a new flea.
12. The Deed of Gift to the British Museum states: 'The Trustees shall so soon as the said collection comes into their hands or is placed at their disposal by Parliament will allow cause to be made and publish a catalogue giving the names and full number of specimens of each species of parasitic insect contained in the said collection'. However, rather characteristically, 30 years later no attempt had been made to fulfill this obligation, hence the compilation of the catalogue was undertaken jointly by these authors. Karl Jordan's attitude was blandly negative, and permission to publish his phylogenetic tree – which they insisted he at last project onto paper – was obtained under duress. Owing to the fact that K.J. was well over 80 and exceedingly deaf, all these importunings and entreaties had to be yelled at the victim, who refused a hearing aid on the grounds it only increased the noises in his own head. Protesting that our knowledge was totally inadequate for the purpose, and that what we required was more collecting and less phylogeny, K.J. ultimately capitulated.
13. When N.C.R. rebuilt the village in 1900 he had several objectives in view: to provide good accommodation for his employees – in fact it was the first village in England to have a bathroom in every house – to promote local crafts such as stone masonry and thatching, and encourage gardening in the village. To further this latter enterprise he organised an annual flower show at Ashton (with champagne but no entrance fee) and awarded prizes for the best gardens. He was immensely proud of the fact that his shepherd, over the years, won 2,000 prizes for his flowers and vegetables. Every visitor to Ashton was taken to see Hill's garden.
14. When Fabian Hirst demonstrated the relationship between plague and the geographical distribution of rat fleas, he wrote: 'The discovery is but further testimony to the essential unity of science and its bearings on the welfare of the human race, for it is the natural outcome of the purely zoological researches of Rothschild and Jordan on the systematics of the Siphonaptera'.
15. W.Holland was an ancient retired cobbler who collected insects as a precarious source of pocket money. He suffered from persecution mania and N.C.R. kept him happy pottering in solitude about the Northamptonshire woodlands. At night reflections from his antediluvian collecting-lamp used to creep round the walls of the nursery – the terrified children convinced they could decipher 'Mene, Mene, Tekel, . . . ', in the flickering shadows.

SCIENTIFIC PUBLICATIONS OF THE DESCENDANTS OF BARON CARL VON ROTHSCHILD* AND LOUISE VON ROTHSCHILD

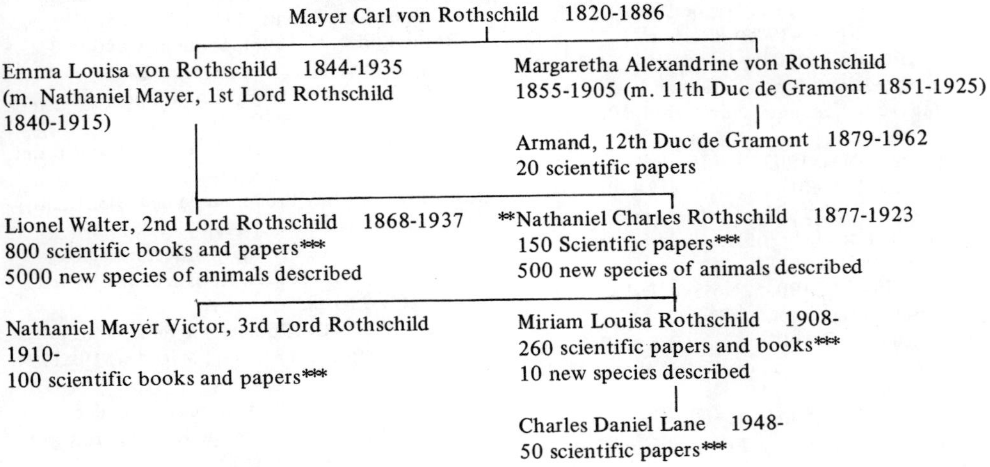

Approximate total: 1380 scientific books and papers

* Baron Henri de Rothschild, grandson of Baron Carl and first cousin of Walter Rothschild, published 127 papers on matters of purely medical interest and these are not included.

** Spouses are omitted from the second generation onwards.

*** Includes joint authorships.

LIST OF THE PUBLICATIONS ON SIPHONAPTERA BY N.C.ROTHSCHILD

Rothschild, N.C. 1895, Casual notes on fleas. Novit. zool. 2(2):66.

Rothschild, N.C. 1897a, A new British flea *(Typhlopsylla pentacanthus).* Ent. Rec. J. Var. 9(3):55.

Rothschild, N.C. 1897b, A new British flea *(Typhlopsylla dasycnema,* sp.nov.*).* Ent. Rec. J. Var. 9(7):159.

Rothschild, N.C. 1898a, A new British flea: *Typhlopsylla spectabilis,* sp.nov. Ent. Rec. J. Var. 10(10):250.

Rothschild, N.C. 1898b, Contributions to the knowledge of the Siphonaptera. Novit. zool. 5(4):533-544.

Rothschild, N.C. 1899, Irish fleas. Irish Nat. 8(12):266.

Rothschild, N.C. 1900a, A new British flea. Ent. Rec. J. Var. 12(1):19-20.

Rothschild, N.C. 1900b, Some new exotic fleas. Ent. Rec. J. Var. 12(2):36-38.

Rothschild, N.C. 1900c, The giant flea: *Hystrichopsylla talpae.* Ent. Rec. J. Var. 12(10): 257-258.

Rothschild, N.C. 1900d, Notes on *Pulex avium* Taschb. Novit. zool. 7(3):539-543.

Rothschild, N.C. 1901a, Notes on *Pulex canis* Curtis, and *Pulex felis* Bouche. Ent. Rec. J. Var. 13(4):126.

Rothschild, N.C. 1901b, A new British flea. Ent. Rec. J. Var. 13(10):284.

Rothschild, N.C. 1901c, A new British flea. Ent. Rec. J. Var. 13(12):362.

Rothschild, N.C. 1902a, Some new Nearctic fleas. Ent. Rec. J. Var. 14(3):62-63.

Rothschild, N.C. 1902b, New British fleas. Ent. mon. Mag. 38(461):225.

Rothschild, N.C. 1903a, A new British flea: *Ceratophyllus londiniensis.* Ent. Rec. J. Var. 15(3):64-65.

Rothschild, N.C. 1903b, New species of Siphonaptera from Egypt and the Sudan. Ent. mon. Mag. 39(467):83-87.

Rothschild, N.C. 1903c, Types of Siphonaptera in the Daleian collection. Ent. mon. Mag. 39(469):144-146.

Rothschild, N.C. 1903d, A new British flea: *Typhloceras poppei,* Wagner. Ent. Rec. J. Var. 15(8):196-197.

Rothschild, N.C. 1903e, A collection of fleas received from Baron Carlo von Erlanger and Mr Oscar Neumann. Novit. zool. 10(2):312-315.

Rothschild, N.C. 1903f, *Ceratophyllus fringillae,* Walker. Ent. Rec. J. Var. 15(12):308-309.

Rothschild, N.C. 1903g, Further contributions to the knowledge of the Siphonaptera. Novit. zool. 10(3):317-325.

Rothschild, N.C. 1903h, Note on *Pulex pallidus* Tasch. Novit. zool. 10(3):542.

Rothschild, N.C. 1903i, A new British flea, *Ceratophyllus dalei,* sp.nov. Entomologist 36(487): 297-298.

Rothschild, N.C. 1904a, New species of Siphonaptera from Egypt. Entomologist 37(488): 1-4.

Rothschild, N.C. 1904b, Description of a new species of Siphonaptera from S.America. Revista chil. Hist. nat. 8(3):147-148.

Rothschild, N.C. 1904c, Further contributions to the knowledge of the Siphonaptera. Novit. zool. 11(2):602-653.

Rothschild, N.C. 1905a, On North American *Ceratophyllus,* a genus of Siphonaptera. Novit. zool. 12(1):153-174.

Rothschild, N.C. 1905b, Some further notes on *Pulex canis* Curtis and *Pulex felis* Bouche. Novit. zool. 12(1):192-193.

Rothschild, N.C. 1905c, Notes on *Stephanocircus dasyuri,* Skuse, and *Stephanocircus simsoni,* sp.nov. Ent. mon. Mag. 41(490):60-62.

Rothschild, N.C. 1905d, Occurrence of *Pulex cheopis,* Rothsch., at Plymouth. Ent. mon. Mag. 41(493):139.

Rothschild, N.C. 1905e, Some new Siphonaptera. Novit. zool. 12(2):479-491.

Rothschild, N.C. 1905f, A new British flea: *Ceratophyllus farreni,* spec.nov. Ent. mon. Mag. 41(498):255-256.

Rothschild, N.C. 1906a, Notes on bat fleas. Novit. zool. 13(1):186-188.

Rothschild, N.C. 1906b, A new British flea: *Ceratophyllus insularis,* spec. nov. Ent. mon. Mag. 42(502):59-60.

Rothschild, N.C. 1906c, A new Egyptian flea. Entomologist 39(515):75.

Rothschild, N.C. 1906d, Note on the species of fleas found upon rats, *Mus rattus* and *Mus decumanus,* in different parts of the world, and on some variations in the proportion of each species in different localities. J. Hyg., Cambridge 6(4):483-485.

Rothschild, N.C. 1906e, Three new Canadian fleas. Canad. Ent. 38(10):321-325.

Rothschild, N.C. 1906f, Some new exotic fleas. Ent. mon. Mag. 42(509):221-224.

Rothschild, N.C. 1907a, A new British flea. Ent. mon. Mag. 43(512):11.

Rothschild, N.C. 1907b, A new British flea. Ent. mon. Mag. 43(513):41-42.

Rothschild, N.C. 1907c, Some new Siphonaptera. Novit. zool. 14(1):329-333.

Rothschild, N.C. 1907d, Instrucciones para la colecta de Pulicidos. Revista chil. Hist. nat. 11(2):99-100.

Rothschild, N.C. 1907e, Some new African Siphonaptera. Ent. mon. Mag. 43(519):175-176.

Rothschild, N.C. 1908a, Notes on a collection of Siphonaptera from the Ruwenzori, Uganda. Ent. mon. Mag. 44(527):76-79.

Rothschild, N.C. 1908b, Siphonaptera. In Y.Sjöstedt, Wissenschaftliche Ergebnisse der Schwedischen Zoologischen Expedition nach dem Kilimandjaro, dem Meru und den Umgebenden Massaisteppen Deutsch-Ostafrikas 1905-1906. Königl. Schwed. Akad. Wissenshaft. 2(11):1-5.

Rothschild, N.C. 1908c, New Siphonaptera. Proc. zool. Soc. Lond. (3):617-626.

Rothschild, N.C. 1908d, Siphonaptera collected by Mr M.P.Anderson in Japan in 1904. Proc. zool. Soc. Lond. (3):627-629.

Rothschild, N.C. 1908e, A new British flea. Ent. mon. Mag. 44(533):231-233.

Rothschild, N.C. 1908f, A new species of bat-flea from Great Britain. Entomologist 41 (547):281-282.

Rothschild, N.C. 1909a, Two new species of *Stephanocircus* from South America. Ent. mon. Mag. 45(536):8-10.

Rothschild, N.C. 1909b, A new flea from Chile. Revista chil. Hist. nat. 13(1):104-106.

Rothschild, N.C. 1909c, Notes on the five-combed bat-fleas forming the genus *Nycteridopsylla,* Oudemans. Entomologist 42(549): 25-28.

Rothschild, N.C. 1909d, A new species of *Stephanocircus* from Chile. Revista chil. Hist. nat. 13(2):181-183.

Rothschild, N.C. 1909e, Some new Siphonaptera. Novit. zool. 16(1):53-56.

Rothschild, N.C. 1909f, Notes on fleas in the K.K.Hofmuseum in Vienna. Novit. zool. 16(1):57-60.

Rothschild, N.C. 1909g, On some American, Australian, and Palearctic Siphonaptera. Novit. zool. 16(1):61-68.

Rothschild, N.C. 1909h, Synonymical note on *Xenopsylla pachyuromyidis* Glink. Novit. zool. 16(1):132.

Rothschild, N.C. 1909i, On *Ctenopsyllus spectabilis* and some closely allied species. Ent. mon. Mag. 45(543):184-186.

Rothschild, N.C. 1909j, Some additional notes on fleas dealt with in previous papers. Novit. zool. 16(2):332.

Rothschild, N.C. 1910a, Synonymical note on *Pulex tripolitanus,* Fulmek. Ent. mon. Mag. 46(549):30.

Rothschild, N.C. 1910b, A new flea from California. Ent. mon. Mag. 46(551):88-89.

Rothschild, N.C. 1910c, *Chiastopsylla,* a new genus of Siphonaptera. Entomologist 43 (563):105.

Rothschild, N.C. 1910d, A synopsis of the fleas found on *Mus norwegicus [=] decumanus, Mus rattus [=] alexandrinus* and *Mus musculus.* Bull. ent. Res. 1(2):89-98.

Rothschild, N.C. 1910e, Two new European Siphonaptera. Ent. mon. Mag. 46(556):207-208.

Rothschild, N.C. 1910f, On some European Siphonaptera. Ent. mon. Mag. 46(558):253-255.

Rothschild, N.C. 1910g, New Chilean Siphonaptera. Revista chil. Hist. nat. 14(1-3):25-28.

Rothschild, N.C. 1911a, Liste des Siphonaptera du Muséum d'Histoire Naturelle de Paris, accompagnée de description des espèces nouvelles. Ann. Sci. nat. (Zool.) (9)12(4-6):203-216.

Rothschild, N.C. 1911b, [Exhibition of some fleas interesting in connection with the spreading of plague.] Proc. zool. Soc. Lond. (1):5.

Rothschild, N.C. 1911c, [Exhibit of two species of flea, *Ctenocephalus canis* and *C.felis.*] Trans. ent. Soc. Lond. (1910)(5):lxvi.

Rothschild, N.C. 1911d, Report on a small collection of fleas from India and China. Rec. Indian Mus. 6(1):43-44.

Rothschild, N.C. 1911e, *Xenopsylla cheopis,* Rothsch., in London. Ent. mon. Mag. 47 (562):68.

Rothschild, N.C. 1911f, A further note on *Xenopsylla cheopis,* Rothsch. Ent. mon. Mag. 47(564):113.

Rothschild, N.C. 1911g, On the bat-fleas described by Kolenati. Novit. zool. 18(1):48-56.

Rothschild, N.C. 1911h, Some new genera and species of Siphonaptera. Novit. zool. 18(1): 117-122.

Rothschild, N.C. 1911i, *Ceratophyllus silantiewi,* Wagner; a 'plague-flea'. Ent. mon. Mag. 47 (565):141.

Rothschild, N.C. 1911j, Notes on the occurrence of *Xenopsylla scopulifer,* Roths., in German East Africa. Ent. mon. Mag. 47(569):234-235.

Rothschild, N.C. 1911k, On a new genus and species of Siphonaptera from Nyasaland. Bull. ent. Res. 2(3):269-272.

Rothschild, N.C. 1912a, A new British flea. Ent. mon. Mag. 48(574):67.

Rothschild, N.C. 1912b, A note on *Ceratophyllus vagabundus,* Boheman. Ent. mon. Mag. 48(574):67.

Rothschild, N.C. 1912c, Description of three new species of Siphonaptera obtained by Capt H.E.M.Douglas, VC, DSO, RAMC, on the Clark expedition in north China. In R.S. Clark & A.de C.Sowerby, Through Shên-kan. The account of the Clark expedition in north China, 1908-9. App. IV:194-203. London and Leipsic.

Rothschild, N.C. 1913a, The host of *Ornithopsylla laetitiae,* Rothsch. Ent. mon. Mag. 49(587):90-91.

Rothschild, N.C. 1913b, Description of the male of *Ceratophyllus borealis,* Roths. (1906). Ent. mon. Mag. 49(591):182.

Rothschild, N.C. 1913c, A new *Listropsylla* and the ♂ of *Ctenophthalmus calceatus,* Waterst. (1912), both from South Africa. Ent. mon. Mag. 49(592):207-208.

Rothschild, N.C. 1913d, A new Palaearctic species of *Xenopsylla.* Ent. Rec. J. Var. 25(10):241.

Rothschild, N.C. 1913e, Two new Palaearctic species of *Rhadinopsylla,* a genus of Siphonaptera. Entomologist 46(606):297-299.

Rothschild, N.C. 1913f, Five new Siphonaptera from Asiatic Russia, collected by W.Rückbeil. Ann. Mag. nat. Hist. (8)12(72):538-544.

Rothschild, N.C. 1914a, Siphonaptera and Clinocoridae. [H.Sauter's Formosa-Ausbeute]. Suppl. ent. (3):117-118.

Rothschild, N.C. 1914b, On three species of *Xenopsylla* occurring on rats in India. Bull. ent. Res. 5(1):83-85.

Rothschild, N.C. 1914c, New Siphonaptera from Peru. Novit. zool. 21(2):239-251.

Rothschild, N.C. 1914d, A further note on Kolenati's bat-fleas. Novit. zool. 21(2):252.

Rothschild, N.C. 1914e, *'Xenopsylla nesiotes':* a correction. Rec. Indian Mus. 10(3):215.

Rothschild, N.C. 1915a, *Stephanocircus pectinipes,* sp.nov. Entomologist 48(621):25-26.

Rothschild, N.C. 1915b, A synopsis of the British Siphonaptera. Ent. mon. Mag. 51(610): 49-112.

Rothschild, N.C. 1915c, Contribution to our knowledge of the Siphonaptera Fracticipita. Novit. zool. 22(2):302-308.

Rothschild, N.C. 1915d, A new African bat-flea. Ent. mon. Mag. 51(618):304-305.

Rothschild, N.C. 1915e, Further notes on Siphonaptera Fracticipita, with descriptions of new genera and species. Ectoparasites 1(1):25-29.

Rothschild, N.C. 1915f, On *Neopsylla* and some allied genera of Siphonaptera. Ectoparasites 1(1):30-44.

Rothschild, N.C. 1916a, Report on the Siphonaptera collected by the British Ornithologists' Union Expedition and the Wollaston Expedition in Dutch New Guinea. Rep. Brit. Orn. Un. & Wollaston Exped. N.Guinea 2 (XII):1-2.

Rothschild, N.C. 1916b, Results of Dr E.Mjöbergs Swedish Scientific expeditions to Australia 1910-1913. 5. Siphonaptera. Ark. zool. 10(1-2)[3]:1-9.

Rothschild, N.C. 1916c, The occurrence of *Xenopsylla cheopis* Roths. in Bristol. Ent. mon. Mag. 52(631):279.

Rothschild, N.C. 1917a, On *Xenopsylla aequisetosus* Enderl. (1901). Ent. mon. Mag. 53 (633):32-33.

Rothschild, N.C. 1917b, The President's Address. Convergent development among certain ectoparasites. Trans. ent. Soc. Lond. cxli-clvi.

Rothschild, N.C. 1919, Siphonoptera collected in Korinchi, West Sumatra, by Messrs H.C. Robinson and C.Boden Kloss. J. Fed. Malay States Mus. 8(3):1-6.

Rothschild, N.C. 1920, A correction. Ectoparasites 1(2):77.

Rothschild, N.C. 1921a, The generic name of the sand-flea. Ectoparasites 1(3):129-130.

Rothschild, N.C. 1921b, Siphonaptera from Juan Fernandez. In C.J.F.Skottsberg (ed.), The natural history of Juan Fernandez and Easter Island. 3(1):48, Uppsala.

Rothschild, N.C. 1922a, Siphonaptera [sic] collected by the Norwegian expedition to Novaya Zemlya in 1921. Rep. sci. Results Norw. Exped. N.Zemblya 1921, Zool. (4):1-9.

Rothschild, N.C. 1922b, Siphonaptères (Pulicides). In M.de Rothschild, Voyage de M.le Baron Maurice de Rothschild en Ethiopie et en Afrique orientale anglaise (1904-1905). Animaux articules (2):791-795. Paris.

LIST OF THE PUBLICATIONS ON SIPHONAPTERA BY K.JORDAN AND N.C.ROTHSCHILD

Jordan, K. & N.C.Rothschild 1906a, Notes on the Siphonaptera from the Argentine described by the late Professor Dr Weyenbergh. Novit. zool. 13(1):170-177.

Jordan, K. & N.C.Rothschild 1906b, A revision of the Sarcopsyllidae, a family of Siphonaptera. Thompson Yates Johnston Labs. Rep. (N.S.) 7(1):15-72.

Jordan, K. & N.C.Rothschild 1908, Revision of the non-combed eyed Siphonaptera. Parasitology, Cambridge 1(1):1-100.

Jordan, K. & N.C.Rothschild 1911a, Katalog der Siphonapteren des Königlichen Zoologischen Museums in Berlin. Novit. zool. 18(1):57-89.

Jordan, K. & N.C.Rothschild 1911b, Some new Siphonaptera from China. Proc. zool. Soc. Lond. (2):365-393.

Jordan, K. & N.C.Rothschild 1912a, List of Siphonaptera collected in Portugal. Novit. zool. 18(3):551-554.

Jordan, K. & N.C.Rothschild 1912b, List of Siphonaptera collected in eastern Hungary. Novit. zool. 19(1):58-62.

Jordan, K. & N.C.Rothschild 1912c, On Siphonaptera collected in Algeria. Novit. zool. 19(2):357-372.

Jordan, K. & N.C.Rothschild 1913a, Siphonaptera. [E.Hartert, Expedition to the central western Sahara.] Novit. zool. 20(1):143-144.

Jordan, K. & N.C.Rothschild 1913b, Siphonaptera collected by Mr Robin Kemp in tropical Africa. Novit. zool. 20(3):528-581.

Jordan, K. & N.C.Rothschild 1913c, On the genus *Amphipsylla,* Wagn. (1909). Zoologist (4)17(203)[869]:401-408.

Jordan, K. & N.C.Rothschild 1914a, On the position of *Notiopsylla* nom.nov., a genus of Siphonaptera. Novit. zool. 21(2):219-223.

Jordan, K. & N.C.Rothschild 1914b, Algerian fleas collected in 1913. Novit. zool. 21(2): 235-238.

Jordan, K. & N.C.Rothschild 1914c, Katalog der

Siphonapteren des Königlichen Zoologischen Museums in Berlin. I. Nachtrag. Novit. zool. 21(2):255-260.

Jordan, K. & N.C.Rothschild 1915a, List of Siphonaptera collected in Algeria in the spring of 1914. Novit. zool. 22(2):308-310.

Jordan K. & N.C.Rothschild 1915b, On some Siphonaptera collected by W.Rückbeil in east Turkestan. Ectoparasites 1(1):1-24.

Jordan, K. & N.C.Rothschild 1915c, Contribution to our knowledge of American Siphonaptera. Ectoparasites 1(1):45-60.

Jordan, K. & N.C.Rothschild 1920a, On the species and genera of Siphonaptera described by Kolenati. Ectoparasites 1(2):61-64.

Jordan, K. & N.C.Rothschild 1920b, On American bird-Ceratophylli. Ectoparasites 1(2):65-76.

Jordan, K. & N.C.Rothschild 1920c, A preliminary catalogue of the Siphonaptera of Switzerland. Ectoparasites 1(2):78-122.

Jordan, K. & N.C.Rothschild 1920d, A new *Ctenophthalmus* from Macedonia. Ectoparasites 1(2):123-125.

Jordan, K. & N.C.Rothschild 1921a, A new species of Sarcopsyllidae. Ectoparasites 1(3): 131-132.

Jordan, K. & N.C.Rothschild 1921b, Four new Palaearctic *Ctenophthalmus.* Ectoparasites 1(3):133-137.

Jordan, K. & N.C.Rothschild 1921c, Two new Palaearctic Siphonaptera. Ectoparasites 1(3): 138-140.

Jordan, K. & N.C.Rothschild 1921d, New genera and species of bat-fleas. Ectoparasites 1(3): 142-162.

Jordan, K. & N.C.Rothschild 1921e, Eight new Ceratophylli. Ectoparasites 1(3):163-177.

Jordan, K. & N.C.Rothschild 1921f, On *Ceratophyllus fasciatus* and some allied Indian species of fleas. Ectoparasites 1(3):178-198.

Jordan, K. & N.C.Rothschild 1922a, New species of Siphonaptera collected by Mr C.Boden Kloss in the Malay peninsula and south Annam. Ectoparasites 1(4):217-222.

Jordan, K. & N.C.Rothschild 1922b, The Siphonaptera collected by Mr E.Jacobson on Sumatra. Ectoparasites 1(4):223-230.

Jordan, K. & N.C.Rothschild 1922c, On *Pygiopsylla* and the allied genera of Siphonaptera. Ectoparasites 1(4):231-265.

Jordan, K. & N.C.Rothschild 1922d, New Siphonaptera. Ectoparasites 1(4):266-283.

Jordan, K. & N.C.Rothschild 1923a, Additions to the catalogue of Swiss Siphonaptera. Ectoparasites 1(5):287-289.

Jordan, K. & N.C.Rothschild 1923b, Further records of Algerian Siphonaptera. Ectoparasites 1(5):290-292.

Jordan, K. & N.C.Rothschild 1923c, On some Siphonaptera from the eastern hemisphere. Ectoparasites 1(5):293-308.

Jordan, K. & N.C.Rothschild 1923d, New American Siphonaptera. Ectoparasites 1(5):309-319.

Jordan, K. & N.C.Rothschild 1923e, On the genera *Rhopalopsyllus* and *Parapsyllus.* Ectoparasites 1(5):320-370.

Jordan, K. & N.C.Rothschild 1926, Siphonaptera from Borneo. Sarawak Mus. J. (3)(10):287-292.

TAXONOMY

ROBERT TRAUB
Department of Microbiology, University of Maryland School of Medicine, Baltimore, USA

NEW GENERA AND SUBGENERA OF PYGIOPSYLLID FLEAS*

ABSTRACT

The names *Stivalius* Jordan & Rothschild, 1922 and *Pygiopsylla* Rothschild, 1906 have been applied to certain species which, according to modern concepts, clearly belong in new genera and subgenera. Such new taxa are named, diagnosed and illustrated herein. The members of Jordan & Rothschild's Group 'B' (1922) are placed in *Bibikovana* n. gen. *Migrastivalius* n. gen. is proposed for the *Stivalius jacobsoni*-group (in nominate subgenus) and for the subgenus *Gryphopsylla* Traub, 1957. *S.mjoebergi* Jordan, 1926 is placed in *Lentistivalius (Destivalius)* n. subgen. *Aviostivalius* n. gen. is established for *S.klossi* Jordan & Rothschild, 1922 and the *S.pomerantzi*-group is placed in *Nestivalius* n. gen. *Afristivalius* n. gen. includes the 15 species in the Ethiopian Region that still had resided in *Stivalius,* while the poorly known *Pygiopsylla celebensis* Ewing, 1924 and *P.sciuri* Ewing, 1924 are both assigned to *Farhangia* n. gen. A key to the known genera and subgenera of pygiopsyllids from Africa and the Asiatic-Pacific area west and northwest of New Guinea, is presented. Some notes on zoogeography and evolution of fleas are included, as pertinent to the new taxa.

The family Pygiopsyllidae is of especial interest in studies on the evolution, zoogeography and classification of fleas, and some of the species are of consequence as potential vectors of plague. Nevertheless, research on this group of fleas has been handicapped by the fact that many of the important species technically reside in two genera that have long been in need of revision, namely *Pygiopsylla* Rothschild, 1906 and *Stivalius* Jordan & Rothschild, 1922, and whose names are still necessarily being applied to species that clearly belong elsewhere. For some years I have been working on a series of long papers to resolve these and other taxonomic problems within the Order and to handle the wealth of new species that have resulted from the field-studies in New Guinea undertaken as a joint project by this Department and the Wau Ecology Institute, B.P. Bishop Museum (Honolulu). However, there has developed a pressing need to refer to some of the new names and concepts,

* This study was supported by Grant No. AI-04242 of the National Institutes of Health, Bethesda, Maryland, USA, with the Department of Microbiology, University of Maryland School of Medicine, Baltimore, Maryland, USA. The field-studies in Ethiopia mentioned in this article were sponsored by Contract N00014-76-C-0393 of the Office of Naval Research and the Navy Medical Research and Development Command, Washington, DC. The opinions and assertions herein are not to be construed as necessarily reflecting the views of the Department of the Navy or the National Institutes of Health.

particularly by other workers doing revisionary studies, and in order to facilitate matters, and make some of our illustrations readily available to colleagues for use in their publications, I have prepared this article to provide new names for the most critical taxa. Because of constraints of time and space, no new species are described herein, and the text is minimal, but supplemented by illustrations of all major features.

One of the difficulties in the taxonomy of pygiopsyllids has been that several species have been described on the basis of only one sex, with the male still remaining unknown. Another complication is that some of the older taxa have never been figured or properly described. Accordingly, I have included here illustrations and diagnostic notes on *Pygiopsylla celebensis* Ewing, 1924 and *P.sciuri* Ewing, 1924, both of which are placed in a single new genus. The hitherto unknown male of *Stivalius mjoebergi* Jordan, 1926; female *S.pomerantzi* Traub, 1951 and relatively little-known taxa like *S.jacobsoni* (Jordan & Rothschild, 1922) and *S.klossi* (Jordan & Rothschild, 1922) have been illustrated in detail and their main features categorized, since each have been placed herein in new genera or subgenera. The dissolution of *Stivalius,* commenced so ably by Holland (1969) and extended by Traub (1972a) is herein completed, whereby students can henceforth at least indicate which group of *'Stivalius'* or *'Pygiopsylla'* they mean when using those names, or the new ones. In order to stress and clarify the main differences between the major taxa involved, a key is presented for the identification of the known genera and subgenera of pygiopsyllids from Africa, Asia, the Philippines and Indonesian islands to and including Sulawesi. (Thus, New Guinea, Australia and South America, per se are excluded.)

Many systematists believe that only non-sexual taxonomic characters are suitable for use at the generic level, either because they prefer features that apply to either sex or because they feel that the genitalia are of value only in determining species. Ewing was convinced of the validity of both of these points (in litt.). Jordan (1933), however, used both approaches in his revision of New World ceratophyllids, but relied more on the non-sexual features, at least for major subdivisions, and avoided employing such structures as the claspers. Holland's sound reclassification of New Guinean *Stivalius* (1969) included both types of characters, and most modern workers have done likewise when describing new genera. There is at least one such student of Siphonaptera who believes that the male genitalic structures by themselves can suffice to establish genera. In the case of the pygiopsyllids, I believe that the various genera can be recognized by careful study of the aedeagus by itself. Nevertheless, I believe that a classification set up on the basis of males alone is undesirable for several reasons, including the fact that certain females could not then be assigned to genus. If one relied solely on the male, *Macrostylophora* Ewing, 1929, for example, could readily be divided into several genera, but thus far it has been impossible to make comparable distinctions in the females. It is difficult to see how a scheme of classification based upon only one sex can properly reflect true evolutionary processes at the generic level.

Regardless, some of the non-genitalic features used in taxonomy must be interpreted with caution because many of them, once regarded as 'fundamental' actual may be adaptive, expressing the habits of the flea or its host. As shown in the accompanying article on adaptations in this volume, congeners can thus differ markedly regarding the presence or absence of striking features like 'crowns of thorns', abdominal ctenidia, supernumerary rows of abdominal bristles, tibial combs and reduction of the pleural arch, etc. Such modifications at times may be at the generic level, but in that case, should be accompanied by other significant changes as well. It is relevant and of interest that in the pygiopsyllids extraordinary variations in the shape and chaetotaxy of the head may have occurred, often resulting in generic designations (and validly so), but with the male genitalia so remarkably uniform as to discourage the taxonomist. In the ceratophyllids, it is the head which shows extremely little evidence of change, whereas the genitalia exhibit bizarre modifications. All these various points and principles have been borne in mind when defining the genera and subgenera in this article, which incorporates both non-genitalic characters as well as some based upon the male and female genitalia.

KEY TO THE GENERA AND SUBGENERA OF KNOWN PYGIOPSYLLIDS FROM AFRICA AND THE ASIATIC-PACIFIC AREA WEST AND NORTHWEST OF NEW GUINEA

(Illustrations of *Stivalius s. str.*, *Medwayella* Traub, 1972 and *Lentistivalius* Traub, 1972, were presented by Traub (1972a) but are not necessary for the use of the key.)

1. Pleural arch of metathorax completely reduced (figs. 83, 95) . . . . *Farhangia* n. gen. (p. 27)

– Pleural arch well-developed (fig. 43, PL.A.) 2

2. With at least one relatively well-developed ctenidium on abdomen (figs. 23, 54). 3

– Abdomen without ctenidia though perhaps some terga with 1-2 apical spinelets. 6

3. With a distinct and characteristic sinus or notch at the oral angle (fig. 30). Comb on 2 T. (fig. 34) with as many spines as pronotal ctenidium. (Borneo) *Migrastivalius (Gryphopsylla)* Traub, 1957, n. gen., n. comb. (p. 20)

– Lacking an oral notch (fig. 18). Abdominal comb(s) with definitely fewer spines than pronotal ctenidium (Borneo and elsewhere) 4

4. With some non-apical, dorsomarginal notches of metatibia (figs. 22, 78) bearing three stout bristles. Protibia similarly modified, or with bristles in a subpectinate row (fig. 20). ♀ tergum 7 with a lower modified bristle (figs. 28, 79, L.M.B.) resembling lower antepygidial bristle (A.B.). Pivotal base of end-chamber either undeveloped (fig. 19) or with at least ventral portion (overlying part of S.I.T.) vestigial or untanned (fig. 76, PIV.B.) 5

– Dorsomarginal notches (non-apical) of metatibia (fig. 49) and protibia (fig. 52) with only two stout bristles. ♀7 T. lacking bristle(s) resembling A.B. (fig. 55, 7 T.). Pivotal base well-developed and extending over S.I.T. (fig. 48, PIV.B.) (Borneo) . . . . *Lentistivalius (Destivalius)* n. subgen. (p. 22)

5. Movable finger (fig. 77, F.) with a patch of small, thin, mesal bristles. F. with distal fringe (D.FR.) limited to area of stiva (STV.). Spermatheca with hilla (fig. 80, H.) with basal half thick-walled and 'submerged' in bulga (B.). Bursa copulatrix (B.C.) flanked by a distinctive rod-like or ovate sclerite which is usually well-tanned. (Ethiopian Region only) . . . . *Afristivalius* n. gen. (partim) (p. 26)

– Movable finger (fig. 18, F.) lacking a patch of mesal bristles; with D. FR. well proximad of stiva, at least in part. Spermatheca (fig. 21) with hilla (H.) not projecting deeply into bulga. B.C. lacking an adjacent sclerite. (Malayan Peninsula and Insular Malayan area) . . . . *Migrastivalius (Migrastivalius)* n. gen. (p. 20)

6. Eye in line with base of procoxa (fig. 1). Metasternum markedly and evenly biconvex, with its lobes extending well over base of metacoxa. With a large patch of mesal bristles on metacoxa. (Known from Australia, New Guinea and Borneo) . . . . *Bibikovana* n. gen. (p. 17)
– Eye somewhat forward of level of base of procoxa (fig. 64). Metasternum lacking a pair of large lobes extending well over base of coxa (fig. 29). Without a patch of mesal bristles on metacoxa (fig. 29) 7

7. Anterodorsal angle of metepisternum acute and upturned. With an unusually well-developed ventral armature (fig. 67, V.AR.) of sclerotized inner tube (S.I.T.) that extends to apex of S.I.T. and terminates in an upcurved crochet-like projection. Movable finger (fig. 69, F.) lacking a stiva. ♀ ventral anal lobe (fig. 74, V.A.L.) relatively well clothed with bristles and hence lacking a gap near anteroventral angle. (Philippines and Sulawesi) . . . . *Nestivalius* n. gen. (p. 24)
– Anterodorsal angle of metepisternum lacking an upturned lobe; broadly rounded or at most blunt (fig. 43, MTS.). S.I.T. lacking a crochet-like process; V.AR. represented at most by a sub-basal spur (figs. 19, 39, 58). F. with a stiva (fig. 59, STV.). ♀ V.A.L. with relatively few marginal bristles and with a conspicuous gap between the basal and apical group of bristles (fig. 62) 8

8. Sclerotized inner tube elongate – about 9 to 30 times as long as broad at middle; arched for at least apical half. Aedeagus lacking crochet-like processes or a caudad-directed arm. B.C. grossly swollen throughout and coiled for most its length or apically folded upon itself; its diameter nearly equal to width (height) of bulga . . . . *Stivalius* Jordan & Rothschild, 1922 *(s.str.)*
– Sclerotized inner tube only about 4 to 5 times as long as broad at middle (figs. 38, 58, S.I.T.); relatively straight. B.C. (figs. 63, 80) unmodified; relatively short, uncoiled; less than half width of bulga 9

9. Metatibia with at least some of dorsomarginal (non-apical) notches bearing three stout bristles 10
– Metatibia with a maximum of two stout bristles in non-apical dorsomarginal notches 11

10. Labial palpi either bearing a small apical sixth segment (fig. 75) *or* fifth segment much longer than fourth. Movable finger (fig. 77, F.) with a group of small mesal bristles, and with distal fringe limited to stiva (STV.). Lacking a crochet-like process extending caudad below S.I.T. (fig. 76). Penis rods making at least half a coil. B.C. (fig. 80) associated with a single sclerotized structure as seen in lateral aspect. (Ethiopian Region only). . . . . *Afristivalius* n. gen. (partim) (p. 26)
– Labial palpi with fifth segment scarcely longer than fourth. F. (fig. 59) lacking a group of small mesal bristles; distal fringe extending well proximad of stiva. With a well-developed crochet process (fig. 58, CR.P.). Penis rods uncoiled. With a small sclerite or tanned patch on both anterior and posterior margins of B.C. (fig. 63) (Southeast Asia to Java) . . . . *Aviostivalius* n. gen. (p. 23)

11. With a ventral spinose lumacaudate process on ♂8 S. Phylax relatively broad and subligulate; less than thrice as tall as broad at middle. Distal arm of 9 S. (D.A.9) not clavate, i.e. with ventral (caudal) margin lacking apical bulge. ♀7 T. with some

bristles modified so as to resemble antepygidial bristles. . . *Medwayella* Traub, 1972
– Lacking a spinose lumacaudate process on ♂8 S. Phylax (fig. 38, PHY.) long and narrow; at least 5 x as long as broad at middle. D.A.9 clavate, ventral margin apically swollen. ♀ 7 T. lacking bristles closely resembling antepygidial bristles. 12

12. Fifth segment of labial palpi twice as long as fourth. . . . . *Afristivalius* n.gen. (partim)(p. 26)
– Fifth segment of labial palpi about 1.5 length of fourth . . . . *Lentistivalius (Lentistivalius)* Traub, 1972

BIBIKOVANA n. gen. (figs. 1-8)

Diagnosis – Comprising the section of *Pygiopsylla* designated as 'Group B' by Jordan & Rothschild (1922) and separable from *Pygiopsylla s.str.* as follows: 1) Maxillary palpi (fig. 1, M.P.) with second segment at least 1.25 times length of first segment instead of being a maximum of 1.23 times (fig. 10). 2) Significantly larger species, viz. despite individual differences and those due to the mounting process or pregnancy, ♂ ranging from 3-4.7 mm, ♀ from 3.5-6.5 mm, whereas in *Pygiopsylla s.str.*, in most species the range is ♂: 2.0-2.6 mm and ♀: 2.5-3.25 mm, but in *P.zethi* (Rothschild, 1904), ♂ reaching about 3 mm and ♀, 3.7 mm. 3) More bristly, e.g. ♂ generally with 2.5-4 rows of bristles on terga 2-6; ♀ with 3.5-5 rows, preceded by some scattered small ones, instead of having 2½ rows in each sex, preceded by 1-3 irregular, very small ones. 4) Metatibia with 8 dorsomarginal, tanned notches (including apical set) bearing one or more stout bristles (fig. 5); instead of 7 (fig. 13). 5) Proximal arm of ♂ sternum 9 (fig. 6, P.A.9) very different from D.A.9 in shape and with apical half or three-fourths much broader than basal portion, the expanded section with its dorsal margin markedly convex, at least near base. In *Pygiopsylla s.str.*, P.A.9 (fig. 14) usually resembles D.A.9 in shape and is unmodified, but if somewhat broader, then dorsal margin flat for apical two-thirds. 6) Distal arm of sternum 9 (fig. 6, D.A.9) with a ventral group of bristles that is larger than the bulk of the other ventromarginals, and generally quite conspicuous. In the other taxon, the ventromarginal bristles on D.A.9 (fig. 14) are uniform, small and thin and scattered. 7) D.A.9 with portions modified instead of being essentially homogeneous, i.e. at least with a ventral thickening but usually consisting of portions that contrast in structure, size and/or shape. In *Pygiopsylla s.str.*, D.A.9 is relatively undifferentiated and uniform in appearance, and the ventral region is unmodified. 8) Aedeagus (fig. 8) with end-chamber covered dorsally by a broad, cowl-like hood (HD.), in contrast to (fig. 16) an essentially single, narrow blade-like median dorsal lobe (M.D.L.) (at times armed with spines as in figure) which may terminate in a conspicuous hook-like projection. 9) With relatively well-developed lateral lobes covering lateroventral portion of end-chamber and fusing with Ford's sclerite (F.SC.). Lateral lobes undeveloped in *Pygiopsylla s.str.* 10) With Ford's sclerite relatively poorly demarcated and obscured by lateral lobes instead of being discrete and in the open. 11) Ventral armature (V.AR.) of sclerotized inner tube (S.I.T.) prominent as a large structure paralleling S.I.T. and projecting distad instead of being restricted to the level of the base of S.I.T. although in some taxa extended ventrad as a hood-like structure (as in fig. 16). 12) With apicomarginal bar-like thickenings (A.TH.AE.) extending dorsocaudad from the floor of the aedeagal pouch to base of Ford's sclerite, presumably representing the pivotal base of typical members of the *Stivalius*-group (figs. 38, 48, PIV.B.). These supporting structures undeveloped in

Pygiopsylla s.str. 13) Spermatheca (fig. 4) with bulga (B.) a large, flabby, sausage-shaped, relatively undifferentiated, sac-like structure, instead of being more tanned, and medially somewhat constricted so that its anterior and posterior portions differ in breadth and/or configuration (fig. 12, B.). 14) Anal stylet (fig. 7, A.S.) essentially completely cylindrical; not somewhat broadened at base (fig. 15). 15) Ventral anal lobe (fig. 7, V.A.L.) with basal ventral portion (proximad of subapical sinus) evenly ovate or semicircular, and fringed with bristles beyond the bend instead of this area being rather angulate, with flat margins, fewer bristles, and usually lacking any bristles basad of angle (fig. 15). 16) With cavity of ♀ eighth spiracle (fig. 3) extended ventrad for some distance along the margin of the eighth tergum. There is no such definitive extension in *Pygiopsylla s.str.* (fig. 11). 17) Bristles of ♀ sternum 8 long, resembling other small bristles instead of being minute. 18) Parasitizing *Rattus,* particularly the more primitive subgenera, whereas *Pygiopsylla s. str.* is primarily found on marsupials. (Diagnostic characters # 1-4 and 16-17 are from Jordan & Rothschild (1922)).

Similarities with Pygiopsylla – In spite of these fundamental differences, *Bibikovana* and *Pygiopsylla s.str.* share some important features. The heads are fundamentally similar (figs. 1:♂, 2:♀, 9:♂, 10:♀) except perhaps that the frontal margin, above the mouthparts, is somewhat straighter and more oblique in *Pygiopsylla* than in *Bibikovana.* The claspers are also built on essentially the same plan (cf. figs. 6 & 14). The characters mentioned by Jordan & Rothschild (1922) for *Pygiopsylla s.lat.* were well chosen and apply to the species described since then, and hence fit both taxa. All known species have unmodified cephalic and thoracic setae, a slightly reduced eye, and a well-developed pronotal comb, a well-developed pleural arch, plantar setae on tarsal segment 5 that are essentially lateral in position, and two antepygidial bristles. The two genera also agree in certain aedeagal features, viz. a large aedeagal pouch (figs. 8, 16, AE.P.); a weakly tanned girdle (G.); a narrow aedeagal apodeme whose margins are parallel most of its length; a long crescent sclerite (C.SC.) with consequently short satellite sclerite (SAT.S.); and a spiculose or delicately fringed area at the apex of the end-chamber, proximad to and below V.AR. A long sac-like bursa copulatrix, equal to or exceeding the length of the spermatheca is also found in both. In each there is a marked reduction of the apical spinelets on the abdominal terga (representing vestigial ctenidia). Most species (and specimens) lack them save for an occasional small one, on 2 T., especially in the ♀, but there are exceptions, e.g. *P. zethi* with 1 or 2 per most abdominal terga.

Comments – *Bibikovana* includes the following species: A) Australia – 1) *arcuata* (Holland, 1971), 2) *colossa* (Rothschild, 1906), 3) *gravis* (Rothschild, 1908), 4) *iridis* (Holland, 1971), 5) *rainbowi* (Rothschild, 1908) and 6) *sinuata* (Holland, 1971); B) New Guinean – 1) *archboldi* (Holland, 1969), 2) *laciniosa laciniosa* (Rothschild, 1908), *l. bismarckensis* (Holland, 1969), 3) *smitiana* (Holland, 1969) and 4) *traubi* (Holland, 1969) and C) Bornean – 1) *tiptoni* (Traub, 1957). The type of the genus is *P. laciniosa.* In *Pygiopsylla s.str.* there remain *hoplia* Jordan & Rothschild, 1922, *hilli* (Rothschild, 1904), *tunneyi* Mardon & Dunnet, 1972 and *zethi* (Rothschild, 1904) from Australia, and *spinata* Holland, 1969, from New Guinea.

Both genera are in need of revision. *Bibikovana* includes species based upon one sex, with the opposite members still undescribed, and it is possible that some males and females may not have been associated properly. The name *P. hoplia* may have been applied to each of two sibling species and/or differences at subspecies level have to be determined. *P. spi-*

nata was described on the basis of the female alone. At the time of description, its affinity with *P.hoplia* was noted, but no direct comparison was made thereto in the absence of the male. Figures 8 to 16 include the first published illustrations of the male, permitting some comments about its status. Although several genera occur both in Australia and New Guinea, among the only fleas (of non-volant hosts) which are known to occur in both areas are commensal species, and *Xenopsylla vexabilis* Jordan, 1925, which really is in that category though also found on native hosts. Even the *Metastivalius* Holland, 1969, of *Rattus* in the two regions are in the category of sibling species rather than subspecies. It is therefore rather to be expected that *P.hoplia* and *P.spinata* represent distinct species, although at least one worker has suggested (in litt.) that they are synonymous. I believe that *P.spinata* is a full species, separable from both 'forms' of *P.hoplia* by the facts that 1) the ventral margin of the head is entire below the eye instead of being biconvex as in Australian *Pygiopsylla* save *P.hilli;* and 2) in the New Guinea species the process P. resembles that of the southern form of *P.hoplia* much more than that of the northern one, but P.A.9 lacks its distinctive hump, thereby resembling that of the northern form. There are other differences in the genitalia, but it is premature to go further until the status of the Australian forms of *P.hoplia* are resolved in writing, and that is beyond the scope of this article.

The name *Bibikovana* is proposed in honour of Professor V.A.Bibikova of the Central Asian Research Anti-Plague Institute at Alma-Ata, U.S.S.R. in token acknowledgement of her outstanding research on Siphonaptera, including studies on the ecology, bionomics and behaviour of fleas; on the ecology of plague and on the complex inter-relationships between the vectors of the infection and the etiological agent.

MIGRASTIVALIUS n. gen. (figs. 17-40)

Diagnosis – Near *Lentistivalius* Traub, 1972 but immediately separable in that: 1) The aedeagus lacks both a distinct caudad-projecting crochet process (cf. fig. 38, CR.P.) and a well-developed pivotal base (fig. 38, PIV.B.). Instead, in *Migrastivalius* there is a fairly discrete, thickened, apical extension of the aedeagal pouch (figs. 19, 39, A.TH.AE.) on each side. This is absent in *Lentistivalius,* where its function presumably is assumed by the phylax (figs. 38, 48, PHY.). 2) Sclerotized inner tube (S.I.T.) unusually narrowed and long, relatively or actually straight, and of uniform width distad of the basal portion. In *Lentistivalius* S.I.T. is short, stout, slightly curved, and of varying thickness (fig. 38). 3) S.I.T. with a distinct ventral spur near the base, which is absent in most *Lentistivalius* (fig. 38). 4) Ford's sclerite (F.SC.) with alpha-portion (ALPH.) terminating in a straight, stiff, hook-like projection. 5) Bursa copulatrix (figs. 21, 40, B.C.) unusually enlarged, i.e. longer and with thicker walls than in most pygiopsyllids (cf. figs. 63, 80). 6) Metatibia with some of stout dorsomarginal bristles in subvertical groups, and hence somewhat pectinate (figs. 22, 31); instead of being primarily in contiguous pairs (fig. 52). 7) Mesotibia even more comb-like and protibia nearly fully pectinate (figs. 20, 32; cf. fig. 49). Other differences: 8) Abdomen with a well-developed comb on tergum 2 (figs. 23, 34, 2 T.). Virtually all *Lentistivalius* lack abdominal combs, but in *Stivalius mjoebergi* Jordan, 1926 (placed below in that genus) there is one on 2 T. and a smaller one on 3 T. (fig. 54). 9) Spermatheca (figs. 21, 40) with bulga (B.) with a straight ventral margin and an evenly sloping dorsal one so that the basal (caudal) third is distinctly broader than the apical one. In *Lentistivalius* the bulga is slightly but evenly arcuate ventrally and the dorsal margin is slightly sinuate. Both halves are sub-equal in breadth or else (fig. 58) the basal half is somewhat narrowed. 10) Metatarsal segment 1 with dorsomarginal stout bristles extending to base instead of only half-way or two-thirds. 11) With at least 1 of the modified bristles of 7 T. resembling the antepygidial bristles (figs. 28, 35, L.M.B.) instead of being scarcely modified.

Generic description – Agreeing with *Lentistivalius* as defined originally except as indicated above and as follows: Dorsal margin of pronotum much longer than half length of dorsal spines of comb. Terga 3-5 with or without an apical spinelet. With 2½ rows of bristles on unmodified abdominal terga. Distal fringe (fig. 37, D.FR.) of movable finger (F.) with a group of 5-6 long bristles. Hood of aedeagus simple (fig. 39, HD.) at times equipped with an apical pair of acuminate uncate projections (fig. 19), and at times terminating in an acute deltoid flap (fig. 39, DEL.FL.). Spiracular fossa of ♀8 T. (figs. 28, 35, 8 SPC.) with vertical portion about 1.5x as broad as base of horizontal one.

Comments – The type of the genus is *Stivalius jacobsoni* (Jordan & Rothschild, 1922), formerly placed in *Pygiopsylla.* There are two subgenera, as described below, at which time the constituent species are designated. The name *Migrastivalius* is derived from the Latin term *migro* meaning to depart from one place to another, and refers to the great distance this genus has wandered from the place of origin of its forebears in the Australian Region, i.e. to the Malayan Peninsula and the Insular Malayan area of Traub (1972c).

MIGRASTIVALIUS (MIGRASTIVALIUS) n. subgen. (figs. 18-29)

Diagnosis – 1) Head (fig. 18) unmodified; of the type representative of *Stivalius s.lat.* 2) Labial palpi not reaching ¾ length of procoxa; with fifth segment unusually long, much longer than fourth. 3) With small, lateral, submarginal bristles on metacoxa (fig. 29) limited to lower two-fifths. 4) With an apical subdorsal spinelet on abdominal terga 3-5. 5) Metatarsus 5 with first pair of plantar bristles lateral in position. 6) First abdominal tergum of usual proportions: ratio of length to height is 1:3. 7) Hood of aedeagus (fig. 19, HD.) apically terminating in left and right fang-like or tooth-like uncate projections. 8) Deltoid flap absent. 9) Clavus of distal arm of ♂ sternum 9 (fig. 25, D.A.9) rather narrowly ovoid rather than broadly so (fig. 36). 10) ♀ clearly with only two antepygidial bristles (fig. 28, A.B.), i.e. lower modified bristle (L.M.B.) displaced towards A.B. but still separate.

Here belongs *M.jacobsoni,* which exists as separate subspecies in Malaya, Java (illustrated here) and Sumatra, and two new species – one on Borneo and some of the Philippines, and one on Sumatra, which will be described elsewhere. All the species infest scansorial or semi-arboreal, nocturnal murines like *Chiropodomys, Pithecheir* and certain *Rattus.*

MIGRASTIVALIUS (GRYPHOPSYLLA) Traub, 1957, n.comb. (figs. 30-37, 39, 40)

Diagnosis – Separable from all known fleas by the conspicuous sinus at the oral angle of the head, producing a snout (fig. 30). Also characterized by 1) the comb on 2 T. (fig. 34) has as many spines, or more, as does the pronotal ctenidium (fig. 30). 2) Labial palpi about ¾ length of procoxa, and with fifth segment scarcely longer than fourth. 3) With group of scattered small, submarginal lateral bristles on metacoxa extending to upper third of coxa. 4) Lacking apical subdorsal spinelets on abdominal terga. 5) Metatarsus 5 with first pair of plantar bristles displaced somewhat ventrad. 6) First abdominal tergum (fig. 34, 1 T.) unusually narrow (short in height); ratio of length to height is 1:5. 7) Hood (fig. 39, HD.) lacking tooth-like apical projections; and with an acute deltoid flap (DEL.FL.). 8) Mace-head shaped clavus of D.A.9 (fig. 36) subrounded. 9) ♀ appearing to have three antepygidial bristles (fig. 35, A.B.) due to 'capture' of lower modified bristle (L.M.B.), which often appears on plate of A.B.

Comments – There is only one species known in this subgenus, *M.(G.) hopkinsi* (Traub, 1957), primarily a parasite of *Rattus (Lenothrix) whiteheadi* and secondarily on *R.(L.) alticola* and only occasionally on other theraphions in the mountains of North Borneo. The major hosts are spiny rats and I suspect that the remarkable notch on the head of this flea serves as a device for gripping the stiff flat bristles characteristic of these rats. Dr Karl Jordan suggested (in litt.), prior to my describing *(Gryphopsylla)* in 1957, that I place it in a separate genus, but it seemed advisable to be rather conservative until more became known about the related taxa. For the time being, at least until we learn whether the snout is adaptive in nature, I prefer to regard *Gryphopsylla* as a subgenus. If the snout is indeed functional for gripping bristly or other hairs, then new species of unrelated fleas, modified in this manner, may be found infesting hosts with similar chaetotaxy or scansorial habits. Likewise, adaptive variations in the number and size of abdominal combs are to be expected in such fleas, correlated with the degree of risk imposed by the traits of the host, as per the companion article in this series.

Notes on the genus Migrastivalius – Certain of the points (#6-8) mentioned in the generic description, namely those dealing with the modifications of the tibiae and with the abdominal comb, are apparently the consequence of adapting to hosts that are nocturnal and scansorial, as per my article on adaptations in this volume. Some of the other unusual morphological features in the genus are also of special interest, such as the development of structures to perform the function ordinarily assumed by the crochets, which are completely vestigial in these species. In both subgenera, the specialized, hook-like spur on Ford's sclerite seem to serve this purpose. It is noteworthy that the Ford's sclerite of *Stivalius s. str.* is specialized in the same way and here too the aedeagus is long and thin, and there are no crochets (Traub, 1972a). As mentioned in that latter reference, such an aedeagus apparently serves as an intromittent organ, instead of the penis rods acting in that capacity, as in the vast majority of fleas. In the subgenus *Migrastivalius* there is a second aedeagal modification that is presumably associated with the loss of the crochets, and that is the extraordinary fang-like projections of the apex of the hood, which I believe must serve as a clasping device in copulation. Moreover, these tooth-like extensions really represent the deltoid flaps of the hood, which ordinarily are far more ventral and anterior in position, overlying S.I.T. in the vicinity of the phylax (figs. 38, 39, 48, DEL.FL.).

In species with a S.I.T. that penetrates the female, the bursa copulatrix is correspondingly modified, and is longer and broader than in the usual case. It is appropriate then, that the females of *Migrastivalius* (figs. 21, 40) are modified along these lines, as compared with 'typical' genera (figs. 63, 80), but not to the level seen in *Stivalius s. str.*, in which the aedeagus is even longer and thinner than in *Migrastivalius.* That this correlation is valid is suggested by the situation regarding *Acanthopsylla richardsoni* Smit, 1953, which was predicted to have an elongate S.I.T. merely because of the characteristics of the female genitalia (Traub, 1972a), and subsequent examination of the male disclosed that this had been a correct assumption.

LENTISTIVALIUS (DESTIVALIUS) n. subgen. (figs. 41-57)

Among the taxa which have not yet been removed from *Stivalius s. lat.* but which obviously require reassignment is *S. mjoebergi* Jordan, 1926. At first glance this species would appear to fit in *Migrastivalius* because of the possession of an abdominal comb on 2 T., etc. but analysis, based upon the hitherto undescribed male, makes it clear that its affinities are with *Lentistivalius,* as a separate subgenus, as follows.

Diagnosis – 1) Separable from all known members of *Stivalius s. lat.* by possessing a comb on 2 T. and, in addition, a smaller one on 3 T. (figs. 45, 54). Distinguishable from the nominate subgenus *Lentistivalius* by the following characters as well. 2) Aedeagus (figs. 47, 48) lacking a well-developed caudad-directed crochet process (cf. fig. 38, CR.P.) which extends to the apex of S.I.T. (although bearing a remnant thereof at ventral portion of pivotal base (PIV.B.)). 3) Phylax (PHY.) much broader subventrally than subdorsally instead of being evenly and narrowly fusiform (fig. 38). 4) Ford's sclerite (F.SC.) with alpha-portion (ALPH.) dorso-apically produced into a rigid, angled, hook-like lobe, with apex opposed to lower lobe of securifer (SEC.) instead of acuminate or broad apex of ALPH. more or less paralleling lower lobe of SEC. (fig. 38). 5) With a sub-basal spur on the ventral margin of S.I.T. instead of the margin being entire or at most bearing a nubbin. 6) Proximal arm of sternum 9 (fig. 44, P.A.9) with dorsal margin evenly and fairly broadly convex from basal fourth to apex instead of paralleling ventral margin or being slightly convex. 7) Distal fringe (fig. 50, D.FR.) of F. consisting of 6-7 long bristles instead of 4-5. 8) Immovable process (P.) of clasper apically ovate instead of being broadly rounded. 9) Spermatheca with bulga (fig. 56, B.) broadest apically; caudad half narrowing due to up-curve of ventral margin instead of being of subequal breadth throughout. 10) ♀ basal sternum with a submedian group of 13-16 bristles and some of these reaching middle of segment (fig. 54). In *L. (Lentistivalius)* there are fewer such bristles and they are restricted to the anterior third.

Separable from *Migrastivalius* as follows: 1) With a comb on 3 T. 2) Pivotal base of end-chamber (fig. 48, PIV.B.) well-developed instead of being inapparent (fig. 39). 3) Correspondingly, lacking conspicuous rod-like extensions (A.TH.AE.) of aedeagal pouch (AE. P.). 4) Metatibia with most dorsomarginal notches bearing a pair of large bristles (fig. 49) instead of having subvertical groups of three stout bristles (fig. 22). 5) Protibia (fig. 52) and mesotibia without a pectinate row of single stout bristles (cf. fig. 20). 6) Movable finger (fig. 50, F.) virtually parallel-sided to near level of stiva (STV.) and straight instead of margins narrowing before apical third. 7) Spermatheca with bulga (fig. 56, B.) narrowing in caudal half instead of broadening (figs. 21, 40). 8) ♀ lacking bristles adjacent to the antepygidial bristles (fig. 55, A.B.) which resemble the latter; instead of having such modified bristles (figs. 28, 35, L.M.B.; fig. 35, U.M.B.).

Comments – The type of the subgenus is the North Bornean *S. mjoebergi,* and no other member is yet known. The subgeneric name *Destivalius* is based on the Latin prefix and

Figure 1 – 8. *Bibikovana laciniosa bismarckensis* (Holland, 1969). PAPUA NEW GUINEA, Western Highlands. 1. ♂ head and prothorax. 2. ♀ preantennal region. 3. Spiracular fossa 8. 4. Spermatheca. 5. ♂ metatibia. 6. Clasper and sternum 9. 7. ♀ anal segments. 8. Apical region of aedeagus.

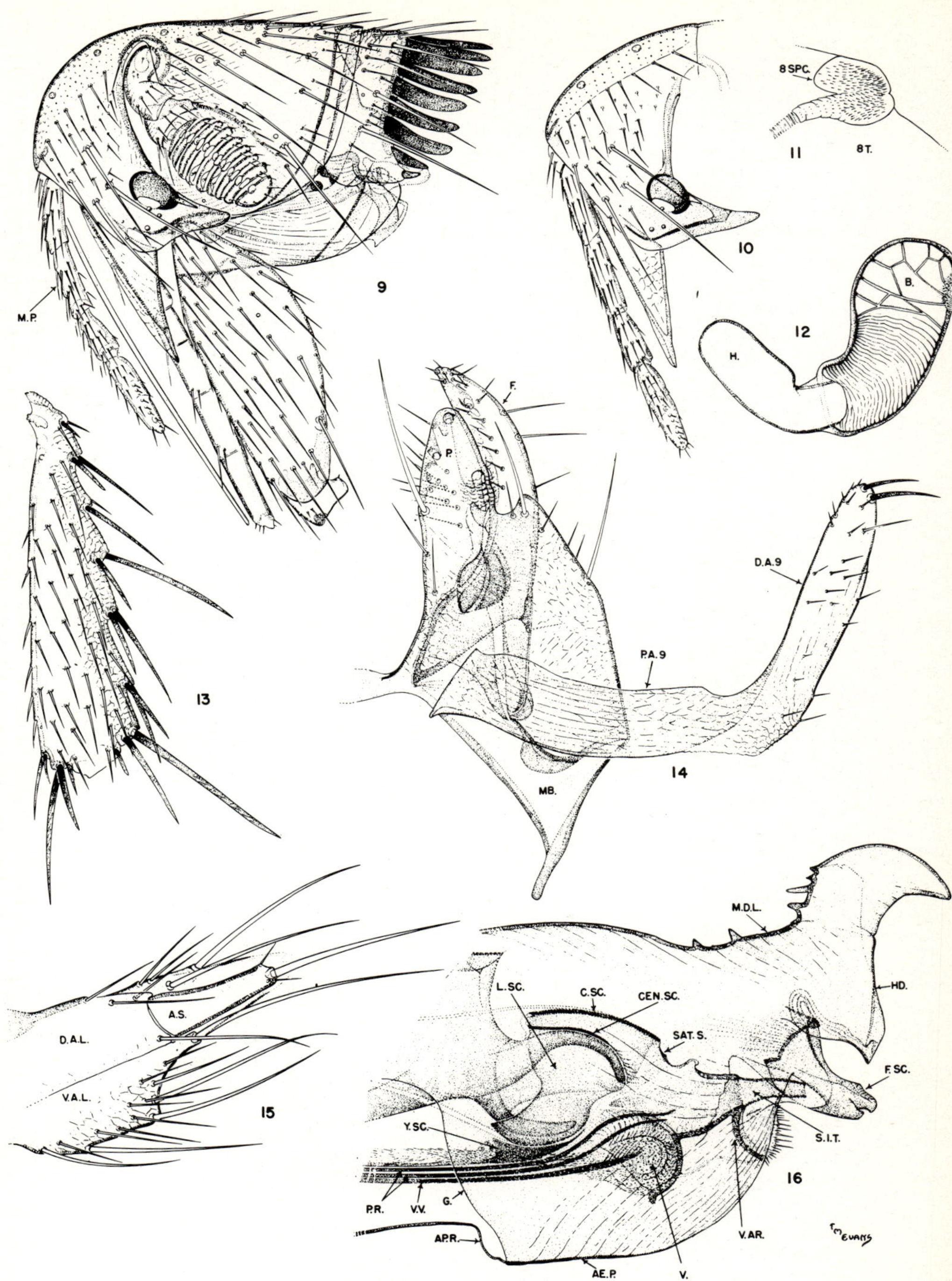

Figures 9 - 16. *Pygiopsylla spinata* Holland, 1969. PAPUA NEW GUINEA. 9. ♂ head and prothorax. 10. ♀ preantennal region. 11. Spiracular fossa 8. 12. Spermatheca. 13. ♂ metatibia. 14. Clasper and sternum 9. 15. ♂ anal segments. 16. Apical region of aedeagus.

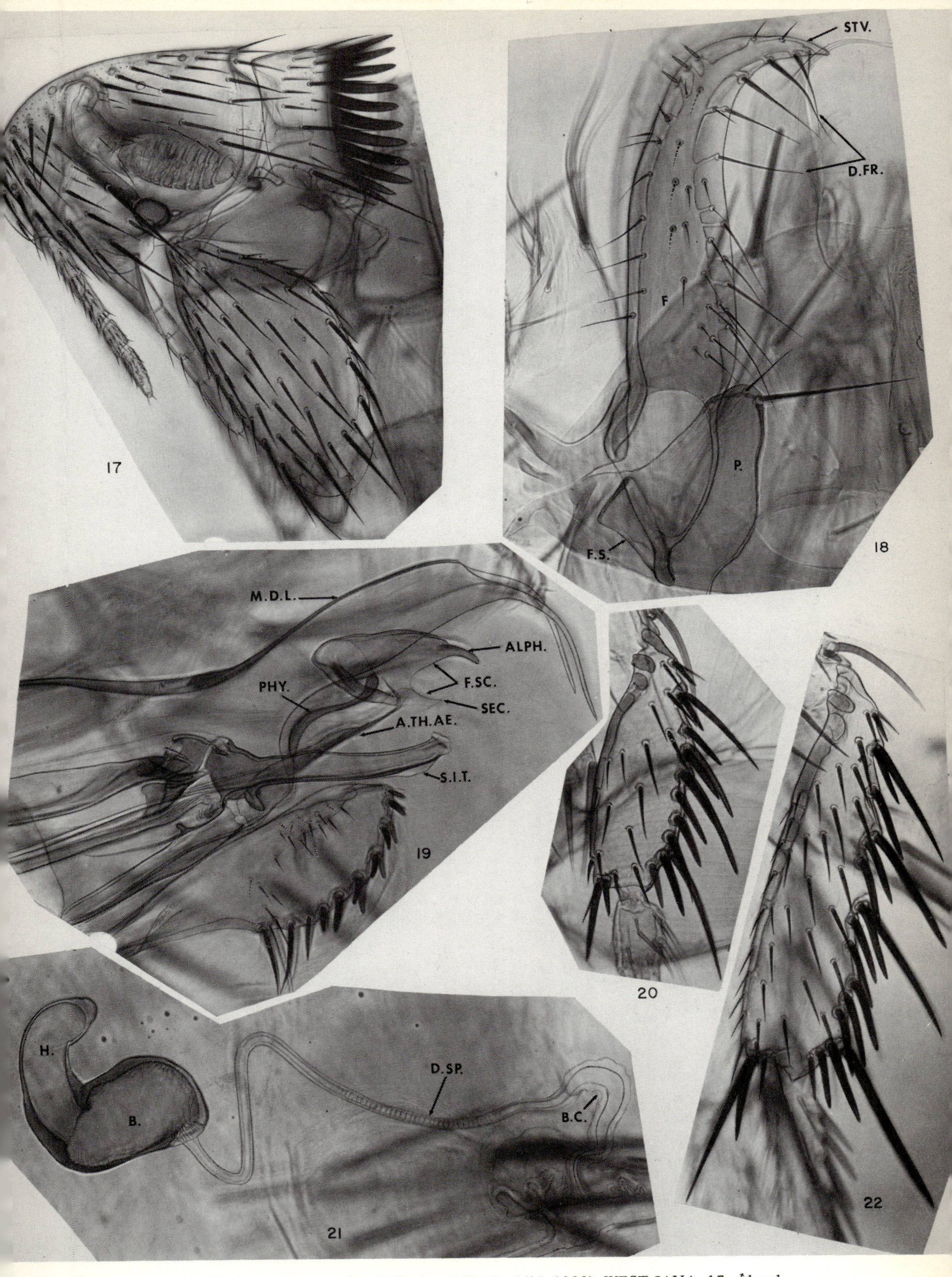

Figures 17 - 22. *Migrastivalius (M.) jacobsoni* (Jordan & Rothschild, 1922). WEST JAVA. 17. ♂ head and prothorax. 18. Processes of clasper. 19. Apex of aedeagus. 20. ♂ mesotibia. 21. Spermatheca and genitalia. 22. ♂ metatibia.

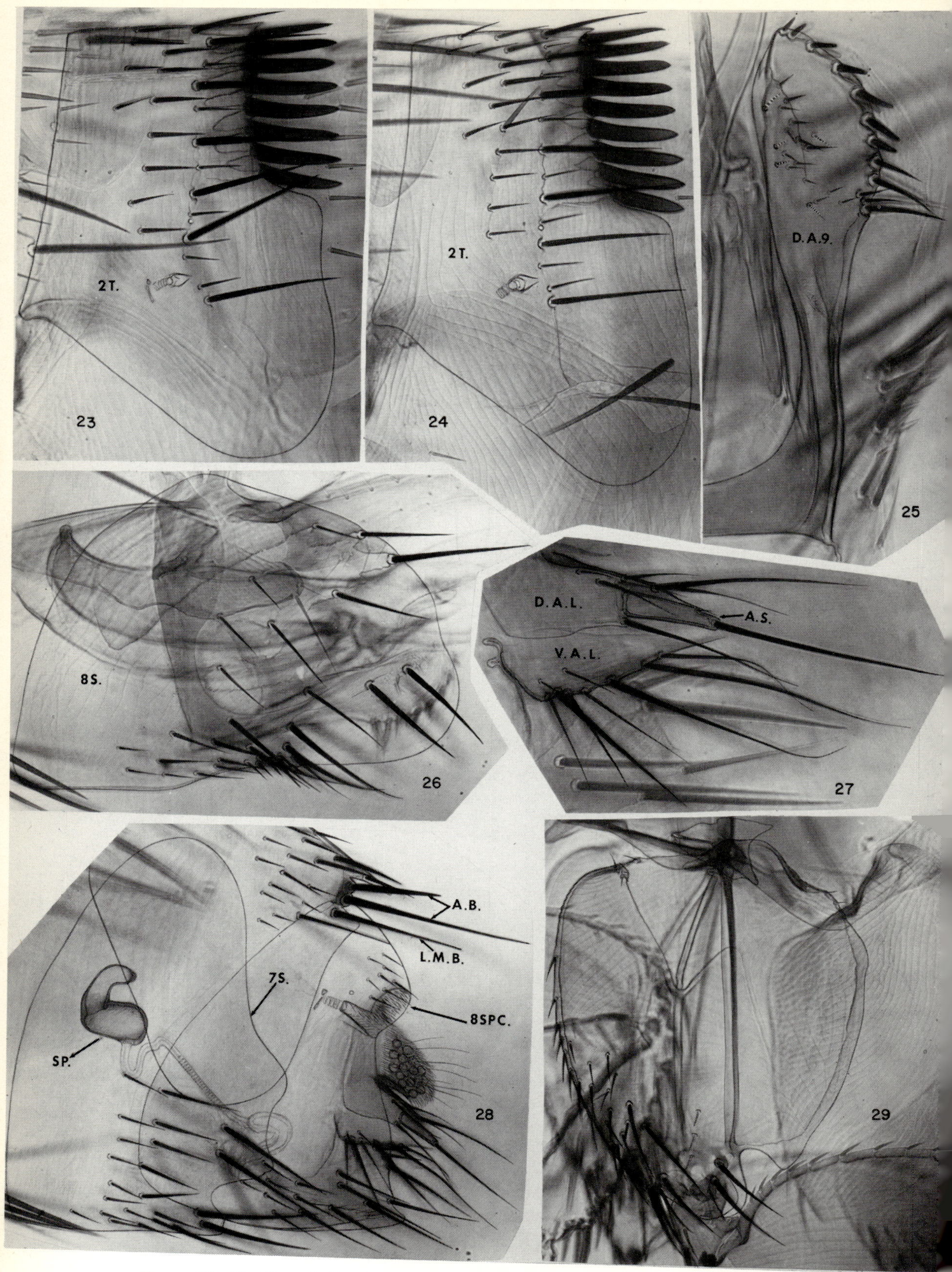

Figures 23 - 29. *Migrastivalius (M.) jacobsoni* (Jordan & Rothschild, 1922). WEST JAVA. 23. ♂ tergum 2. 24. ♀ tergum 2. 25. Distal arm of sternum 9. 26. ♂ sternum 8. 27. ♀ anal segments. 28. ♀ modified abdominal segments. 29. ♂ metacoxa.

Figures 30 - 34. *Migrastivalius (Gryphopsylla) hopkinsi* (Traub, 1957). NORTH BORNEO. 30. ♂ head and prothorax. 31. ♂ metatibia. 32. ♂ protibia. 33. ♂ anal segments. 34. ♂ terga 1-2.

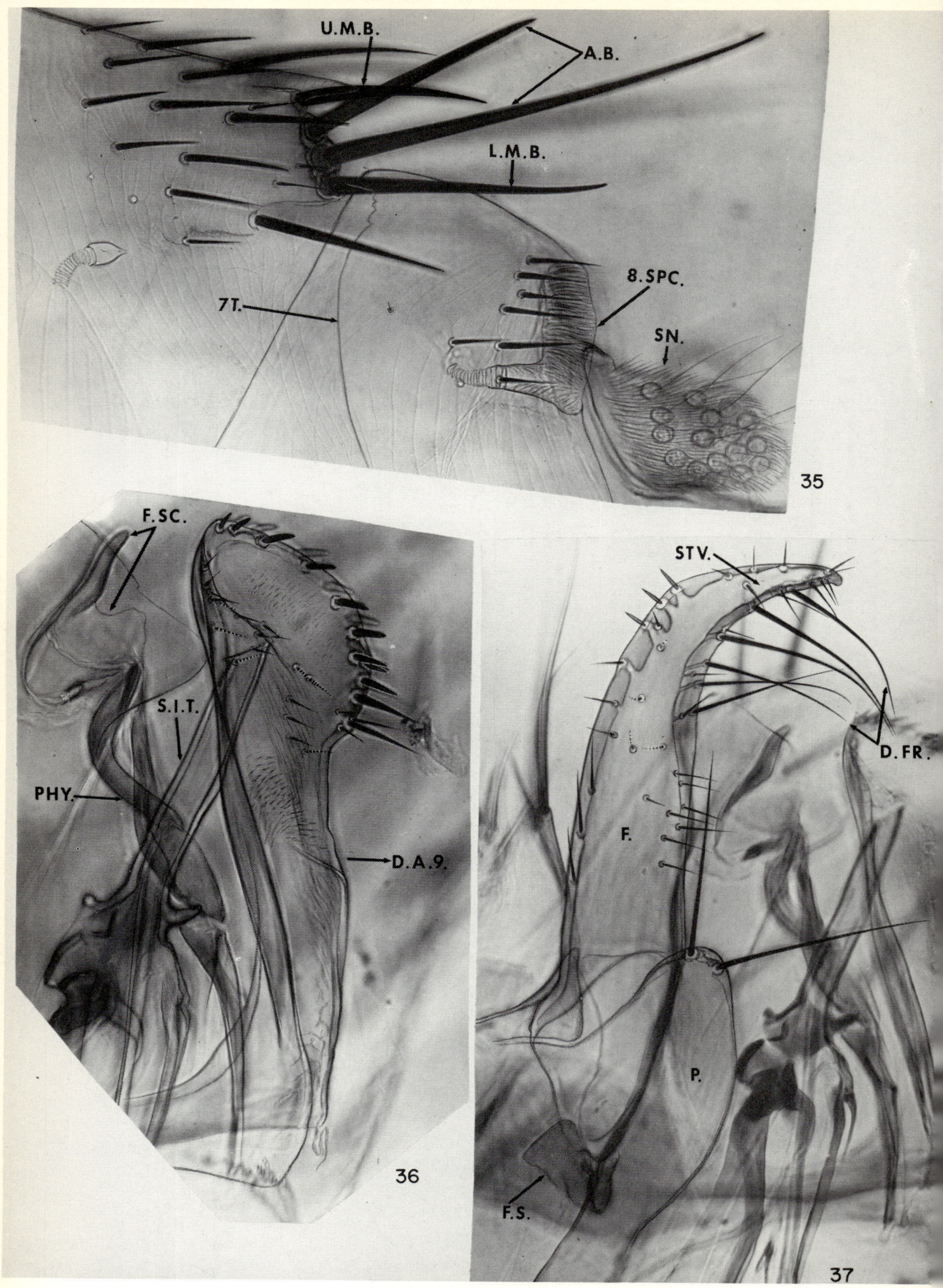

Figures 35 - 37. *Migrastivalius (Gryphopsylla) hopkinsi* (Traub, 1957). NORTH BORNEO. 35. ♀ antepygidial region. 36. Distal arm of sternum 9. 37. Processes of clasper.

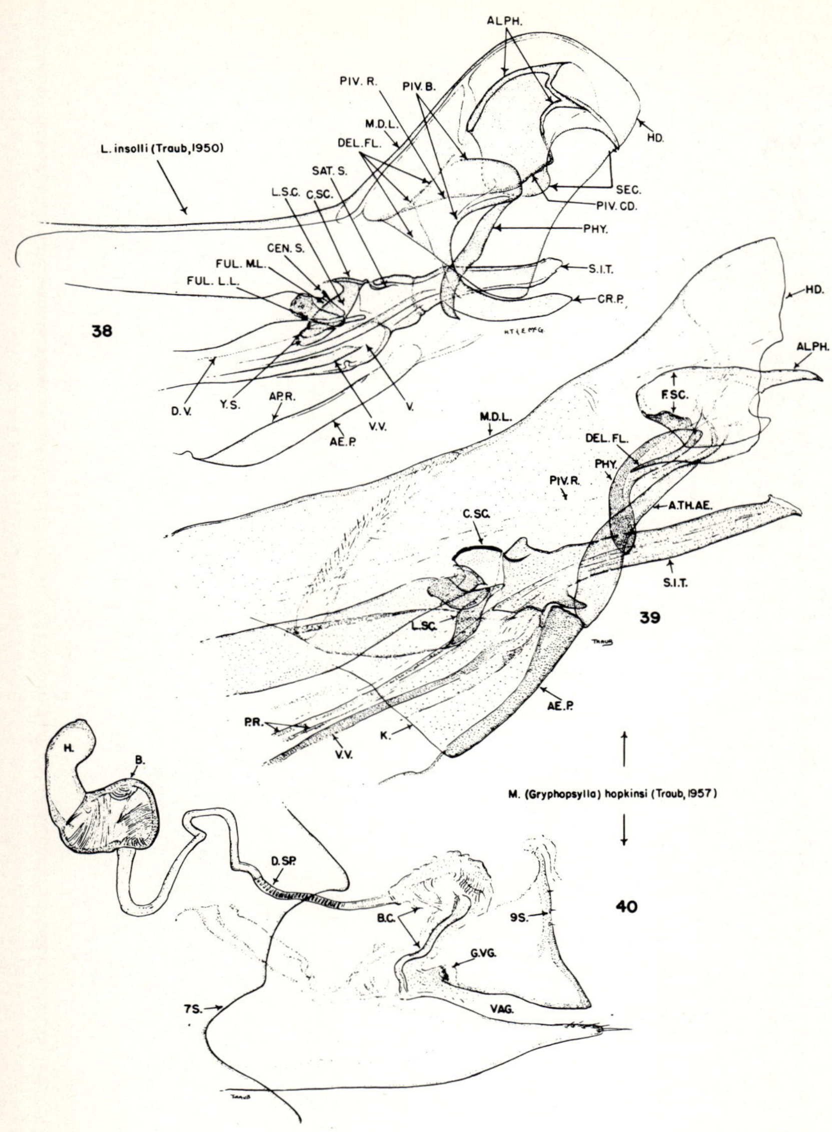

Figure 38. *Lentistivalius (L.) insolli* (Traub, 1950). MALAYA. Apical region of aedeagus.

Figures 39 - 40. *Migrastivalius (Gryphopsylla) hopkinsi* (Traub, 1957). NORTH BORNEO. 39. Apical region of aedeagus. 40. Spermatheca and genitalia.

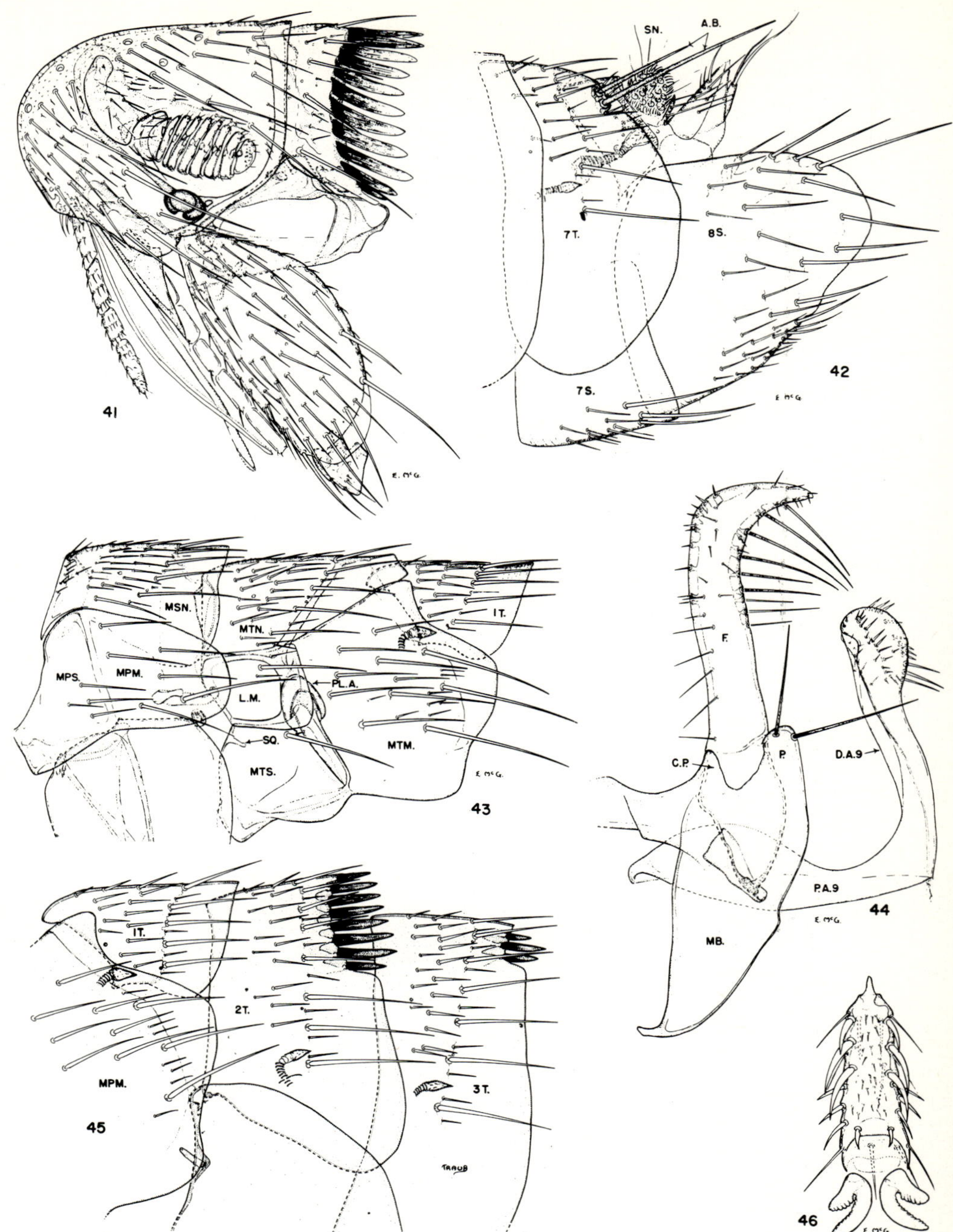

Figures 41 - 46. *Lentistivalius (Destivalius) mjoebergi* (Jordan, 1926). NORTH BORNEO. 41. ♂ head and prothorax. 42. ♂ modified abdominal segments. 43. ♂ meso- and metathorax. 44. Clasper and sternum 9. 45. ♂ terga 1-3. 46. Tarsal claw.

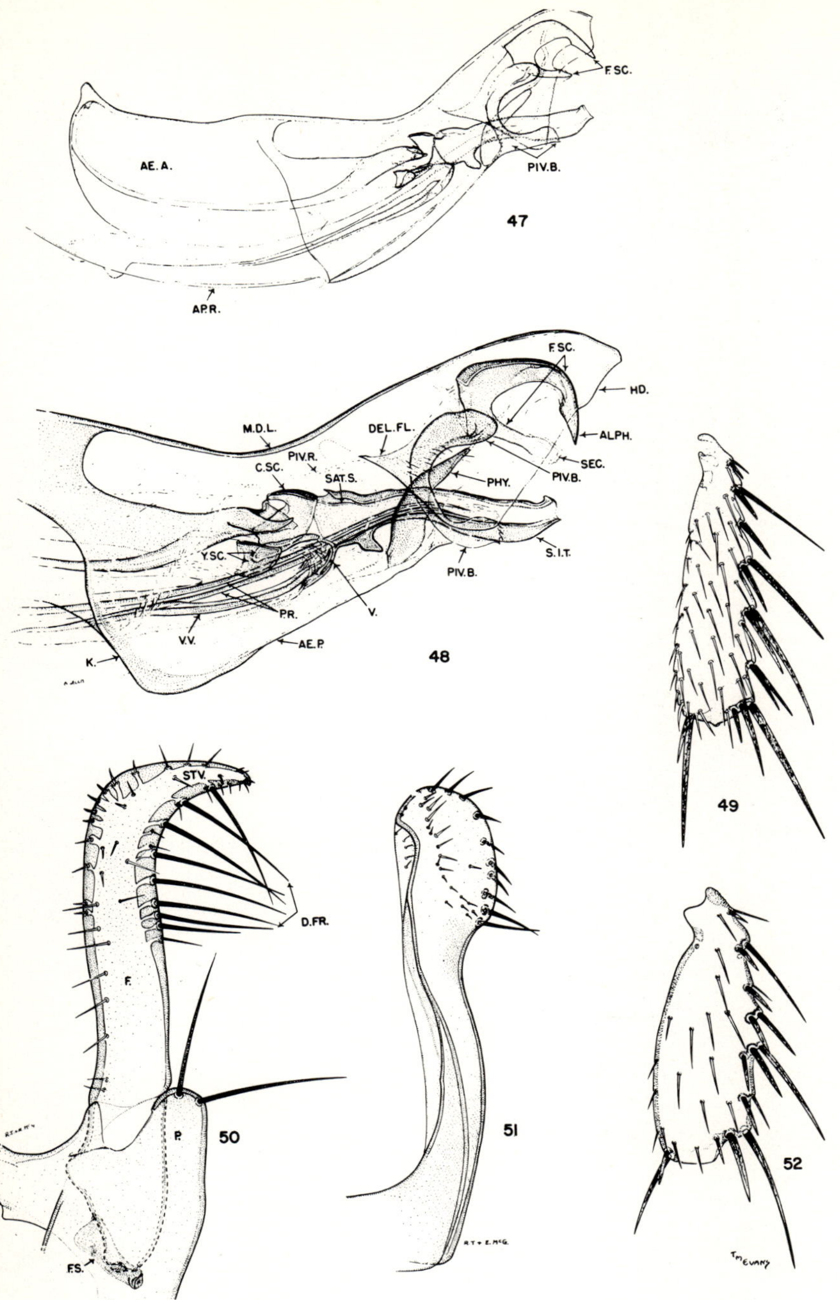

Figures 47 - 52. *Lentistivalius (Destivalius) mjoebergi* (Jordan, 1926). NORTH BORNEO. 47. Aedeagus. 48. Apical region of aedeagus. 49. ♂ metatibia. 50. Processes of clasper. 51. Distal arm of sternum 9. 52. ♂ protibia.

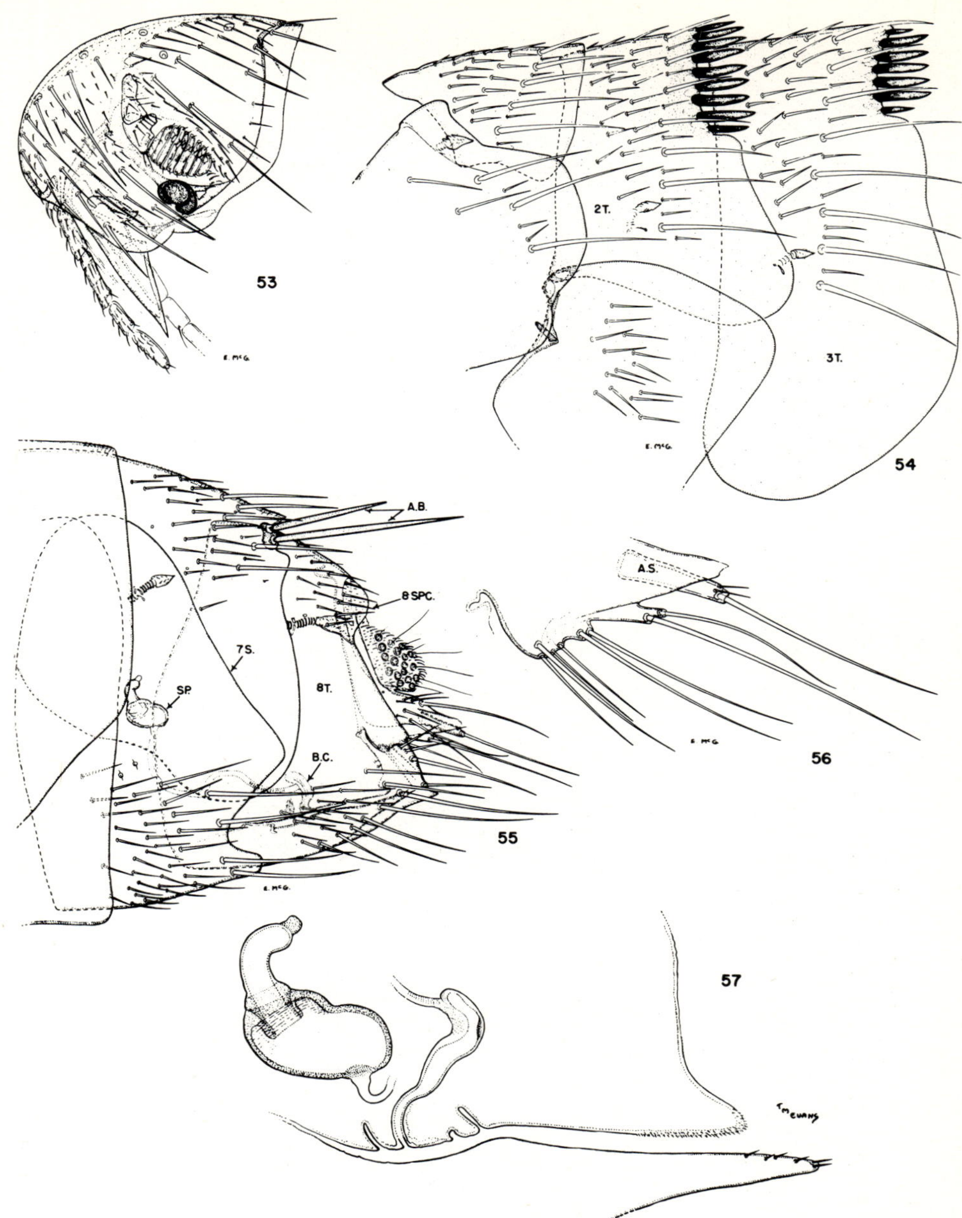

Figures 53 - 57. *Lentistivalius (Destivalius) mjoebergi* (Jordan, 1926). NORTH BORNEO. 53. ♀ head. 54. ♀ terga 1-3. 55. ♀ modified abdominal segments. 56. ♀ anal segments. 57. Spermatheca and bursa copulatrix.

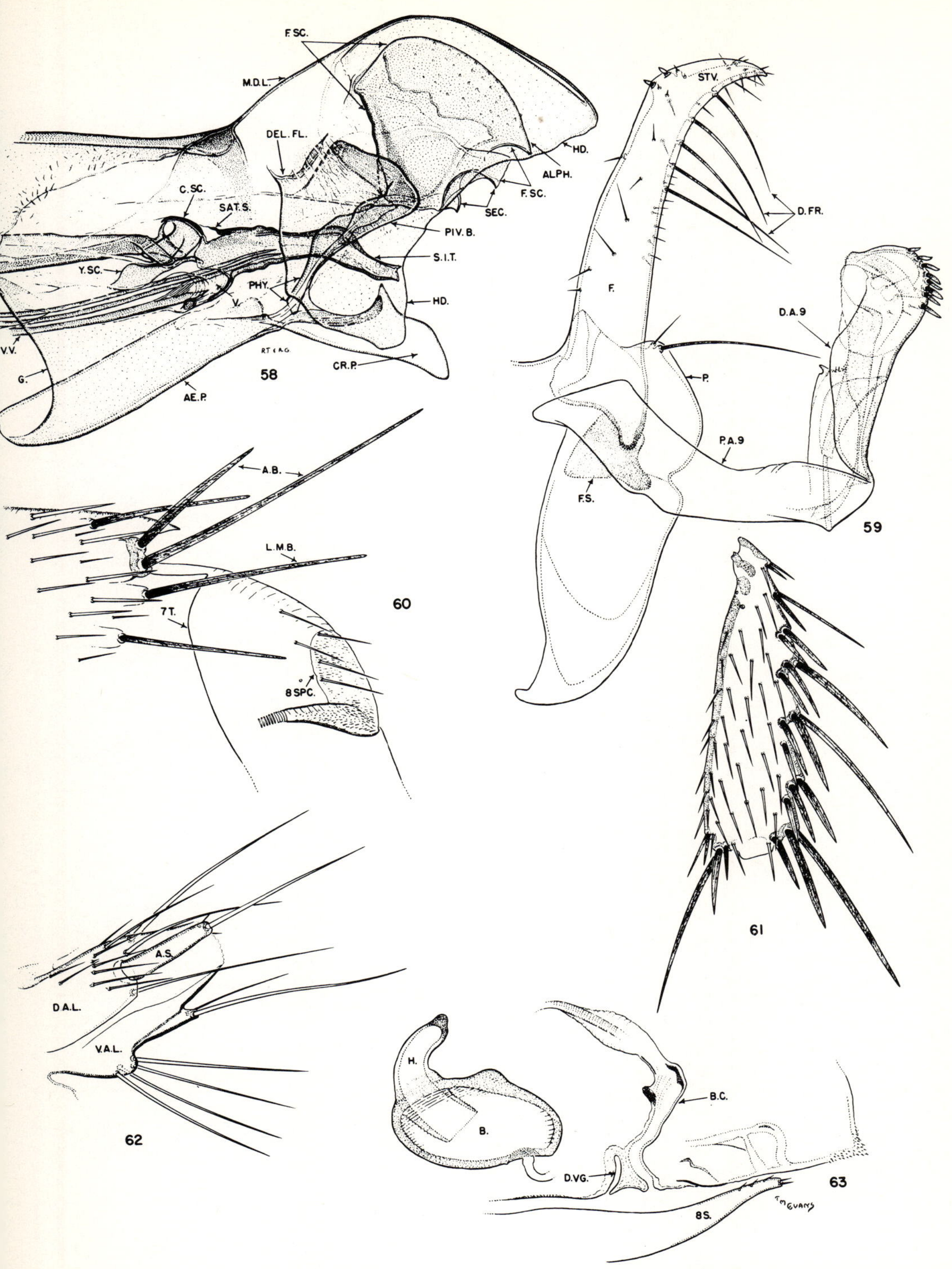

Figures 58 - 63. *Aviostivalius klossi* (Jordan & Rothschild, 1922). MALAYA. 58. Apical region of aedeagus. 59. Clasper and sternum 9. 60. ♀ antepygidial region. 61. ♂ metatibia. 62. ♀ anal segments. 63. Spermatheca and bursa copulatrix.

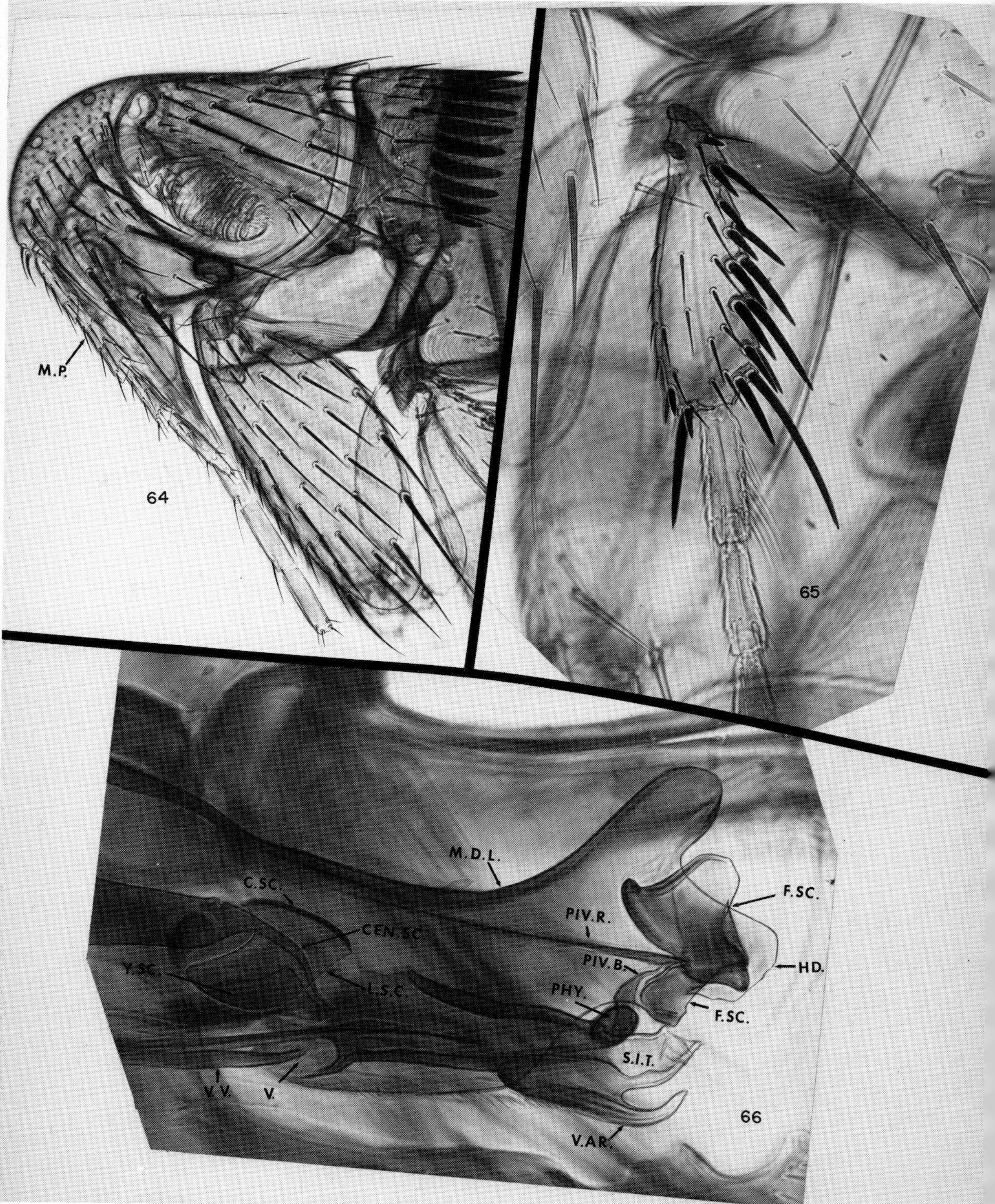

Figures 64 - 66. *Nestivalius pomerantzi* (Traub, 1951). PHILIPPINES, Agusan, Davao Province, Mindanao. 64. ♂ head and prothorax. 65. ♂ protibia. 66. Apical region of aedeagus.

Figures 67 - 70. *Nestivalius pomerantzi* (Traub, 1951). PHILIPPINES, Agusan, Davao Province, Mindanao. 67. Apex of aedeagus. 68. Distal arm of sternum 9. 69. Processes of clasper. 70. ♂ metatibia.

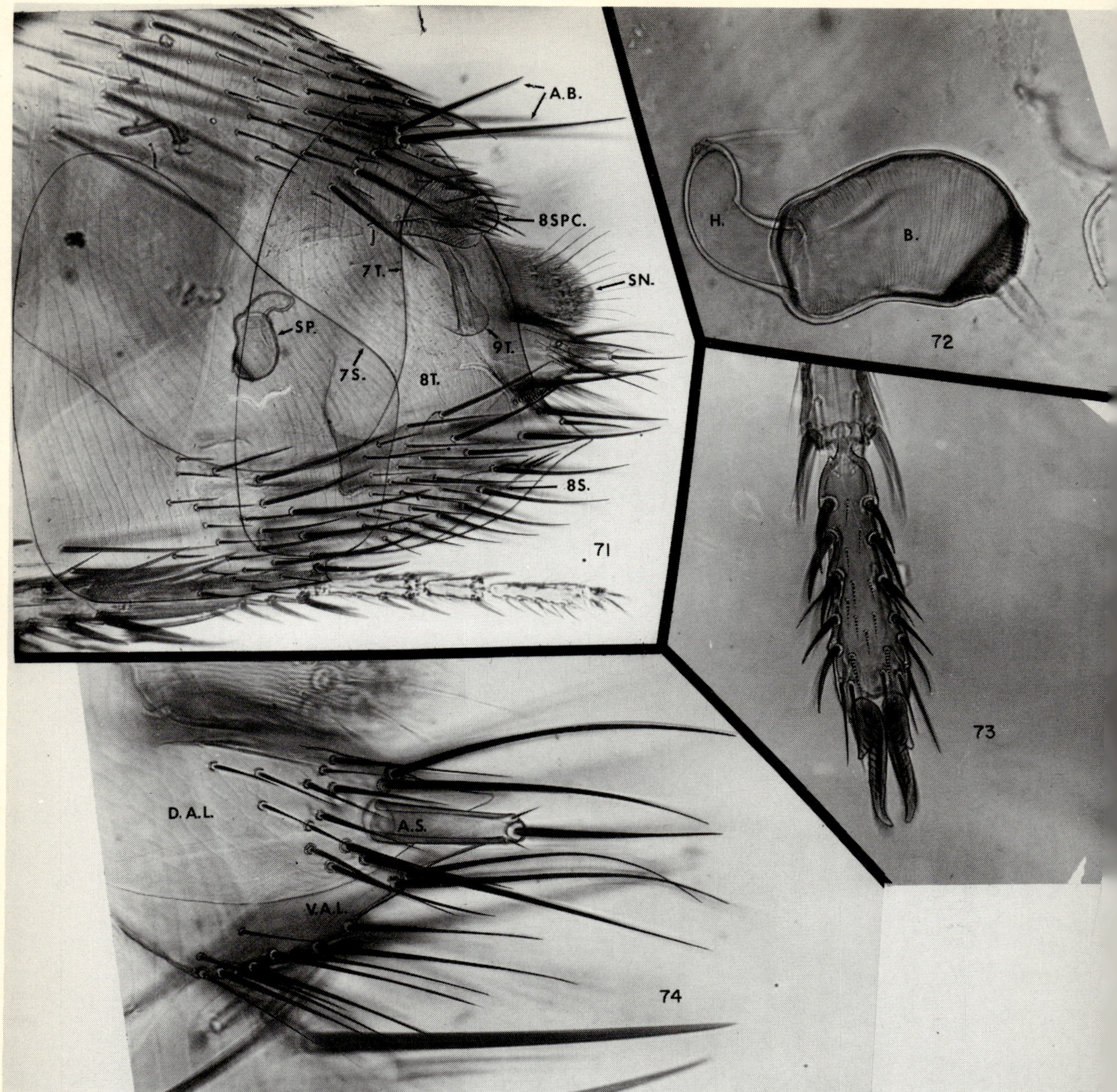

Figures 71 - 74. *Nestivalius pomerantzi* (Traub, 1951). PHILIPPINES, Agusan, Davao Province, Mindanao. 71. ♀ modified abdominal segments. 72. Spermatheca. 73. ♂ metatarsus. 74. ♀ anal segments.

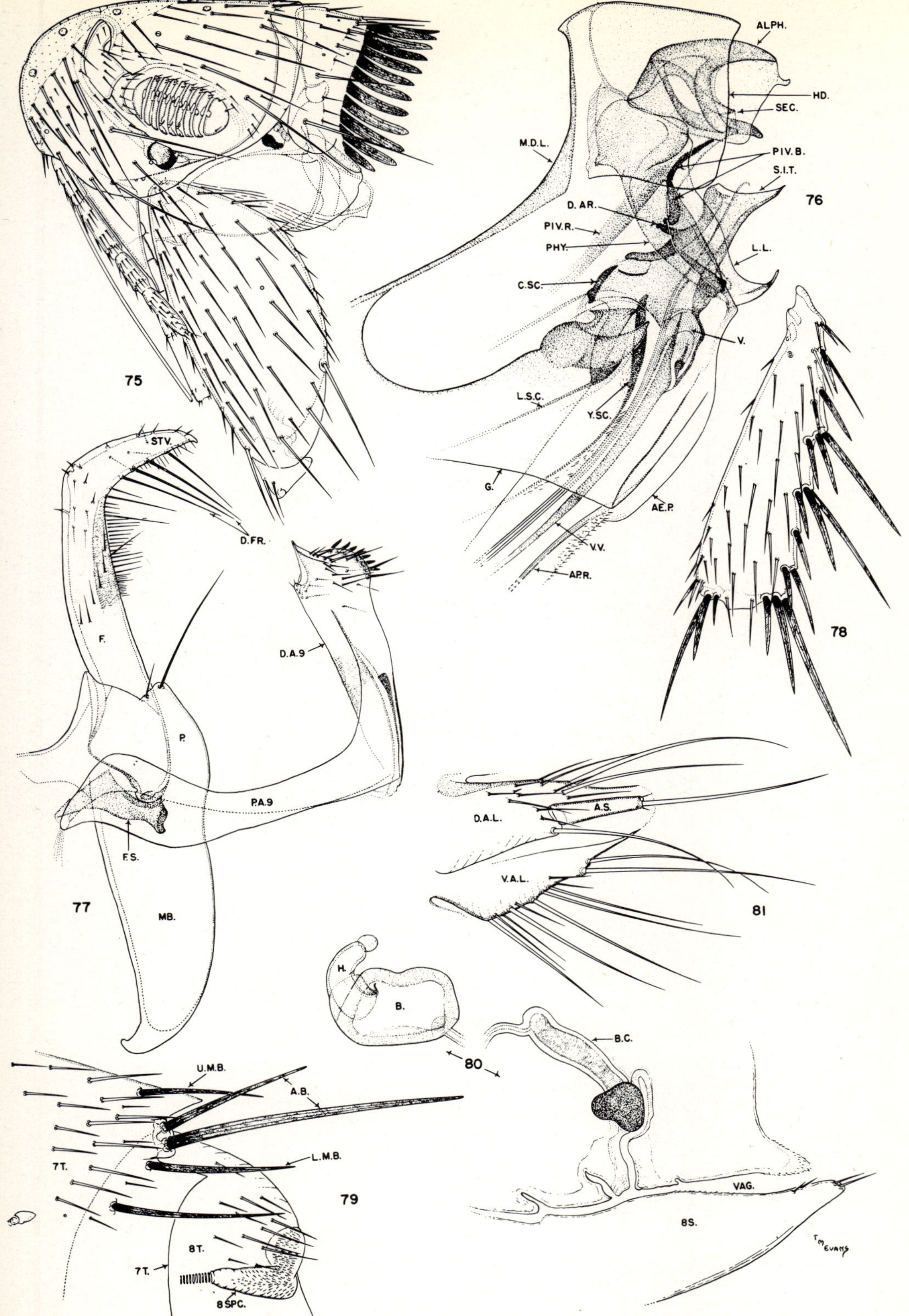

Figures 75 - 81. *Afristivalius torvus* (Rothschild, 1908). CONGO. 75. ♂ head and prothorax. 76. Apical region of aedeagus. 77. Clasper and sternum 9. 78. ♂ metatibia. 79. ♀ antepygidial region. 80. Spermatheca and bursa copulatrix. 81. ♀ anal segments.

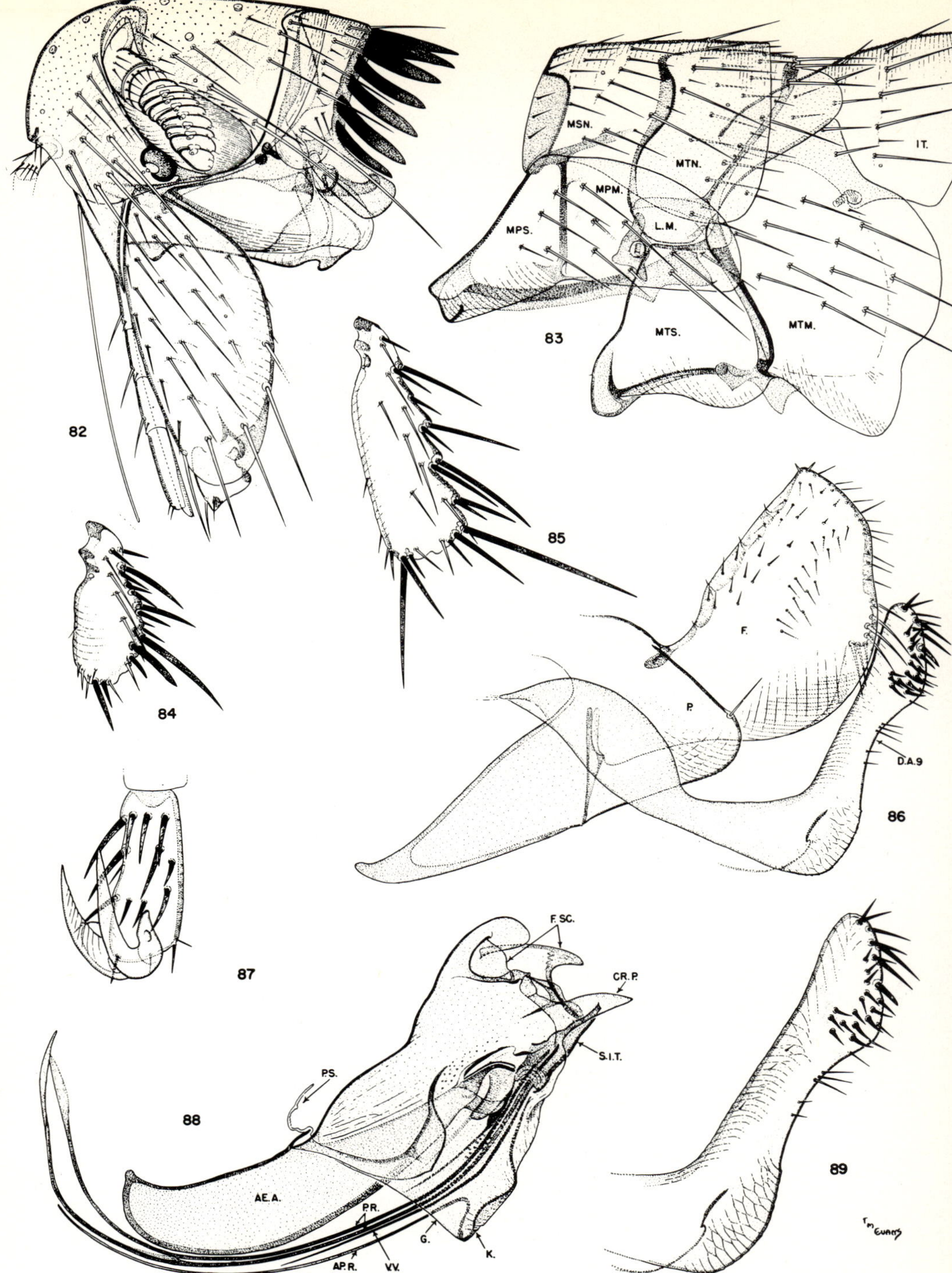

Figures 82 - 89. *Farhangia celebensis* (Ewing, 1924). SULAWESI (Celebes). 82. ♂ head and prothorax. 83. ♂ thorax. 84. ♂ protibia. 85. ♂ metatibia. 86. Clasper and sternum 9. 87. ♂ metatarsus 5 (inner aspect). 88. Aedeagus. 89. Distal arm of sternum 9.

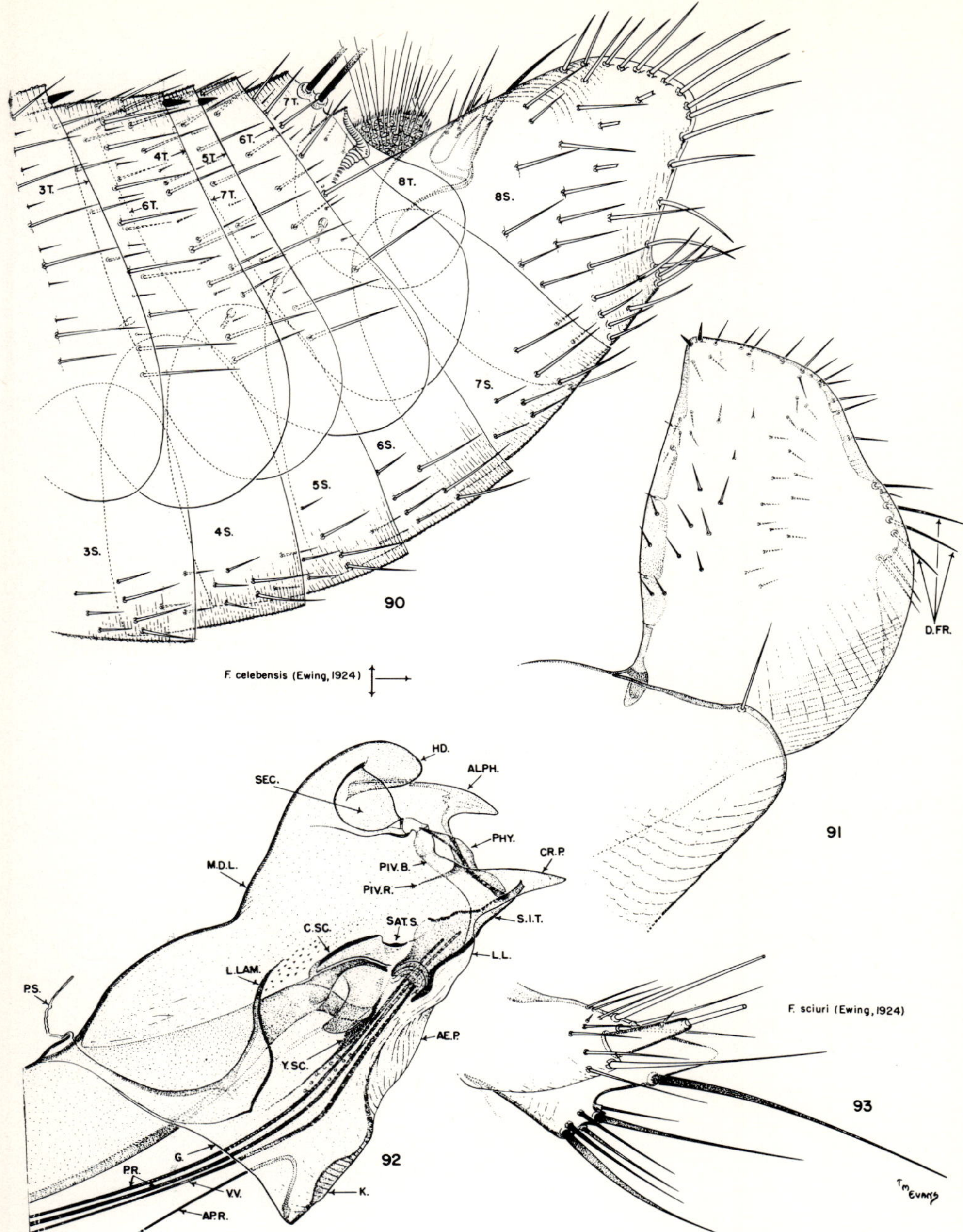

Figures 90 - 93. *Farhangia celebensis* (Ewing, 1924). SULAWESI (Celebes). 90. ♂ abdomen. 91. Processes of clasper. 92. Apical region of aedeagus. 93. *Farhangia sciuri* (Ewing, 1924). INDONESIAN BORNEO, E. Sungei Merah. ♀ anal segments.

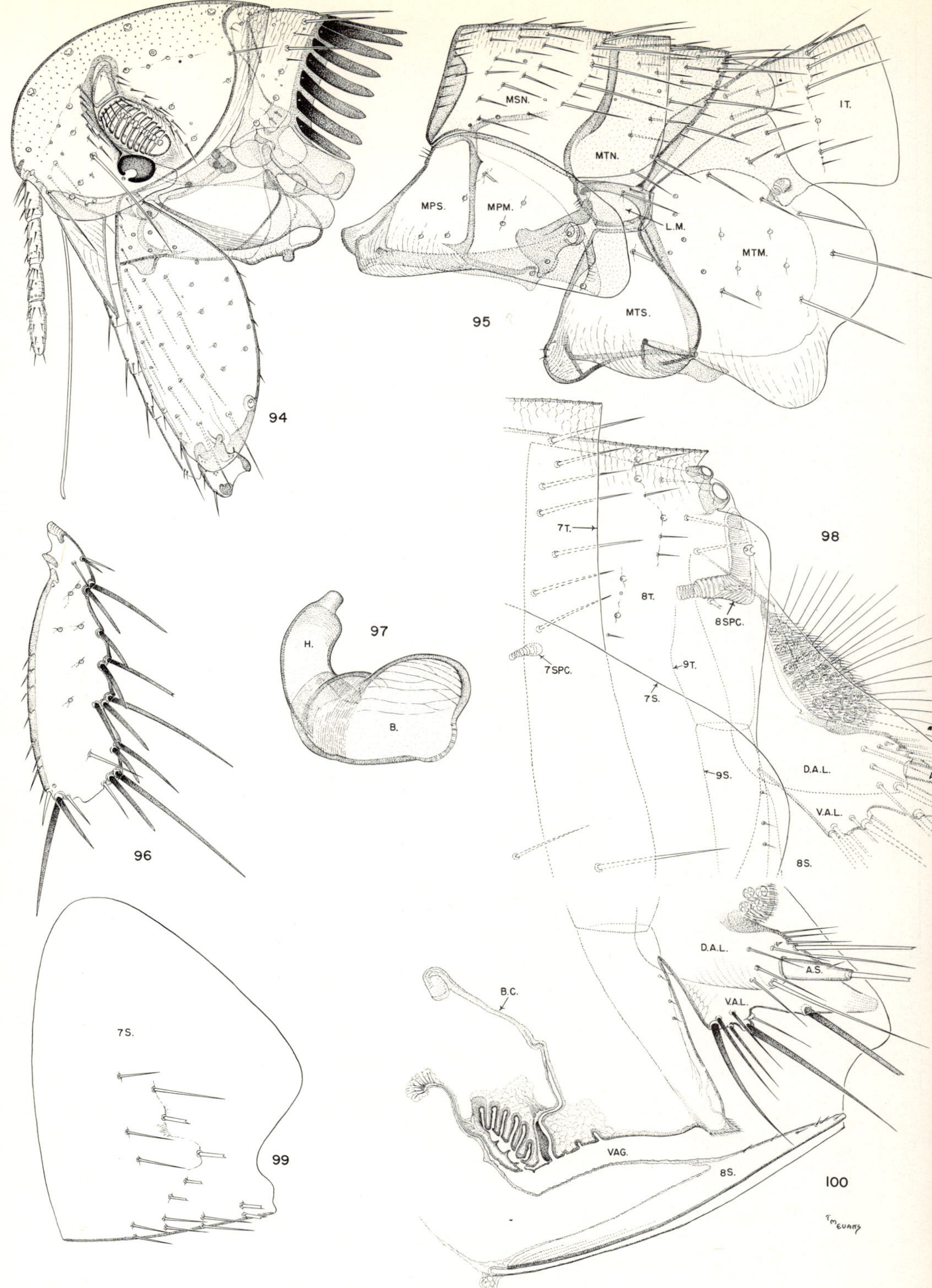

Figures 94 - 100. *Farhangia sciuri* (Ewing, 1924). INDONESIAN BORNEO, E. Sungei Merah. 94. ♀ head and prothorax. 95. ♀ thorax. 96. ♀ metatibia. 97. ♀ sternum 7. 98. ♀ modified abdominal segments. 99. Spermatheca. 100. Bursa copulatrix.

preposition *de*, in the sense of 'derived from' and refers to the descent from the *Stivalius*-group.

The abdominal combs (and nature of the pronotal comb as well) in my opinion, reflect the hazards posed by the scansorial habits of *Tupaia baluensis,* which is the prime host of this flea and which provided almost all of our records on Mt. Kinabalu and Mt. Trus Madi. (In contrast, *L.vomerus* Traub, 1972, in the same precise habitat is common on ground-squirrels, various rats, *Hylomys* and the tree-shrew *Dendrogale,* all of which spend most of their time on the ground.) The comb on 3 T. is better developed in the ♀ (fig. 54) than in the ♂ (fig. 45) and this is true of the other combs and general chaetotaxy, as is often the case in fleas. It is interesting that this species, which has a vestigial crochet process, possesses a rather elongate, relatively narrow S.I.T. (fig. 48) as compared to the nominate subgenus, and a hook-like Ford's sclerite, suggestive of the path taken by *Migrastivalius* but not as far along. The length and girth of the bursa copulatrix (fig. 56, B.C.) are also intermediate in that regard.

AVIOSTIVALIUS n. gen. (figs. 58-63)

Diagnosis – Near *Lentistivalius* but readily distinguishable as follows: 1) Labial palpi reaching to apex of procoxa instead of clearly falling short thereof. 2) With some of the dorsomarginal notches on metatibia bearing a group of three stout bristles (fig. 61) instead of a maximum of two (fig. 49). 3) Apical group of dorsomarginals on metatibia consisting of three stout bristles. In *Lentistivalius* there are two stout bristles, and if there is a third, it is thin. 4) First segment of metatarsus with stout dorsomarginal bristles along entire length instead of being restricted to subapical region. 5) Base of movable finger (fig. 59, F.) subtruncate, i.e. produced only into a short lobe; this ventrocaudal in position and its length only about one-third that of adjacent fulcral sclerite (F.S.); not with base produced into a long lobe submedian in position and about 1.5x length of F.S. (fig. 44). 6) Fulcral sclerite with height almost equalling length instead of being only one-third to one-half as tall as long. 7) Aedeagus (fig. 58) with crochet process (CR.P.) produced into a huge lobe with a median dorsal spur; not with CR.P. ligulate and lacking a spur (figs. 38, 48). 8) ♂ sensilium unusually tall for its length, the ratio being 1:3 instead of only about 1:1 (fig. 42, SN.). 9) ♀ with spiracular fossa of 8 T. (fig. 60, 8 SPC.) with vertical arm scarcely broader than base of horizontal arm; instead of being at least 1.5x (fig. 55) to 2.5x. 10) ♀ with lower modified bristle (fig. 60, L.M.B.) resembling an A.B.; not distinctly shorter, thinner and lighter (fig. 55). 11) ♀ basal sternum with 1-3 bristles on lower ventro-caudal quadrant. These missing in *Lentistivalius.* 12) ♀ with base of A.B. recessed between dorsal and ventral lobes of 7 T. and ventral lobe with fairly acute apex (fig. 60) instead of with lower lobe undeveloped or blunt. 13) Spermatheca (fig. 63) with bulga (B.) narrowing caudad at level of median dorsal bulge, instead of being essentially subequal throughout or, if somewhat narrowed caudally (fig. 57), then lacking a dorsal median convexity.

Generic description – Agreeing with *Lentistivalius* as defined in 1972 except as indicated above and as follows: Typical abdominal terga with three rows of bristles, but first incomplete. Fulcral sclerite (fig. 59, F.S.) broader (longer) at base than tall; longest ventrally. Girdle of aedeagal pouch (fig. 58, G.) fairly well sclerotized. Phylax (PHY.) in part blending with pivotal base (PIV.B.) and hence not fully discrete as a separate sclerite.

Comments – The type of this monotypic genus is *S.klossi* (Jordan & Rothschild, 1922), a species that ranges from Java and Sumatra through the Malayan Peninsula to Indo-China and at least southeast China (Fukien). Various subspecies exist, but only one, *A.k.bispinosus* (Li & Wang, 1958) (*NEW COMBINATION)* has been described. The generic name is derived from the Latin *avius*, meaning remote, wandering or far from home, and refers to the great distance this taxon is found from its ultimate place of origin in the Australian Region.

NESTIVALIUS n.gen. (figs. 64-74)

Diagnosis – Close to, and probably a derivative of, *Orthopsylloides* Holland, 1969, but distinguishable as follows: 1) Anterodorsal angle of metepisternum acute and upturned, instead of being broadly rounded, or at most, blunt. 2) With metatibia bearing at most two bristles in dorsomarginal notches (fig. 70) and lacking any adjacent single stout bristles instead of some bearing three bristles or a single modified one near a pair. 3) Protibia (fig. 65) likewise at most with two dorsomarginals in notches instead of three. 4) Maxillary palpi (fig. 64, M.P.) with second (especially) and third segments significantly broader at apex than at base, i.e. tapering towards base instead of being only slightly broader and segment not tapered. 5) Crescent sclerite of aedeagus (fig. 68, C.SC.) long, its length exceeding breadth of middle of S.I.T., whereas in *Orthopsylloides* C.SC. scarcely exceeds half the diameter of S.I.T. 6) Satellite sclerite undeveloped. In *Orthopsylloides* it is well sclerotized, as in other taxa with a short C.SC. 7) Dorsal margin of aedeagal apodeme with a conspicuous hump instead of being straight or only slightly modified. 8) Ventral armature (fig. 67, V.AR.) extending to apex of S.I.T. and terminating in an acute upturned projection, whereas in the other genus, V.AR. is at most subapical and lacks a hook-like extension. 9) Hood (HD.) with a large sinus at level of Ford's sclerite, instead of being entire here or slightly sinuate. 10) Phylax (PHY.) reduced to a small spherical mass in the dorsal, subapical region of S.I.T., not a distinct rod-like sclerite. 11) ♂ ventral anal lobe (fig. 69, V.A.L.) with 2-4 (rarely only one) short marginal bristles below the long apical one. These are missing in *Orthopsylloides.* 12) Spermatheca with bulga (fig. 72, B.) about 1.8-2x as long as broad at maximum level and caudal region not appreciably broader than anterior half; instead of being at most 1.5x as long as broad, and dilated caudally. 13) ♀ with base of antepygidial bristles (fig. 73, A.B.) recessed between well-developed dorsal and ventral lobe-like extensions of 7 T.; the ventral one acute. In *Orthopsylloides* there is scarcely any dorsal lobe and the ventral one is at most subacute. 14) ♀ V.A.L. (fig. 74) relatively well-clothed with marginal bristles so that there is no conspicuous gap at basal fourth which is comparable to the subapical one, instead of having two marginal gaps. 15) V.A.L. with a group of 13-15 small subdorsal bristles that extend unto the basal half; instead of having only 2-5 such bristles and not reaching dorsoapicad of middle of V.A.L.

Generic description – Agreeing with *Orthopsylloides* as defined by Holland, except as diagnosed above and as follows. Pronotal spines subovate or bluntly rounded (fig. 64). D.A.9 only slightly curved dorsad apically (fig. 68).

In addition to being unusual in lacking a stiva on the movable finger (fig. 69, F.; cf. figs. 50, 59), *Nestivalius* shares some characteristic aedeagal features with *Orthopsylloides* (which have not yet been cited or illustrated) viz: 1) S.I.T. (figs. 66, 67) relatively short and thick and lacking dorsal modifications or a sub-basal spur. 2) With a very large ventral armature (V.AR.) which is relatively or actually apical in position. 3) Pivotal ridge (PIV. R.) of Ford's sclerite (F.SC.) unusually well-developed, often quite broad and rod-like. 4) Phylax (PHY.) relatively reduced and not presenting as a distinct sclerite flanking entire side of S.I.T. 5) Pivotal base (PIV.B.) also reduced; not developed sufficiently to reach, or serve as a support for, the anteroventral portion of F.SC. (cf. fig. 48).

Comments – The type of the genus is *pomerantzi* (Traub, 1951) described as a *Stivalius,* as was *ralius* (Traub, 1951), which also belongs here, as does a new species from Sulawesi. Both *N.pomerantzi* and *N.ralius* were described from the Philippines, Mindanao, Davao Province on Mt. McKinley; the former from 2 ♂ ex '*Rattus*'; the latter from 1 ♀ taken at a 70 m lower elevation ex '*Apomys*' by the same team under the direction of H.Hoogstraal. Although *Apomys* is now considered a synonym of *Rattus,* the fact that both names were used simultaneously indicates that two different species of *Rattus* were involved. At the time of description it was pointed out that *N.ralius* might turn out to be the ♀ of *N.pomerantzi.* The ♀ of *N.pomerantzi* is now illustrated for the first time, and there are few differences from *N.ralius* that may be significant, e.g. the narrower base of the hilla (fig. 74, H.) in the former. However, no subsequent material has been taken from Mt. McKinley and hence the status of *N.ralius* still cannot be evaluated since 1) females of species of *Stivalius s. lat.* often show little if any difference from one another and 2) there are marked and valid variations at the subspecies level of *N.pomerantzi* (e.g. the numbers of bristles on ♀ 8 T. near the spiracular fossa). One therefore cannot extrapolate from the characters seen in females from other mountains on Mindanao and determine the status of *N.ralius.*

The unusual position, size and talon-like modification of the ventral armature of S.I.T. (fig. 66, V.AR.) apparently is in response to the loss of the crochet as a structure for clasping the female. Such hyperdevelopment is an example of the Principle of Structural Compensation (Traub, 1972b), paralleling the situation in *Parastivalius.*

The close relationship of *Nestivalius* to *Orthopsylloides* is of interest in evolution and zoogeography. The latter is essentially a New Guinean genus, described on the basis of two species, and I know of an additional four new species from that island and one from the Solomon Islands. The vast majority of the genera of pygiopsyllids that occur on New Guinea are either bioendemic or else are restricted to Australia and New Guinea (Traub, 1972c; and in press, this volume). The fact that a genus that is apparently derived from *Orthopsylloides* exists on Sulawesi and the Philippines makes sense zoogeographically. All the *Orthopsylloides* are parasites on native species of the genus *Rattus,* a relatively youthful taxon which I believe arose in Southeast Asia and moved back and forth between the islands of the Oriental Region and New Guinea and Australia in the recent geological past (though not as recently as the *Rattus rattus*-group) (op. cit.). *Nestivalius* and *Bibikovana* are also *Rattus*-fleas, and the latter is found in Australia, New Guinea and Borneo. Significantly, its host on Borneo is a primitive species of *Rattus.* As mentioned in the references cited, as *Rattus* moved towards Australia and New Guinea, it carried the Palaearctic-derived *Sigmactenus* with it, and then acquired and transported in the return direction some pygiopsyllids, such as these. Species of these pygiopsyllids have evolved on the various islands en route, e.g. the new *Nestivalius* on Sulawesi, six species of a new genus on Sulawesi and West-Irian, and the *Orthopsylloides* in the Solomons, as *Rattus* moved northeast.

AFRISTIVALIUS n.gen. (figs. 75-81)

There are two genera of pygiopsyllids that are found in Africa (both occurring only south of the Sahara): 1) *Lentistivalius,* represented there by only one known species, *L.alienus* (Smit, 1958) and 2) a component of *Stivalius s.lat.,* constituting a distinctive genus, *Afristivalius* n.gen., including 15 known species and categorized as follows.

Diagnosis – Immediately separable from typical *Stivalius s.lat.* (including *Lentistivalius,* to which it is related and with which it is being directly compared) in that the movable finger (fig. 77, F.) bears a patch of small, thin, mesal bristles below the stiva (STV.); and 2) the distal fringe (D.FR.) is limited essentially to the stiva area (cf. figs. 50 and 59). Further characterized as follows: 3) Labial palpi (fig. 75) 5- or 6-segmented and not extending beyond apex of procoxa; generally reaching only ¾ its length. 4) Labial palpi with apparent fifth segment about twice the length of the fourth, but apical one-sixth generally set off as a fairly distinct segment (#6) and if not truly divided here, then still bearing the small apical bristles indicating the joint. In *Lentistivalius* and most *Stivalius s.lat.* (but not *Migrastivalius* n.gen.), #5 is slightly longer or subequal. 5) With some dorsomarginal notches of metatibia (fig. 78) usually with three stout bristles. 6) Anterodorsal angle of mesepisternum squarish, not up-curved. 7) Aedeagal apodeme with a conspicuous pre-apical sinus or concavity; markedly convex caudad of this sinus. 8) Ford's sclerite (fig. 76, F.SC.) massive and bearing at least one hook or acuminate projection. 9) Lacking a true crochet or caudad-directed crochet process (cf. figs. 38 and 58). 10) With a lateral lobe (L.L.) covering part of the sclerotized inner tube (S.I.T.) in the vicinity of the well-developed phylax (PHY.). L.L. at times becoming narrowed or acuminate apically. 11) S.I.T. without a ventral spur though bearing a dorsal spur near middle and often an apical dorsal thickening. 12) Pivotal base well-developed, at least at base of F.SC. and often untanned more ventrad. 13) Penis rods making at least half a coil, and generally with 1-2 convolutions. 14) ♀ ventral anal lobe (fig. 81, V.A.L.) unusually long, clearly longer than anal stylet (A.S.) and with anteroventral margin squarish, not rounded. 15) Spermatheca (fig. 80) with bulga (B.) thick-walled; lacking internal striae except near orifice to duct. 16) Hilla (H.) unusually thick-walled and hence with basal half appearing 'submerged' in bulga. 17) With a tanned sclerite or spot on the left and right sides of the bursa copulatrix (B.C.); this generally conspicuous and characteristic in shape in various species, but poorly developed in *A.curtiductus* (Smit, 1958). 18) In both sexes, antepygidial bristles (fig. 79, A.B.) with base 'recessed' somewhat within short lobes of seventh tergum (7 T.).

Generic description – Agreeing with *Lentistivalius* as defined originally except as indicated above and as follows. Genal margin entire below eye. Dorsal margin of pronotum generally subequal to dorsalmost spines of comb. Lateral metanotal area usually subquadrate, without one axis greatly exceeding the other. Third segment of metatarsus long, generally about 1.7x #4 and 1.4x #5. Abdominal terga 2-4 or 5 usually with a subdorsal apical spinelet but at times bearing a short comb of longish spines on some terga (i.e. 3 species with 3 comblets). Unmodified abdominal terga with 2½-3 rows of bristles. ♀ with 1 or 2 of lower modified bristles (fig. 79, L.M.B.) resembling A.B. Immovable process of clasper (fig. 77, P.) fairly broad or subovate. Fulcral sclerite (F.S.) usually longer than tall but with maximum breadth at anterior (dorsal) margin. Crescent sclerite (fig. 76, C.SC.) short. Satellite sclerite (SAT.S.) subequal to C.SC. but less tanned. Caverna spiculosa not developed. ♀ 8 S. usually flask-shaped apically but at times subovate.

Comments – Type species: *S.torvus* (Rothschild, 1908), which has been well described and figured by Smit (1958). Included are: *afer* (Rothschild, 1908); *azevedoi* (Ribiero, 1974) (♀ only); *curtiductus* (Smit, 1958) (♀ only); *fallociosus* (Smit, 1958); *perilis* (Smit, 1958); *pirloti* (Smit, 1958); *rahmi* (Beaucournu, 1969); *richardi* (Jordan, 1936); *sellatus* (Jordan & Rothschild, 1923); *smitianus* (Beaucournu, 1966); *timanus* (Jordan, 1938) and *vancanneyti* (Berteaux, 1947). *S.nigeriensis* (Jordan, 1938) is tentatively placed here but presumably is modified as a shrew-flea and hence is somewhat anomalous, with a narrow pronotum, generally only two pairs of bristles in the notches of the metatibiae as in *Lentistivalius,* etc.

The generic name is derived from the Latin word *Afri,* meaning the dwellers in Africa and refers to the bioendemic nature of the taxon. The genus, however, does not occur throughout the continent and has not been taken in Saharan Africa, Ethiopia, nor in the southern parts (Rhodesia and southwards). However, it is represented in Angola by several species, and the others are scattered across the breadth of the continent between those limits. *Lentistivalius alienus* has been collected in Mozambique and *Afristivalius* probably occurs there as well. The South African Siphonapteran fauna has been well studied and the absence of *Afristivalius* in the records may represent the true state of affairs. The aridity there and in the Saharan areas may be a limiting factor for *Afristivalius.*

The typical species are apparently parasites of murines, i.e. native rats formerly considered *Rattus* but now placed in a variety of genera, and there are nearly always 22-24 straight spines in the genal comb, suggesting the condition in other groups of rat-fleas (vide my article on adaptations in this volume). In *A.rahmi,* however, the comb has 28-30 spines. This species infests a swamp-rat, and the hyperdevelopment of the ctenidium is considered a consequence of parasitizing a host whose traits place the flea at special risk (op. cit.).

FARHANGIA n.gen. (figs. 82-100)

From the original descriptions it has been impossible to correctly determine the true status of the fleas described as *Pygiopsylla celebensis* Ewing, 1924 and *P.sciuri* Ewing, 1924. Each is known only from the type, the former a ♂ 'from the skin of *Sciurus evidens*' collected in Sulawesi, the latter a ♀ 'from the skin of *Sciurus atricapillus*' taken on Borneo. Both of the types are in poor shape, devoid of most of the limbs and many bristles and somewhat distorted, with parts obscured, but it is apparent at a glance that they are not *Pygiopsylla* and instead represent a highly distinctive new genus, designated in the above caption and categorized as follows.

Diagnosis – While agreeing with *Pygiopsylla s.lat.* in that the sensilium is not as markedly convex as in *Stivalius s.lat., Farhangia* n.gen. is immediately separable from '*Pygiopsylla*' in that 1) the eye is in line with the base of the procoxa (figs. 82, 94) instead of being more forward; 2) the metasternum (figs. 83, 95) is only slightly biconvex, with both lobes extending only slightly over the metacoxa, instead of being highly biconvex; 3) there is no large patch of mesal bristles on the metacoxa, whereas 4) there is a lateral group of bristles on the ♀ basal sternum, instead of vice versa regarding 3) and 4). Closer to *Stivalius s.lat.* than *Pygiopsylla* but with no known close relatives and highly unusual in the family in that: 5) there has been a complete loss of the pleural arch (figs. 83, 95; cf. fig. 43, PL.A.). 6) Lateral metanotal area (L.M.) relatively very narrow, much longer than tall, instead of being subquadrate or with one dimension only slightly exceeding the other, as in *Pygiopsylla* and *Stivalius s.lat.* 7) Metanotum (MTN.) more than 1.4x as tall (at midline) as dorsal margin is long (excluding flange) instead of being subequal. 8) Dorsal margin of MTN. scarcely longer than ventral one instead of being much longer due to posterior margin diverging at an angle of about 45-55° from horizontal. 9) Squamulum absent. 10) Metatarsal segment 5 (fig. 87) with a displaced pair of plantar bristles lying between the apparent first pair, and with a total of only four laterals and presumably also with a displaced pair between the last lateral pair; instead of bearing the usual six pairs of plantar bristles, of which the first and/or third or fourth are slightly displaced (e.g. fig. 46). (Note: only one metatarsus 5 on specimens extant, and it is out of alighment and partially obscured, as figured.) 11) With only two rows of bristles on metatibia, and those incomplete (figs. 85, 96); instead of with approximately three or four irregular rows (e.g. figs. 13, 49, 61). Further separable from *Pygiopsylla* and *Stivalius s.lat.*, or characterized, as follows: 12) With two rows of abdominal bristles (fig. 90). 13) Metatibia with seven dorsomarginal notches (including apical one) bearing stout bristles, but with only a maximum of two stout ones each. 14) Base of antepygidial bristles (fig. 90, 98) not recessed within margin of 7 T., i.e. lacking any dorsal or ventral lobes (cf. figs. 42, 60, 71). 15) Immovable process (fig. 86, P.) very broadly rounded; not produced even into a short lobe (cf. figs. 18, 44). 16) Movable finger (F.) massive; portion distad of P. shaped somewhat like half an oblate spheroid, thus lacking a stiva completely; with small, scattered, thin bristles near anterior margin (cf. figs. 18, 44, 50). 17) F. with distal fringe (D.FR.) represented by a group of relatively thin bristles near middle of exposed caudal margin. 18) Fulcral sclerite not developed. 19) D.A.9 (figs. 86, 88) relatively simple; lacking long sclerotizations, subdivisions, or structures like a dorsal subapical spur or supramedial flap. 20) D.A.9 bearing a clavus armed with spiniforms and subspiniforms. 21) Aedeagus with a well-developed keel (figs. 88, 92, K.). 22) Alpha portion (ALPH.) of Ford's sclerite (F.SC.) falcate. 23) With a well-developed crochet process (CR.P.) proximally intimately associated with the pivotal base (PIV.B.). 24) Phylax (PHY.) small but distinct. 25) S.I.T. basally broad and apical half narrow; lacking distinctive spurs. 26) Crescent sclerite (C.SC.) long. 27) Very unusual in family in possessing a proximal spur (P.S.) arising from dorsal apex of girdle (G.). 28) Penis rods (P.R.) uncoiled. 29) ♂ 8 S. (fig. 90) very large; caudal margin adorned with long bristles, as is apical half of ventral margin, but those more widely spaced; with no such dorsomarginals. 30) Spermatheca (fig. 97) with bulga (B.) longer than broad; basal half (near duct) much broader than apical one. 31) Hilla (H.) partially within bulga; with an apical papilla. 32) Bursa copulatrix (fig. 100, B.C.) long and narrow; flanked (right and left sides) near vagina (VAG.) by a small well tanned sclerite, adjacent to the series of glands at the usual site of the duplicature vaginalis (cf. fig. 63, D.VG.). 33) ♀ ventral anal lobe (fig. 93, V.A.L.) of type common in *Stivalius s.lat.*, i.e. with a sinus between basal group of long bristles at anteroventral angle and the subapical long bristle. (Further generic description unwarranted because of the poor condition of the specimens extant.)

Comments – The type of the genus is *P.celebensis* Ewing, 1924, and this unique genus is named for Dr A.Farhang-Azad in appreciation of his outstanding studies on Siphonaptera and other aspects of medical zoology and his splendid assistance in our investigations on the ecology of murine typhus in Ethiopia.

Examination of the skins of the reported hosts of *F.celebensis* and *F.sciuri* provide the following information. The host cited by Ewing for the former is today known as *Prosciu-*

rillus m.murinus and was collected on 13 January 1916; that for *F.sciuri* is *Callosciurus prevosti,* Indonesian Borneo: E.Sungei Merah (0°50′N, 116°49′E), H.C.Raven, 15 March 1914. However, from Ewing's statement 'from the skin of', it is not clear as to whether the fleas were actually collected on the bodies of the hosts at the time, in Sulawesi or Borneo, or if they were later found on the skin of the preserved specimens at the US National Museum. Today this question apparently cannot be resolved, but if they were collected from study-skins, the possibility of error in the data must be considered.

When the male of *F.sciuri* is discovered, it is possible that it will prove to be not congeneric with *F.celebensis,* but the two species agree on such fundamental characters, that also are exceptional in the family, that this seems highly unlikely. The fact that one species is from Sulawesi, and the other from Borneo does not militate against their belonging to the same genus. *Medwayella* Traub, 1972, another *Callosciurus*-parasite has such a distribution, and *Bibikovana* n.gen. (supra) probably does as well since it is known from the Australian Region and Borneo.

ACKNOWLEDGEMENTS

As designated on the figures, the following illustrators prepared various drawings used in this article: T.M.Evans, E.McGoodwin and A.Green. Some were drawn by the author or represent composites by him and another artist, as indicated. The photomicrographs were prepared by J.Navarro and retouched by him and the author. Miss Helle Starcke provided editorial assistance. To all these colleagues I express my thanks.

REFERENCES

Holland, G.P. 1969, Contribution towards a monograph of the fleas of New Guinea. Mem. ent. Soc. Canad. (61), 77 pp.

Jordan, K. 1933, Fleas collected by Dr Max Bartels in Java. Novit. zool. 38: 352-357.

Jordan, K. & N.C.Rothschild 1922, On *Pygiopsylla* and the allied genera of Siphonaptera. Ectoparasites 1: 231-265.

Smit, F.G.A.M. 1958, The African species of *Stivalius,* a genus of Siphonaptera. Bull. Br. Mus. (Nat. Hist.) Ent. 7: 41-76.

Traub, R. 1972a, Notes on zoogeography, convergent evolution and taxonomy of fleas (Siphonaptera), based on collections from Gunong Benom and elsewhere in South-east Asia. I. New taxa (Pygiopsyllidae, Pygiopsyllinae). Bull. Br. Mus. (Nat. Hist.) Zool. 23: 201-305.

Traub, R. 1972b, Notes on zoogeography, convergent evolution and taxonomy of fleas (Siphonaptera), based on collections from Gunong Benom and elsewhere in South-east Asia. II. Convergent evolution. Bull. Br. Mus. (Nat. Hist.) Zool. 23: 306-388.

Traub, R. 1972c, Notes on zoogeography, convergent evolution and taxonomy of fleas (Siphonaptera), based on collections from Gunong Benom and elsewhere in South-east Asia. III. Zoogeography. Bull. Br. Mus. (Nat. Hist.) Zool. 23: 389-450.

(Submitted for publication August, 1977)

EVOLUTION AND ZOOGEOGRAPHY

ROBERT TRAUB
Department of Microbiology, University of Maryland School of Medicine, Baltimore, USA

SOME ADAPTIVE MODIFICATIONS IN FLEAS*

ABSTRACT

In addition to serving as a means of protection and to facilitate passage through fur, the combs and bristles of fleas help maintain a hold on the host. Thus, in many parts of the world, unrelated fleas have independently developed various specializations like 'crowns of thorns' or other modifications of spines or spiniforms on the head which are absent in their kin, or even congeners, which infest other kinds of hosts in the area. By some sort of 'co-ordinated evolution' these adaptations have been associated with a series of other developments serving the common function of maintaining a hold on the hairs as the flea rests or feeds, e.g. a keel-like narrowing of the head. As with fleas with other manifestations of hyperdevelopment of spines and bristles, such modifications are restricted to Siphonaptera whose hosts have attributes that are unusually hazardous for flea-survival, where a detached flea would have little chance of finding a new host, or else its larva would emerge in a hostile environment. Such hosts include mammals that are volant or gliding, or are both nocturnal and scansorial, or have vast home-ranges, and the like. Observations are presented indicating that fleas with a well-developed genal comb are also in such a category regarding their host. Additional observations are presented indicating how the bristles and spines of the fleas may be tailored to fit the vestiture of the host. For example, hosts that have true spines, like hedgehogs, tenrecs and porcupines, have fleas whose bristles are spaced unusually far apart and/or are exceptionally thickened, and which may bear highly modified combs.

Photomicrographs, including dorsal and anterior aspects of the head of fleas, are presented to illustrate these points. In those few taxa of fleas in which both the eye and the genal combs are well-developed, the ctenidium also serves to protect the eye from contact with hairs or feathers. In fleas without such a comb, the eye is recessed, and thus is shielded by the sides of the head in addition to the usual 'eye-bristle'.

* This study was supported by Grant No. AI-04242 of the National Institutes of Health, Bethesda, Maryland, USA, with the Department of Microbiology, University of Maryland School of Medicine, Baltimore, Maryland, USA. The field-studies in Ethiopia mentioned in this article were sponsored by Contract N00014-76-C-0393 of the Office of Naval Research and the Navy Medical Research and Development Command, Washington, DC. The opinions and assertions herein are not to be construed as necessarily reflecting the views of the Department of the Navy or the National Institutes of Health.

1 INTRODUCTION

Fleas are remarkably adapted for their particular mode of existence. The specializations may be physiological, e.g. the maturation of the sex glands may be linked, via hormones, to that of the host (M.Rothschild, 1965a, 1965b), or there may be complex enzymal relationships affecting the clotting of host-blood within the flea (Bibikova & Klassovsky, 1974; Cavanaugh, 1971). Such fascinating attributes are beyond the scope of this article, and instead we will be concerned with external morphological features such as ctenidia, chaetotaxy and pertinent modifications of the head and eyes of fleas, all of which illustrate, to some degree, the moulding effect of the environment upon the flea, in the course of evolution.

After a brief review of salient points recently made about such adaptations, some new and original observations along these lines are reported herein. Additional observations are presented to support the view that the spines of the pronotal and genal combs may be tailored to fit the major hairs of the host and thus, among other functions, retard sideways or backward motion of the flea as the beak, teeth or claws of the host, or other forces, threaten to dislodge it. Thus, it is known that even the *bristles* of the flea can be modified in an analogous fashion, as in the case of fleas of spiny hosts like porcupines, tenrecs and echidnas. Data are provided indicating that the various helmets, frontal combs, crowns of spiniforms or marginal groups of spiniforms of fleas are adaptive in a number of respects. Some presumably serve to temporarily anchor the flea to the hairs of the host, while the parasite is resting or feeding, but they nevertheless can immediately be unhooked by mere forward motion on the part of the flea. Regardless, the presence of such helmets and crowns, etc. is primarily if not wholly limited to fleas which are subject to unusual survival pressure because of the specialized habits or hazardous environs of the host, or because of some unusual attributes of the fleas themselves. The situation is therefore analogous to that reported for fleas bearing supernumerary or hyperdeveloped combs and bristles in general (Traub, 1972b, 1972c), as in the case of bat-fleas, bird-fleas, etc. Tibial combs, and well-developed genal ctenidia, are also shown to be associated with fleas at special risk because of environmental or other constraints.

It will be shown that the development of frontal combs of spines or crowns of spiniform bristles, etc. has invariably been accompanied by an extreme narrowing of the anterior margin of the head from side to side, with the degree of thinness corresponding to the length and marginal nature of the comb or crown. Such changes, and the modifications mentioned above, have been accomplished irrespective of the provenance or geographical distribution of the flea involved, and represent examples of convergence. Among other uses, when present in fleas with well-developed eyes, the genal comb presumably serves to protect the eyes from injury by the hairs of the host.

In this article, as in others in which I have discussed setal modifications and ctenidia (e.g. Traub, 1968, 1972b), an important distinction is made between: 1) what are commonly referred to as spines or ctenidial spines and 2) spine-like bristles (also termed spiniforms), even though the former type are, fundamentally, highly modified bristles themselves. As pointed out in Rothschild & Traub (1971), true spines, which are multicellular outgrowths of the cuticle, do not occur in Siphonaptera. The ctenidial outgrowths (on combs on the gena, helmet, pronotum or abdomen of certain fleas) which are generally called spines originate histologically in the pre-imago as a setal process arising from a socket. In the imago, however, all external signs of articulation have essentially, or completely, disappeared. The ctenidial spines are parallel-sided at the base and generally appear to arise directly from the cuticle. However, in some cases, particularly in the helmet-comb of Stephanocircidae (figs. 4, 5), vestiges of the articulation may be discerned on close study. In marked contrast even to such exceptions, in spine-like bristles (spiniforms) the articulation is readily visible, and the base of the bristle is narrowed where it is inserted into the socket. The bristle-like nature of the structure is apparent, even if the apex is broadly rounded (as in the false helmet shown in figs. 1 and 2). The contrast between spines on the ctenidium and spiniforms on the frons are well shown in figs. 6 and 8. In fleas, the distinctions between spines and spine-like bristles are usually clear-cut, unlike in the Dipteran bat-parasites Nycteribiidae and Streblidae, where all sorts of intergrades occur.

It is not surprising that the spines of the helmet-comb of stephanocircids are of an intermediate type, since, as pointed out by Traub (1968), false combs (if present) are found only on those parts of the body of the flea where true combs are not known to occur, at least in that phylogenetic group of fleas. Thus, there are fleas with false combs on the metepimere, on the seventh tergum, etc., but the only Siphonaptera with what has been regarded as a bona fide comb on the anterior part of the head (in contrast to an upright, displaced, genal comb) are the helmet-fleas, Stephanocircidae and Macropsyllidae. However, the helmet is a secondary modification of the head, even though occurring in an ancient group of fleas (Traub & Dunnet, 1973), and the accompanying comb cannot be a primitive feature. It is therefore to be expected that such a ctenidium would be evolutionarily or histologically derived from modified bristles, and hence be of a different type than the combs of spines on gena, pronotum or abdomen.

2 REVIEW OF THE LITERATURE

Combs of short flat spines occur in a variety of ectoparasites, including polyctenids (Hemiptera), streblids and nycteribiids (Diptera) and even certain Mallophaga, but they have been so intimately associated with fleas that when Ritsema described the unique beaver parasite *Platypsyllus castoris* in 1869, he placed it with the fleas, as a separate family, because of the ctenidium it bore. Only later was it correctly assigned to the Coleoptera by Leconte (Ritsema, 1880: 185). Because such external modifications of ectoparasites, especially fleas, are so strikingly specialized, they served as the theme for a brilliant Presidential Address of N.C.Rothschild to the Royal Entomological Society of London 60 years ago (1917). In it he pointed out (p. 9) 'the function of those bristles which are directed away from the body or legs is to rest on the hairs of the host when the parasite is not moving. The parasite by this means can hang in the pelt without slipping out, as a broken-off many-branched twig remains hanging in a bush'. N.C.Rothschild also stressed that these supporting bristles point backwards so as not to impede forward motion, and that fleas cannot move backwards. Another important observation he made was that bristles also serve to fend off the hairs of the host and thus prevent their slipping into sutures and entangling the flea. Rothschild noted (p. 13) that 'the "parrying" bristles are most conspicuous at those joints where the greatest flexibility obtains between parts of the insects, (such as) . . . at the joints between . . . prothorax and metathorax in those cases where the former is separated . . . .' Other functions were 'to strengthen the body and render it slippery so that it can withstand pressure more successfully and glide more easily through the fur . . . The combs . . . have a similar function to the bristles'. Even more than a century ago, however, Kolenati (1863) believed that the pronotal comb assisted the flea in maintaining a grip on the pelage.

N.C.Rothschild (1917) did not discuss the genal ctenidium as such but it had long been assumed that this comb likewise aided the flea in remaining on the host (Kolenati, 1856; Taschenberg, 1880). According to Smit (1972), Roesel in 1749 claimed that the genal comb facilitated passage through the hairs, but Smit felt that the function of that ctenidium was to protect the base of the mouthparts and/or the basal joint of the forecoxae. However, Smit did not say how this was accomplished in fleas lacking a genal comb.

Little was written about the functions of combs and bristles in the four decades following N.C.Rothschild's imaginative paper, although in 1929 Ioff ascribed the unusual, rounded genal spines of a Palaearctic species of *Ctenophthalmus* Kolenati, 1856 (I.3)* which infested *Spalax* mole-rats (Spalacidae), to the 'extremely thick soft silky fur' of the host. Indeed, we now know that in the Ethiopian Region several *Ctenophthalmus* of a different subgenus, parasitizing *Tachyoryctes* mole-rats (Rhizomyidae), share these and other features, as indicated by Hopkins & Rothschild (1966) and by Smit (1976), who made no comments about possible functions or associations. Smit (1958) noted that the pronotal spines of shrew-fleas have a characteristic appearance, and Traub & Barrera (1966) pointed out that both genal and pronotal ctenidia of shrew-fleas regularly differ significantly from those of mole-fleas belonging to the same genus. They also demonstrated that these modifications represented convergence since the combs of unrelated fleas infesting the same class of host tended to resemble one another. Such uniformly and consistently distinctive genal and/or pronotal spines occur in shrew-fleas representing three families and four hystrichopsyllid subfamilies in different parts of the world (Traub, 1972c). Congeners of such fleas, occurring on moles rather than shrews, consistently and characteristically have narrower, straighter genal and pronotal spines, and more spines in the pronotum. In fact, I now point out that even *Callopsylla seminovi* Ioff, 1936 (III.1), the sole ceratophyllid infesting an insectivore (a mole), rather than typical hosts like rodents, has unusually narrow and more numerous spines in the pronotal comb than do its relatives on other hosts. Additional examples of such developments were provided by Traub & Evans (1967) and Traub (1968, 1969). Thus, Traub (1968) and Holland (1969) independently observed that the fleas of peramelid bandicoots have stiletto-like pronotal spines, while their relatives infesting rats (by far in the majority) possessed bluntly rounded spines. The former also stated that the gaps between the spines correlated well with the nature of the major hairs of the respective hosts. Traub (1972b) alluded to several additional new species of *Metastivalius* Holland, 1969 (II.1) and *Papuapsylla* Holland, 1969 (II.1) which supported those views.

Smit (1972) and Traub & Dunnet (1973) independently noted that the helmet of stephanocircid fleas is an acutely narrowed, keel-like extension of the head. The former stated that the vertical helmet-comb 'acts like the bow of a ship and also protects the sensitive cephalic organs that are situated posterior to this ctenidium'. The latter agreed that the helmet-comb (like the other ctenidia) aided the flea in gliding through the fur, with the knife-like edge of the helmet parting the fine hairs as the flea moved. However, they also believed that the helmet spines could also be latched onto the closely knit hairs of the

* The family and subfamily of the various taxa are indicated in this way, following the system of Traub (1972b).

host, particularly since the helmet is movable. In that way the flea could remain hooked as it fed or rested, a procedure facilitated by the short mouthparts characteristic of helmet-fleas (and also of 'crested' fleas and those with marginal groups or crowns of spiniforms, as mentioned below (p. 51)). This means of attachment, reinforced by the genal and pronotal ctenidia, was regarded by them as an adaptation against dislodgement by the host, since movement in any direction but forward was impeded. In support of their view, Traub & Dunnet (1973) reported that in their new species, *Stephanocircus harrisoni* (VII.1), the acutely pointed spines of the helmet of this bandicoot-flea were apparently 'adapted' to fit the coarse flattened hairs of its host, while in contrast, the close-fitting, bluntly tipped helmet-spines of *Coronapsylla jarvisi* (Rothschild, 1908) (VII.1) jibed with the somewhat bristly fur of its host *Antechinus*. Further, the helmet-spines of the murid-infesting stephanocircines were consistently found to be of yet another pattern.

Even though it was conceded that the pronotal combs of fleas had other functions as well, it therefore appeared that such modifications were adaptive, associated with a particular type of host pelage, serving to prevent dislodgement by the beak, tooth or claw of the host. Indeed, Humphries (1966, 1967) reported specific instances in which the gap between the pronotal spines of certain fleas corresponded with the diameter of the major hairs of the host, and believed that the comb would 'lock' onto the hairs if the flea was pulled backwards. Traub (1972b) stated that in squirrel-fleas, representing 17 genera in three families and subfamilies, from areas in Africa, Eurasia and the New World, one or two of the subventral spines of the pronotal comb were uniformly of an unusual breadth. Moreover, when allied species, infesting other kinds of hosts, occurred in the same genus, they lacked such modified spines.

Not just the spines and combs, but even the bristles of fleas were noted as becoming modified in the evolutionary quest for security in the fur of the host, as vide the development of marginal groups of spiniform bristles and even of 'false combs' of modified bristles on various parts of the body (Traub, 1968, 1969, 1972b, 1972c). For example, it was prognosticated that pygiopsyllid fleas like *Idiochaetis* Jordan, 1937, *Ernestinia* Smit, 1953 and *Striopsylla rugata* (Jordan, 1937) which possess flat heads and marginal or submarginal spiniforms or spinose bristles, press their heads against the skin of the host and cling to the hairs via the spiniforms, aided by the exceptionally stout forecoxae (Traub, 1968). Field observations reported by Traub, (1969) confirmed that this actually was the case in *Muesebeckella* Traub, 1969 (II.1) and *Striopsylla vandeuseni* Holland, 1969 (II.1), which possess such spiniforms, and especially for *Idiochaetis illustris* Jordan, 1937 (II.1) with its well-developed tiara. The mechanism was found to be very effective in inhibiting all induced motion save for forward movement. Traub & Dunnet (1973) noted that the occipital and thoracic bristles of helmet-fleas of peramelid bandicoots were exceptionally short and thick as compared to other helmet-fleas and suggested that this was an adaptation to the coarse flat hairs of the host. It was shown (Traub, 1972b) that possession of a tibial comb of modified bristles was correlated with the presence of both a narrowed head and the possession of a frontal row of bristles (usually spiniform) on the head. The modifications were considered as adaptive features associated with the habit of clinging to the host in a sedentary but temporary manner.

The false combs of spine-like bristles on the abdomen and thorax of certain fleas were deemed as adaptive, associated with enhancing the chances of latching onto, and remaining on, the host (Traub, 1968). It was shown that such false combs, like the supernumerary true ones, were characteristic of hosts like bats, which were volant, or which otherwise posed special hazards for the survival of the flea (Humphries, 1966; Traub, 1968, 1972b, 1972c). This point is discussed in the paragraph below. It was stressed that false combs were known to occur only on those parts of the body where true combs were absent in the respective taxa (Traub, 1968). These various points seemed sufficiently well based to permit that author to predict that even fleas in a family like the Rhopalopsyllidae, which completely lack combs, could be expected to include a member with a false comb of spiniforms on the pronotum provided it infested a host like a bird or arboreal, nocturnal mammal. A new genus of rhopalopsyllid with such a comb was found shortly thereafter (Mendez, 1968), and it was suggested (Traub, 1972b) that its true host may be the type previously predicted.

More than the vestiture of the host is involved in the evolutionary forces that account for the modifications of spines and bristles seen in fleas, for in many instances the crucial factors seem to be the *habits* of the host or the flea. Thus, hyperdevelopment of spines and combs is characteristic in fleas which would have inordinately difficult problems of survival if they were dislodged, or failed to attach to their host when leaping thereon, e.g. fleas of volant animals like birds or bats, or of gliding mammals such as flying-squirrels or flying-phalangers, or of hosts which are arboreal and/or nocturnal.

Another evolutionary trend in chaetotaxy which is unrelated to host vestiture, is seen in true nest-fleas, which act like bedbugs and hide in crevices, emerging to feed when the host is sleeping in the nest, and hence are rarely found on the mammals themselves (Traub, 1972b). There the tendency is towards reduction of combs, bristles and the size of the eye, and lessened ability to jump, associated with loss of the pleural arch and the resilin which provides the motive power for leaping (Traub, 1972b; M.Rothschild et al., 1973). Another modification imposed by the environment occurs in desert-fleas, where regardless of affinity of the parasite, the pronotal comb flares away from the sides of the body to a greater degree than in fleas from other areas (even if in the same genus, or species, e.g. *Orchopeas sexdentatus* (Baker, 1904) (III.1) from different habitats) (Traub & Evans, 1967).

All these observations have therefore suggested that the evolutionary significance of the correlations between the pattern of the Siphonapteran combs and bristles and the attributes of the host (its vestiture or its habits), or else with the special traits or environment of the flea, often transcend the phylogenetic and taxonomic relationships of the flea (Traub, 1977). However, it must be emphasized that so eminent an authority as Smit does not agree with these concepts. Instead, in an opus (1972) in which he superbly deals with adaptations in the tarsal claws, including differences in bird-fleas and mammal-fleas, Smit stated that he believes that the function of the pronotal comb is essentially wholly protective, serving to shield the intersegmental membrane between head and thorax, and to fend off the hairs of the host. He denies that the pronotal comb can grasp hairs in any way or at any time, and explains the correlation between shape of spine and girth of hair as being 'of preventive nature, obviating the slight, virtually hypothetical, chance of hairs getting caught between them'. These points are rebutted at length by Traub (1977). Suffice it to say here that 1) Smit's hypothesis would not account for the function of genal or abdominal combs or for their marked convergent modifications; 2) A stylized simple type of pronotal comb would serve to protect the intersegmental membrane and hence there would be no selective pressure to account for such striking convergent or parallel developments in shape and numbers of spines in genal and pronotal ctenidia; 3) He overlooks the hyperdevelopment of combs and spines in fleas at unusual risk; 4) There would

be no way to account for the fact that the *bristles* of fleas at times are also obviously tailored to fit the hairs or spines of the host. In support of the last point, it is only necessary to glance at the fleas of spinose mammals like porcupines, echidnas, tenrecs and hedgehogs, as mentioned in the section on new observations reported below (p. 59).
5) The ctenidial spines are not special extensions of the nota designed to cover the base of the following segment, and do not extend beyond the flange. Instead they are derived from the flange of the nota itself. As N.C.Rothschild (1917) put it, the combs 'usually appear to be exaggerated serrations of the edges of the segments'. It would seem that whatever membrane exists under the nota would also occur under a comb, although in modified form. Rather than the comb serving to protect the membrane, it may be that the latter is the consequence of the development of the ctenidium (Traub, 1977) and serves to protect bodyparts that would otherwise be uncovered, while permitting some flexibility in the comb itself. However, according to M.Rothschild (1976), J.Schlein believes that 'combs . . . are essentially devices for the protection of intersegmental arthrodial membrane', thereby agreeing with Smit. It is difficult to see how the need to protect an underlying membrane can result in the truly remarkable convergence seen in the pronotal comb of all bird-fleas in nine genera in three families, and which entails the three different features pointed out by Traub (1969). Further: 1) where there are congeners infesting rodents, they lack such modifications and 2) the name *Ceratophyllus* is applied to some fleas that actually belong to another genus but which have been placed in *Ceratophyllus* solely because they happen to have become adapted to birds and developed the type of comb characteristic of bird-fleas. Nevertheless, we must admit that we do not know if (or how) this comb of supernumerary, relatively narrow, close-set, straight spines serves to help maintain a hold on the feathered host, or if it has yet another function.

3 DISCUSSION AND NEW OBSERVATIONS

There is no doubt that the combs and bristles have a multiplicity of functions in fleas, including parrying the hairs of the host and facilitating passage through its vestiture, and in the case of ctenidia, protecting delicate structures beneath, and providing rigidity to the body and hence lessening the chances of the flea being crushed by the host. Some of the bristles are tactile or otherwise sensory. Like N.C.Rothschild (1917), I believe that both types of structures also serve to maintain the flea's hold upon its host, but go further in opining that at times the spines and bristles may be especially modified to readily latch onto the hairs if the host attempted to drag the flea in any direction but forward. Also, as mentioned above, frontomarginal tiaras of spines or spiniforms (and perhaps other groups of spines and modified bristles) may act as a temporary hold-fast while the flea feeds or rests. Recent and new observations indicate that the concomitant adaptive modifications are even more fundamental than has been suggested in the articles reviewed above, in that they exceed such features as tailoring of spines to fit the hairs of the host, and also accurately reflect the degree of hazard faced by the parasite if it fails to grasp a new host or remain on the body of a current one.

3.1 *Hyperdevelopment of spines and bristles*

Before dealing with these points, it is desirable to discuss supernumerary or hyperdeveloped combs and bristles in fleas. The selective pressure upon the course of evolution which are exerted by the hazards faced by fleas of volant hosts and those which are both nocturnal and arboreal must be immense, and this statement also applies to fleas which are otherwise exposed to environmental risks. This is attested by the frequency with which such modifications like concomitant hyperdevelopment of true and false combs, or spiniforms on the head, etc. have arisen in the Order, often without regard to geographical distribution or to the phylogeny or taxonomy of the fleas involved, and by the fact that the phenomenon is restricted, apparently without exception, to fleas in two major categories: 1) species whose hosts have characteristics that place the fleas at unusual risk regarding survival or 2) fleas whose own life patterns or physiological requirements make failure to locate or remain on the host exceptionally hazardous. In the first group are fleas which infest birds, bats, flying-squirrels, flying-phalangers, hosts like martens, tree-climbing nocturnal rodents and marsupials or other hosts whose habits similarly produce risks for errant fleas, such as leaping mammals like elephant-shrews or rabbits, or those with a vast territory. Examples in the second group include fleas that emerge only in the winter, or which are restricted to the environmental conditions of the hosts' burrows or runways; or which are extremely host-specific and cannot survive or breed if on another species of bird or mammal; or whose larvae can successfully develop only within the nest of the host. Further, I cannot recall any fur-fleas of such hosts which do lack modifications which appear to enhance their chances of survival in such special straits. In contrast, none of the fur-fleas of the ordinary, ground-dwelling mammals like *Citellus* and allies, *Cynomys,* marmots or hyrax possess supernumerary combs or large genal or pronotal ctenidia, nor do they have tiaras of spiniforms, crests of spines or tibial combs. For example, the true host of the combless *Xenopsylla cheopis* (Rothschild, 1903) is a diurnal African murine, *Arvicanthis* (Traub, 1972d).

Even the apparent exceptions to the principle that fleas at special risk possess adaptive chaetotaxic devices for maintaining their hold on the host, can be readily explained, e.g. the combless *Echidnophaga* Olliff, 1886 (IX.1) on bats have anchoring mouthparts and hence have no need for supplemental aids. At first thought, the frontal row of spiniforms on the fleas of pikas (*Ochotona*) seem anomalous in this regard, for those lagomorphs are often diurnal and are terrestrial, not scansorial. However, it must be borne in mind that they inhabit high mountains and are active in the winter, where and when a cast-off or newly emerged flea is in dire straits in those frigid climes if it does not soon find a host (vide p. 54). Similarly, the fleas like *Stenoponia* Jordan & Rothschild, 1911 (I.9) or *Hystrichopsylla* Taschenberg, 1880 (I.1) which bear one or more extra combs and which may infest ground-dwelling (but nocturnal) hosts, also are winter-fleas and hence are at special risk. Some, like certain *Stenoponia,* are desert-denizens and others are montane forms. Thus, while the presence of multiple combs in fleas of non-volant hosts has been regarded as a primitive character (Jordan, 1947; Traub, 1968, 1972b) and while such fleas tend to parasitize primitive hosts (Traub, 1972b; Traub & Dunnet, 1973), I now stress that there are other factors involved, namely the degree of hazard to survival posed by the traits of the host. Several Australian fleas, such as the macropsyllid helmet-flea *Macropsylla hercules* Rothschild, 1905 (VIII) and *Stephanopsylla thomasi* (Rothschild, 1903) (VIII) and the stephanocircid helmet-flea *Coronapsylla jarvisi* (VII.1) illustrate this point. These fleas, which bear one or more combs on the head and abdomen as well as an exceptionally well-developed pronotal comb, primarily or wholly infest scansorial or arboreal nocturnal marsupials or rats. Pygiopsyllids which bear an abdominal comb in addition to a large pronotal one (but which lack a genal ctenidium) include the Southeast Asian *Migrastivalius (Grypho-*

psylla) hopkinsi (Traub, 1957), *Migrastivalius (Migrastivalius) jacobsoni* (Jordan & Rothschild, 1922) and *Lentistivalius (Destivalius) mjoebergi* (Jordan, 1926) and the members of the Neotropical *Ctenidiosomus* Jordan, 1931. *M.(G.)hopkinsi* is known from sundry *Rattus,* tree-squirrels, tree-shrews and viverrids, all of which are arboreal or scansorial, and some of which are nocturnal. *M.(M.)jacobsoni* is virtually invariably from tree-nesting or semi-arboreal, nocturnal mammals like the murids *Chiropodomys, Pithecheir* and *Rattus cremoriventer* and the *R.rajah*-group and the gymnuran pen-tailed shrew *Ptilocercus,* while *L.(D.)mjoebergi* is a parasite of tree-shrews. Data are available on one of the three African '*Stivalius*' which possess abdominal combs, namely *Afristivalius pirloti* (Smit, 1958) and its host, *Praomys morio,* is semi-arboreal and active at night. *Ctenidiosomus* are from marsupial or rodent hosts with similar habits.

The trend towards hyperdevelopment of the pronotal comb of fleas at such environmental risk is well illustrated by the genera *Corypsylla* (I.8) and *Nearctopsylla* Rothschild, 1915 (I.8). In each there is one species which infests carnivores (mustelids), where the large home-range and occasional scansorial habit impose extra hazards for the flea's survival (and that of its larvae), and these species have a far greater number of spines in the pronotal comb, and the ctenidium covers a greater area, than in the case of the vast majority of fleas in these genera, which infest insectivores. Another illuminating example is provided by the pygiopsyllid *Uropsylla* Rothschild, 1905 (II.3) which bears a particularly well-developed pronotal comb (and rows of unusually stiff abdominal bristles). This is a flea of the far-ranging native marsupial 'cat', *Dasyurus,* which lacks a permanent den, and is somewhat scansorial, and I believe that these points are factors pertaining not only to the chaetotaxy of the flea but also the fact that the larvae of *Uropsylla* are almost unique in the Order in being parasitic on the host of the adult, burrowing into its superficial tissues. This principle is also illustrated by *Euhoplopsyllus glacialis* (Taschenberg, 1880) (IX.3), a flea of the Arctic hare, in which the larva is likewise parasitic, as pointed out by Freeman & Madsen (1949) based upon observations in Greenland. The hare has no permanent nest or resting place, and in such rigorous climes, and under such conditions, the survival value of this adaptation is obvious. Fleas at somewhat lesser risk are not so well adapted but exhibit this same trend. Murines are usually nocturnal and are excellent climbers, and their fleas tend to have more spines in the pronotal comb (e.g. *Neopsylla* Wagner, 1903, *Sigmactenus* Traub, 1950, various *Stivalius s.lat.*) than do their relatives on other hosts. Thus, the murine *Afristivalius* have 22-24 pronotal spines but the species of *Lentistivalius* Traub, 1972 on African shrews (crepuscular and surface-dwelling) and on Asian tree-shrews (rather diurnal but scansorial) have about 20 pronotal spines. Of special interest here is *A.rahmi* Beaucournu, 1969 which has 28-30 pronotal spines. Its host is *Malacomys,* a swamp-dweller whose feet are modified for its sodden habitat, a hostile environment for a stray flea or its larva.

This last point emphasizes that in holometabolous insects, the larval stage must be considered as a possible contributing factor in the evolution of adaptations seen in the adult. The precise requirements of the larvae have been mentioned as a possible reason why sibling species of *Medwayella* Traub, 1972 (II.1), occurring in the same locus, show different host relationships, with some restricted to ground-dwelling squirrels and another occurring on arboreal hosts (Traub, 1972a). Certain species of fur-fleas, e.g. *Orchopeas howardi* (Baker, 1895) (III.1) glue their eggs to debris in the nest of the host-squirrel instead of laying them loosely in the hairs of the host like cat-fleas (Traub, 1972b), and this implies special environmental conditions are necessary for the larvae. *Nosopsyllus fasciatus* (Bosc,

1801) (III.1) larvae obtain host-blood by sucking it from the anus of the replete flea stimulated by the larva grasping the pygidium (Molyneux, 1967). Host-blood ingested via the dried excrement of fleas are reported as constituting the main source of nutrition for larval cat-fleas (Strenger, 1973). It, therefore, seems likely that where flea-larvae have specific requirements available only or mainly in the nest, such as those mentioned above, or certain conditions of temperature and humidity, any structural modifications which facilitate the flea's remaining on its host successfully and reaching the nest, would also have selective value insofar as concerns the larval stage as well. The vertical frontal comb of certain fleas, or other features, may thus also reflect such an evolutionary force by the larval requirements.

3.2 *The fleas of saltatorial hosts*

Among the habits of hosts that present problems of survival for the various stages of fleas may be the trait of leaping or jumping, since either there are no specific fur-fleas infesting such mammals, or else the fleas that do, tend to exhibit hyperdevelopment of spines or bristles, or are semi-sessile or true sticktight fleas that can remain attached despite the activities of the host. The dearth or literal absence of records of characteristic fur-fleas (or even any at all) from saltatorial hosts is surprising. In part this is due to limited collecting, as in the case of certain Australian marsupials, but the data nevertheless seem sufficiently valid to substantiate this impression, especially since the non-saltatorial relatives generally do have fleas, as do other mammals in those areas. There are no records of fleas at all from eight of the ten genera of kangaroos (macropodids) in Australia and New Guinea, and for the remainder, only a few obvious strays have been recorded, as in the works by Holland (1969) and Dunnet & Mardon (1974). Three of the five genera of rat-kangaroos progress by leaping, and fleas have been listed only from *Potorous,* and these are all non-specific save for the little-known *Austropsylla* Holland, 1971 (II.1), which is apparently somewhat of a nest-flea. Some *Bettongia* may be saltatorial, and it seems significant that only sticktight fleas (*Echidnophaga*) and helmet-fleas have been reported from it.

Some bandicoots have a leaping or galloping gait, viz. some *Perameles* and *Peroryctes,* and records from the former include helmet-fleas and *Acanthopsylla* Jordan & Rothschild, 1922 (II.1) which are modified along pertinent lines, and the catholic *Pygiopsylla hoplia* Jordan & Rothschild, 1922 (II.1) and allies, which are not (as is typical of such fleas). *Peroryctes* carry *Striopsylla* Holland, 1969 (II.1), *Parastivalius* Holland, 1969 (II.1) and other pygiopsyllids with specialized bristles and pronotal spines, as mentioned elsewhere (pp. 48, 60). *Macrotis* (formerly called *Thylacomys*) canters and hops and is infested with *Echidnophaga.* Some *Sminthopsis* and *Antechinomys,* which are marsupial 'mice', bound or gallop. *Echidnophaga* and *Acanthopsylla* are known from the former, while there are no records for the latter. Specific fleas have not been found on *Notomys,* the Australian hopping mouse (a murid). (Certain murids like *Mesembriomys* and *Conilurus* (Australian) and *Lorentzimys* (New Guinean) were formerly believed to be saltatorial because of their large hind-feet, but are now regarded as scansorial.)

The five genera of macroscelids (elephant-shrews) are excellent jumpers and are intimately associated with chimaeropsyllid fleas with a prominent vertical genal comb and a pronotal one as well. They also may carry a *Caenopsylla* Rothschild, 1909 (IV.2), a taxon with a genal comb, and sticktight fleas. Fleas of the genus *Cratynius* Jordan, 1933 (IV.1), infest the gymnuran *Hylomys,* an active jumper, and bear a frontomarginal group of spiniforms. The rabbits and hares are among the most conspicuous of saltatorial mammals. Of their fleas, *Spilopsyllus* Baker, 1905 (IX.3), *Cediopsylla* Jordan, 1925 (IX.3), *Nesolagobius* Jordan & Rothschild, 1922 (IX.2) and one species of *Hoplopsyllus* Baker, 1905 (IX.3) have a relatively well-developed genal comb and pronotal one, and they, and the other rabbit-fleas, are specialized in other ways as well (p. 56). The Palaearctic dipodids *Allactaga, Jaculus,* etc. are saltatorial, desert-denizens and carry fleas like *Mesopsylla* Dampf, 1910 (IV.2) and *Desertopsylla* Argyropulo, 1946 (IV.2) which have a short genal comb, and the latter bears supernumerary bristles on the head. (These rodents also are infested with *Ophthalmopsylla* Wagner & Ioff, 1926 (IV.2), which lack these special features, but this taxon perhaps may be primarily burrow-fleas rather than fur-species.) The Nearctic heteromyids *Dipodomys* are called kangaroo-rats because of their habitus and leaping habits, and are infested with *Meringis* Jordan, 1937 (I.7), which bear a small genal ctenidium and a well-developed pronotal one. (The *Meringis* on *Perognathus,* which is non-saltatorial and occurs in the same foci, resemble those on *Dipodomys.*) The zapodid jumping-mice in the New-World (*Zapus* and *Napaeozapus*) do not have any specific fleas but the Palaearctic *Sicista* (partly scansorial) is parasitized by a *Leptopsylla* Jordan & Rothschild, 1911 (IV.1) which has frontal spiniforms and well-developed genal and pronotal combs.

In all probability it is not the saltatorial habit per se (and any consequent risk to dislodged fleas) that is the main factor making life on these hosts especially precarious for flea-survival, but, rather, the selective pressure is due to traits associated with it, such as the relatively extended home-range, or leaving the territory when seeking food or when alarmed, and ending up in a biotope unfavourable for fleas. Another likely reason is that these hosts coincidentally have attributes that are detrimental to Siphonaptera. The following illustrate some of these points. The gymnuran *Hylomys* lives in a fringe-habitat of logs and fallen brush that is narrowly restricted and conditions just nearby may be completely different and unfavourable for fleas so highly specific to this host or for their larvae. Rabbits, bandicoots and kangaroos travel over a large area, and fleas falling off the host or emerging from the cocoon in such places, would have a difficult time finding a food source or the eggs may drop from the host in habitats unsuitable for the larvae. These particular mammals do not return to a previous nest-site or breeding-place, so if fleas do hatch out where the last brood of young were born, they are unlikely to find a host. (These factors apparently account for the absence of fleas on herbivores in herds in most parts of the world, and macropodids belong in that category. Their adaptive radiation, like that of equids and ruminants, coincided with the development of grasses in the Miocene, and they have been major occupants of grazing and browsing niches in Australia ever since (Tyndale-Biscoe, 1973).) The elephant-shrews (except *Elephantulus*) and dipodids (save *Jaculus*) tend to live singly or in pairs, and this lessens the chance of their fleas finding new hosts. Many of these rodents (dipodids, sicistines) frequently travel a long distance to find food and may enter new habitats, and a majority hibernate or aestivate for long periods, during which time the fleas must remain safely on their hosts or face death by scratching, or perhaps end up in an extremely adverse environment without a source of blood. The macroscelids and jumping-mice generally lack an individual burrow system, and the former and many of these rodents do not have permanent nest-sites. Most of these mammals are residents of xeric areas, where it is essential for fur-fleas to remain on this host to avoid rigorous or fatal conditions off it.

Under those conditions, it is to be expected that hosts in these categories would either lack specific fur-fleas or would be parasitized only with species that are especially adapted, e.g. possess a heightened means of remaining attached, whether by supernumerary spines or bristles, by anchoring mouthparts, or else be unusually specialized, such as possessing parasitic larvae or having their reproductive behaviour synchronized with the hormones of the host (p. 56).

3.3 *Spines and spiniforms on the heads of fleas*

It is remarkable that so many fleas, representing a variety of genera, subfamilies and families, have developed what I here term '*crowns of thorns*', i.e. 1) purely marginal rows or groups of backward-directed spiniform bristles on the head, or 2) a subvertical genal comb that approximates the anterior cephalic margin, or 3) a comb of spines on a special helmet. However, there are several groups of fleas which approach the condition seen in Siphonaptera with crowns of thorns. In these, the first row of frontal bristles is displaced appreciably towards the margin of the head, but is not quite submarginal, and/or there are somewhat thorny bristles or subspiniforms on the frons, varying in number from a few to many. Examples (all pygiopsyllids) include certain *Rectidigitus* Holland, 1969, *Striopsylla* and *Ernestinia* and the *Metastivalius mordax*-group, as well as *M.shawmayeri* (Jordan, 1933). Such Siphonaptera are here termed as 'quasi-thorny fleas'. In many of these, the head is flattened in varying degrees and fairly broad laterally, and the coxae are unusually stout, as discussed elsewhere (Traub, 1968, 1969), and below, when other intermediate forms are mentioned (p. 51).

Fleas with an upright genal comb on the *posterior* part of the preantennal region which does not approximate the frontal margin, are not considered in the category of crowns of thorns unless they also bear frontomarginal spiniforms as pointed out below (p. 47). Some fleas with vertical ctenidia approach this condition, depending upon how close part of the comb comes to the front of the head.

Fleas in which the crown of thorns consists of an apical tiara of spiniforms, include the pygiopsyllids *Smitella* Traub, 1968, *Idiochaetis, Muesebeckella,* a new and undescribed genus, *Metastivalius anaxilas* (M.Rothschild, 1934), *Lentistivalius ferinus* (Rothschild, 1908) (II.1) and allies; the hystrichopsyllid *Stenistomera alpina* (Baker, 1895) (I.2); the leptopsyllids like *Leptopsylla (Pectinoctenus) pamirensis* (Ioff, 1946) (IV.1); certain *Ctenophyllus* (IV.2); and *Cratyniius* (IV.1). Notably, in some genera (e.g. *Metastivalius, Lentistivalius, Stenistomera* Rothschild, 1915 and *Ctenophyllus*) only certain species are so modified, their congeners being less so or even completely unaffected. Other taxa have a small group of marginal frontal spiniforms, such as a second undescribed genus of pygiopsyllid, the species of *Acanthopsylla, Metastivalius tamberan* Holland, 1969, a new genus near *Rectidigitus* (II.1), *Peromyscopsylla* I.Fox, 1939 and *Leptopsylla* and allies (IV.1), etc., with the number of spiniforms ranging from 1-6. All the stephanocircids have a true comb of vertical spines on the helmet. Chimaeropsyllids with a submarginal frontal upright comb include *Epirimia* de Meillon, 1940, *Macroscelidopsylla* de Meillon & Marcus, 1958 and *Demeillonia* Hopkins & Rothschild, 1956 (XII.3). Such 'crested' fleas also occur in the Hystrichopsyllidae, e.g. *Corypsylla* (I.8) where the combs are even closer to the anterior margin. In *Sigmactenus* Traub, 1950 (IV.1) only the upper part of the huge sigmoid ctenidium is near the frontal margin.

This taxonomically diverse group of fleas with a crown of thorns have a surprising number of features in common: 1) a series of backward-directed barb-like structures or spines on the head which are presumably useful in maintaining a grip on the hairs of the host; 2) the hooking devices are in a subvertical row or parallel the frontal margin, and 3) they are anteromarginal or submarginal in position, or at least partly so, with at least one spine or spiniform near the anterodorsal angle. The presence of such a series of modifications is impressive enough, but the power of evolutionary forces involved is indicated by the variety of ways in which these results have been achieved. Some of the head combs are genal ctenidia which have assumed a vertical position as shown in figs. 18, 24, 26. The helmet of stephanocircids (fig. 4), of which there are three different types, and which may also represent convergence at the subfamily level (Traub & Dünnet, 1973) bears a special ctenidium. A wondrous development is exhibited by the pygiopsyllid *Smitella* (fig. 1), which is a helmet-flea in which the comb consists of modified spine-like bristles, not spines, as witness their setal bases (Traub, 1968). Despite that basic difference, the helmet of *Smitella* strikingly resembles that of the stephanocircids in the possession of a large canal and many submarginal canaliculi, displacement of certain sensory placoids, etc. Even the spiniform bristles of fleas are not uniform in origin. In some instances, they are formed during the life of the adult flea, as the deciduous attenuated tips of the bristles are shed, leaving the barb, as in *M.anaxilas* (fig. 12) and *I.illustris* (fig. 6) (Traub, 1968; Jordan, 1937). This is also the case of the leptopsyllid *Ctenophyllus,* as reported by Ioff & Scalon (1954). In other instances, there is no sign of a deciduous tip, and the spiniform is apparently the result of evolutionary processes.

Another remarkable feature shared by fleas with a crown of thorns is that 4) the frons of the head is narrowed appreciably, or even conspicuously, from side to side, producing a keel-like edge, whereas related taxa (and others) without such spines or spiniforms have a broad head. Moreover, the degree of acuity of the margin varies directly with the size of the row or group of spiniforms or spines, and with its proximity to the margin. 5) Fleas with crowns of thorns all presumably face special hazards if they fail to remain on the host when it is out of the nest. Further, the degree of the risk posed to the flea is reflected in the level of development of the modified spiniforms or ctenidium. 6) In these fleas, the mouthparts are unusually short, usually extending only about half-way down the fore-coxae, and only rarely reaching ¾ of the length of the coxae. Fleas that lack such vestiture have definitely longer mouthparts, even if in the same genus. Once again, the species that are partially modified regarding development of spiniforms tend to have mouthparts that are intermediate in size. 7) The majority of fleas with such adaptations on the head tend to have a short frons that is taller than long (as seen in the lateral aspect), a character which is generally coupled with the presence of false combs of bristles on the tibia (Traub, 1972b).

Table 1 compares fleas with crowns of thorns with examples of fleas with a horizontal genal ctenidium or which differ otherwise. Data are summarized therein as to whether the hosts are scansorial, etc., and pertinent figure references are cited. From this table it is apparent that, in addition to the critical points made above, the taxa with horizontal combs on the head lack a wafer-thin anterior cephalic margin, nor do they possess apico-frontal spiniforms.

In order to illustrate the keel-like anterior margin of the head in fleas with crowns of thorns and other modifications, dissections were prepared and photographed, showing the anterior aspects of the frons (e.g. figs. 2, 10, 11, etc.) and the dorsal aspect of the head and pronotum (e.g. figs. 3, 5, 8, 9, etc.). When present, the keel can be readily seen. Also, the degree of approximation of the spines, or of the left and right bristles of the first row

TAXA	CLASSIFICATION	HEAD-SPINIFORMS			GENAL CTENIDIA				HELMET COMB	FRONTAL MARGIN ACUTE	FIGURE REFERENCES	MAJOR HOSTS			OTHER HAZARDS
		NO. OF TAXA	TRUE TIARA	NO. OF MARGINALS	VERTICAL		HORIZONTAL	LENGTH				SCANSORIAL /ARBOREAL	NOCTURNAL	VOLANT	
					ANTERIOR	MEDIAN									
STEPHANOCIRCIDS (9 Gen.)	VII	0/9 Gen.							9/9	+	4,5	+	+		
SMITELLA	II.1	1/1	+						(False)	+	1,2,3	+	+		
IDIOCHAETIS	II.1	2/2	+							+	6,8,10	+	+		
METASTIVALIUS ANAXILAS-GROUP	II.1	6/6	+							+	12,14,16	+	+		
(OTHERS)		0/13	0							0	13,15,17	0	±		
NEW GENUS	II.1	1/1	+							+		+	+		
ACANTHOPSYLLA	II.1	19/19		1,2,4						+		+	+		
MUESEBECKELLA	II.1	3/3		5						+		+	+		
PAPUAPSYLLA	II.1	0/23								0		0	±		
HOOGSTRAALIA	II.1	0/4				+		± L		0	46,48,51	0	0	+	
HYSTRICHOPSYLLA	I.1	0/16					+	L		0		0	+		+
STENOPONIA	I.9	0/16				+	(Part)	L		0		0	+		+
CTENOPHTHALMUS	I.3	0/110					+	S		0		0	±		
PALAEOPSYLLA	I.2	0/33				+		S-L		0		0	+		+
DORATOPSYLLINAE (6 Gen.)	I.5	0/6					+	S		0		0	+		+
STENISTOMERA MACRODACTYLA	I.2	0/1		0						0		0	±		+
(OTHERS)		2/2		4-5						+		+	+		+
CORYPSYLLA	I.8	0/3			3/3			± L		+	19,21,23		+		+
RHADINOPSYLLA & KIN	I.8	0/55				+		S		0					
NEOPSYLLA	I.7	0/43					+	S		0	7,9,11	0	0 or ±		
CERATOPHYLLUS - EX BIRDS	III.1	0/53								0	41,43,45			+	
MACROSTYLOPHORA	III.1	0/28								0		+	0		
ORCHOPEAS	III.1	0/9								0		+ (Some)	0 or ±		
LEPTOPSYLLA (LEPTOPSYLLA)	IV.1	9/9	+ in 1	2-6		+		S		+		+	+		+
L. (PECTINOCTENUS)	IV.1	5/6		0-6		+		L		0 or ±	31,33,35	(Some)	+		+
PEROMYSCOPSYLLA	IV.1	18/18		2		+		S		±		(Some)	+		+
SIGMACTENUS	IV.1	0/7		0- ±4	(Dorsal)	+		L		+(Dorsal)	18,20,22	+	+		+
CTENOPHYLLUS (S. LAT.) CT. & ACONOTHOBIUS	IV.2	9/9	±							±			+ or 0		+
(OTHERS)		0/6	0							0			(?)		(?)
APHROPSYLLA	IX.1	0/2				+		± L		0				+	
CTENOCEPHALIDES	IX.1	0/9					+	S-L		0	47,49,50,52	(Some)	(Some)		+
CEDIOPSYLLA & SPILOPSYLLUS	IX.1	0/5				+		S-L		0					+
CENTETIPSYLLA	IX.1	0/1				+		S		0	38				+
ARCHAEOPSYLLA	IX.1	0/1				+		S		0	32,34,36				+
PULEX	IX.1	0/6								0	40,42,44				
EPIRIMIINES (3 Gen.)	XII.3	0/3			±			L		±	24-27,29	1/3	+ or 0		+

Table 1. Comparison of certain fleas regarding spiniforms or combs on the head.
Numerator = Number with character; Denominator = Total number in taxon; + = Feature applies; Blank or 0 = Absent; ± = Somewhat; S = Small; L = Large.

serves to indicate the degree of acuity of the keel or anterior margin at the various levels.

The series of convergent developments in the true helmet-fleas and in the pygiopsyllid *Smitella,* with its comb of spiniforms, mentioned above, is paralleled by the wafer-thin nature of the helmet in both taxa (cf. figs. 1-3 and 4-5). Because the helmet is so extremely flattened, the spines (or spiniforms) of the frontal comb on the left side closely approximates those on the right, as shown in the figures. In this way, the keel of the helmet can slip between the hairs, and the hooklets on each side presumably can latch onto a contiguous pair of bristles. *Idiochaetis* has a greatly narrowed frontal margin (figs. 6, 8, 10) and the entire dorsal and anterior margin of the preantennal region is here festooned with thorny barbs, which again are closely appressed. A new genus at hand is similar in both respects. *Metastivalius anaxilas* (figs. 12, 14, 16), like its five sibling species, has a tiara of spiniforms along the frontal margin, and here too the narrowness of that region is striking. In contrast, however, the closely related *M.mordax* (Rothschild, 1908) (figs. 13, 15, 17) lacks a crown of thorns, and, in comparison, the frons of the head is much broader (from left to right) and thus the bristles on the left side are far more distant from those on the right than in the preceding taxa. Nevertheless, as a quasi-thorny flea, *M.mordax* approaches *M.anaxilas* to a greater degree regarding these modifications than do some other members of the genus in which the bristles are unmodified – and where the head is even much broader apically than in *M.mordax.* This condition is also seen in *Lentistivalius,* where the apical margin is acute in *L.ferinus,* which bears a tiara of spiniforms, and broad in all the other species, which have unmodified bristles. Pygiopsyllids like *Papuapsylla, Traubia* Smit, 1953 and *Pygiopsylla* Rothschild, 1906, which lack cephalic spiniforms or combs, likewise have a broad head.

In fleas where the vertical comb borders the *front* of the head, the anterior margin is greatly narrowed, as in *Corypsylla* (figs. 19, 21, 23). Where the head comb does not quite approximate the margin, as in the epirimiine *Epirimia* (figs. 24, 25, 27) and *Demeillonia* (figs. 26, 29), the frontal margin is not so keel-like, but, nevertheless, is greatly narrowed, as indicated by the narrow gap between the bristles of the first row on the left and right sides of the frons. *Sigmactenus* (figs. 18, 20, 22) is instructive in that the bases of the left and right spines are closest together when at the dorsal regions, where the comb turns anteriorward and approaches the margin. Thus, the frontal part of the head is much more narrowed dorsally than ventrally. The correlation between the position and number of spiniforms, and the breadth of the frons, is similarly demonstrated by the fleas with just a few thorn-like anteromarginal bristles, such as *Leptopsylla* (figs. 31, 33, 35), *Ctenophyllus* and *Acanthopsylla.* The frontal margin is significantly narrowed in *Leptopsylla* at the area of insertion of the spiniforms, especially in the species with a group of about six anteromarginal spiniforms as in the figures. It is less narrowed in species of *Ctenophyllus* where the row of thorny bristles is submarginal, but, notably, the members of *Ctenophyllus s.lat.* which lack spiniforms have broad heads. In *Acanthopsylla,* where there are 1, 2 or 4 spiniforms at the anterodorsal angle, the degree of narrowing reflects the number of thorny bristles involved.

These observations suggest that it is not the mere presence of vertical combs or groups of spiniforms on the head that determine the acute nature of the anteromarginal portion of the head, but rather that the proximity of the spines or thorns to the frontal margin is the critical factor. Thus, *Dinopsyllus* C.Fox, 1925 (I.4) bears a large, vertical genal ctenidium which in the main is on the posterior portion of the frons and only dorsally does it approach the frontal margin, and at best is never even quite submarginal. In this genus, the anterior part of the head is quite broad from left to right, albeit somewhat narrowed subdorsally, and strictly speaking it does not meet the above criteria for a crown of thorns though obviously evolving in that direction. *Palaeopsylla* Wagner, 1903 (I.2) also has a vertical comb on the head, but it is shorter and further back than in *Dinopsyllus,* and the frontal margin of the head is broader yet.

The same principle is followed by the spiniforms – the head is narrowed apically only if the hook-like bristles are wholly frontomarginal. For example, in quasi-thorny fleas like certain *Striopsylla* and *Metastivalius shawmayeri* (Jordan, 1933), the head is adorned with stout barb-like bristles, but the frons is broad (and flattened) apically. This distinction in the head of *Idiochaetis* versus *Ernestinia* and *Striopsylla* lend support to my contention that, like the spines, the bristles of the flea can bear the imprint of the host's vestiture, and likewise function as a holding-device. As mentioned above, *Striopsylla* and *Ernestinia* are believed to ram the head against the body of the host (via stout forecoxae) and latch onto the hairs (Traub, 1968, 1969; Traub & Dunnet, 1973). Fleas like *Idiochaetis* effectively hook onto the hairs by means of the keel-like anterior margin and the apical crown of spiniforms, and can dangle securely in that position (*op. cit.* and new observations in the field by this author). The difference is presumably because *Striopsylla* and similarly adorned fleas primarily infest peramelid bandicoots, which have coarse fur, and hence an ultra-fine keel with barbs that are extremely close together (left and right sides) could not operate efficiently against such hairs, whereas such a tiara of spiniforms would function well if the host had fine hairs that were close together, as do the giant rats which *Idiochaetis* parasitize. The modifications seen in *M.shawmayeri* lend credence to this contention. This species is essentially a quasi-thorny one, with a short broad head, and infests bandicoots. In *Obtusifrontia* Holland, 1969, the head is also flattened and broad, and there is a frontomarginal group of downward-directed, stout bristles or subspiniforms. This taxon is another bandicoot-flea and acts like *Striopsylla.* The helmet-combs of stephanocircids likewise provide corroborative evidence. The peramelid-flea, *Stephanocircus harrisoni* has a helmet that is much more flattened (lateral and anterior aspects) than in the rat-stephanocircids, and the spines are hence further apart laterally and, as already mentioned, each spine is relatively distant from those above and below it. Both the helmet and the spines, then, are fitted to the coarse hair of the bandicoot. The rats infested by stephanocircids have fine hair, and the helmets (which have a long, very narrow keel) and the spines of the flea correspond to that vestiture. The same trend re a shorter, broader helmet-keel is shown by *S.dasyuri* Skuse, 1893, but to a lesser degree, and it makes sense that this flea infests both peramelids and rats. The bristles on the occipital region of these bandicoot-stephanocircids are also characteristically specialized and differ from those on their rat-infesting congeners, as mentioned below (p. 60).

Like anteromarginal spiniforms, vertical genal combs, then, are associated with a marked lateral narrowing of the front portion of the head, provided the comb itself is subapical. What then is the case when the upright comb is restricted to the caudal part of the gena, or when the genal comb is horizontal? If the taxon likewise has frontomarginal spiniforms in addition to a short subcaudal vertical comb, as in *Leptopsylla* (figs. 31, 33, 35), then the head is markedly narrowed anteriorly, as in other fleas with crowns of thorns. When species with such a genal comb lack marginal spiniforms, as in *Hoogstraalia* Traub, 1950 (II.1) (figs. 46, 48, 51), *Cediopsylla* (IX.3), *Chimaeropsylla* Rothschild, 1911 (XII.1), etc., the head is frontally broad.

In taxa with a horizontal genal comb, the front of the head is broad, as in *Ctenocepha-*

lides felis (Bouche, 1835) (IX.2) (figs. 47, 49, 50), where the ventral aspect (fig. 52) is also illustrated to emphasize this point, and as in *Doratopsylla* Jordan & Rothschild, 1912 (I.5), *Ctenophthalmus* (I.3) and *Archaeopsylla* Dampf, 1908 (IX.2) (figs. 32, 34, 36), etc. The same is true for taxa which possess a ventral or subventral caudal genal comb consisting of two crossed spines, e.g. *Chiastopsylla* Rothschild, 1910 (XII.2), and the neopsylline *Epitedia* Jordan, 1938 (I.7), *Meringis, Phalacropsylla* Rothschild, 1915, *Rothschildiana* Smit, 1952 and *Neopsylla* Wagner, 1907 (figs. 7, 9, 11). (As can be seen from the illustrations, the comb does not consist of two 'characteristic' crossed or overlapping spines, but instead, the spines are at different levels and give an erroneous impression in the mounted flea. It is clear from the true position that the comb could indeed function to grasp hairs, but could not do so if the apparent position were real.)

The nature of the preantennal region of *Macropsylla* Rothschild, 1905 emphasizes some of these points. The head-comb of this monotypic, highly distinctive genus is anomalous in that part of it is horizontal, extending from the very front of the head to the level of the middle of the second antennal segment, at which point it swings abruptly dorsad and runs dorso-caudad at an angle of about 75°, terminating at the rudimentary eye, which has been pushed to a height near the base of the antenna. Hopkins & Rothschild (1956) regarded the angulate comb of the Macropsyllidae as 'an exceptionally well-developed genal comb and presumably represents both the genal comb and helmet-comb of the Stephanocircidae'. Dunnet & Mardon (1974) refer to it as a genal comb. If my theories are correct about the frontal margin seen in helmet-fleas and those with horizontal genal combs or caudally placed vertical combs, then one would expect the following modifications in *Macropsylla:* The anterior and anterodorsal margin would be greatly narrowed as in bona fide helmet-fleas, but the head would be broad at the rear of the horizontal comb (and the level of the vertical comb). This is exactly what is seen in the dissected flea (prepared too late for photography for inclusion here) although the frontal keel bears a ridge on each side of the razor edge. The head as seen dorsally is an acute triangle, tapering towards the front; as seen from the frontal aspect, it is narrowest at the anteroventral angle, and closest to the first spine. This too is in accord with these concepts. The keel-like frontal margin suggests that at least that portion of the head represents a true helmet. This view is further supported by the fact that the mouthparts are inserted well caudad from the tip of the head, especially in *Stephanopsylla* Rothschild, 1911, the only other genus in the family Macropsyllidae. In fleas with an ordinary genal ctenidium, like *Stenoponia,* the mouthparts are near the front of the head. The upright part of the comb in *Macropsylla* may indeed be derived from the genal comb since the vestigial eye has obviously been forced dorsad as the ctenidium assumed the vertical position in the course of evolution (as in *Sigmactenus,* etc.).

The fifth point in common in Siphonaptera with crowns of thorns is the intimate and apparently invariable association with hosts whose habits or environment present special hazards to the fleas, especially mammals which are both nocturnal and arboreal or scansorial. The relationship is well-illustrated by the genus *Leptopsylla* and its allies, which also demonstrate how the degree of development of spiniforms and spines is correlated with the degree of risk faced by the flea. In the nominate subgenus of *Leptopsylla* the number of genal spines varies from three to six, and the fronto-marginal spiniforms range from two to six, while one species, *L.putoraki* Ioff, 1950, has the entire anterior margin of the head festooned with spiniforms and bears six genal spines. The greatest number of both genal spines and spiniforms occurs in the fleas at greatest risk – e.g. *L.putoraki* specifically infests *Diplomesodon,* a shrew which is not only a desert-denizen but is solitary, frequently changes its burrows and is apparently rare under the best of conditions. The intermediate condition regarding spines and spiniforms occurs in fleas of dormice, which are arboreal and nocturnal; those of *Sicista,* the birch mouse, which is a good leaper and climber, nesting off the ground, and of hamsters in the mountain-deserts of Central Asia, where there are severe environmental hazards. The fewest spines and spiniforms are found in fleas of ground-dwelling murines, which, as nocturnal hosts in semideserts, do exert pressure for the evolution of combs and spiniforms in their fleas. *L.sexdentata* (Wagner, 1930) with six genal spines and four spiniforms, is primarily associated with *R.rattus,* which often nests in trees and is an excellent climber, as suggested by the name 'roof-rat' for the commensal form. In many parts of the world, this rat carries *L.segnis* (Schönherr, 1811), another spinose flea, and one that despite the name 'house-mouse flea', may prove to be a characteristic parasite of *Rattus* rather than *Mus.* In our studies on murine typhus in Ethiopia and Baltimore we have observed that *L.segnis* occurs in patches on the vertex of the head of its host, or deep within the hairs of the cheek. About 90% of the *L.segnis* on the host, at times comprising as many as 90-95 individuals, have been collected from those sites on freshly-caught rats. Notably, even after etherization, these fleas have often remained in the fur of the host, fixed by the spiniforms. Some may be found loose in the fur, and the number seems to reflect how long the rats have remained agitated in traps.

The correlation between spiniforms and genal spines, with environmental problems posed by the host, is particularly well portrayed by the subgenus *L.(Pectinoctenus)* Wagner, 1928 (figs. 31, 33, 35) and by the allied genus *Sigmactenus.* In the former, the number of genal spines ranges from 6 to 20, and it is noteworthy that the species with the greatest number of spines lacks spiniforms. The maximum ctenidial development occurs in fleas on nocturnal hosts (e.g. *Phodopus*) in arid areas and deserts, or in far north-eastern Asia under equally rigorous climatic conditions, where fleas off the host may have little chance of survival. In *Sigmactenus* (figs. 18, 20, 22) the genal comb consists of from 7 to 15 spines and the dorsal part of the ctenidium is placed nearer the anterior margin of the head, and there are no highly modified anterofrontal spiniforms. The pronotal ctenidium and tibial combs are unusually well-developed. These are parasites of nocturnal scansorial murines, inhabiting moss-forest or cloud-forest, where environmental conditions would probably be unsuitable for larval development (and perhaps for the flea as well) outside the nest of the host.

The combination of characters found in *Sigmactenus* is so firmly correlated with hosts of that type that I was certain that even though *Epirimia* belongs to another family (XII.3) and occurs thousands of miles distant, it too must infest a nocturnal, arboreal host because of the nature of its genal and pronotal combs (figs. 24, 25, 27) and its exceptionally specialized tibial combs (figs. 28, 30). In fact, so closely do these taxa resemble one another regarding head, pronotum and tibiae that, if one relied solely on such structures, he could readily believe they belonged to the same family. Recent collections from a series of dormice, made by Dr D.Schlitter in a joint project in South Africa, substantiate the belief about the traits of the true host of *Epirimia.*

The question may be asked why *Lentistivalius ferinus* is the only member of the genus (7 species known to me) with a crown of thorns, while even another shrew-infesting member lacks one. In reply, it is stressed that in addition to the regular problems associated with infesting shrews is the difficulty of survival in terrain that is either wholly arid or is subject to monsoon conditions of prolonged drought alternating with extensive flooding. (There are relatively few species of native fleas to be found on ground-dwelling hosts in such terrain anywhere.)

The sixth feature shared by Siphonaptera with crowns of thorns (Table 1), namely the short mouthparts (e.g. figs. 1, 4, 6), is not limited to such fleas, but it is significantly scarce in the Order to be a notable characteristic. *Metastivalius* illustrates the principles involved. The laciniae and palpi are very short in *M.anaxilas* (fig. 12) and only slightly longer in the quasi-thorny *M.mordax* (fig. 13) and allies, but they are much longer in members of the genus which lack modified bristles. Other quasi-thorny fleas like *Striopsylla* etc., and fleas which also tend to have shortened, flat heads and stout forecoxae, e.g. *Obtusifrontia*, likewise have short mouthparts. However, not all fleas with mouthparts of this type have flat or foreshortened heads, e.g. certain ischnopsyllids (bat-fleas). Where we do have observations, such a feeding apparatus is associated with the habits of the flea when imbibing blood, and the subject is treated further below (pp. 54, 55).

The seventh point common to fleas with crowns of thorns pertains to the foreshortened nature of the head, a characteristic, which, as pointed out just above and elsewhere, occurs in other Siphonaptera where spines and/or bristles have been modified so as to enhance attachment (pp. 37, 55). Fleas with tibial combs often have such flattened heads, but a notable exception is *Ctenoparia* Rothschild, 1909 (I.1).

Just as fleas have developed hold-fast devices in various ways, so are there transitional forms in the evolution of crowns of thorns. Some have been mentioned, but others not so advanced are scattered throughout the Order. Some have supernumerary bristles on the head and exhibit a degree of forward displacement of the first row of frontal bristles, but the bristles are not particularly spiny. Such fleas invariably seem to be parasites of mammals that are somewhat scansorial and facultatively crepuscular rather than nocturnal, or else are scansorial but somewhat diurnal, or which pose some element of environmental hazard to the dislodged or newly emerged flea. In short, the degree of morphological modification and risk are intermediate between those of the nocturnal/arboreal group and the ground-dwelling, routine-type of host. Examples of such transitional fleas are the pygiopsyllids *L.(D.)mjoebergi, Lentistivalius vomerus* Traub, 1972, *M.(M.)jacobsoni,* the subgenus *Gryphopsylla* Traub, 1957 (all of which have a complete abdominal comb, or one or more partially developed ones), *Traubia* Smit, 1953, as well as representatives in other families, such as *Jellisonia* Traub, 1944 (III.1), *Pleochaetis* Jordan, 1933 (III.1), *Kohlsia* Traub, 1950 (III.1), certain *Neopsylla* (I.7) and *Amphipsylla* Wagner, 1909 (IV.2), etc. Certain of these also have pectinate bristles on the tibia. These fleas are broad-headed, but there may be some shortening in those where the frontal row is displaced anteriorwards to a greater degree and/or there is a tendency towards the development of subspiniforms.

The various stages in the position and numbers of spines in the *Leptopsylla-Sigmactenus* complex serve to stress the occurrence of intermediate forms in the modification of the upright genal ctenidium, including the correlation of the acuity of the frontal margin with the proximity of the comb. Thus, in some *L.(Pectinoctenus)* the comb is too far caudal to appreciably affect the narrowness of the head, and the same is true for chimaeropsyllines as compared to the kindred epirimiines, *Nearctopsylla* versus *Corypsylla,* and certain species of *Corypsylla* as compared to *C.kohlsi* Hubbard, 1940 (figs. 9, 21, 23).

Although fleas are not supposed to be capable of moving backwards in the fur (N.C. Rothschild, 1917; Smit, 1972), it should not be inferred that they are incapable of changing direction, for they readily can do so. Such flexibility is facilitated by the soft connection between the pronotum and mesonotum, although, thanks to the suture and membrane between head and pronotum, there is more mobility to the head itself than is generally appreciated. (The helmet is movable in most stephanocircids, as pointed out by Wagner (1934) and Traub & Dunnet (1973), and I believe that in fleas with a crown of thorns and in quasi-thorny fleas there is more movement possible than in ordinary species, and that this trait is likewise associated with their special habits and adaptations.) By turning only slightly to one side, a flea with a crown of thorns can no doubt hook a hair with its anteromarginal spiniforms or spines. If the host has very fine hairs, such a flea probably can simultaneously latch onto two contiguous ones by inserting its highly acute anterior margin between them and then merely stopping short. All these fleas have to do to detach immediately is to merely move forward.

3.4 *The frequency of occurrence of crowns of thorns*

It is edifying to discuss the frequency of occurrence of crowns of thorns in the Order and also to what extent such fleas share their hosts with other fleas. Only a relatively small proportion of fleas bear a crown of thorns. Thus, of the 241 genera known to me, only 27 (11%) have species in this category. Many of the affected genera consist of but one or two species altogether and only in the stephanocircids are all nine genera and 38 species so modified. All 19 of the species in *Acanthopsylla* and the 18 *Peromyscopsylla* have crowns of thorns, as do six of 19 *Metastivalius* and 15 of 16 *Leptopsylla.* Altogether, these 27 genera include 149 species of which 130 (87%) bear these modifications.

As already indicated, the hosts of fleas with crowns of thorns invariably seem to possess traits which are obviously hazardous regarding flea-survival, and the vast majority of them are both nocturnal and arboreal/scansorial. There are two somewhat surprising and apparently new generalizations that can be made about specific fur-fleas that infest hosts in that last-named group. Firstly, in any one area there generally is only one species of characteristic flea parasitizing such a host, even though, if the entire host-range were considered, a rich Siphonapteran fauna may be represented even at the generic level, as in the case of the flying-squirrels *Hylopetes* and *Petaurista.* (That unique species of flea may often lack spines or spiniforms on the head, but, regardless, it invariably bears supernumerary bristles, spines or combs, as noted by Traub (1972b) and discussed above.) Secondly, when there is more than one specific fur-flea on the nocturnal/arboreal host, both or all are apt to possess a crown of thorns or cephalic spiniforms.

The following hosts of this scansorial or arboreal/nocturnal category are examples of mammals infested with single species of fleas as specific or prime parasites in part of the range of the host. (Fleas with crowns of thorns are marked '#', those that are quasi-thorny are denoted '†'). Dormice, with a *Myoxopsylla* Wagner, 1927 (III.1) or a *Leptopsylla* #; flying-squirrels, with an *Opisodasys* Jordan, 1933 (III.1), *Hollandipsylla* Traub, 1953 (III.1), *Smitipsylla* Lewis, 1971 (III.1) or a single species of several other genera; the scansorial *Apodemus,* with *Peromyscopsylla bidentata* # (Kolenati, 1863) (IV.1) or *Leptopsylla; Gymnobelideus* opossum with *Stephanocircus domrowi* # Traub & Dunnet, 1973 (VII.1); *Petaurus,* with a new species of *Papuapsylla; Antechinus* and *S.greeni* # Traub & Dunnet, 1973; and martens with *Ceratophyllus lunatus* Jordan & Rothschild, 1920 (III.1). Some fleas associated with nocturnal hosts are more characteristic of the arboreal biotope rather than being specific to any one kind of host and hence can occur on a number of nocturnal mammals occupying that ecological niche, e.g. *Acanthopsylla* #

(II.1) on *Dactylonax* and *Dactylopsylla* phalagerids; *Metastivalius tamberan* # (II.1) on *Melomys* and marsupial mice; *Sigmactenus cavifrons* # Smit, 1975 (IV.1) on *Pogonomelomys* and arboreal *Pogonomys; Sigmactenus toxopeusi* # Smit, 1953 and *Metastivalius anaxilas* # on various *Rattus; M.(M.)jacobsoni* (II.1) on the pen-tailed shrew and other Malaysian hosts already mentioned; *Acanthopsylla rothschildi* # (Rainbow, 1905) on *Sminthopsis* and *Antechinus* in Australia and *Acanthopsylla* # spp. and certain *Rectidigitus* † (II.1) on various marsupial mice and murines in New Guinea.

Examples of the second point, concerning scansorial or arboreal/nocturnal hosts which in one locus carry several specific fur-fleas, include the following pygiopsyllids: *Melomys* with *Acanthopsylla* #, *Smitella* # (II.1) and/or a new genus #; *Pseudocheirus* with *Acanthopsylla* # and *Muesebeckella* # (II.1); New Guinean giant rats like *Anisomys* and *Mallomys* with *Idiochaetis* #, a second new genus #, certain *Rectidigitus* † and *Papuapsylla corrugis* (Jordan, 1933).

As for fleas without cephalic modifications occurring on this type of host along with those bearing crowns and crests, it must be stressed that more is involved than just the risks posed by the habits of the host. The phylogeny and behaviour of the flea, and the length of the association between parasite and host are among the factors to consider. A species without a history of a true comb, or one that has lost a ctenidium in the course of evolution, cannot produce a new one (although it may evolve a false comb of spiniforms) (Traub, 1968, 1972b). The combs, of course, have a variety of functions, and one flea may utilize the ctenidium while resting or feeding, whereas another may feed quickly and repeatedly, without a hold-fast device. Since we know of relatively recent (geologically speaking) changes of host, e.g. rat-fleas switching to marsupials, as in some *Metastivalius* and *Papuapsylla,* with a concomitant change in the pronotal comb (Traub, 1972b), and rodent-*Monopsyllus* adapting to birds (Rothschild & Clay, 1952), it is expected that younger changes exist which have not yet produced overt morphological modifications. Another point is that there are alternative mechanisms for facilitating remaining on an arboreal/nocturnal host, such as the supernumerary spines, combs and bristles, and false ctenidia, mentioned above, and the downward extension of the pronotal comb so as to cover the vinculum, etc. (Traub, 1969, 1972b). Similar functions are ascribed to the spinose heads of the quasi-thorny fleas. These adaptations on the thorax, abdomen or legs may serve in lieu of crowns of thorns and may supplement them. In the paragraph immediately above, the examples cited which lack the symbol '#' and '†' are species which possess such alternative devices.

Even so, as we have seen, there are remarkably few instances where fur-fleas lacking marginal spines or spiniforms infest the same host as species with crowns of thorns. The cited *Papuapsylla* from the giant rats is the only possible pertinent example I can recall, and it seems significant that this species is often also found on a ground-dwelling host like *Peroryctes* and certain *Rattus.* It, therefore, may not be a true *Mallomys/Anisomys* flea.

There are a very few hosts of fleas with crowns of thorns which are not in the category of nocturnal/arboreal (although they are generally nocturnal or crepuscular), e.g. the elephant-shrew, with *Demeillonia* (figs. 26, 29) and *Macroscelidopsylla;* some moles and shrews, with *Corypsylla* (figs. 19, 21, 23); and the lesser gymnuran, *Hylomys,* with *Craty-nius.* As mentioned above, all of these hosts are associated with environmental conditions potentially or actually adverse to fleas, while the elephant-shrews, like *Hylomys,* also progress by leaping, which, as indicated, somehow seems to be another trait associated with special hold-fast modifications in fleas. *Hylomys* is also partly scansorial. Another example, the shrew-flea *Lentistivalius ferinus,* has already been discussed (p. 50). All of these hosts seem to have no other taxa of fleas infesting them in any one area.

Bearing in mind that fleas with crowns of thorns are essentially or wholly restricted to risk-posing hosts, it seems clear that adapting to such hosts is a difficult evolutionary adjustment for fleas to make. This is indicated by 1) less than six per cent of the total flea fauna at the species level are so modified, and 2) the fact that even where such fleas occur, generally only one or two species per taxon of host are encountered. In contrast, for example, regarding ground-dwelling, facultatively or fully diurnal hosts, as many as eight species of fleas per individual host have been noted for *Peromyscus* (Holland, 1950), and eight species (representing four genera) have been found in a single burrow of *Rhombomys* in Iran (Farhang-Azad & Neronov, 1973). In the southern Pre-Balkash Region, USSR, alone, 27 species have been recorded from *Rhombomys* (Mikulin, 1956). The characteristic fur-fleas of *Rhombomys, Meriones* and other 'terrestrial' diurnal hosts either lack ctenidia altogether or have an unspecialized pronotal comb, e.g. *Xenopsylla* Glinkiewicz, 1907, *Nosopsyllus* Jordan, 1933, etc. In this category are well over 96% of the specimens (e.g. *Callopsylla tiflovi* Wagner, 1936) from *Ochotona* in Iran, Baluchistan, etc., where the pikas are active during the day. In those habitats, *Ctenophyllus* with cephalic spiniforms are rare, suggesting that there is something specialized about the behaviour of those species which are true parasites of *Ochotona,* or about their particular biotopes.

3.5 *Co-ordinated evolution of components of crowns of thorns*

The evolutionary mechanism that can result in such modifications as the crowns of thorns in fleas staggers the imagination, for it involves initiating, maintaining and synchronizing several nominally independent developments for a single, complex purpose. Moreover, several different methods are employed in various fleas to achieve the same ultimate result. Among other changes mentioned above, the process called for the development of 1) hooking devices, whether true cuticular spines or spiniform bristles; 2) arrangement of these barb-like structures in a vertical row; 3) migration of the spiny outgrowths towards the anterior margin of the head; 4) narrowing the frontal region to an extreme so as to bring into immediate play the spines on each side of the head; and 5) unusually short mouth-parts, rendering possible the particular stance adapted when feeding, etc. The basic device may be a genal comb that has become exaggerated in size and vertical in position (and pushing up the eye in the process, as noted by Jordan (1950) and Traub (1969, 1972b)) or else it may be a pectinate group of articulated modified bristles (again developed by a variety of means). Another example of what I term 'co-ordinated evolution' in fleas is provided by the coupled presence of a flattened but laterally broad head, an upright comb on the posterior part of the gena (or as an alternate, spiniforms on the frons), short mouth-parts and stout coxae, all again functioning together to permit temporary attachment while feeding on the host. (These fleas also have special concomitant buccal features which are adaptive but are too complex for treatment here.) It is difficult to comprehend how four or more apparently unrelated steps could evolve and become integrated for a common function, although the process probably is based upon such genetic phenomena as major genes, super-genes and switch-genes such as Ford (1975) mentions in connection with mimicry in the butterfly *Papilio.* He points out that 'if any linked major genes interact usefully, structural interchanges bringing them on the same chromosome will be favoured' (p. 118), and refers to 'bringing together the loci of co-adapted genes so as to form a super-

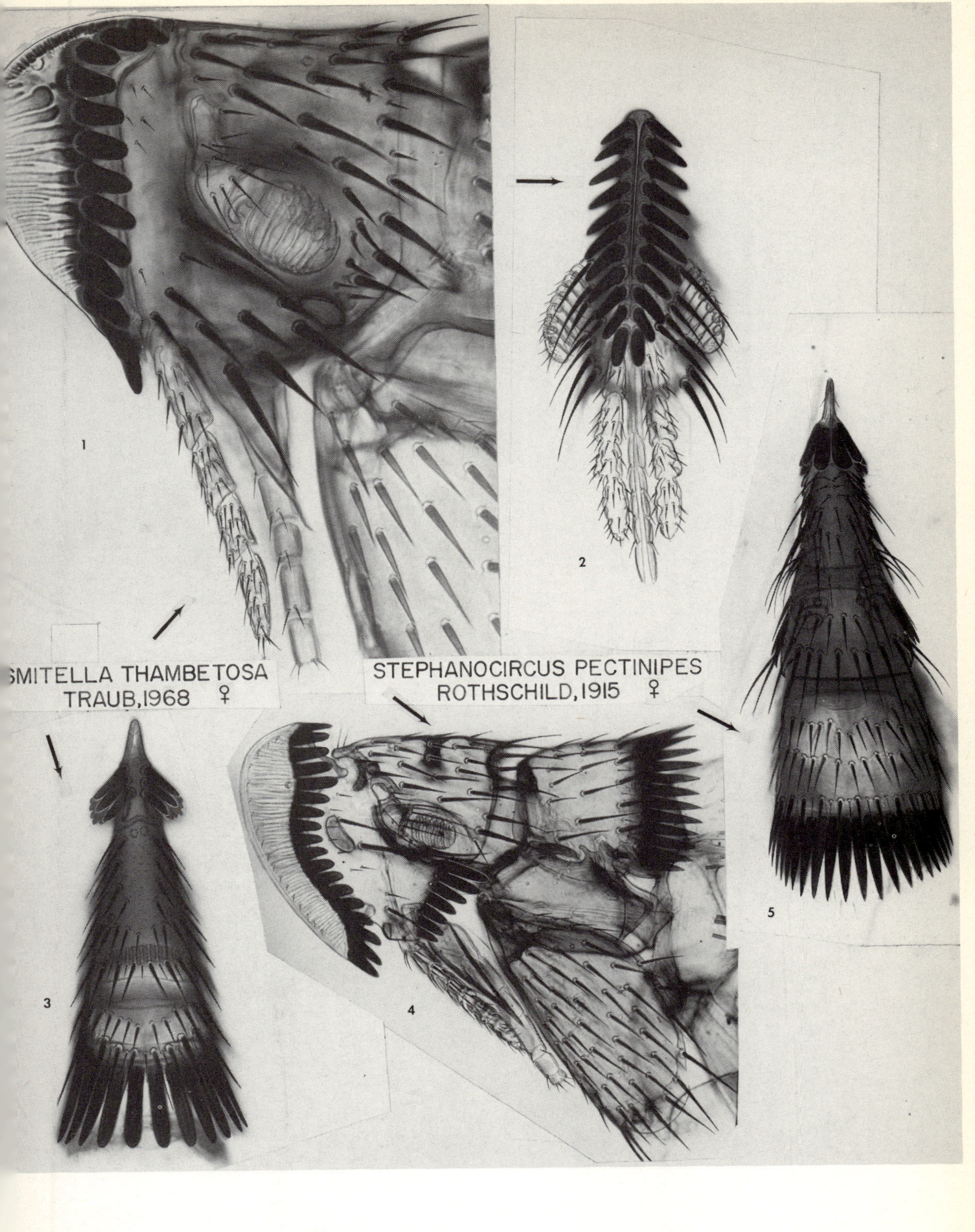
1
2
SMITELLA THAMBETOSA
TRAUB, 1968 ♀
STEPHANOCIRCUS PECTINIPES
ROTHSCHILD, 1915 ♀
3
4
5

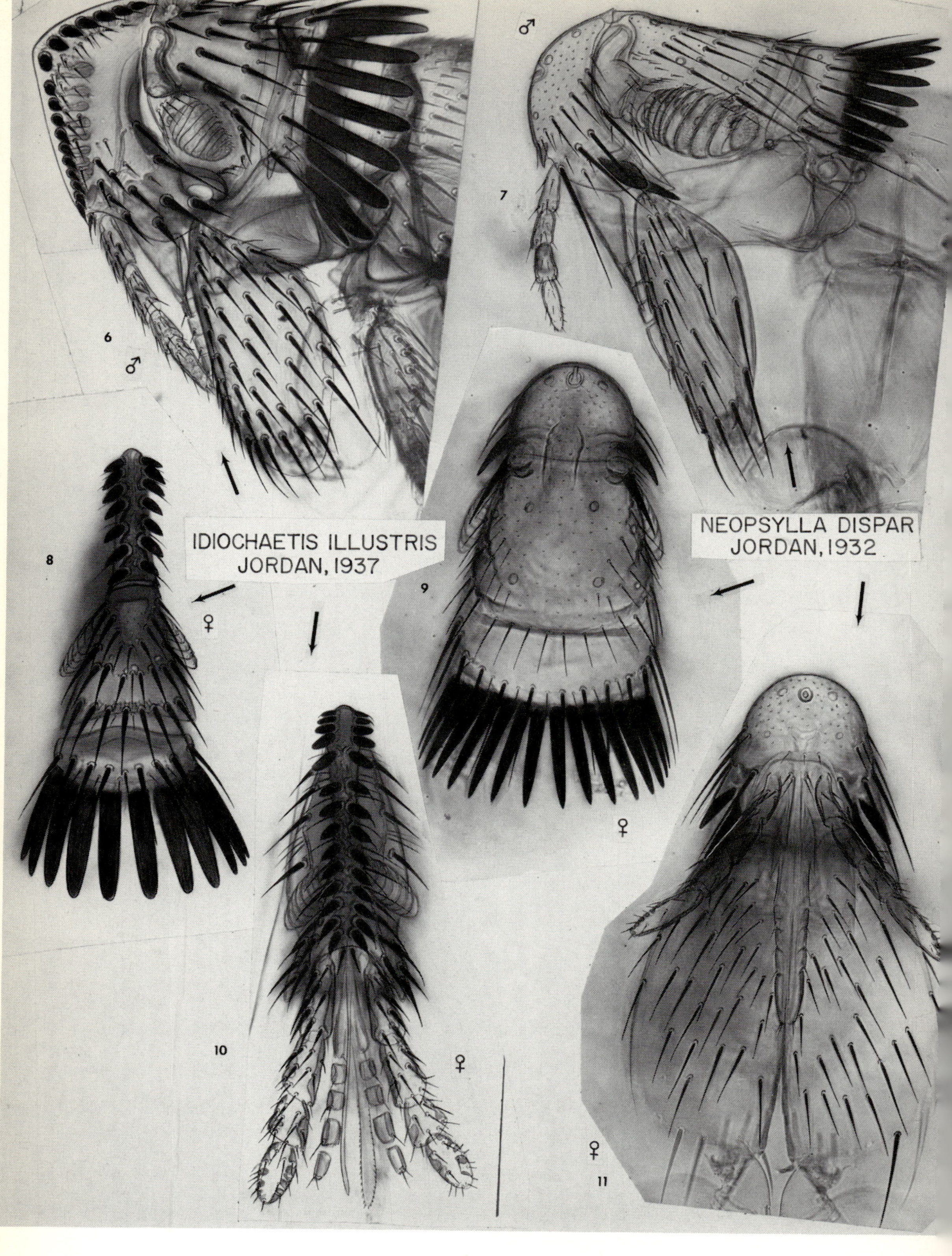

♂
6
♂
7
IDIOCHAETIS ILLUSTRIS
JORDAN, 1937
NEOPSYLLA DISPAR
JORDAN, 1932
8
♀
9
♀
10
♀
♀
11

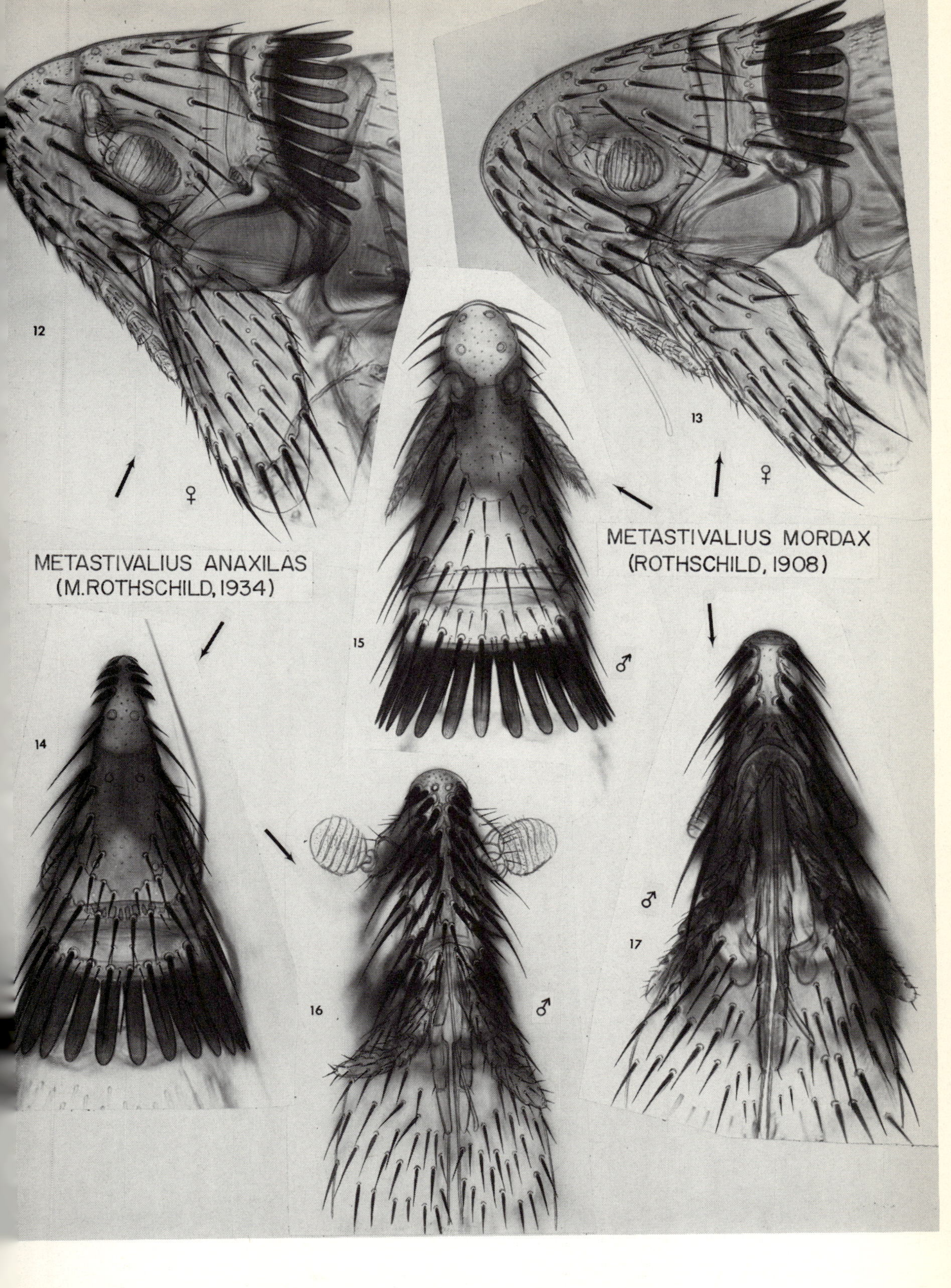
12
♀
METASTIVALIUS ANAXILAS
(M.ROTHSCHILD, 1934)
13
♀
METASTIVALIUS MORDAX
(ROTHSCHILD, 1908)
15
♂
14
16
♂
♂
17

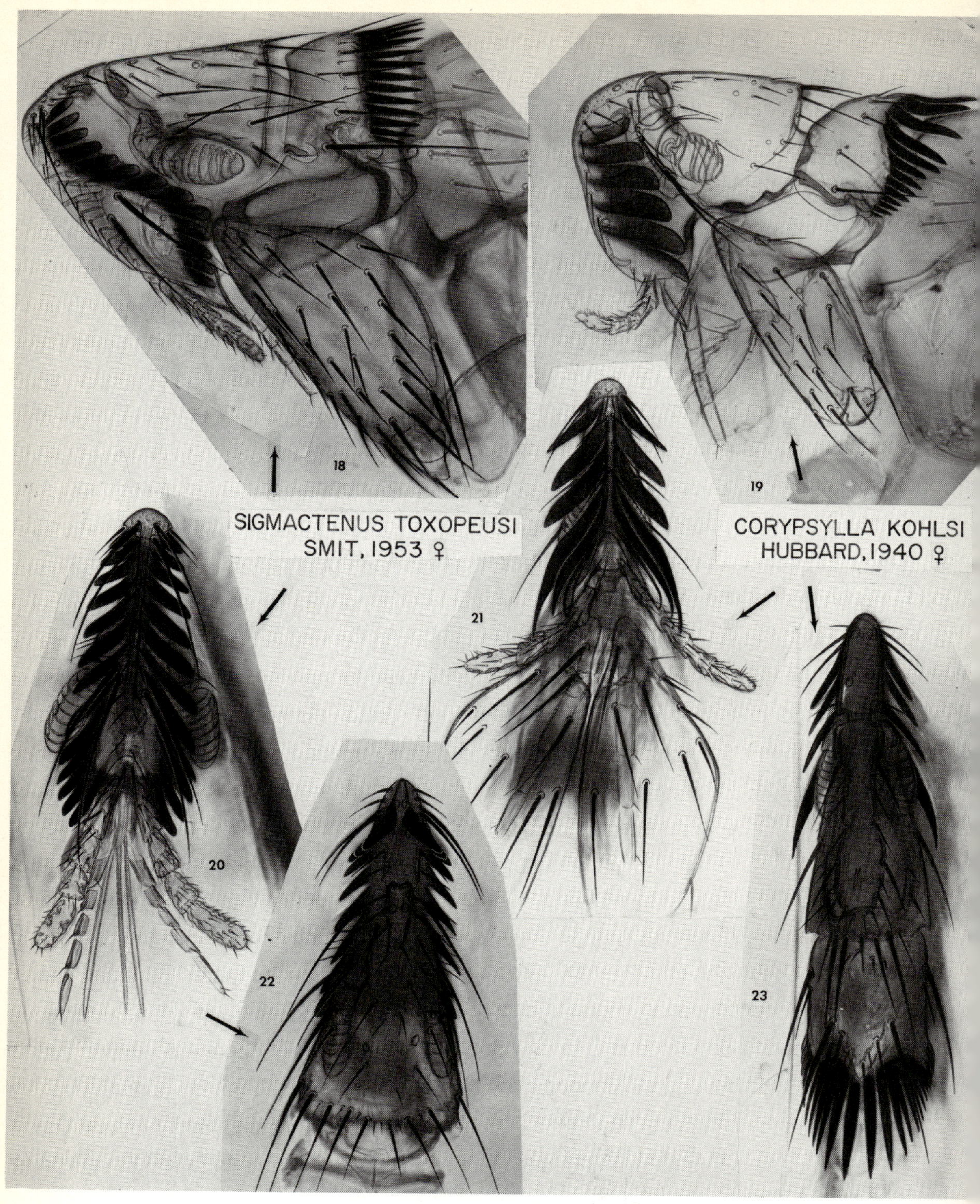
18
19
SIGMACTENUS TOXOPEUSI
SMIT, 1953 ♀
CORYPSYLLA KOHLSI
HUBBARD, 1940 ♀
20
21
22
23

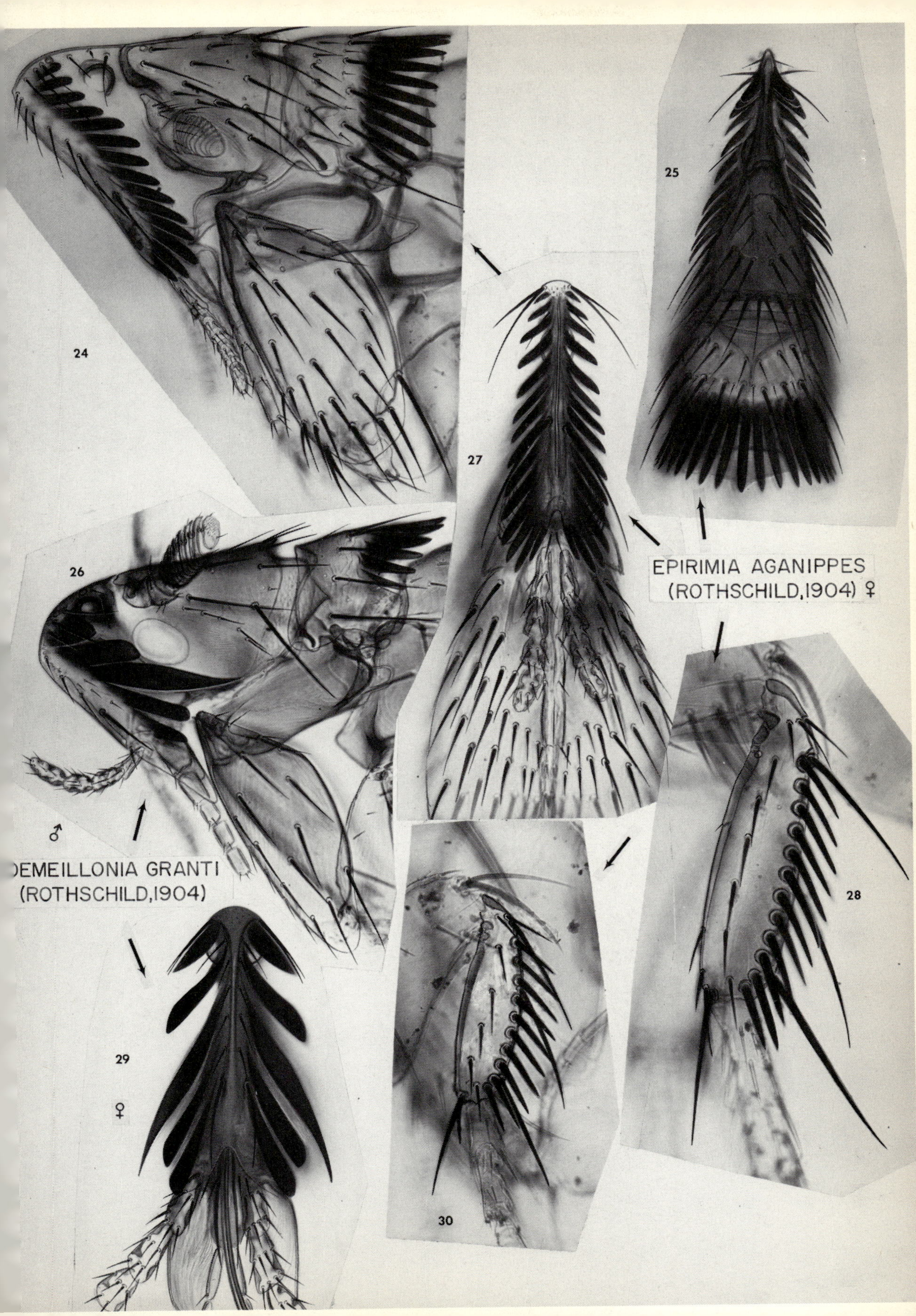
24
25
26
27
EPIRIMIA AGANIPPES
(ROTHSCHILD,1904) ♀
♂
DEMEILLONIA GRANTI
(ROTHSCHILD,1904)
28
29
♀
30

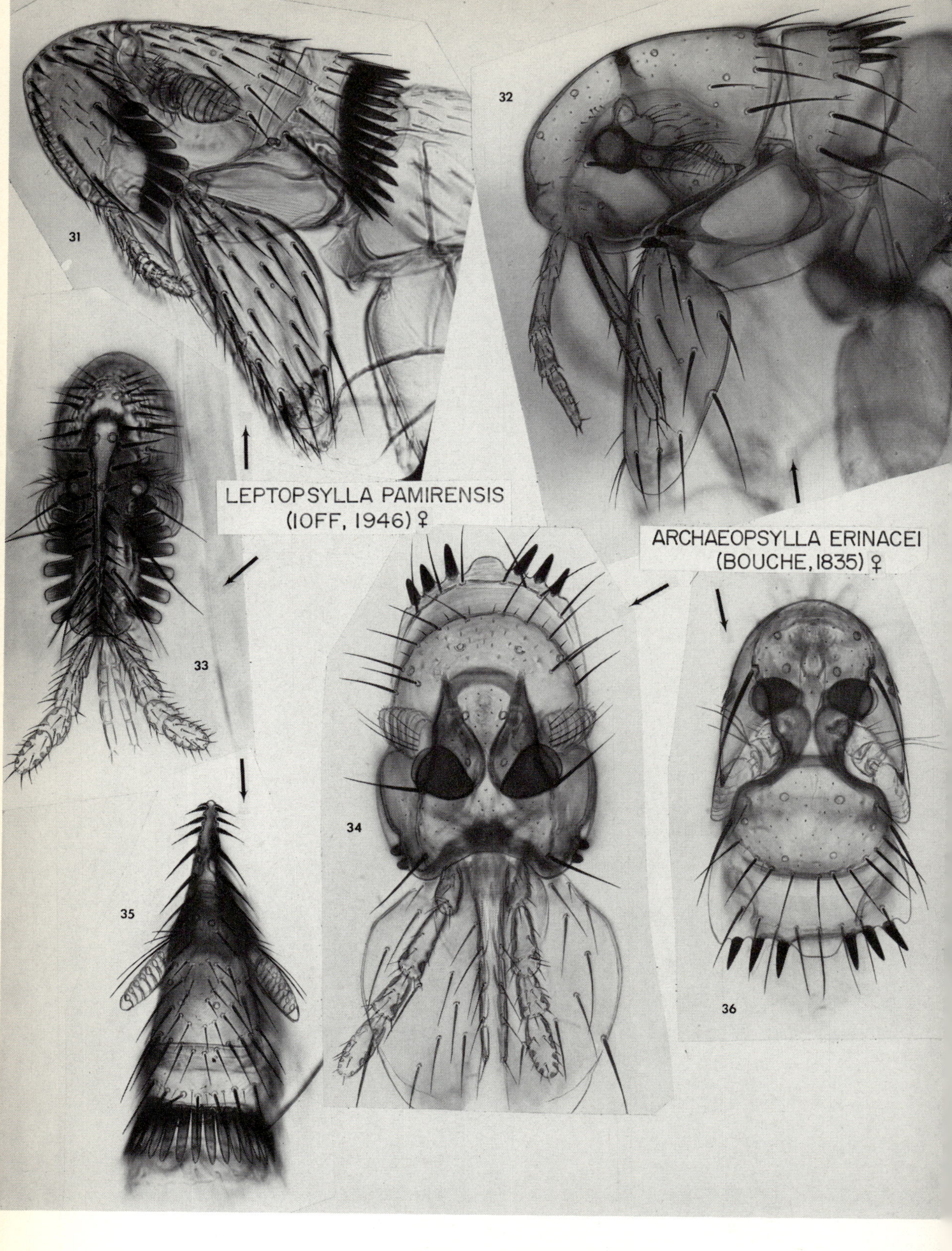
31
32
LEPTOPSYLLA PAMIRENSIS
(IOFF, 1946) ♀
ARCHAEOPSYLLA ERINACEI
(BOUCHE, 1835) ♀
33
34
35
36

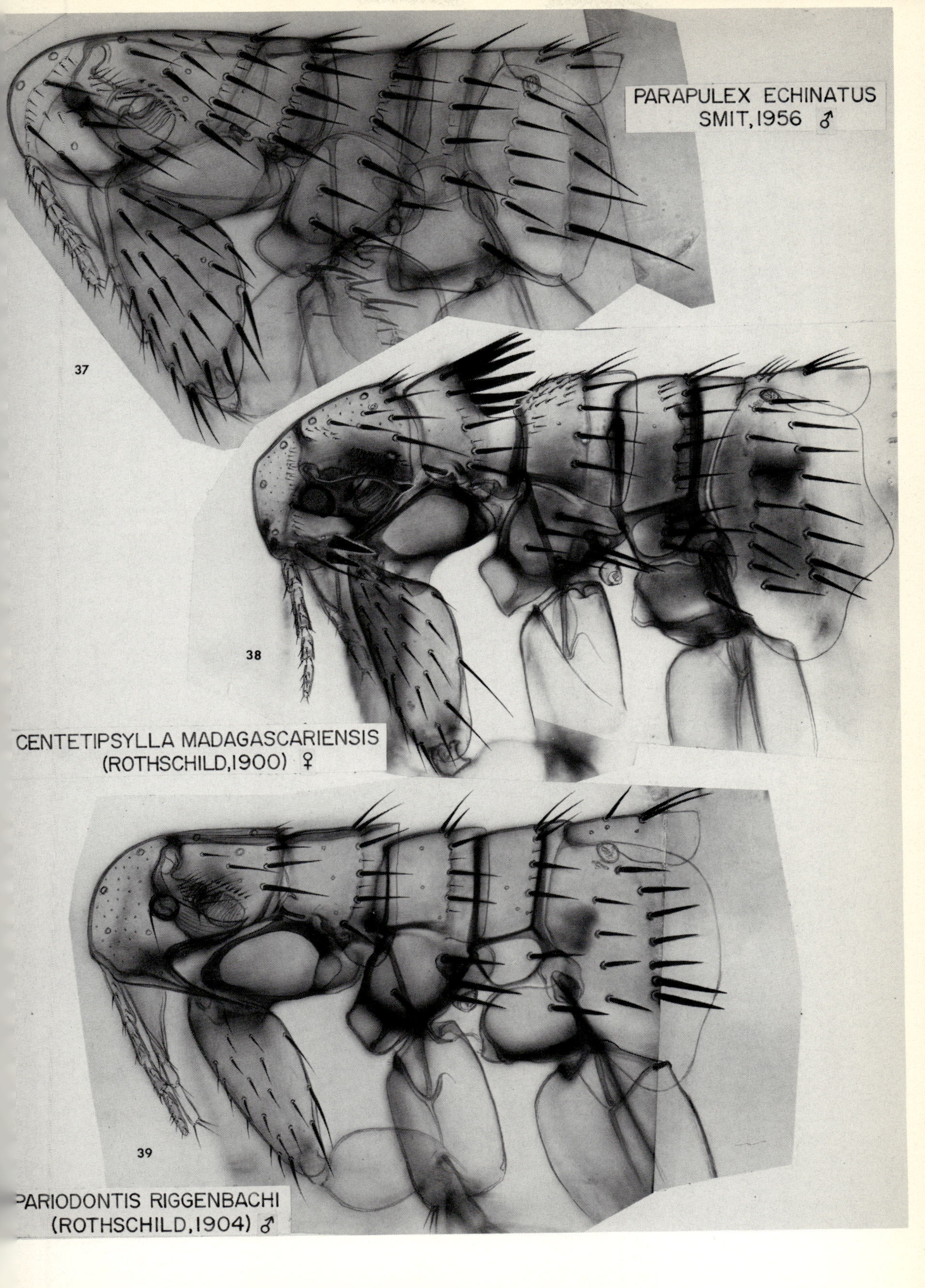
PARAPULEX ECHINATUS
SMIT, 1956 ♂
37
38
CENTETIPSYLLA MADAGASCARIENSIS
(ROTHSCHILD, 1900) ♀
39
PARIODONTIS RIGGENBACHI
(ROTHSCHILD, 1904) ♂

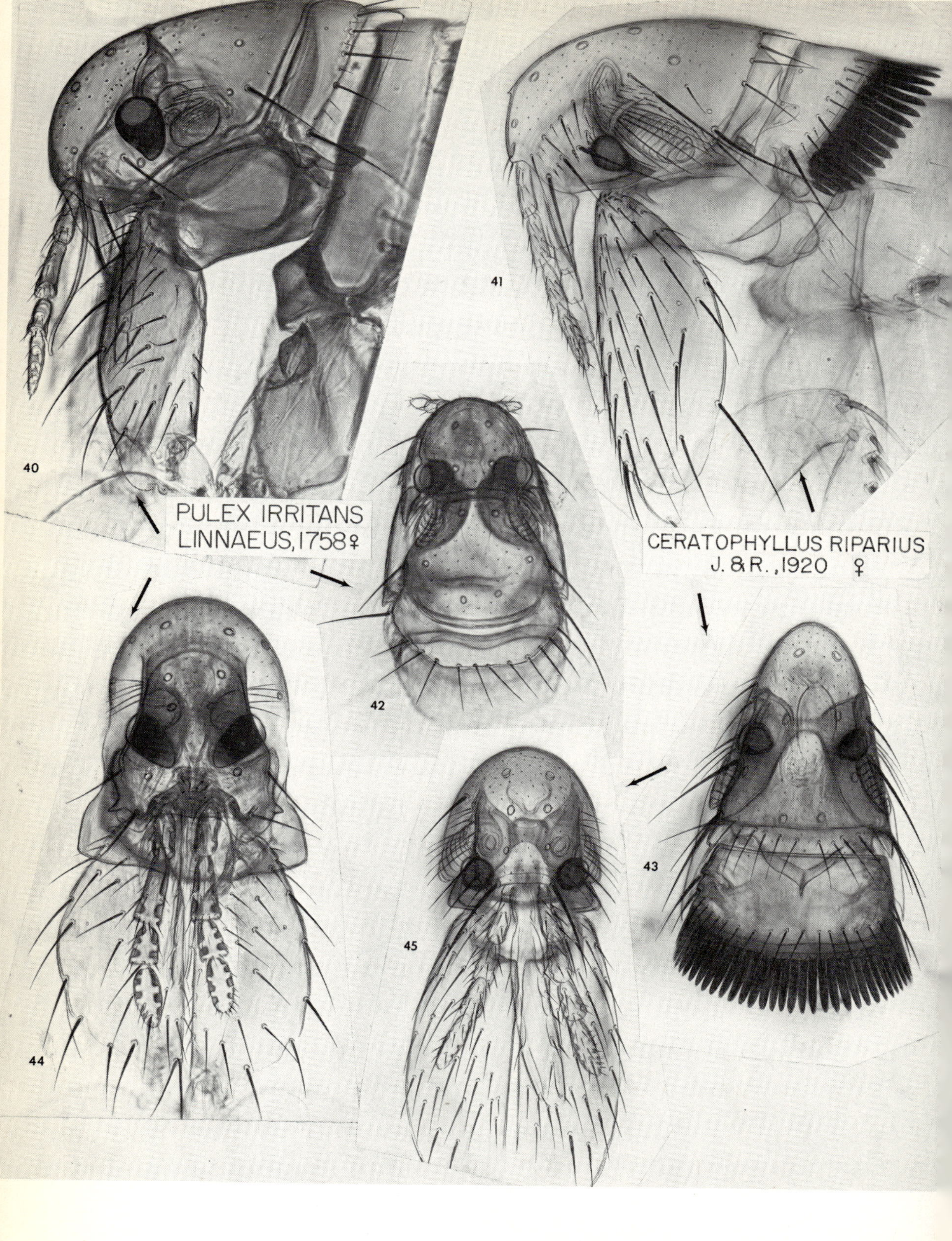

40
41
42
43
44
45
PULEX IRRITANS
LINNAEUS, 1758♀
CERATOPHYLLUS RIPARIUS
J. & R., 1920 ♀

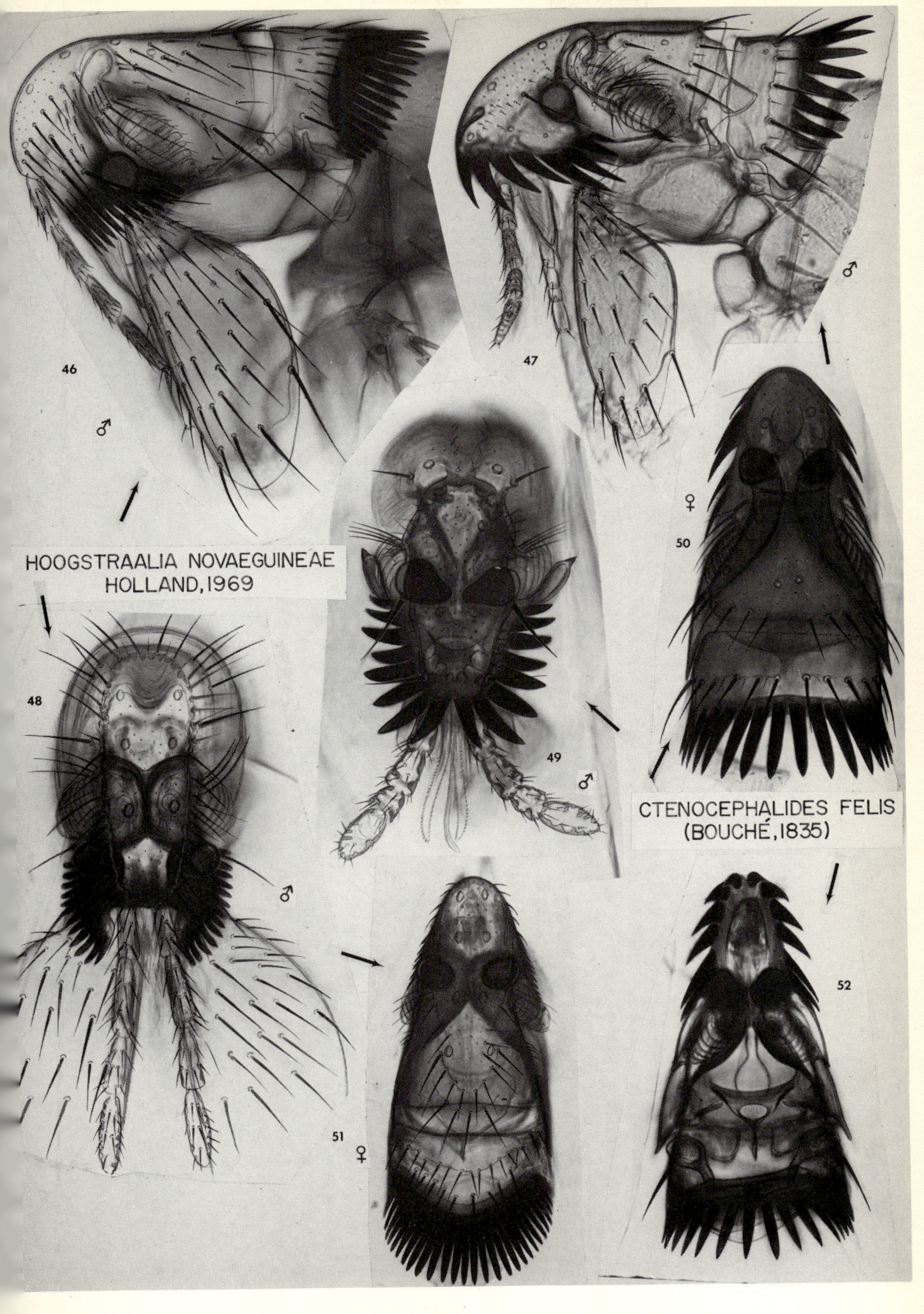
46
♂
47
♂
HOOGSTRAALIA NOVAEGUINEAE
HOLLAND, 1969
48
♂
49
♂
50
♀
CTENOCEPHALIDES FELIS
(BOUCHÉ, 1835)
51
♀
52

gene'. Clearly, more must be involved in the evolution of a crown of thorns than the linkage of genes which result in the appearance of marked family resemblances in humans that are known to occur for generations, because the latter do not seem to be associated with function. This situation in fleas differs even more from the 'correlated responses' mentioned by Mayr (1963) which occur because a common gene is modified and which may entail haphazard or coincidental changes, involving widely separated parts of the body. Instead, we are concerned with a series of modifications that somehow are directed towards a combined operation. Moreover, there have been really two different evolutionary processes at work in the production of crowns of thorns. The first three points above must have been progressive developments but the other associated features probably were concurrent with item 3), ensuring proper performance. The situation is more intricate than, and not quite analogous with, the 'coaptions d'accrochage' in which, as Simpson (1953) states: 'two anatomical parts arise separately in the embryo and subsequently fit together and operate in a common function', such as the modification of femur and tibia in mantids to form a device for grasping prey.

Examples of co-ordinated evolution abound in nature, e.g. the poison-apparatus in snakes, which entailed modification of saliva into a potent neurotoxin or hemolytic toxin; formation of a structure to contain the poison; changing a tooth into a hollow fang (complete with orifices) or grooved structure; developing a means of forcibly injecting the toxin into the victim; and behavioural changes, such as following the prey until the poison takes effect, etc. Another instance is provided by the development, in mammals, of milk glands from modified sweat glands; with concomitant storage facilities and disposal mechanism; and the appearance in the progeny of suitable enzymes for the digestion of milk.

3.6 *The adaptive nature of the genal comb*

The genal comb, like the pronotal comb, has a variety of functions. In addition to being protective and assisting passage through the hairs, in at least some instances, it is presumably adapted to enabling the flea to secure a grip on the hairs of the host while feeding or resting. It has been pointed out above how some fleas apparently employ frontomarginal genal combs as temporary hold-fast devices. As we have seen, I also believe that with certain types of pelage, upright combs on the posterior part of the gena can also be brought into similar play, especially since such ctenidia are nearly always accompanied by a short, often flattened head, stout mouthparts (not exceeding ¾ the length of the forecoxae and usually much shorter) and stout forecoxae. These adaptations, working together, apparently serve as a mechanism for temporarily adhering closely to the host while feeding. Those who believe that ctenidia in fleas are solely protective, stress that the combs are invariably associated with mobile joints, apparently placing the bases of the mouthparts in that category. It is difficult to see how a long, vertical genal comb, regardless of position, protects mobile joints, or even the region of the oral cavity (cf. figs. 18, 19, 41, 47). Smit (1972) believes the upright genal comb guards delicate sensory parts of the head, but there are no apparent structural differences in the antennal margins of such combed and combless fleas. Further, if a horizontal genal comb has been developed to shield the bases of the mouthparts, then how do the feeding organs of combless fleas differ, and how do the latter species manage?

As examples of fleas with such posteriorly placed, upright genal combs and corollary modifications are *Cediopsylla* (IX.3), *Spilopsyllus* (IX.3), *Chimaeropsylla* (XII.1), *Chilio-*

psylla Rothschild, 1915 (I.3), *Palaeopsylla* (I.3) and *Nearctopsylla* (I.8), representing a span of three families. As already noted, some fleas substitute tiaras or masses of spiniforms (e.g. the pygiopsyllids *Idiochaetis* and *Striopsylla*) for the vertical genal comb to meet these objectives. That these particular genal ctenidia (and frontal spiniforms) actually function in this way is further suggested by the facts that: 1) the only short-headed fleas with long mouthparts are the true sticktight fleas, and these are in a completely different category in that they penetrate the host until even their thorax is imbedded, and remain attached for long periods – for weeks in the female. 2) Fleas with long palpi lack a genal comb whether or not they are sticktight species. 3) Those sticktight fleas which lack a genal comb, use other mechanisms to hold onto the host, such as elongate styli well armed with serrations as in *Echidnophaga* and *Rhynchopsyllus* Haller, 1880 (X.2), etc. or else utilize specialized tibiae and tarsi for this purpose, as in the malacopsyllids. 4) *Cediopsylla, Spilopsyllus* and *Chimaeropsylla* are semi-sedentary species, but lack the extreme specializations mentioned in 3) and hence any support provided by the ctenidia would be useful.

Moreover, ordinary horizonal or oblique genal ctenidia could function to latch onto hairs of the host if fleas with such combs assumed a subvertical position, when feeding, analogous to that seen in *Ctenocephalides felis* as illustrated by Balashov & Daiter (1973: 173). In that posture, the spines would point caudad like the pronotal spines do when the flea is oriented in the usual way. In some cases the genal comb serves to shield the large eye from contact with hairs, as mentioned below.

Regardless of such functions, the foregoing remarks about the correlation between fleas at special risk and the possession of ctenidia are applicable to this structure to a surprising degree, and the evolutionary selective pressure of this factor seems to outweigh phylogenetic considerations. A good example is provided by pulicid archaeopsylline *Ctenocephalides* Stiles & Collins, 1930, in which most members bear a full genal (and pronotal) comb, and which fundamentally are all parasites of carnivorous mammals with vast ranges and, generally, nocturnal habits. Some of these even infest scansorial hosts. Notably, however, the two species of *Ctenocephalides* on hyrax (Hyracoidea, Procaviidae) possess a genal comb that is greatly reduced, consisting of only 1 to 3 spines, of which 1 or 2 may even be vestigial or aborted. Hyraxes are diurnal terrestrial animals living in groups, often in greatly circumscribed areas, where a newly emerged or 'lost' flea would have little difficulty in encountering a host.

Representatives of three families and four subfamilies of fleas infest shrews or moles. All but one of these fleas possess a genal comb, and there are four or more such spines in all these genera except *Ctenophthalmus,* e.g. *Corypsylla* (fig. 19). The hosts of these fleas are crepuscular or nocturnal and added to this hazard is the fact that all of these taxa are extremely host-specific, suggesting that they cannot survive on other mammals, a factor that must also exert selective evolutionary pressure. The mole-fleas are probably restricted to environmental conditions in the talpid-burrows, and similar constraints seem to limit the shrew-fleas to the micro-habitats of the host. Another potential limiting factor is that of the requirements of the flea larvae for successful development in the nest. Rabbit-fleas are particularly instructive regarding the probable adaptive value of the comb, since they face a multiple hazard regarding individual and species survival, e.g. their hosts are saltatorial, lack a permanent nest-site or resting-area and cover a large territory. In *Spilopsyllus* Baker, 1905 (IX.3) and at least some *Cediopsylla* (IX.3), maturation of the sex glands of the flea is dependent upon their feeding on pregnant rabbits. This hormonal association synchronizes the breeding of both fleas and host so that the larvae develop in favourable sites and the newly emerged fleas have an appropriate source of food (M.Rothschild, 1965a, 1965b). These spilopsylline rabbit-fleas have a well-developed genal comb (and pronotal one as well), and are semi-sedentary in nature, a feature that is primarily dependent upon the flat head and stout, well-serrated mouthparts of these fleas but probably is assisted by the ctenidia as well. The feeding habits and appearance of these fleas (shape of head and ctenidia) resemble those of *Chimaeropsylla* (XII.1), an African taxon in another family, but which, like the rabbit-fleas, infests saltatorial hosts.

Hoplopsyllus pectinatus Barrera, 1967 (IX.3), a flea of the relict volcano-rabbit *Romerolagus* in the mountains of Mexico, also has a genal ctenidium, unlike *H.anomalus* (Baker, 1904), a flea of diurnal ground-squirrels. Since *H.pectinatus* and a *Cediopsylla* restricted to this rabbit have been found only when the *Romerolagus* are pregnant and have young (Barrera, 1967), these too may be bound to the hormonal cycle of its host. The closely related genus *Euhoplopsyllus* Ewing, 1940, infests hares and lacks a genal ctenidium, but, significantly, the absence of the comb seems compensated for by the facts that the pronotal comb is much better developed than in *Hoplopsyllus* and that the larvae may be parasitic on the same host (as at least is the case in *E.glacialis*). It also is noteworthy that the archaeopsylline *Nesolagobius* which infests a highly bio-endemic rabbit *(Nesolagus)* in Sumatra not only possesses a well-developed genal ctenidium but morphologically resembles *H.pectinatus,* which belongs to another subfamily. Since the leptopsyllid rabbit-flea *Odontopsyllus* Baker, 1905 lacks a genal comb, it might appear that this invalidates this hypothesis, but not only are well-developed genal ctenidia unknown in that subfamily, but there are many supernumerary spines in the pronotal comb of *Odontopsyllus* and the species are heavily bristled – modifications which provide an enhanced facility of attachment.

As already mentioned, none of the fur-infesting fleas of diurnal, ground-dwelling hosts have well-developed genal combs (i.e. consisting of four or more spines). Instead, such ctenidia are generally found in fleas whose hosts' mode of life is more hazardous to the flea. Examples not already treated in this section include fleas of hosts that are: 1) nocturnal and/or arboreal (or scansorial) such as the fleas *Epirimia* (XII.3) (fig. 24) (also in xeric areas); *Adoratopsylla* Ewing, 1925 (I.5); many *Dinopsyllus* (I.4); *Chiliopsylla* (I.3); the stephanocircid helmet-fleas of Australia (VII.1) and South America (VII.2); and the leptopsyllids *Sigmactenus* (fig. 18), *Leptopsylla* and *L.(Pectinoctenus)* (fig. 31) (IV.1). 2) Nocturnal and occuring in an adverse habitat – *Ctenoparia* (I.1), in Chile; winter-species of *Hypsophthalmus* Jordan & Rothschild, 1913 (XII.1) on *Otomys* in the mountains of South Africa; some *Dinopsyllus* in xeric zones; *Leptopsylla putoraki* on a desert-shrew. 3) Birds – *Hoogstraalia* (II.1) (fig. 46) and *Aphropsylla* Jordan, 1932 (IX.2). 4) Carnivores with vast home-ranges – *Idilla* Smit, 1957 (I.5) and *Acedestia* Jordan, 1937 (I.5), infesting marsupials in Australia. 5) Live in ground saturated with water, where fleas and their larvae would probably find it difficult to survive – *Paratyphloceras* Ewing, 1940 (I.8) and *Hystrichopsylla schefferi* Chapin, 1919 (I.1) on *Aplodontia,* the mountain beaver, a host that is also active on and under snow. Epirimiine fleas on saltatorial hosts in xeric areas have been mentioned above.

Dinopsyllus is particularly instructive regarding the genal comb. All members of this large genus have five large, conspicuous genal spines with the exception of the monotypic subgenus *Cryptoctenopsyllus* Wagner, 1939 (I.4), in which the comb is at most represented by a single tiny spine. Significantly, this flea is a specific parasite of *Bathyergus,* the South

African sand-mole or mole-rat (Bathyergidae), which is a burrowing wholly subterranean species. Reduction of the comb is characteristic of fleas of such hosts (as is the increase in the numbers and length of bristles) (Traub, 1972b). In such a restricted habitat there ordinarily cannot be a problem of finding a new host, if need be, nor are there likely to be other pressures to remain attached to the host.

Limits of time and space preclude discussion of fleas with short genal combs, but many pertinent examples exist, and some (*Meringis* and *Mesopsylla,* etc.) have already been mentioned. Others include xerophilic *Caenopsylla* (IV.2) on macroscelids, or on desert-carnivores, or on *Ctenodactylus,* which climbs subvertical rock walls; *Chiastopsylla* (XII.2) in arid parts of the Ethiopian Region; *Carteretta* C.Fox, 1927 (I.3) on *Perognathus* in US deserts; and the neopsyllines *Phalacropsylla* in those habitats or atop mountains in Mexico; and the *Neopsylla* (figs. 7, 9, 11) with tibial combs, on nocturnal scansorial rats, etc.

There is a group of fleas which are more commonly found in the nests of rodents and insectivores than on the hosts themselves when out of their homes, and yet which lack the dramatic reduction in chaetotaxy and other special features associated with true nest-fleas as defined by Traub (1972b). In this former division are *Rhadinopsylla* Jordan & Rothschild, 1912 and allies, and probably *Stenischia* Jordan, 1932, etc., bearing genal combs with about five spines. These seem like admirable candidates for the category of fleas whose larvae are wholly dependent upon the environmental conditions provided by the nest, and have been regarded as partial nest-fleas (Traub, 1972b).

A genal ctenidium, then, particularly one that is well-developed, seems to be more adaptive and functional than has been generally realized, especially insofar as concerns a potential means of furthering a hold on the host. However, in some instances, this comb seems to serve in yet another capacity, namely protecting a large eye which protrudes from the side of the head, thereby apparently accounting for the anomalous situation wherein a small minority of fleas are conspicuous in having both a large eye and a well-developed genal ctenidium, whereas the vast majority of such combed fleas are eyeless or have greatly reduced eyes.

In order to discuss this topic, it is first necessary to consider the condition in combless fleas with large eyes, as in *Pulex* Linnaeus, 1758 (IX.1) (figs. 40, 42, 44) and *Ceratophyllus* Curtis, 1832 (III.1) (figs. 41, 43, 45). Such fleas use a variety of mechanisms to protect the eye against contact with the hairs of the host, e.g. the well-named 'eye bristle' which is a large bristle immediately in front of the eye. In *Chaetopsylla* Kohaut, 1903 (XI) there is a row of unusually stout and long bristles in that position. As can be seen from the illustrations of dissected specimens, far more than bristles are involved in protecting the eye. In some cases, the eye is recessed in position (figs. 42-45) so that the eyes scarcely bulge beyond the walls of the head in that area, while other parts of the head extend even more laterad. It is thus impossible for hairs to brush the eye. In others the ventral margin of the head protrudes laterad of the eye, either directly in front of it or behind it, serving to achieve the same purpose, since, even by itself, the caudal extension is sufficiently great to part the hairs as the flea moves. In *Hoogstraalia* (II.1) (figs. 46, 48, 51), which is a bird-flea and is the only representative of the family to bear a genal comb, the eye is not so recessed in the head, and it is the ctenidium which provides most of the shield for the eye. *Aphropsylla* (IX.2), another bird-flea, is likewise protected by the comb. *Ctenocephalides* (IX.2) is the only common flea with both a large eye and a well-developed genal comb, and as can be seen from figs. 47, 49, 50 and 52, the ctenidial spines (and gena) extend sufficiently laterad to parry the hairs. In a spiny host like hedgehogs, protecting the eye must pose special problems, and it is of interest that in such a flea, *Archaeopsylla* (IX.2) (figs. 32, 34, 36), even though the comb is greatly reduced, it, and parts of the head, extend well laterad of the eye, and that while there are only two bristles on the preantennal region, one is the 'eye bristle'.

3.7 *Modifications in vestiture in fleas of spiny and bristly hosts*

In fleas of spiny hosts like porcupines, hedgehogs, tenrecs, echidnas, etc. the pronotal spines (if present) are highly modified, being reduced in number and consisting of well-separated very short dorsal or subdorsal spines which are broad at the base and apically pointed (Holland, 1964; Traub, 1972b). Also, as pointed out by the latter author, the broad gaps between the flea's spines correspond with the spines of the host, and the shortened comb is apparently similarly adaptive, because a long (deep) ctenidium could not bear spines divergent to that degree, since they would either overlap or else be non-functional for other physical reasons. Moreover, in those fleas of spiny hosts, the bristles of the thorax and part of the abdomen obviously also show the influence of the host's vestiture since they generally are much stouter, shorter, and much more widely spaced than are their relatives infesting ordinary hosts, while at times some of the bristles are in vertical groups of two or three, forming a sort of comb (Traub, 1977). Examples of such 'spiny' fleas of spiny hosts are the species of *Pariodontis* Jordan & Rothschild, 1908 (IX.4) (fig. 39) (ex porcupines), *Centetipsylla* Jordan, 1925 (IX.2) (fig. 38) (ex tenrecs) and *Archaeopsylla* (IX.2) (figs. 32, 34, 36) (ex hedgehogs), all pulicids, and the pygiopsyllid *Bradiopsylla* Jordan & Rothschild, 1922 (II.1) (ex echidnas). In the fleas whose hosts bear real spines, only the caudad-directed (horizontal) bristles are affected, not the subvertical ones on the forecoxae. Thus, the modifications are restricted to those bristles that could function effectively against the host's spines (figs. 38, 39). (As can be seen, the features of the small genal combs of *Centetipsylla* and *Archaeopsylla* are in accord with these concepts of fitting the host's spines. Figures 34 and 36 show that the pronotal spines of *Archaeopsylla* could also span the hedgehog's spines.) The pectinate group of three modified bristles on the lateral metanotal area of *Centetipsylla* and on the mesepimeron of *Pariodontis* are highly unusual, if not unique in the Order. Even the combless *Parapulex* Wagner, 1910 (IX.1) (two species), parasitizing the 'spiny mice' *Acomys,* have bristles strikingly modified along these lines, as shown in fig. 37. *Synosternus* Jordan, 1925 (IX.4) and its allies are devoid of combs, and it is instructive to note that in *S.pallidus* (Taschenberg, 1880), which infests hedgehogs, the thoracic and abdominal bristles are twice as far apart (and half the number) as in *S.cleopatrae* (Rothschild, 1903), which is a gerbil-flea in the areas. The unusually broad gap between the bristles of the former clearly seems to be an accommodation to the spines of its host.

Since the fleas with 'spinose' and well-spaced bristles are limited to hosts with spiny hairs or true spines, it is difficult to conceive how such unusual chaetotaxy could have evolved if not in response to the host's vestiture, or how they could fail to operate as a mechanism against dislodgement. The conspicuous hook-like genal lobe of *Pariodontis* also seems like an adaptation for holding onto the host's spines.

The paucity of Siphonapteran fauna on truly spiny mammals may also be related to the nature of the host's vestiture. Certain animals, like the Nearactic porcupines, apparently lack fleas, while others have a geographical range that apparently far exceeds that of their fleas, e.g. hedgehogs in North Africa, *Acomys* in Pakistan and Iran, and, presumably, echidnas in New Guinea, all of which have characteristic fleas elsewhere. Moreover, none of the spiny hosts seem to bear more than one species of Siphonapteran parasite. These points suggest that adaptation to these types of hosts is a difficult step for fleas and that development of the vestiture characteristic of spiny fleas is part of the process, indicating in turn that the pattern of spines and bristles is indeed functional.

Parallel chaetotaxic modifications are shown by fleas infesting mammals with bristles or coarse hairs, but the structural changes are to a lesser degree, as one would expect. The fleas of peramelid bandicoots are good examples of this principle. In bandicoot-stephanocircids the ventralmost bristles on the occiput of the head are far stouter than are other bristles on those species, while other helmet-fleas lack such modifications (Traub & Dunnet, 1973). The distinctive nature of the helmet, genal and pronotal spines has been mentioned above (p. 48) and elsewhere (op. cit.; Traub, 1972b), as has the trend for New Guinean bandicoot-fleas like *Parastivalius* (II.1), etc. to have sharply pointed pronotal spines (Traub, 1968, 1972b; Holland, 1969). These spines also tend to be divergent, and not subparallel like they are in the rat-pygiopsyllids. The sharply pointed genal and pronotal spines of the hystrichopsyllid *Idilla chera* Jordan, 1937, another parasite of peramelids, fit this picture of spines adapted to the coarse fur of bandicoots. In addition, the New Guinean peramelid-fleas, which include a minimum of 12 species belonging to at least four genera, also tend to have thorny bristles on the sides of a very broad, short, flattened head, or else they have thoracic and abdominal bristles that are spaced further apart than in the pygiopsyllids on rats and other hosts. In some species these bristles are unusually stout. The fact that such thorny frontal bristles frequently are irregular instead of being in rows lead me to suspect that they match the spacing of the coarse hairs of the host, and thus have enhanced value as a hold-fast mechanism. The unique spur at the anteroventral angle of the head of (*Gryphopsylla*) (II.1) may prove to be an adaptation for securing the bristly hairs of *Rattus whiteheadi* and *R.alticola,* its most common hosts.

3.8 *The numbers of spines in the comb of bird-fleas*

The extraordinary convergence exhibited by bird-fleas of three or four families extend to at least four features involving the combs and bristles (Traub, 1969, 1972b), and other adaptations on the legs (Smit, 1972). The feature of supernumerary narrow, parallel, straight spines in a pronotal comb that does not reach the vinculum, is diagnostic, but the number of surplus spines is actually significantly greater than that generally reported in the literature. This is shown in the photographs of the dissected specimens and is due to the fact that there are more dorsal spines in the comb than can be envisaged from the lateral aspect, which of course, is how fleas have been studied in taxonomy (compare figs. 41 with 43, and 46 with 51). Bird-fleas tend to be stout and the dorsal margin, which is broader from side to side than is ordinarily appreciated, is foreshortened as seen in the mounted specimen. Hence the number of spines cited in descriptions seem to have been low by about four to eight spines. For example, on the basis of the lateral aspect, one would estimate that the comb of this *Hoogstraalia* (fig. 46) consists of a total of about 34 spines, but from the dorsal aspect alone (fig. 51) one can see 28 spines, excluding the lower ones. For the *Ceratophyllus,* the comparable figures are 36 (fig. 41) and 32 (fig. 43) spines.

3.9 *The attachment value of the modified spines and bristles*

There are thus myriads of examples where the hyperdevelopment or modification of ctenidia, spines, spiniforms and bristles clearly seems adaptive, while the absence of such specialized structures in those fleas whose hosts do not present unusual problems for flea-survival, is equally impressive. The marked correlation between type of flea-vestiture and that of the host, or that with the environment or behaviour of the host, are surely too great for spines and bristles of the fleas to lack any function in maintaining a hold on the host, be it either while the flea is feeding, resting, mating, or in general, to reduce the chances of dislodgement. On the basis of what we have noted regarding *L.segnis* in Ethiopia, and my observation on *Idiochaetis* in New Guinea, these species may remain in a sessile position for hours. They can also suddenly move forward and disappear into the fur. Krampitz (1965) considered *L.segnis* to be 'sessile' and stated that it did not leave the host voluntarily, but did not say how long it remained attached. These points suggest that the main feature of the crown of thorns indeed is to provide a means of attachment, and that these fleas behave differently on the host than do those which lack such modified bristles and move actively about in the pelage, and which may leave and return to the host frequently. Thus, large numbers of *engorged* male and female *Orchopeas* Jordan, 1933 may be observed in squirrel nests, indicating that they had just recently left the host, and gravid ones seen gluing the eggs to the debris (Traub, unpublished). *L.segnis* in contrast, does not leave the host to lay its eggs, nor are they deposited loosely in the hairs. Instead the sticky eggs are expelled by force and apparently adhere to the surface where they fall (ibid.), which is apt to be in the nest, and where there are suitable conditions for the development of the larvae. Fleas with crowns of thorns probably copulate on the host, as I have observed for *L.segnis,* whereas fleas unmodified thusly may behave like *Orchopeas,* and some bird-fleas like certain *Ceratophyllus* and *Geohollandia solida* (Rothschild, 1916), and mate in the nest of the host, shortly after emerging from the pupal cocoon. Hyper-development of bristles and spines and consequent enhancement of the capability of remaining on the host, which as we have seen is a relatively uncommon phenomenon, thus can reduce the contact with the hostile external environment, but the selective pressure of such hazards can of course lead to other types of evolutionary developments in fleas, as in the case of those having to cope with the high temperatures of the desert. There, as Bibikova, Il'inskaya et al. (1963) have pointed out, contact with the hot sands can kill *Rhombomys*-fleas within seconds. Such fleas move from deep within the burrow system to the opening and there await a passing or entering mammal, thereby reducing to a minimum the chance of missing the host when leaping. Adaptations for an existence ensconced in the nest is another such mechanism (Traub, 1972b). It has been indicated herein that the degree of modification of spines and bristles is correlated with the extent of the environmental risk, as illustrated by the species of *Leptopsylla* (p. 49). The allied genus *Peromyscopsylla* apparently is another example. Here there are but two spines in the genal comb, which is vertical, and there are merely a few frontomarginal spiniforms (with concomitant narrowing of the head from side to side). These fleas are seasonal, e.g. *P.scotti* I.Fox, 1939 is essentially restricted to the autumn (Jackson & DeFoliart, 1976), a feature which must entail some survival pressure, but nothing like that experienced by fleas infesting wholly nocturnal, arboreal hosts, as witness the complete (but false!) helmet of *Smitella* (fig. 1).

Inasmuch as the frontomarginal spiniforms in certain leptopsyllid and pygiopsyllid fleas are actually bristles which have lost their attenuated tips during the early life of the adult (figs. 6, 12), as pointed out above (p. 45), it seems axiomatic that such modifications are functional. That this is so is suggested by the observations of Kosminsky (1959), who noted sequential changes in the acuity of the spiniform tips as *Leptopsylla* fleas aged, and who used the appearance of the spiniforms and of the genal spines as indicators of the comparative age of the specimens. For example, he believed that fleas with blunt or rounded spiniforms, or with genal spines that were broken at the tip, were more than a week old. It seems logical to assume that such changes in these fleas of non-burrowing hosts are the result of 'wear and tear' of devices used in attachment, and are not due to trauma caused by sand and debris as I believe occurs in the *Ctenophthalmus* infesting fossorial hosts like 'mole-rats'. Ioff (1929) has remarked on the great frequency of injured spines in the genal comb of *Spalax-Ctenophthalmus,* and I have often noticed similar deformities on members of that genus from Ethiopian *Tachyoryctes.*

When affirming that the ctenidia and tiaras of spiniforms serve to help maintain a grip on the host and to discourage the flea from being moved in any direction but forward, I am not implying that their function is to entangle hairs – least of all to the extent that fleas in collections would exhibit entrapped hairs in the ctenidia. If the combs did latch onto hairs to the point where one or more were actually snared, then such a flea would be handicapped in moving on its host, not facilitated. Consequently, such live fleas would be easy prey, and hence rarities, and hyperdevelopment of combs and bristles would not have much survival value. Moreover, if the host captured a flea which did snag hairs while being pulled backwards, then the flea would not survive to end up in a specimen-tube. Nevertheless, Smit (1972) mentions the failure to observe entrapped hairs on preserved fleas as an argument against ctenidia functioning as an attachment device. It is, of course, true that fleas in alcohol-tubes often have hairs snagged onto their tarsal claws, but it does not follow that 1) those objects had necessarily been attached when the flea was actually placed in the tube, since hairs are frequently brushed off the host when a person is collecting fleas, and such loose hairs often end up in the vial in a tangle of tarsi as the fleas struggle in the alcohol, and 2) the claws (six pairs) are designed by nature to actively grasp hairs regularly, whereas the combs at best are rather passive agents which oppose movement in the direction of the hairs.

There are some points about true nest-fleas that seem to have been generally overlooked but which have bearing on the attachment-value of combs and bristles. A soft-bodied flea that has lost the power to leap, is blind, lacks combs, and has markedly reduced chaetotaxy, may readily find perfect environmental conditions within the nest of its host, crawling out to feed when the host is asleep. However, there obviously are problems in the perpetuation of such a species if the flea were obliged to find a new nest when the host abandons the old one, for it is not adapted to remaining on the host and accompanying it, or to facing the rigors of the outside world. It therefore seems highly significant that such greatly modified nest-fleas (e.g. *Anomiopsyllus* Baker, 1904, *Megarthroglossus* Jordan & Rothschild, 1915, *Conorhinopsylla* Stewart, 1930, *Callistopsyllus* Jordan & Rothschild, 1915, etc.) seem to be restricted to hosts that use the same nest continuously, and/or add to it, over a period of years. A prime example, and indeed, a host for all these genera, is the Nearctic, cricetid pack-rat, *Neotoma,* whose nests are occupied for years, decades, or even centuries. Such mammals are most common in arid areas or live in caves or rock-ledges, but other types of habitats are associated with long-term nests, such as hollow trees and tree-holes in mesic climes, and these often harbour such true nest-fleas. Birds nesting in banks provide similar conditions, and when such swallows return in the spring they find nest-fleas awaiting them (Traub, 1972b). In contrast, the nest-fleas whose hosts build new burrows or nests annually or within a relatively short period as compared to the above type, must find new quarters and food when their hosts depart, and accomplish this either by crawling to it or adhering to the host long enough to make the journey. Fleas like *Stenischia, Rhadinopsylla, Corypsylla* and others that have only some of the above-mentioned attributes and have been regarded as 'partial nest-fleas' (Traub, 1972b) seem to have hosts of this type. The fact that these fleas bear genal and ctenidial combs, coupled with the pattern of evolutionary development just outlined, suggest that the ctenidia are functional and enhance chances for survival of the individual – and the species.

3.10 *The need for more data*

There still is a great deal to be learned about the functions of the combs and bristles of fleas and the relationships of such structures to the vestiture and habits of the host. Perhaps the greatest need is for direct observations in the field and in the laboratory on the behaviour of fleas on the host, such as when feeding, resting and moving, and such research is difficult and time-consuming. Equally urgently required are more data on the host-relationships of fleas, especially insofar as concerns certain taxa and geographic areas, so that we can obtain more information on the correlations between the chaetotaxy of the fleas and the traits of the host. For example, nothing seems to be known about the fleas of the spiny dormouse (Platycanthomyidae), a scansorial rodent in southern India, which is so abundant that at times it is an agricultural pest. Such fleas would face two somewhat antagonistic evolutionary pressures: 1) reduction of combs because of the host's spines and 2) development of supernumerary bristles and spines associated with the nocturnal arboreal habits of the host. Does such a flea (if it exists) have an abbreviated subdorsal comb (or an unusually long one) of widely separated, short spines, coupled with well-separated stout bristles and a tibial comb and/or spiniforms on the head? What are the fleas of the group of *Perognathus* which have bristly fur, and how are they adapted? Existing records pertain only to the fine-furred species. Does the absence of reports of true macropodid fleas represent a phenomenon, or inadequate collecting? It is hoped that some of the audience and readers will undertake investigations or collect data along these lines.

3.11 *Adaptive modifications in other ectoparasites*

All ectoparasites are of course highly modified for their particular mode of life (N.C. Rothschild, 1917), and one example will suffice to support the general theme of this article. The proposed survival value of the modified spines, spiniforms and bristles in fleas recalls the case of some of the morphological variations seen in Mallophaga, so well treated by Rothschild & Clay (1952). In the bird-lice there are many examples of both convergent evolution (i.e. in unrelated taxa) and parallel evolution (those with a common ancestry). For example, 'Mallophaga living on the head and neck, where they are out of reach of the bird's beak, have less of a need for rapid movement, and have become adapted to a comparatively sedentary life on the feathers'. Such lice are relatively short and broad, with claws adapted to clinging to feathers; etc. 'Such fat-bodied forms on other parts of the body would easily be picked off by the bird or crushed by its bill . . . On the wings and back, where the louse is always in danger from the bill, a flattened, elongated type is found, which is able to move rapidly'. Certain *Philopterus,* sedentary species infesting motmots and nine different families of Passeriform birds, show unusual, yet similar morphological modifications of the head, and Rothschild & Clay believe that these are in response to a similarity in structure of the feathers of the head of their particular hosts.

4 CONCLUSIONS

The spines and bristles of fleas must serve to help maintain a hold on the host, and not merely act for protection, or for facilitating passage through the fur or feathers of the host, or for parrying hairs, etc. This is indicated by the facts that: 1) a large number of unrelated fleas in various parts of the world, infesting taxonomically varied hosts, have developed convergent chaetotaxic modifications for the same apparent function. 2) Allied taxa, even congeners, living on other kinds of hosts in the same area nearly always lack such structures. 3) A variety of methods have been used by different fleas in evolving structures which produce the same effect. 4) Several nominally independent evolutionary changes have been somehow co-ordinated in developing structures like tiaras of spiniforms or crests of genal combs, etc., in any one taxon. 5) These have been integrated with concurrent, lateral narrowing of the head, and modifications of the mouthparts and legs for the common function of permitting temporary, selective attachment to the host. 6) Field observations verify that in at least some taxa these adaptive cephalic spiniforms are used as hold-fast devices while the flea is feeding or resting. 7) Such modifications like tiaras or masses of spiniforms and frontal vertical ctenidia on the head, are restricted to fleas whose hosts present special problems for survival of fleas (or their larvae), e.g. they are nocturnal and either arboreal and scansorial, have vast home-ranges, inhabit hostile terrain, are saltatorial, etc. 8) Other chaetotaxic manifestations of supernumerary bristles, spines, true combs, and false combs are likewise associated with hosts which place fleas at unusual risk. 9) In contrast, fur-fleas of ground-dwelling diurnal hosts invariably lack crowns of thorns, large genal ctenidia, abdominal combs or other supernumerary spines or bristles. 10) Bristles and spines are often tailored to fit the vestiture of the host. Thus, spiny mammals have fleas that are spiny themselves or have extraordinarily special combs (or lack fleas altogether), and bristly hosts also tend to have fleas with unusual and characteristic chaetotaxy. 11) The well-developed horizontal genal comb is also adaptive and presumably serves in part to enhance attachment for the flea. Hosts of such fleas are in the category of presenting hazards to their parasites.

The specialized chaetotaxy thus has selective value, increasing the chances of survival, as do other morphological features of the flea which reflect the traits of the host and its environment and those of the flea as well, e.g. reduction of the eye in fleas of fossorial hosts, loss of leaping ability in nest-fleas, etc. (Traub, 1972b; M.Rothschild, 1965a). The overall association is so profound that it is now possible to merely glance at a new genus or species of flea and make correct statements about some characteristic attributes of its host.

In the small minority of fleas which possess both a large eye and a genal comb, the ctenidium protects the eye from contact with the hairs of the host. In eyed fleas lacking such a comb, the eye is recessed or otherwise protected by lateral expansions of the sides of the head.

There is an urgent need for more data on the functions of combs and bristles in fleas, and on the relationships of such structures with the vestiture and habits of the host.

5 ACKNOWLEDGEMENTS

The photomicrographs were prepared by J.Navarro and were retouched by the photographer and the author. Specimens of South African fleas, which proved very useful in this study, were kindly collected by Dr Duane Schlitter of the Carnegie Museum, Pittsburgh. The dissections were made by R.Traub and mounted by Phuangthong Malikul. Dr A.Farhang-Azad very kindly reviewed the manuscript and corroborated my views about the host-relationships of Iranian fleas. Editorial assistance was provided by Helle Starcke. To these colleagues, I extend my thanks.

REFERENCES

Balashov, Yu.S. & A.B.Daiter 1973, Blood-sucking arthropods and rickettsiae. Acad. Sci. USSR, Zool. Inst. 170-191, 213-250.

Barrera, A. 1967, Redefinicion de *Cediopsylla* Jordan y *Hoplopsyllus* Baker. Nuevas especies, comentarios sobre el concepto de relicto y un caso de evolucion convergente. Rev. Soc. mex. Hist. nat. 27:67-88.

Bibikova, V.A., Il'Inskaya, V.L., Kaluzhenova, Z.P., Morozova, I.V. & Shmuter, M.F. 1963, Biology of fleas of the genus *Xenopsylla* in Sary-Ishikotrau desert. Zool. Zh. 42(7): 1045-1051.

Bibikova, V.A. & Klassovsky, L.N. 1974, Transmission of plague by fleas. Meditsina 190 pp. Bibliography (sep. vol.) 166-187.

Cavanaugh, D.C. 1971, Specific effect of temperature upon transmission of the plague bacillus by the oriental rat flea, *Xenopsylla cheopis.* Amer. J. trop. Med. Hyg. 20(2): 264-273.

Dunnet, G.M. & Mardon, D.K. 1974, A monograph of Australian fleas (Siphonaptera). Austral. J. Zool., Suppl. Ser. No. 30, 271 pp.

Farhang-Azad, A. & Neronov, V. 1973, The flea fauna of the great gerbil (*Rhombomys opimus* Licht.) in Iran. Folia Parasit. (Praha) 20:343-351.

Ford, E.B. 1975, Ecological genetics. London, Chapman & Hall. 442 pp.

Freeman, R.B. & Madsen, H. 1949, A parasitic flea larva. Nature 164(4161): 187-188.

Holland, G.P. 1950, Notes on some British Columbian fleas, with remarks on their relationships and distribution. Proc. ent. Soc. B.C. 46:1-9.

Holland, G.P. 1964, Evolution, classification, and host relationships of Siphonaptera. Ann. Rev. Ent. 9: 123-146.

Holland, G.P. 1969, Contribution towards a monograph of the fleas of New Guinea. Mem. ent. Soc. Canad. (61), 77 pp.

Hopkins, G.H.E. & Rothschild, M. 1956, An illustrated catalogue of the Rothschild collection of fleas (Siphonaptera) in the British Museum. Vol. II, 445 pp. London, Brit. Mus. nat. Hist.

Hopkins, G.H.E. & Rothschild, M. 1966, An illustrated catalogue of the Rothschild collection of fleas (Siphonaptera) in the British Museum. Vol. IV, 549 pp. London, Brit. Mus. nat. Hist.

Humphries, D.A. 1966, The function of combs in fleas. Ent. mon. Mag. 102:232-236.

Humphries, D.A. 1967, Function of combs in ectoparasites. Nature 215: 319.

Ioff, I.G. 1929, Material for the study of the ectoparasite fauna of southeast USSR. VI. Fleas of mole-rats (Spalacidae). Izv. gosud. mikrobiol. Inst. Rostove (8): 29-43. German summary pp. 56-59.

Ioff, I.G. & Scalon, O.I. 1954, Keys to the fleas of eastern Siberia, the Far East and adjacent districts. In: Handbook for the identification of fleas of East Siberia. Moscow, Medgiz, 275 pp.

Jackson, J.O. & DeFoliart, G.R. 1976, Relationships of the white-footed mouse, *Peromyscus leucopus,* and its associated fleas (Siphonaptera) in southwestern Wisconsin. J. med. Ent. 13(3): 351-356.

Jordan, K. 1937, Three new bird-fleas from

Kashmir. Novit. zool. 40: 299-306.
Jordan, K. 1947, On some phylogenetic problems within the order of Siphonaptera (= Suctoria). Tijdschr. Ent. 88: 79-93.
Jordan, K. 1950, On characteristics common to all known species of Suctoria and some trends of evolution in this order of insects. 8 Int. Congr. Ent.: 1-9.
Kolenati, F. 1856, Die Parasiten der Chiropteren. Brünn, Rohrer 8: 1-51.
Kolenati, F. 1863, Beiträge zur Kenntniss der Phthirio-Myriarien. Versuch einer Monographie der Aphanipteren, Nycteribien und Strebliden. Horae Soc. ent. Ross. 2: 9-109.
Kosminsky, R.B. 1959, Determination of the age of fleas of the species *Leptopsylla segnis* Schönh. and *L.taschenbergi* Wagn. 10 Conf. parasitol. Prob. & nat. Focal Dis. 2: 76-77.
Krampitz, H.E. 1965, Beobachtungen an einer Laboratoriumszucht von *Leptopsylla segnis* Schönherr, 1811 (Insecta, Siphonaptera). Z. f. Parasitenk. 26(3): 197-214.
Mayr, E. 1963, Animal species and evolution. Cambridge, Mass., Belknap Press, Harvard University, 797 pp.
Mendez, E. 1968, *Scolopsyllus columbianus,* a new genus and species of the family Rhopalopsyllidae (Siphonaptera) from Columbia. J. med. Ent. 5(3): 405-410.
Mikulin, M.A. 1956, Data on the flea fauna of Asia. 2. Fauna and some feature of geographic distribution of fleas of the great gerbil in deserts of the Southern Prebalkhasch region. In: Transactions of Central Asian Anti-Plague Research Institute, vol. 2, Alma-Ata. pp. 95-107.
Molyneux, D.H. 1967, Feeding behaviour of the larval rat flea *Nosopsyllus fasciatus* Bosc. Nature 215: 779.
Ritsema, C. 1880. Versuch einer chronologischen Uebersicht der bisher beschriebenen oder benannten Arten der Gattung *Pulex* Lin. mit Berücksichtigung ihrer Synonymen. Z. Gesammt. Naturwiss. 5: 181-185.
Rothschild, M. 1965a, Fleas. Sci. Amer. 213: 44-53.
Rothschild, M. 1965b, The rabbit flea and hormones. Endeavour 24: 162-168.
Rothschild, M. 1976, Notes on fleas (Part II): The internal organs: can they throw any light on relationships within the Order? Proc. Brit. ent. Nat. Hist. Soc. 9: 97-110.
Rothschild, M. & Clay, T. 1952, Fleas, flukes and cuckoos, a study of bird parasites. London, 304 pp.
Rothschild, M., Schlein, Y., Parker, K., Neville, C. & Sternberg, S. 1973, The flying leap of the flea. Sci. Amer. 229(5): 92-100.
Rothschild, M. & Traub, R. 1971, A revised glossary of terms used in the taxonomy and morphology of fleas. In: Hopkins, G.H.E. & Rothschild, M., 'An Illustrated Catalogue of the Rothschild Collection of Fleas (Siphonaptera) in the British Museum (Natural History)'. London: British Museum (Natural History), Vol. V, pp. 8-85.
Rothschild, N.C. 1917, Convergent development among certain ectoparasites. (Presidential Address) Proc. ent. Soc. Lond. 1916: 141-156.
Simpson, G.G. 1953, The major features of evolution. New York, Columbia Univ. Press, 434 pp.
Smit, F.G.A.M. 1958, The African species of *Stivalius,* a genus of Siphonaptera. Bull. Br. Mus. nat. Hist. (Ent.) 7: 41-76.
Smit, F.G.A.M. 1972, On some adaptive structures in Siphonaptera. Folia Parasit. (Praha) 19: 5-17.
Smit, F.G.A.M. 1976, Two new east African mole-rat fleas (Siphonaptera, Hysterichopsyllidae). Rev. zool. Afr. 90(1): 46-52.
Strenger, A. 1973, Zur Ernährungsbiologie der Larve von *Ctenocephalides felis felis* B. Zool. Jb. Syst. 100(1): 64-80.
Taschenberg, O. 1880, Die Flöhe. Die Arten der Insektenordnung Suctoria nach ihrem Chitinskelet monographisch dargestellt. Halle, 120 pp.
Traub, R. 1968, *Smitella thambetosa,* n.gen. and n.sp., a remarkable 'helmeted' flea from New Guinea (Siphonaptera, Pygiopsyllidae) with notes on convergent evolution. J. med. Ent. 5: 375-404.
Traub, R. 1969, *Muesebeckella,* a new genus of flea from New Guinea (Siphonaptera, Pygiopsyllidae). Proc. ent. Soc. Wash. 71: 374-396.
Traub, R. 1972a, Notes on zoogeography, convergent evolution and taxonomy of fleas (Siphonaptera), based on collections from Gunong Benom and elsewhere in South-east Asia. I. New Taxa (Pygiopsyllidae, Pygiopsyllinae). Bull. Br. Mus. nat. Hist. (Zool.) 23: 201-305.
Traub, R. 1972b, Notes on zoogeography, convergent evolution and taxonomy of fleas (Siphonaptera), based on collections from Gunong Benom and elsewhere in South-east Asia. II. Convergent evolution. Bull. Br. Mus. nat. Hist. (Zool.) 23: 307-387.
Traub, R. 1972c, The Colloquium on the zoogeography and ecology of ectoparasites, their

hosts and related infections at the Second International Congress of Parasitology, 1970. 23. The relationships between the spines, combs and other skeletal features of fleas (Siphonaptera) and the vestiture, affinities and habits of their hosts. J. med. Ent. 9(6): 601.

Traub, R. 1972d, Ibid. 27. Notes on fleas and ecology of plague. J. med. Ent. 9(6): 603.

Traub, R. 1977, *Tiflovia,* a new genus of pygiopsyllid flea from New Guinea, with notes on convergent evolution and zoogeography (Siphonaptera). J. med. Ent. 13(6): 653-685.

Traub, R. & Barrera, A. 1966, New species of *Ctenophthalmus* from Mexico, with notes on the ctenidia of shrew-fleas (Siphonaptera) as examples of convergent evolution. J. med. Ent. 3(2): 127-145.

Traub, R. & Dunnet, G.M. 1973, Revision of the Siphonapteran genus *Stephanocircus* Skuse, 1893 (Stephanocircidae). Austral. J. Zool., Suppl. Ser., No. 20:41-128.

Traub, R. & Evans, T.M. 1967, Descriptions of new species of hystrichopsyllid fleas, with notes on arched pronotal combs, convergent evolution and zoogeography (Siphonaptera). Pacif. Insects 9: 603-677.

Tyndale-Biscoe, H. 1973, Life of marsupials. (Contemporary Biology Series). New York, Elsevier Publ. Co. Inc. 254 pp.

Wagner, J. 1934, Weitere Beiträge zur Auffassung des sogenannten 'Caput fractum' bei Insekten. (Uber den Kopfbau der 'helmtragenden' Flöhe.) Zool. Anz. 106: 7-15.

PETER W. PRICE
Department of Entomology, University of Illinois, Urbana, Ill., USA

THE EXTENT OF ADAPTIVE RADIATION IN FLEAS

ABSTRACT

Many parasite taxa have undergone extensive adaptive radiation, making parasites in general more abundant than other kinds of organisms. Rapid evolution seems to be typical of many parasite taxa, but fleas appear to be anomalous because gross morphology is quite uniform throughout the order, evolutionary rates appear to be slower than in their hosts, and adaptive radiation has not been very extensive. A comparative analysis of numbers of parasites in families of fleas, the dipterous families Oestridae, Hippoboscidae and Streblidae, the mallophagan family Philopteridae, and the family Ichneumonidae (Hymenoptera), indicates that, when the number of potential hosts available to parasite taxa, and the specificity of the parasites are considered, the fleas fit the trends seen in the other parasite families. Considering that fleas are principally adapted to exploiting mammals, and that fleas show a relatively low specificity of host utilization, they appear to have evolved at rates comparable to the other parasite families.

1 INTRODUCTION

As a way of life among organisms on this earth parasitism is the commonest. At least half the organisms on earth are parasites for they live in or on another living organism, obtain from it part or all of their organic nutriment, they exhibit some degree of adaptive structural modification, and cause some degree of real damage to the host. The British fauna is probably the most thoroughly studied in the world and the checklist of British insects by Kloet and Hincks (1945) enables an evaluation of the abundance of predators, non-parasitic herbivores such as bees and grasshoppers, parasitic insects on plants, parasitic insects on animals, and saprophagous insects (Price, 1977). Parasitic insects comprise 72 per cent of British insects while predators account for only four per cent, and non-parasitic herbivores only 2.4 per cent of the fauna. Adaptive radiation among parasitic taxa has been spectacular when compared to predatory and non-parasitic herbivore taxa. When this preponderance of parasitic insects is added to the very large numbers of parasitic mites, nematodes, flatworms, fungi, bacteria and protozoa it is clear that a majority of organisms is parasitic.

Judging by the British fauna, fleas have not entered into adaptive radiation to the same extent as the parasitic taxa in other orders of insects. Fleas number only 47 in Britain whereas the parasitic Hymenoptera number 5 342. The fleas do appear to be anomalous in this respect.

Adaptive radiation in the largest families of parasitic insects is impressive when compared to predatory families (Price, 1977). In the ten largest families of parasitic insects in the British fauna in the categories predators, parasites on plants, and parasites on animals, family sizes in the parasitic categories (mean sizes 329 and 478 for parasites on plants and on animals respectively) are many times larger than those in the predatory category (mean size 45 species per family).

These kinds of observation, which are summarized in Price (1977) have led me to generate some simple general concepts on the evolutionary biology of parasites. The most important are: 1) Parasites are very specialized relative to predators. 2) Evolutionary rates and speciation rates are high. 3) Adaptive radiation is extensive, the extent depending on several factors, the most important being: a) the number of hosts available for potential exploitation, b) the selective pressure for co-evolutionary modification (i.e. for specialization or specificity), c) evolutionary time available for colonization of hosts. 4) Types of speciation other than through geographic isolation are at least as important as allopatric speciation. Particularly, sympatric speciation is likely to be common because host shifts will frequently effectively isolate populations (Bush, 1975a, b).

2 ARE FLEAS ANOMALOUS?

The Siphanoptera appear to be exceptions to these general concepts. Gross morphology is strikingly uniform throughout the order, evolutionary rates appear to be slower than in their hosts (Jordan, 1942), and adaptive radiation in the order, thought to be represented by some 3 000 species and subspecies (Holland, 1964), is not spectacular for one which is exclusively parasitic in the adult stage.

There are two major considerations which are cause for skepticism about the apparently exceptional status of the Siphonaptera. Both require a biological species concept rather than the typological species concept (Mayr, 1963).

1. Slow morphological change does not necessarily indicate slow genetic change. We know of the enormous selective pressures imposed by host topography and grooming or preening for conformity to a certain set of characters. There is probably more evidence for convergent evolution in fleas than in any other group of insects (Traub, 1969, 1972 and this volume for references), and convergence of characters may be seen even between fleas and other mammal-infesting ectoparasites such as streblid flies in the evolution of strikingly similar combs (cf. Wenzel, Tipton & Kiewlicz, 1966; Traub; Marshall, this volume). Kethley and Johnston (1975) also emphasize the strictures imposed by the topographic features of hosts on the morphology and ecology of ectoparasites. With such strong forces operating to conserve morphology, we should regard morphological change as a very poor indicator of genetic change and therefore evolutionary change.

2. Recent studies indicate that many groups of mammals, which also provide the majority of hosts for fleas, are evolving and speciating more rapidly than other vertebrates and invertebrates examined so far (Wilson et al., 1975; Bush et al., 1977). Social structuring produces small effective populations enabling karyotypic changes to become fixed rapidly, resulting in reproductive isolation from the parental group and thus rapid speciation following the stasipatric or parapatric model of White (1968) and Bush (1975b). Such

chromosomal evolution has resulted in species flocks of fossorial rodents such as pocket gophers and mole rats studied by Nevo and associates (Nevo & Shaw, 1972; Nevo & Shkolnik, 1974; Nevo et al., 1974; Nevo & Bar-El, 1976) with each species in proximity to the others.

Such rapid speciation and the existence of species flocks must affect the population structure and evolution of fleas exploiting these hosts. Where hosts speciate parapatrically, there is a strong probability of isolation developing between their fleas and subsequent genetic divergence through the founder effect and genetic drift in the initial very small populations on new species. Therefore, we should anticipate considerable genetic difference between 'populations' of fleas on different host species in a species flock. In fact there appears to be even considerable morphological difference. The genus *Dactylopsylla* has as true hosts fossorial rodents in the family Geomyidae (Lewis, 1975). As we should expect from Nevo's detailed studies among the hosts, it is by no means clear how many biological species exist and many of the described subspecies probably deserve biological species rank. Of the fleas on these hosts, Lewis (1975) says 'Our knowledge of the systematics of the genus *Dactylopsylla* is in a similar state of disarray, and the number of valid taxa accepted in the following discussion is purely arbitrary'. He treats 15 species of which three have subspecies and there are eight subspecies in all; not an unusual number of subspecies for genera in the Ceratophyllidae as treated by Lewis. However, where 'our knowledge of the species and subspecies is quite complete' (Lewis, 1975) due to a monograph by Stark (1970) on the genus *Thrassis,* species of which utilize mainly burrowing sciurids, the 11 species listed have six with subspecies and 24 subspecies in all, a very high proportion of species with subspecies for genera in the Ceratophyllidae. It is my contention that a detailed study of isozyme patterns of these subspecies would indicate biological species status in many cases and many new species of *Dactylopsylla* would emerge with such a study. Speciation in fleas can probably be quite as rapid as in their hosts.

A similar situation may exist among the Mallophaga on hyraxes in South Africa, 'for nearly every form of *Procavia capensis* and *Heterohyrax syriacus* found in South Africa has its own specific forms of *Procavicola* s.str. Moreover, these lice are species, not subspecies, for though the females are inseparable, the differences in the male genitalia are often such as to render cross-mating difficult or even impossible' (Hopkins, 1949).

Therefore it would be extremely interesting to look at the isozyme variation of the fossorial rodent hosts, and their fleas, lice and other parasites, in order to determine the extent to which host population structure and speciation influences that of their parasites.

3 ANALYSIS

Because of my skepticism about the reality of conservative evolution in fleas, I will now ask, and attempt to answer the question – Have fleas evolved at a significantly slower rate than other parasitic taxa? For an approach to this question I return to the general concepts on parasites, specifically to factors influencing the extent of adaptive radiation in a taxon of parasites. Two working principles are available as follows: 1) The extent of adaptive radiation of a parasite taxon should be positively correlated with the number of hosts offering similar resources. 2) The extent of adaptive radiation of a parasite taxon should increase with increasing specificity to hosts and to host parts. That is, as resources are subdivided more species can utilize them and still co-exist. The third major factor influencing adaptive radiation, the evolutionary time available for colonization of hosts, cannot be used for parasites since these times are inadequately known.

I have tested if these principles are upheld in five families of parasitic insects on vertebrates representing several different modes of parsitism: Hippoboscidae (Diptera), Hystrichopsyllidae (Siphonaptera), Oestridae (Diptera), Philopteridae (Mallophaga) and Streblidae (Diptera). The hippoboscid adults (louse flies) are parasitic on birds mainly, and are usually winged, at least until a host is discovered. The larvae are reared internally and leave the mother just before pupariation, and the puparia normally fall to the ground. Hystrichopsyllid adults are parasitic on small mammals, apterous but with good jumping powers, and the larvae are free-living in host nesting material. Oestrid adults are free-living, strong flying, and the larvae are parasitic, most commonly on the larger mammals. Philopterids spend their whole life on the host, adults being apterous, and birds are utilized as hosts. Streblid adults are parasitic on bats, they are usually winged and the larvae are as in the Hippoboscidae, whereas the puparia are fixed on the roof of the bat roost. In addition,

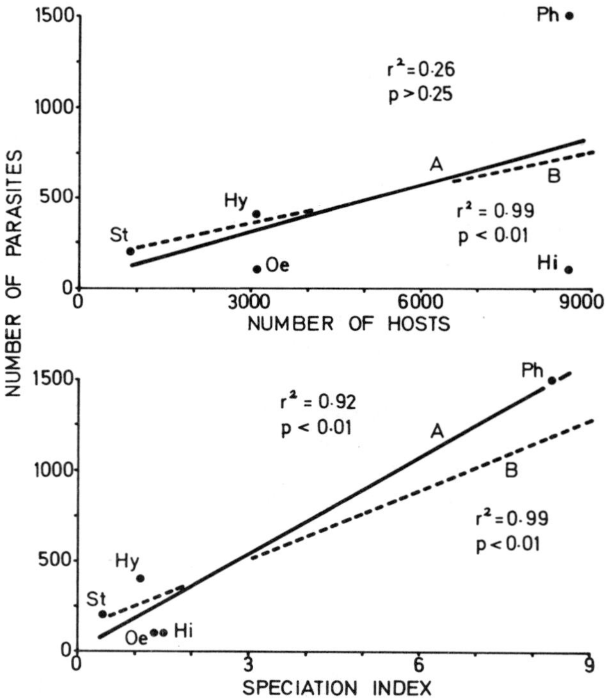

Figure 1. The number of parasite species in each family in relation to the number of hosts available (above) and the speciation index (below). The speciation index is the product of numbers which relate to the number of hosts available and the specificity of the parasites. The regression line A is for the five families shown, and the regression line B includes data on the Ichneumonidae with an estimated 215,000 hosts available, and 14,800 described species in the family. St = Streblidae, Hy = Hystrichopsyllidae, Oe = Oestridae, Ph = Philopteridae, Hi = Hippoboscidae. Slopes of the regression lines for A and B above are b = 0.086 and 0.068 respectively, and for A and B below b = 177.0 and 128.9 respectively.

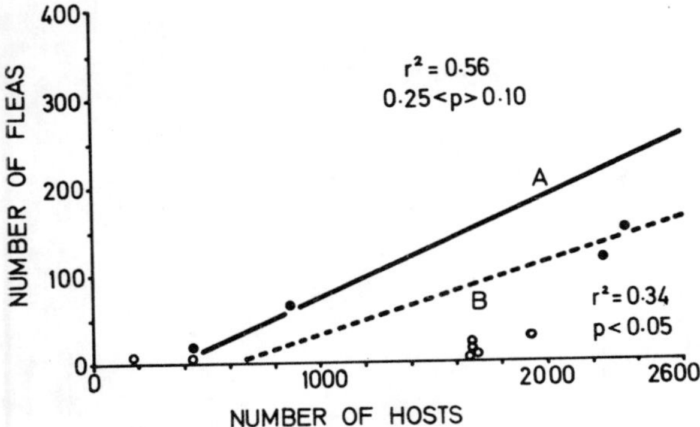

Figure 2. The number of flea species in each family treated by Hopkins and Rothschild (1953-71) in relation to the number of hosts available in the taxa already utilized by members of the family. Regression line A is for the five major families (○) which are from left to right Vermipsyllidae, Ischnopsyllidae, Pulicidae, Leptopsyllidae and Hystrichopsyllidae. Regression line B is for data on all flea families. Slopes of the regression lines for A and B are $b = 0.114$ and 0.082 respectively.

data have been used from the Ichneumonidae (Hymenoptera) which parasitize insects, for this family has undergone perhaps the most extensive adaptive radiation of any insect family, being thought to be represented by about 60 000 species (Townes, 1969). Adults are free-living and the larvae are parasitic as in the Oestridae but on arthropods.

The first principle was tested by correlating the number of hosts available in the world fauna and the number of species in each parasite family in the world (fig. 1). For the five families on mammals and birds alone, the correlation is not significant, largely because the Hippoboscidae is a small family considering that there are about 8 600 birds available as hosts. When the Ichneumonidae is included in the correlation, with 14 800 described parasite species and about 215 000 potential hosts, the correlation is highly significant and the slope of the regression line is very similar. Numbers of potential hosts appear to be valuable in predicting numbers of parasites, but other factors clearly influence the difference between numbers in the Philopteridae and Hippoboscidae, both of which utilize largely birds as hosts.

A similar approach may be taken for the families of fleas thus far treated by Hopkins & Rothschild (1953, 1956, 1962, 1966, 1971). The five major families therein, Pulicidae, Vermipsyllidae, Ischnopsyllidae, Hystrichopsyllidae and Leptopsyllidae are represented in much of the world and the number of hosts available can be crudely estimated from a world list of the mammals (Anderson & Knox Jones, 1967). The correlation accounts for 56 per cent of the variance (fig. 2) but is not significant at the ten per cent level, but the trend is clearly defined. When the small families from the Hopkins & Rothschild catalogue are included, most of which have very restricted geographic distributions, the total number of potential hosts in a world list grossly overestimates availability, but the correlation including large and small families is significant at the 5 per cent level. A detailed study of the biogeography of parasites and hosts available is warranted and would no doubt improve the correlation between number of hosts and number of fleas.

In the first analysis, the family Hystrichopsyllidae indicates that the number of species in the family is comparable to numbers in families of other parasitic groups. In the second analysis, the regression lines have a comparable slope to those in fig. 1 indicating that fleas

NUMBER OF HOSTS	PHILOPTERIDAE	STREBLIDAE	OESTRIDAE	HYSTRICHOPSYLLIDAE	HIPPOBOSCIDAE
1	97	56	49	37	17
2	2	22	19	20	9
3	<1	13	7	9	15
4	<1	5	6	5	13
5	<1	2	6	5	7
6	<1	1	4	6	4
7	0	1	4	2	4
8	<1		0	3	4
9			4	3	2
10-19			2	9	4
20-29				<1	7
30-39				<1	2
40-49					0
50-59					2
60-69					0
70-79					0
80-89					7
NUMBER OF PARASITES	1446	135	53	172	46

Table 1. Percentage of species in families of parasitic insects on mammals and birds in each class of number of hosts attacked. Data from sources given in the text.

have responded by adaptive radiation to an extent similar to other families of parasites.

The second principle was tested using an analysis of specificity for each of the families examined above. Check lists and catalogues for major faunal regions were analysed to find the number of hosts utilized by each species of parasite for which records existed and the family was then summarized by a frequency distribution of numbers of parasite species in each host number class. The host-parasite lists analysed were as follows: Hippoboscidae (Bequaert, 1956), all Siphonaptera (Hopkins & Rothschild, 1953, 1956, 1962, 1966, 1971, using only records from European and Mediterranean subregions of the Palaearctic and the Nearctic), Oestridae (Zumpt, 1965), Philopteridae (Hopkins & Clay, 1952), Streblidae (Wenzel, Tipton & Kiewlicz, 1966; Wenzel, 1976) and Ichneumonidae (Townes & Townes, 1951).

The degree of specificity seen in each parasite family examined differed enormously from a high of 97 per cent of species specific to a single host in the Philopteridae down to only 17 per cent of species host specific in the Hippoboscidae (table 1). These differences are clearly correlated to the degree to which a species of parasite is associated throughout its life cycle with the host and the degree to which hosts occupy discrete habitats. Specificity in the Streblidae is probably largely attributable to the occupation of discrete habitats by their bat hosts, while that in the Philopteridae more to the close asssociation between parasite and host throughout the life cycle of the parasite. Specificity in the Oestridae is caused more by a shortage of alternative hosts in many parts of the Old World than by a selection pressure for specificity in the presence of alternative hosts as in the more specialized families.

Specificity in the Philopteridae is overestimated in table 1 because mainly type hosts are given in the world list by Hopkins & Clay (1952) but this is the only list available for a

NUMBER OF HOSTS	TUNGIDAE (WORLD)	PULICIDAE (WORLD)	PULICIDAE (HOLARCTIC)	VERMIPSYLLIDAE	ISCHNOPSYLLIDAE	HYSTRICHOPSYLLIDAE	LEPTOPSYLLIDAE
1	19	35	12	20	21	37	30
2	38	20	16	20	25	20	20
3	25	11	8	30	14	9	9
4	6	6	4	10	7	5	7
5	0	6	8	0	4	5	5
6	0	1	0	20	4	6	5
7	0	1	0		7	2	5
8	6	2	12		4	3	7
9	0	1	4		4	3	5
10-19	6	9	20		11	9	9
20-29		3	8			<1	
30-39		1	4			<1	
40-49		1	4				
>50		3					
TOTAL SP. IN HOLARCTIC	1	25	25	10	28	172	44
TOTAL SP. IN CAT.	18	114	114	23	68	399	150

Table 2. Percentages of species in families in the Siphonaptera in each class of number of hosts attacked. Host records are from Hopkins & Rothschild (1953-71), and are restricted to the better studied regions of the world: the European and Mediterranean subregions of the Palaearctic and all subregions of the Nearctic. Data for the Pulicidae are given for the world and the above regions of the Holarctic to show how the estimate of specificity declines with increased collecting. Data on Tungidae of the world, not well represented in the Holarctic, are also provided, and together with data on the Pulicidae indicate the apparent low specificity of these fleas.

large faunal region. However, there is no doubt that the lice are much more specific than fleas. For Israel, Theodor & Costa (1967) provide host information which indicates that lice are 82 per cent host specific while the fleas are only 46 per cent host specific.

If the principle that specificity increases the potential for adaptive radiation is valid then the relationship between number of hosts and number of parasites should be modified by an index of specificity. The simplest index is to use the proportion of species in the family that utilize only one host. The index of specificity for the Philopteridae is therefore 0.97. Then a relative index of the extent of potential speciation in a family, the speciation index, can be obtained from the product of the number of hosts available and the index of specificity. For the Philopteridae, with 8 600 bird species available as hosts, the speciation index would be $8.6 \times 0.97 = 8.3$. For the Hippoboscidae with the same number of potential hosts, the index would be $8.6 \times 0.17 = 1.5$. The speciation indices for the five families of parasites correlate well with the number of parasites in each family and account for 92 per cent of the variance (fig. 1). The slope of the regression line is similar when the Ichneumonidae, which has a speciation index of 114.0, is included in the analysis. The improvement in regressions is largely due to the displacement of the Hippoboscidae far to the left due to low specificity in the family which results in a low specia-

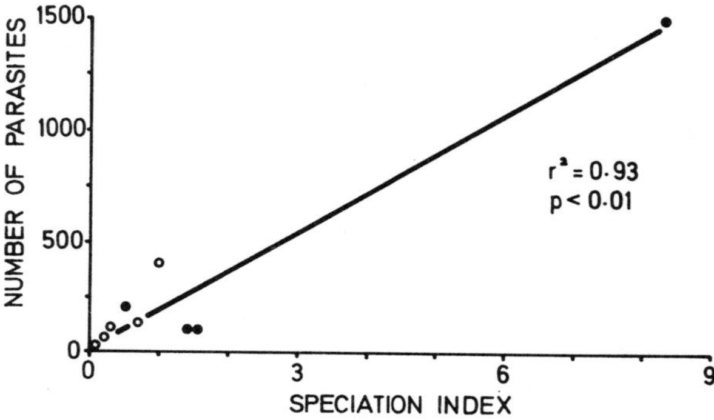

Figure 3. The number of parasite species in each family in relation to the speciation index. The flea families (○) are in the same order as in figure 2 and other families (●) are in the same order as in figure 1 (lower). The slope of the regression line is $b = 173.9$.

tion index. However, the value of the analysis is weakened by a shortage of data on parasite families with speciation indices above 1.5.

The major families of the Siphonaptera so far catalogued by Hopkins & Rothschild also show differences in specificity (table 2), although to what extent they are real or due to differences in the degree to which their host associations are fully known cannot be ascertained at this time. However, a very tentative analysis deriving speciation indices from the specificity indices provided in table 2 shows that the fleas fit well into the general trend seen in the four other families (fig. 3), that there is a trend within the flea families themselves, and that the pattern in the fleas is consistent with that of the other parasite families.

4 CONCLUSIONS

In conclusion, fleas seem to have undergone adaptive radiation to a degree similar to that in other groups of parasites when the number of hosts and the specificity of fleas are taken into account. No anomalous patterns have emerged. Given that there are relatively few hosts available to fleas since most species utilize mammals, and that they show relatively low specificity, we should expect a small number of species in any taxon when compared to parasites on larger groups of hosts such as birds or insects, and parasites which are more specific. It is interesting to note that the Siphonaptera and Mallophaga have members on mammals and birds. The former order is estimated to contain about 3 000 species (Holland, 1964) and the latter more than 10 000 species (Hopkins & Clay, 1952). We should expect this kind of difference in numbers simply from the differences in specificity particularly since the lice specialize on body parts of the host much more than the fleas so that species packing can be even tighter than the specificity index would indicate.

The apparently conservative nature of morphological evolution in the fleas that Jordan (1942) identified probably results from the strictures on morphology imposed by the topography of hosts. Hosts may speciate but unless the topography of the related species diverges, there is no selective pressure for flea morphology to change. However, where such selective pressures do exist and flea populations on the new host species are effectively isolated, the potential for rapid evolution appears to exist as seen in the fleas on fossorial rodents.

5 ACKNOLWEDGEMENTS

I am very grateful to Dr Miriam Rothschild for her invitation to present this paper at the International Conference on Fleas and to Dr Theresa Clay for pointing out that the specificity index for the Philopteridae is an over-estimate of real specificity in the family.

REFERENCES

Anderson, S. & J.Knox Jones, Jr. (eds.) 1967, Recent mammals of the world. A synopsis of families. New York, Ronald Press, 453 p.

Bequaert, J.C. 1956, The Hippoboscidae or louse flies of mammals and birds. Part II. Taxonomy, evolution and revision of American genera and species. Ent. Amer. 36:417-611.

Bush, G.L. 1975a, Sympatric speciation in phytophagous parasitic insects. In: P.W.Price (ed.), Evolutionary strategies of parasitic insects and mites. New York, Plenum. p. 187-206.

Bush, G.L. 1975b, Modes of animal speciation. Ann. Rev. Ecol. Syst. 6:339-364.

Bush, G.L., S.M.Case, A.C.Wilson & J.L.Patton 1977, Rapid speciation and chromosomal evolution in mammals. Proc. Nat. Acad. Sci. USA 74:3942-3946.

Holland, G.P. 1964, Evolution, classification and host relationships of Siphonaptera. Ann. Rev. Ent. 9:123-146.

Hopkins, G.H.E. 1949, The host-associations of the lice of mammals. Proc. Zool. Soc. London 119:387-604.

Hopkins, G.H.E. & T.Clay 1952, A check list of the genera and species of Mallophaga. London, British Museum (Natural History), 362 p.

Hopkins, G.H.E. & M.Rothschild 1953-71, An illustrated catalogue of the Rothschild collection of fleas (Siphonaptera) in the British Museum (Natural History), Vols. 1-5. London, British Museum (Natural History).

Jordan, K. 1942, On *Parapsyllus* and some closely related genera of Siphonaptera. Rev. Esp. Ent. 18:7-29.

Kethley, J.B. & D.E.Johnston 1975, Resource tracking patterns in bird and mammal ectoparasites. Misc. Pub. Ent. Soc. Amer. 9:229-254.

Kloet, G.S. & W.D.Hincks 1945, A check list of British insects. Stockport, Kloet & Hincks. 483 p.

Lewis, R.E. 1975, Notes on the geographical distribution and host preferences in the order Siphonaptera. Part 6. Ceratophyllidae. J. Med. Ent. 11:658-676.

Mayr, E. 1963, Animal species and evolution. Cambridge, Mass. Belknap Press of Harvard Univ. Press. 797 p.

Nevo, E. & H.Bar-El 1976, Hybridization and speciation in fossorial mole rats. Evolution 30:831-840.

Nevo, E., Y.J.Kim, C.R.Shaw & C.S.Thaeler, Jr. 1974, Genetic variation, selection and speciation in *Thomomys talpoides* pocket gophers. Evolution 28:1-23.

Nevo, E. & C.R.Shaw 1972, Genetic variation in the subterranean mammal, *Spalax ehrenbergi.* Biochem. Genet. 7:235-241.

Nevo, E. & A.Shkolnik 1974, Adaptive metabolic variation of chromosome forms in mole rats, *Spalax.* Experientia 30:724-726.

Price, P.W. 1977, General concepts on the evolutionary biology of parasites. Evolution 31: 405-420.

Stark, H.E. 1970, A revision of the flea genus *Thrassis* with observations on ecology and relationship to plague. Univ. Calif. Pub. Ent. 53:1-184.

Theodor, O. & M.Costa 1967, A survey of the parasites of wild mammals and birds in Israel. Part 1. Ectoparasites. Israel Acad. Sci. Human. Sect. Sci. pp. 5-117.

Townes, H. 1969, The genera of Ichneumonidae. Part 1. Mem. Amer. Ent. Inst. 11:1-300.

Townes, H. & M.Townes 1951, Family Ichneumonidae. In: C.F.W.Muesebeck & K.V.Krombein (eds.), Hymenoptera of America north of Mexico. Synoptic catalog. p. 184-409. US Dept. Agric., Agric. Monogr. 2.

Traub, R. 1969, *Muesebeckella,* a new genus of flea from New Guinea, with notes on convergent evolution (Siphonaptera: Pygiopsyllidae). Proc. Ent. Soc. Wash. 71:374-396.

Traub, R. 1972, Notes on zoogeography, convergent evolution and taxonomy of fleas (Siphonaptera), based on collections from Gunong Benom and elsewhere in South-East Asia. II. Convergent evolution. Bull. Brit. Mus. Nat. Hist. (Zool.) 23:307-387.

Wenzel, R.L. 1976, The streblid batflies of Venezuela (Diptera: Streblidae). Brigham Young Univ. Sci. Bull. Biol. Ser. 20(4):1-177.

Wenzel, R.L., V.J.Tipton & A.Kiewlicz 1966, The streblid batflies of Panama (Diptera Calypterae: Streblidae). In: R.L.Wenzel & V.J.Tipton (eds.), Ectoparasites of Panama. p. 405-675. Chicago, Field Museum of Natural History.

White, M.J.D. 1968, Models of speciation. Science 159:1065-1070.

Wilson, A.C., G.L.Bush, S.M.Case & M.-C.King 1975, Social structuring of mammalian populations and rate of chromosomal evolution. Proc. Nat. Acad. Sci. USA 72:5061-5065.

Zumpt, F. 1965, Myiasis in man and animals in the Old World. A textbook for physicians, veterinarians and zoologists. London, Butterworths, 267 p.

ADRIAN G. MARSHALL
University of Aberdeen, UK

THE FUNCTION OF COMBS IN ECTOPARASITIC INSECTS

ABSTRACT

Combs (ctenidia) are a prominent feature in a number of insects ectoparasitic upon mammals, but their function has been disputed. Observations on living and dead ectoparasites from bats in West Malaysia, including Polyctenidae (Hemiptera), Nycteribiidae (Diptera) and Ischnopsyllidae (Siphonaptera), suggest that their function is not to prevent dislodgement from the host, but rather to protect highly mobile joints and their associated membranes which are not otherwise protected from damage.

1 INTRODUCTION

Combs (ctenidia) are rows of stout, non-articulated spines* which occur only in insects ectoparasitic upon warm-blooded vertebrates, particularly mammals. They are a prominent feature in members of the Polyctenidae (Hemiptera), Nycteribiidae and Streblidae (Diptera), Siphonaptera, and Platypsyllidae (Coleoptera), and analogous structures may be seen in certain Ischnocera (Phthiraptera), Hippoboscidae (Diptera), and Pyralidae (Lepidoptera) (Baer, 1952; Theodor, 1957; Dr T.Clay, pers. comm.). Combs are clearly a fine example of convergent evolution, but their function has long been in dispute. In the past, suggestions have included rudimentary wings, spiracular coverings or organs of hearing (for references, see Mukerji & Dasgupta, 1953), but currently two theories are extant: that they are protective devices associated with mobile joints (Smit, 1972; Schlein in Rothschild, 1976), or that they are organs of attachment (Humphries, 1966, 1967; Traub, 1972a, b and papers by the same author cited therein). Indeed, Humphries suggested that the distance between the spines of the comb in a number of ectoparasites was closely correlated with the diameter of the host's hair, the comb thus being able to trap hair and so prevent dislodgement.

This controversy has so far centred largely upon studies of preserved fleas, although both Humphries (1967) and Theodor (1957, 1967) believed that the combs of Nycteribiidae functioned as organs of attachment. Studies upon insects ectoparasitic upon bats in West Malaysia, in which particular emphasis was placed upon identifying true host relation-

* I follow Rothschild and Traub (1971) in the use of the term 'spines', although ctenidial spines, at least in fleas, are not true spines but rather modified setae.

ships (Marshall, 1970, 1971, in press; Maa & Marshall, in prep.) have allowed me to make observations upon the function of combs in living and preserved Polyctenidae, Ischnopsyllidae (Siphonaptera), and particularly Nycteribiidae, the second largest group of comb-bearing insects.

2 MATERIALS AND METHODS

Observations were made sporadically between 1966-1976 on living insects in the field in various parts of West Malaysia and in the laboratory at the University of Malaya. Some insects were sectioned at 8 μ and stained with Mallory's Triple Stain. Measurements of bat hair were made at the University of Malaya and the British Museum (Natural History), and of ectoparasites in my own collection. All measurements were made using a standard compound microscope and ocular micrometer, and were recorded to $\pm 1.5\ \mu$. Host names are taken from Medway (1969) with subsequent amendments by Medway (pers. comm.).

The following number of measurements were attempted but not always achieved: (a) Host hair diameter: from one male and female bat of each species two hairs measured from between shoulder blades, one from back, and two from rump above tail giving total of 10 hairs. Hairs were exceedingly variable in diameter, and an attempt was made to measure the maximum width of each hair. (b) Comb spacing: measurements of distance (gaps) between spine tips near centre of each comb made on two males and two females of each insect species. The following are the total number of gaps measured for each comb: Polyctenidae – for each comb 32; Nycteribiidae – thoracic 20, abdominal 40; Siphonaptera – genal 4, prothoracic 16. Certain combs, notably the thoracic combs of Nycteribiidae, are very three-dimensional in structure and thus difficult to measure. (c) Thoracic length: measurements made along mid-line of two males and two females of each insect species to give indication of relative size.

3 OBSERVATIONS

3.1 *Polyctenidae*

Three species occur in West Malaysia, all host specific to different species of bat: *Eoctenes intermedius* (Speiser) on *Taphozous melanopogon* Temminck, *E.spasmae* (Waterhouse) on *Megaderma spasma* (L), and *Polyctenes molossus* Giglioli on *Megaderma lyra* Geoffroy (Maa, 1959, 1964; Maa & Marshall, in prep.). Maa (1964) records a total of nine types of comb on the 25 known species of Polyctenidae, but no species possesses them all; they are well-illustrated in Ferris & Usinger (1939) and Maa (1964). Comb spacing and host hair diameter for adults of the three species are given in table 1.

Observations were made on living *E.spasmae* (fig. 1A, B). All stages of this viviparous ectoparasite remain permanently on the host, living in the fur through which they can move rapidly due to the action of the meso- and metathoracic legs. The prothoracic legs face forwards below the greatly expanded pronotum and are virtually without claws. The genal comb occurs anterior to them and presumably guides hair away from them. The other four combs are all associated with highly mobile body joints; however, the two thoracic combs are absent in nymphal *E.spasmae.* At rest or feeding, these insects tended to

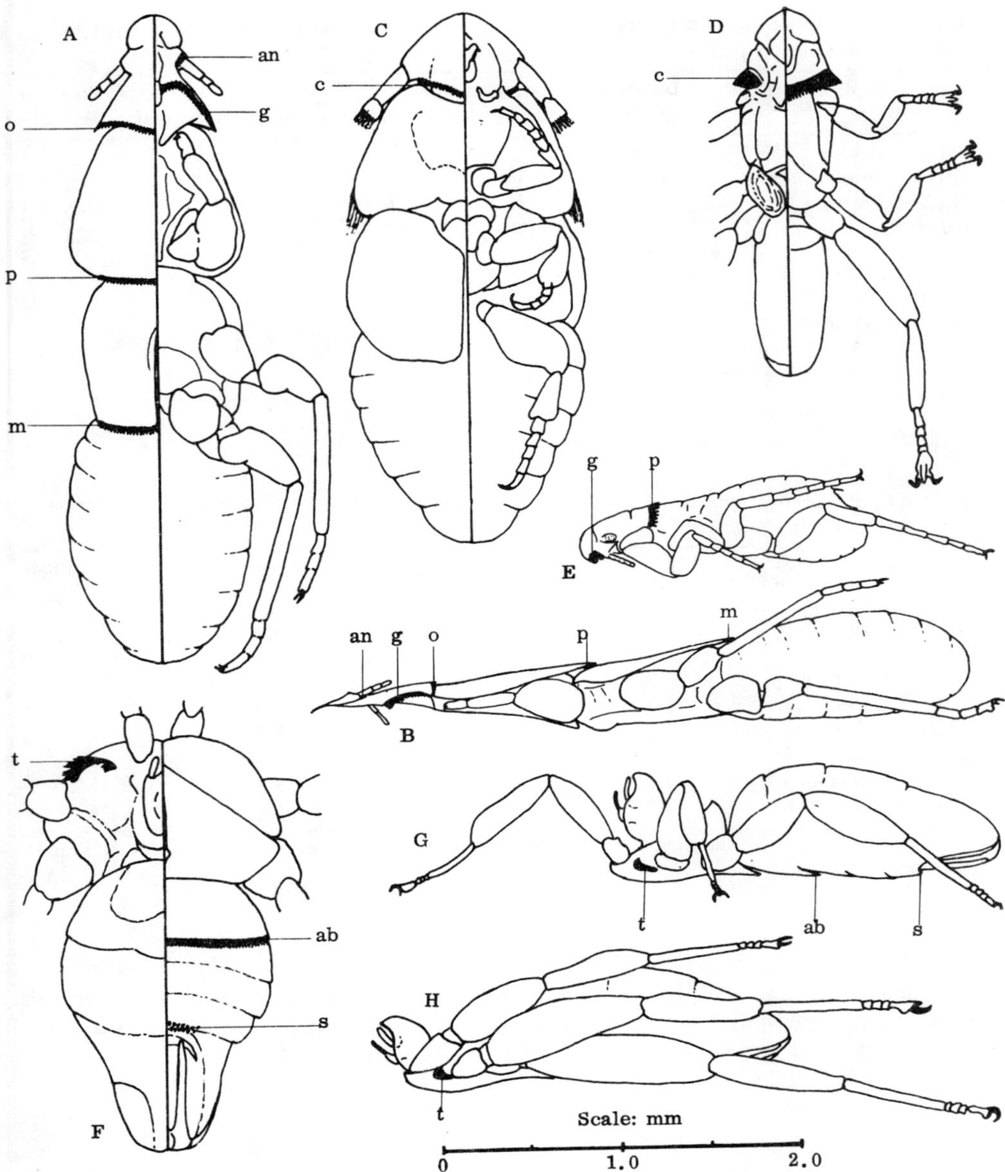

Figure 1. The position of combs and of posture in ectoparasitic insects.
A. Polyctenidae: *Eoctenes spasmae,* adult ♀; dorsal (left) ventral (right).
B. The same, lateral view, locomotion in fur.
C. Platypsyllidae: *Platypsyllus castoris,* adult ♀; dorsal (left), ventral (right).
D. Streblidae: *Metelasmus pseudopterus,* adult ♂; dorsal (left), ventral (right) (after Jobling, 1936).
E. Ischnopsyllidae: *Lagaropsylla mera,* lateral view, locomotion in fur.
F. Nycteribiidae: *Basilia hispida,* adult ♂; dorsal (left), ventral (right).
G. The same, lateral view, locomotion on surface of fur.
H. The same, lateral view, feeding posture.
ab – abdominal comb, an – antennal comb, c – cephalic comb, g – genal comb, m – mesonotal comb, o – occipital comb, p – pronotal comb, s – sternite 5, t – thoracic comb.

Table 1. Comb spacing and host hair diameter in adult Polyctenidae and Ischnopsyllidae from West Malaysia (for hosts, see text).

	Distance between spine tips Antennal	Occipital	Genal	Pronotal	Mesonotal	Host hair dia. (μ)
Polyctenidae						
Eoctenes intermedius	18.4	15.9	19.9	16.7	17.0	17.2
Eoctenes spasmae	21.8	17.0	23.0	17.7	17.7	18.5
Polyctenes molossus	26.7	20.8	27.6	18.9	*	17.3
Ischnopsyllidae						
Lagaropsylla mera	–	–	14.0	21.2	–	17.5/19.5
Lagaropsylla turba	–	–	14.0	23.8	–	**
Thaumapsylla breviceps	–	–	14.0	31.3	–	25.1/26.7
Thaumapsylla longiforceps	–	–	16.2	32.4	–	26.7

* *Polyctenes* species do not possess a mesonotal comb
** *C.torquatus* is virtually hairless

Table 2. Comb spacing, thoracic length and host hair diameter in Nycteribiidae from West Malaysia (for hosts, see text).

	Distance between spine tips (μ) Thoracic	Abdominal	Length of thorax (μ)	Host hair dia. (μ)
(a) *Basilia pudibunda*	16.8	14.2	664	14.7
Basilia hispida	18.3	16.0	807	16.1/19.8
Basilia marshalli	18.0	15.6	819	13.5
Basilia roylii	26.0	16.8	821	16.8
(b) *Nycteribia parvuloides*	21.6	12.4	533	16.2
Nycteribia allotopoides	20.8	12.5	643	16.2
Penicillidia sumatrensis	22.8	(30.4)*	948	16.2
Penicillidia actedona	27.5	*	1 361	16.2

* Abdominal comb reduced in both sexes of *P.sumatrensis,* and on *P.actedona* reduced in male and absent in female

remain with their heads close to the skin and bodies at an angle. Live adults could grip tenaciously, whereas dead insects offered little resistance when pulled backwards through the host's fur. The dorsal combs were usually closely adpressed to the body. At no time was hair observed caught in any comb.

3.2 *Nycteribiidae*

Forty-four species of Nycteribiidae have been recorded from Malaysia, 28 of these being associated with a single host species, 12 with two or three hosts of the same genus and four with unknown hosts (Maa & Marshall, in prep.). Three combs occur in members of the family (fig. 1F, G, H; see Theodor (1967), for illustrations):

1. Paired thoracic combs, present in all but two of the approximately 200 known species and subspecies, lie on the anterior-lateral part of the thorax. Each consists of a movable sclerite, part of the mesothoracic coxa (Schlein, 1970), which bears a row of curved spines reminiscent of the fingers of a hand held palm upwards with fingers somewhat apart and bent. At rest, the comb lies within a groove and the spines point anteriorly; but it can be everted so that the spines point dorsally or even posteriorly.

2. The abdominal comb lies on the posterior margin of sternite 1 + 2, and is generally flat and well-defined, although in a few species it is reduced or absent.

3. In many species, a third comb is present in males only, lying on the posterior margin of abdominal sternite 5.

Table 2 gives comb spacing, thoracic length and host hair diameter for certain Nycteribiidae from West Malaysia: (a) for four of the seven *Basilia* species: *B.pudibunda* Schuurmans Stekhoven from *Myotis haseltii* (Temminck), *B.hispida* Theodor from *Tylonycteris pachypus* (Temminck) and *T.robustula* Thomas, *B.marshalli* Maa from *Pipistrellus javanicus* (Gray), *B.roylii* (Westwood) from *Scotophilus kuhlii* Leach; (b) for the species parasitising the bat *Miniopterus medius* Thomas & Wroughton: *Nycteribia parvuloides* Theodor, *N.allotopoides* Theodor, *Penicillidia sumatrensis* Theodor, *P.actedona* Theodor.

Observations were made on many species of living Nycteribiidae, notably *Basilia hispida* and those species found on *Miniopterus medius* (Marshall, 1970, 1971). These viviparous insects virtually never leave their hosts, spending their whole life in the fur over which they can run with great rapidity and agility. Smaller species, such as *Nycteribia* species, can also push through the fur, all legs being directed posteriorly. All species feed head downwards with legs directed posteriorly, and will adopt this posture if disturbed (fig. 1H). At least in dead insects, the thoracic combs are everted when the legs are directed posteriorly. All legs possess powerful claws and live insects could grip tenaciously with these. Dead insects offered little resistance when pulled backwards through the host's fur. The abdominal comb always remained closely adpressed to the abdomen, which is highly expandable in females (see Marshall, 1970, plate 2). At no time was hair observed caught in any comb.

3.3 *Ischnopsyllidae*

Four species occur in West Malaysia associated with the following species of bats: *Lagaropsylla mera* Jordan & Rothschild with *Tadarida mops* (de Blainville) and *T.plicata* (Buchannan); *L.turba* Smit with *Cheiromeles torquatus* Horsfield; *Thaumapsylla breviceps* Rothschild with *Eonycteris spelaea* (Dobson) and *Rousettus amplexicaudatus* (Geoffroy); and *T.longiforceps* Traub with *R.amplexicaudatus* (Maa & Marshall, in prep.). Comb spacing and host hair diameter for these fleas are given in table 1. Fleas may possess combs on the head (helmet and genal), on the pronotum, and in some families including Ischnopsyllidae on the metathoracic notum or epimeron and up to six dorsal abdominal sclerites (Smit, 1972). Of the 2 000 described species and subspecies about 20 per cent have no combs, 40 per cent have only a pronotal comb and 40 per cent have both head and pronotal combs. These various combs are excellently illustrated in Hopkins & Rothschild (1956), Rothschild & Traub (1971), and in other papers by Traub cited herein.

Observations were made on living *L.mera* on *T.mops*, *L.turba* on *C.torquatus* and *T. breviceps* on *E.spelaea.* On the virtually hairless *C.torquatus, L.turba* walked and stood in normal insectan fashion, but when feeding stood on their heads, their bodies held vertically and no legs in contact with the host's skin. The hosts of the other two species have dense fur through which the fleas could move rapidly with all legs directed posteriorly (fig. 1E); when feeding their posture resembled that of Nycteribiidae. All species walked rapidly and with agility on their long legs, and could grip tenaciously with well-developed tarsal claws. Dead *L.mera* offered little resistance when pulled backwards through the host's fur. The most mobile body point was that between the pro- and mesothorax; the pronotal comb associated with this was normally closely adpressed to the body. At no time was hair observed caught in any comb.

4 DISCUSSION

Combs occur in a wide variety of insects associated with the fur of mammals, and it is reasonable to believe that they have evolved for a specific purpose. Most discussion has centred upon fleas for these are easily the largest group of comb-bearing ectoparasites. In a number of interesting papers, Traub and others (Traub, 1966, 1968, 1969, 1972a, b; Traub & Barrera, 1966; Traub & Dunnet, 1973; Traub & Evans, 1967) have attempted to correlate the structure of flea combs with the biology of both parasite and host, believing that they show adaptations to both the pelage of the host and the host's environment. Throughout these discussions the authors assumed that the function of combs is to prevent dislodgement from the host, and Traub has indeed observed attachment between the spiniform setae on the head of certain rodent fleas and the hairs of their host (Traub, 1969 and pers. comm.). Humphries (1966, 1967) also believed that combs are organs of attachment, noting that the gaps between the tip of the spines was about 1.75 times greater than the average diameter of the host hair in 15 species of fleas, one Nycteribiidae and one Platypsyllidae.

In contrast, Smit (1972) believed that flea combs were mainly protective devices as Rothschild (1917) had suggested. The pronotal comb, which occurs in 80 per cent of fleas and is thus presumably the most important, is associated with the main body joint and its greatly developed intersegmental membrane. Smit did not believe that the structure of combs would allow hair to become trapped between the spines, and in hundreds of thousands of fleas examined had never seen a hair so caught.

My observations on Polyctenidae, Nycteribiidae and Ischnopsyllidae, made upon a diversity of species whose true host associations are known, support the view of Smit (1972) and Schlein (in Rothschild, 1976) that combs are protective. I do not believe that combs are organs of attachment for the following reasons:

1. Combs could only function as organs of attachment if the insects were dragged *backwards* through the fur by their hosts. This is exceedingly unlikely with such agile insects, and on the rare occasions when it does occur is likely to result in serious damage. Well-developed tarsal claws or the mouthparts are the major organs of attachment in ectoparasites, and it is difficult to see how any comb-like structure could significantly improve attachment or prevent dislodgement. The resistance felt when dead insects are drawn backwards through the fur is due to the backward-pointing setae which are a prominent feature of many ectoparasitic insects.

2. The position of certain combs (e.g. thoracic in Nycteribiidae) and the form of others in which the spines are closely adpressed to the succeeding sclerite (e.g. abdominal in Nycteribiidae, thoracic in Polyctenidae and certain fleas; see stereoscan photographs in Rothschild & Traub (1971, plate 7A), Smit (1975, fig. 6)) would prevent them acting as organs of attachment.

3. The gaps between spine tips and the form of the spine, from pointed to blunt, varies within a comb and between combs of a given species (tables 1, 2). Furthermore, hair diameter within a single host species varies from one type of hair to another and along the length of a given hair. For example, a hair of *Tylonycteris pachypus* varied from 4 μ in diameter at the base to a maximum of 17 μ. Bat ectoparasites are active insects and will thus come in contact with hair of widely varying thickness. Thus the relationship proposed by Humphries (1966, 1967) does not hold true, and it is unlikely that an examination of comb structure will help elucidate problems of host-specificity, as Marshall (1976) hypothesised.

4. Hair was never seen to become caught in or below a comb, nor was broken hair ever found in the combs of many score of Polyctenidae or hundreds of Nycteribiidae examined.

5. Those advocating the role of combs as organs of attachment have not explained why combs are located where they are, which generally do not appear to be optimal sites for such an organ. Nor do they suggest why these sites are usually associated with mobile joints or organs with their attendant delicate intersegmental membranes which are not otherwise protected from hair (e.g. by the legs).

That combs are usually associated with such sites is clear:

a) Polyctenidae: genal with mouthparts, prolegs; others with body joints, antennae.

b) Nycteribiidae: thoracic with mesothoracic coxal joint; abdominal with heavily sclerotised sternite 1 + 2 and abdomen; in males sternite 5 combs with genitalia.

c) Streblidae – Streblinae: certain species possess a comb on the lateral and ventral surface of the head associated with the mobile head/thorax joint (Jobling, 1929, 1936, 1949; fig. 1D). It is likely that these species are unusual amongst Streblidae in not remaining upon the membranous areas of their hosts, but entering the fur (Wenzel & Tipton, 1966).

d) Siphonaptera: head combs with cephalic organs and mouthparts; pronotal with pro-/meso-thoracic joint; others possibly with mobile joints on active, flexible fleas.

e) Platypsyllidae – *Platypsyllus castoris* (Ritsema) possesses a cephalic comb associated with the mobile head/prothorax joint (fig. 1C). This beetle is a permanent ectoparasite of the beaver *Castor canadensis* (Kuhl), spending all its life in the fur through which it can move with great agility (Janzen, 1963).

It is suggested that combs give good protection to mobile joints from the abrasive action of hair, for they are more dense than thin dispersed setae, and more flexible than a chitinous collar. Furthermore, Schlein (pers. comm. and in Rothschild, 1976) has noted that comb spines tend to be longer than the intersegmental membranes with which they are associated, and that fleas with a more flexible thorax have longer, stouter and more numerous spines than those with a more compact and rigid thorax. This may explain why fleas with a narrow pronotum (i.e. more flexible) have long spines and those with a broad pronotum short spines (Traub & Evans, 1967; Traub, 1969).

However, some aspects of the distribution of combs still remain puzzling. For example within Polyctenidae the presence or absence of combs varies greatly between the five known genera, within the genus *Eoctenes,* and even between different instars of *E.spasmae* (Maa, 1959, 1964). In the Nycteribiidae *Penicillidia* species parasitizing *Miniopterus* species have either lost all their combs (i.e. *P.progressa* (Scott)), or the abdominal comb is reduced or absent, whereas the much smaller *Nycteribia* species from the same hosts all possess well-developed combs. Indeed all Nycteribiidae with reduced or absent combs are large species (Theodor, 1967) and it is possible that these insects, living more on the surface of the fur and with thicker exoskeletons, have less need of the protection afforded by combs. Amongst fleas, Smit (1972) believes that there is a correlation between combless-ness and avian hosts or nest-dwelling habit or permanent attachment to the host. However, certain Ceratophylloidea infesting birds actually have more spines in the pronotal combs than mammal-infesting relatives (Hopkins, 1957) although these species may be losing their combs by densation (Smit, 1972). The fact that bat fleas, and those of other arboreal animals, have such well-developed combs (Traub, 1972b) may be because they spend most of their lives in the host's fur, jumping little but being highly flexible and agile walkers.

In conclusion, I believe that the major function of combs is to protect highly mobile joints and their associated membranes which are not otherwise protected from damage in those insects living in mammalian fur. Further studies, particularly upon the habitat and mobility of living insects, are now needed to test this hypothesis. However, no doubt problems will still remain for as Dr Karl Jordan once remarked to his successor as doyen of British Siphonapterists, Mr Frans Smit: 'Nature alive is nature illogical'.

5 ACKNOWLEDGEMENTS

I wish to thank Dr Miriam Rothschild for her hospitality at the Conference, and her unfailing kindness and enthusiasm over many years. The following have also generously offered assistance with this paper: Prof Dr Mr A.H.Benton, T.Clay, G.M.Dunnet, J.E.Hill, T.C.Maa, D.K.Mardon, B.C.Nelson, J.Schlein, F.G.A.M.Smit, R.Traub and J.K.Waage.

REFERENCES

Baer, J.G. 1952, Ecology of animal parasites. Urbana, University of Illinois Press.

Ferris, G.F. & R.L.Usinger 1939, The family Polyctenidae (Hemiptera: Heteroptera). Microentomology 4:1-50.

Hopkins, G.H.E. 1957, Host associations of Siphonaptera. In: J.G.Baer (ed.), First symposium on host specificity among parasites of vertebrates, p. 64-87. Université de Neuchâtel, Institut de Zoologie.

Hopkins, G.H.E. & M.Rothschild 1956, An illustrated catalogue of the Rothschild Collection of Fleas (Siphonaptera) in the British Museum (Natural History) Vol. II. London, Brit. Mus. (Nat. Hist.).

Humphries, D.A. 1966, The function of combs in fleas. Ent. Mon. Mag. 102:232-236.

Humphries, D.A. 1967, Function of combs in ectoparasites. Nature 215:319.

Janzen, D.H. 1963, Observations on populations of adult beaver beetles, *Platypsyllus castoris* (Platypsyllidae: Coleoptera). Pan-Pacific Ent. 39:215-228.

Jobling, B. 1929, A comparative study of the structure of the head and mouth parts in the Streblidae (Diptera Pupipara). Parasitology 21:417-445.

Jobling, B. 1936, A revision of the subfamilies of the Streblidae and the genera of the subfamily Streblinae (Diptera Acalypterae) including a redescription of *Metelasmus pseudopterus* Coquillet and a description of two new species from Africa. Parasitology 28:355-380.

Jobling, B. 1949, Host-parasite relationship between the American Streblidae and the bats, with a new key to the American genera and a record of the Streblidae from Trinidad, British West Indies (Diptera). Parasitology 39: 315-329.

Maa, T.C. 1959, The family Polyctenidae in Malaya (Hemiptera). Pacif. Insects 1:415-422.

Maa, T.C. 1964, A review of the Old World Polyctenidae (Hemiptera: Cimicoidea). Pacif. Insects 6:494-516.

Maa, T.C. & A.G.Marshall (in prep.), The ectoparasitic insect fauna of bats (Chiroptera) in the Malesian subregion.

Marshall, A.G. 1970, The life cycle of *Basilia hispida* Theodor 1967 (Diptera: Nycteribiidae) in Malaysia. Parasitology 61:1-18.

Marshall, A.G. 1971, The ecology of *Basilia hispida* (Diptera: Nycteribiidae). J. Animal Ecol. 40:141-154.

Marshall, A.G. 1976, Host specificity amongst arthropods ectoparasitic upon mammals and birds in the New Hebrides. Ecol. Ent. 1:189-199.

Marshall, A.G. (in press), The comparative ecology of insects ectoparasitic upon bats in West Malaysia. In: Proceedings of the Fifth International Bat Research Conference, Alberqueque, USA, 1978.

Medway, Lord 1969, Wild mammals of Malaya and offshore islands including Singapore. London, Oxford University Press.

Mukerji, D. & B.Dasgupta 1954, Some aspects of the anatomy of the bat-fly *Cyclopodia sykesi* Westwood (Diptera: Pupipara: Nycteribiidae). Proc. Zool. Soc. Bengal 7:1-25.

Rothschild, M. 1976, Notes on fleas (Part II): the internal organs: can they throw any light on relationships within the order? Proc. Brit. Ent. and Nat. Hist. Soc. 8:97-110.

Rothschild, M. & R.Traub 1971, A revised glossary of terms used in the taxonomy and morphology of fleas. In: G.H.E.Hopkins & M.Rothschild, An illustrated catalogue of the Rothschild Collection of Fleas (Siphonaptera) in the British Museum (Natural History) Vol. V, p. 8-85. London, Brit. Mus. (Nat. Hist.).

Rothschild, N.C. 1917, Convergent development among certain ectoparasites. Proc. Ent. Soc. London 1916, p. 141-156.

Schlein, J. 1970, A comparative study of the thoracic skeleton and musculature of the Pupipara and the Glossinidae (Diptera). Parasitology 60:327-373.

Smit, F.G.A.M. 1972, On some adaptive structures in Siphonaptera. Folia Parasit. 19:5-17.

Smit, F.G.A.M. 1975, A new bat flea from Australia. J. Ent. (B) 42:283-288.

Theodor, O. 1957, Parasitic adaptation and host-parasite specificity in the pupiparous Diptera. In: J.G.Baer (ed.), First symposium on host specificity among parasites of vertebrates, p. 50-63. Université de Neuchâtel, Institut de Zoologie.

Theodor, O. 1967, An illustrated catalogue of the Rothschild Collection of Nycteribiidae (Diptera) in the British Museum (Natural History). London, Brit. Mus. (Nat. Hist.).

Traub, R. 1966, Some examples of convergent evolution in Siphonaptera (abstract). Proc. Roy. Ent. Soc. London (C) 31:37-38.

Traub, R. 1968, *Smitella thambetosa,* n.gen. and n.sp., a remarkable 'helmeted' flea from New Guinea (Siphonaptera, Pygiopsyllidae) with notes on convergent evolution. J. Med. Ent. 5:375-404.

Traub, R. 1969, *Muesebeckella,* a new genus of flea from New Guinea, with notes on convergent evolution. Proc. Ent. Soc. Wash. 71:374-396.

Traub, R. 1972a, The Gunong Benom Expedition 1967. 12. Notes on zoogeography, convergent evolution and taxonomy of fleas (Siphonaptera), based on collections from Gunong Benom and elsewhere in South-east Asia. II. Convergent evolution. Bull. Brit. Mus. (Nat. Hist.) (Zool.) 23:309-389.

Traub, R. 1972b, The relationships between the spines, combs and other skeletal features of fleas (Siphonaptera) and the vestiture, affinities and habits of their hosts. J. Med. Ent. 9: 601.

Traub, R. & A.Barrera 1966, New species of *Ctenophthalmus* from Mexico, with notes on the ctenidia of shrew-fleas (Siphonaptera) as examples of convergent evolution. J. Med. Ent. 3:127-145.

Traub, R. & G.M.Dunnet 1973, Revision of the Siphonapteran genus *Stephanocircus* Skuse, 1893 (Stephanocircidae). Austral. J. Zool., Suppl. Ser. 1973, 20:41-128.

Traub, R. & T.M.Evans 1967, Descriptions of new species of Hystrichopsyllid fleas with notes on arched pronotal combs, convergent evolution and zoogeography (Siphonaptera). Pacif. Insects 9:603-677.

Wenzel, R.L. & V.J.Tipton 1966, Some relationships between mammal hosts and their ectoparasites. In: R.L.Wenzel & V.J.Tipton (eds.), Ectoparasites of Panama, p. 677-723. Chicago, Field Museum of Natural History.

GEORGE P. HOLLAND
Biosystematics Research Institute, Ottawa, Ontario, Canada

THE WONDERFUL ONE HOST FLEA

The *Thrassis* looked at the *Catallagia*
and said 'Look here, you dog!
You're supposed to be on a deer-mouse,
and this here's my groundhog!'

'Silly goon' giggled the mouse-flea,
'Shut up, and don't make a fuss.
You must be blind if you can't see
That this is a *Peromyscus'*.

The *Thrassis* groaned in blank dismay
and said 'My Gawd! I'm lost
It was late when I came home last night.
Guess I got my burrows crossed.'

'That's the trouble with you monozoid fellows'
The *Catallagia* started to say
'Take away your special chosen host
and you cry like a fool all day'.

'Now I feed on mice, I don't care what kind
I really don't give a hoot.
I even like squirrels, and chipmunks are fine,
and my Grandma once bit a coot!'

'But you, with your smelly old marmots,
You don't live a fine life like me.
Crawling all day in a rockslide –
I'm ashamed to admit you're a flea!'

The *Thrassis* started to howl and weep
Got his palpi all stuck up with goo
and said 'I'll never be happy until
I'm a flea of the world, like you'.

The mouse-flea was sorry, and said 'There there,
Please don't be quite so sad.
Try a bite while you're sitting on this mouse.
You'll find it isn't too bad'.

So the *Thrassis* twiddled his mouthparts
and shoved them in up to the hilt
all the time glancing sheepishly sideways,
For he was conscious of criminal guilt.

For thousands of years all his fathers
Had fed on one genus of hosts
'Twas tradition, and now that he'd broken it
He was afraid of their spirits and ghosts.

His head capsule blushed rosy red
and he stopped and said 'Listen to this!
I'll starve ere I'm false to my breed,
Or my name's not *acamantis*'.

'Go feed on your squirrels, your chipmunks and mice
And take this old mouse back, goldarn it!
For I'll have you know there is nothing so nice
As a fat yellow-bellied marmot!'

'Oh *Marmota flaviventris avara*
Listen to me as I cry
I'm a one-hoster. If I can't have you,
I'll pine away 'til I die.'

Then he turned from the *Catallagia,*
Jumped off the white-footed mouse,
Ignored all the mites and the other fleas,
And thumbed his nose at a louse.

He started to search for his woodchuck
He tramped up hill and down dale.
Day after day, without any luck –
He became very thin and pale.

He stumbled on and on and on
His breath was short and gasping
When he spied a big rock, in the sun
With a marmot on it, basking.

He kicked up his tarsi in joy,
And put on a slight burst of speed

'Here's a meal that I can enjoy'
He muttered and started to feed.

Contented at last – Oh happy bliss
He fed as he scurried around
Through the fur. But Hark! What was this?
Surely he heard a sound!

He popped out both his antennae
And turned on his radar full strength
North by east through the pelage he followed
The high frequency call of a wench!

A moment later he saw her,
The cutest blond little flea
His proventriculus up in his pharynx
He gasped 'Won't you belong to me?'

He pursued through the forest of hairs
The slim little *Thrassis* girl.
Oh gone were his troubles and cares –
His brain was all awhirl.

At last he caught up to her
Out on the marmot's ear.
And there he begged and cajoled her
Until she said, 'Yes, my dear'.

'Let me take you to dinner' he offered
'There's a nice tender place back here'.
Side by side they prodded their beaks
In the poor old groundhog's rear.

The 'chuck woke up with a whistle
And must have been ever so sorry
To be bitten. However, this will
Have to be the end of our story.

Yes, we now leave our *Thrassis* in comfort
With best wishes and sincere felicity,
For ever and aye we must marvel at
His wondrous host specificity!

G.P.Holland
1945

ROBERT TRAUB
Department of Microbiology, University of Maryland School of Medicine, Baltimore, USA

THE ZOOGEOGRAPHY AND EVOLUTION OF SOME FLEAS, LICE AND MAMMALS

CONTENTS

ABSTRACT

Data on the zoogeography, phylogeny and evolution of fleas and lice support the concepts of austral faunal relationships, transatlantic connections and other aspects of the theory of continental drift. Primitive hosts tend to have primitive fleas and lice, while the most evolutionarily youthful ectoparasites are associated with the most advanced mammals. Fleas generally parasitize the hosts with which they evolved or else those which developed later, rather than infest hosts lower on the evolutionary scale. Primitive hosts and their fleas and lice tend to be conservative and change very slowly at the generic, or even species level, especially as compared to relatively recently evolved forms such as murids and their fleas. The close correlation between the kind of ectoparasite occurring on tetrapods and the geological time the hosts first arose extends to class and order, with only mites occurring as true parasites of Amphibia; mites and ticks on reptiles, etc., extending to a gamut of ectoparasites on rodents. Thus, hosts arising before the Paleocene lack Anoplura, e.g. the bats.

The Siphonaptera demonstrating southern affinities (e.g. stephanocircids, doratopsyllines, pygiopsyllids and pulicids) are the more primitive groups of fleas. The main families on the boreal continents (Ceratophyllidae and Leptopsyllidae) are essentially northern in distribution and are clearly more youthful in evolutionary development than the preceding ones. The siphonapteran relationships among the austral continents are at the level of subfamily or family, not the genus.

The marsupials presumably arose in South America in the Late Jurassic or Early Cretaceous and dispersed to Australia, at least in the Early Cretaceous, via Antarctica, carrying stephanocircid and doratopsylline fleas and amblyceran Mallophaga. Other marsupial elements moved into North America and, eventually, into Europe via transatlantic connections. Later, a marsupial travelled to South America from Australia or Antarctica, transporting the forebears of a genus of pygiopsyllid. The Insectivora arose in Asia and entered North America without fleas or lice. The monophyletic order Siphonaptera must date back to the Jurassic and existed on parts of Pangea, at least as ancestral hystrichopsyllids, and certain families or subfamilies arose in components of Gondwanaland and others in Laurasia, as indicated, e.g. Pygiopsyllidae in Australia, Pulicidae in Africa, etc. The Anoplura arose in North America, perhaps in Early Paleocene.

There were African/South American faunal connections, by rafting, in the Early Eocene, involving 'hystrichomorph' rodents and ancestral ceboid monkeys and their ectoparasites, e.g. polyplacid lice of phiomorph and caviomorph rodents; ctenophthalmine fleas; and perhaps *Pediculus* Anoplura. Virological and other parasitological evidence also suggest the possibility of such faunal relationships. The data on Anoplura indicate that the ultimate roots of the murids go back to the Asian mainland, even if rats as such arose in Wallacea or southeast Asian islands. Regardless, *Rattus* is a relatively youthful taxon and moved from Southeast Asia towards Australia, and transported some fleas of Palearctic derivation. On movements in the opposite direction, they carried pygiopsyllids. Penetration by sciurids into Borneo, Sulawesi, etc. was subsequent to that of some murids. Madagascar was originally much further north than at present, probably opposite Somalia, as indicated by their cricetid rodents and leptopsyllid fleas of Palearctic origin, and by the dearth of native murids and their ectoparasites. The distribution of fleas like *Odontopsyllus* and some hystrichopsyllids suggest that direct transatlantic connections formerly existed between Europe and North America, but the absence of such data for the ceratophyllids, a more recently evolved group, indicates the corridor was terminated after the Eocene. The *Pediculus* and *Pthirus* lines of human lice may antedate the divergence of the humanoid and anthropoid branches.

1 INTRODUCTION

Studies on the faunal affinities and zoogeography of ectoparasites like fleas, lice and trombiculid mites have led to the demonstration of the occurrence of chigger-borne rickettsiosis and tick typhus in wholly unexpected geographical and ecological areas, and have contributed significantly to our knowledge of the ecology and distribution of plague and murine typhus (213, 214, 215, 216). Inasmuch as all fleas are obligate, blood-sucking ectoparasites of mammals and birds; are often quite specific in their host relationships; and have been associated with their hosts for countless eons – so much so that the morphological features of these insects often reflect the vestiture or habits of the host (200, 204, 209) – it is only to be expected that data on the phylogeny, systematics, host relationships and zoogeography of fleas (and lice, etc.) could advance our understanding of the historical biogeography and evolution of their hosts. Such evidence on fleas, denoting faunal connections among the southern continents (205, 206), has since been supported by recent findings on plate tectonics and continental drift. Traub (1968b, 1972c, 1972d) reported that data on fleas strongly militated against the widely accepted hypothesis of Simpson (1961) that marsupials entered Australia after penetrating the Australian archipelagos from Asia, to which they allegedly had emigrated from North America. The Asian route 'is now almost universally rejected' (Cracraft, 1974a). Data summarized in this paper extend those ideas of faunal relationships among the austral continents. However, as I noted in 1972 (205), studies on Siphonaptera were in accord with Simpson's views that *Rattus* reached Australia and New Guinea via the Indonesian islands, and new data, herein reported, substantiate that contention.

There are cogent reasons why the Siphonaptera are useful tools in zoogeography. Thus, there is unanimous agreement that the order arose only once, and since all fleas have a common ancestry, clues are available regarding where and when the various families of fleas originated, e.g. Australian and South American fleas could not have arisen de novo, completely independently. Also, the current scheme of classification is accepted as fundamentally sound at the family and critical subfamily levels. Fleas are so complex morphologically that it is relatively easy to decide which features represent convergence and which are due to parallel evolution. Mammalogy and paleontology, in contrast, have been plagued with such problems, and worse, key fossils may be represented by only a few teeth and hence even their assignments to order may be controversial, e.g. the necrolestids of South America. Generalizations about the zoogeography and history of fleas can therefore be based on a relatively firm foundation and can lead to similar deductions about their hosts, despite the fact that there is yet much to be learned about the fleas of certain parts of the world, e.g. Sulawesi and certain critical hosts like African phiomorph rodents and giant tree-rats of Flores and the Philippines.

In this paper, efforts are made to summarize available data on Siphònaptera in order to show that continental drift has indeed affected the distribution of some contemporary mammals, and that there have been faunal connections between the southern continents, and between eastern North America and western Europe, even though the vast majority of mammalogists and paleontologists have decried or denied such a possibility. Because of unavoidable and severe constraints of time and space, it is impossible to present herein details on all the data or to provide background material on plate tectonics, continental drift and the distribution of fleas and their hosts. Such material will be treated at greater length in an article in preparation. There have been no articles on Siphonaptera that discuss zoogeography of fleas per se since my papers of 1972 (205, 206), although Lewis (1972, 1973, 1974a, b, c, 1975) has compiled a very useful outline of the geographic and host distribution of Siphonaptera. There have been some excellent and stimulating recent papers reviewing continental drift, the historical geography of marsupials, austral relationships, faunal dispersal routes, etc. (17, 18, 21, 25-31, 64, 89-92, 121, 122, 131, 132, 151-154, 161, 184, 188, 190, 192).

It will be shown herein that the data on fleas suggest that: 1) the order Siphonaptera is an ancient one, probably going back to the Jurassic, more than 140 million years ago (m.y.a.); 2) as an order, fleas may antedate the families, and perhaps even some orders of mammals existing today, while some Siphonapteran families are older than some of their current families of hosts; 3) representatives were probably present on at least some of the major 'continents' i.e., on components of Pangea and the super-continents Laurasia and Gondwanaland before rifting occurred; 4) such fleas were already differentiated geographically into 'families', or else evolved independently after the breakup of Gondwanaland and Laurasia; 5) some of the contemporary (modern) families had been established as such while there were still faunal connections, at least by island-hopping or drifting, between the austral continents; 6) some of the existing families of fleas (and their hosts) arose in the Southern Hemisphere; 7) the faunal interchange between the austral continents was primarily at the subfamily level, not at the tribal or generic one. Observations on the zoogeography and host relationships of lice will be cited to support some of these contentions on the zoogeography of mammals, and the historical geography of pertinent hosts are briefly reviewed. Examples will be given where such evidence has been ignored or denigrated because of the widespread conviction that Earth had been static regarding the position of the continents, or else, if there had been any movement, it all occurred too long ago to have affected the distribution of mammals and birds. Data are presented indicating that

1) marsupials entered Australia from South America; 2) the marsupials arose in South America and 3) the 'hystrichomorph' rodents of South America had African roots. There are brief incursions into such topics as the origin of the marsupials, the birthplace of the Insectivora and Rodentia, comparative rates of evolution of host versus parasite; and on the faunal affinities of Madagascar.

It should be borne in mind that, unless otherwise specified, the Siphonapteran data pertain to indigenous fleas that infest non-volant hosts. Commensal and introduced forms are ordinarily excluded, as are fleas from birds and bats because these hosts can transport fleas vast distances and thus confuse the overall picture.

The terminology and concepts of the zoogeographical areas as used in this text are classical ones but as modified by Traub (1972c) in order to treat high altitude extensions of one region into another. Revised systems have been published subsequent to the preparation of the tables in this paper, but it would take too much time and space to incorporate the changes (nearly all of which correspond to those I had independently developed). As in previous papers, the scheme of classification of mammals which has been followed considers the Cricetidae and Muridae as two separate families.

2 BACKGROUND

At least until the revolution in thinking recently wrought by the theory of plate tectonics, there had been general acceptance of the concepts of Matthew (1915, 1928), as extended and espoused by Simpson (e.g. 165, 166, 168, 169, 171-174) and others (37-39, 46, 82-84, 146-148), who hypothecated, or accepted, that the mammals of the southern continents were all descended from migrants from Eurasia or North America. Additional references were cited by Traub (1968, 1972c, 1972d), who stressed that the data on fleas strongly indicated otherwise. Actually, views contrary to those of Matthew and Simpson go back at least as far as the turn of the century insofar as concerns the origin of Australian mammals (183-185). According to Baker (1931), in 1898, studies on tapeworms were used to demonstrate the common origin of Australian and South American marsupials. Unfortunately, theories of austral origin, or of southern faunal connections, have often been deprecated, overlooked or ignored, especially when based upon parasitological evidence. Thus, the extensive works of Harrison (e.g. 50, 53, 55, 56), which date back as far as 1915, were not even mentioned in Darlington's classic opus (1963) on zoogeography. The large paper by Schwarz (1924) on the evolution and radiation of mammals, which includes distributions based upon continental drift, seems to have completely escaped attention. Wolfson's contentions (1948, 1955) about the effect of continental drift upon bird migration and on the distribution of birds likewise made little or no impression. Although Harrison (1914), on the basis of entomological data, claimed that the flightless birds, ostriches (Africa), rheas (South America) and emus and cassowaries (Australia) were related and that they arose in the Southern Hemisphere, they were placed in separate orders by the majority of ornithologists decades ago and remained there. Other parasitologists later provided evidence concerning lice, mites, tapeworms and nematodes that firmly suggested that the ostrich and rhea were closely allied (7, 8, 11, 160). Mayr (1957) was sufficiently impressed thereby to reconsider their taxonomic status and propose these two groups of birds had a common ancestor and that the ostriches reached Africa via North America and Asia. Cameron (1960) suggested an Antarctic route. However, in

1972 Mayr reverted to his former conviction that there were no direct faunal connections between the birds of Africa and South America; stated that it was 'a moot point' as to whether the rhea and ostrich are related, and claimed that there was 'no evidence whatsoever for any relationship' between the South American and Australian avifauna (139). Keast (1977) also expressed reservations about such possible affinities. However, the resurrection of Wegener's theory of continental drift (1912), effected by the new data on plate tectonics, and the consequent re-examination of the question of the ratite birds, led several ornithologists to propose that the ostriches, rheas, emus and cassowaries did have a monophyletic origin, and that perhaps the kiwis (New Zealand); the fossil moas (New Zealand) and aepyornithids (Madagascar) were of the same stock (27, 29, 154, 164), and utilized Gondwanaland means of dispersal (Rich, 1975; Cracraft, 1973a). It is ironic that these ornithologists reached those conclusions without considering, or at least referring to, the parasitological evidence.

Wagner (1932) claimed that the potential of fleas in studies on zoogeography was not being achieved, and alluded to his observations that there was a common fauna of desert-rodents and fleas (e.g. *Xenopsylla* and *Coptopsylla*) in the vast area ranging from North Africa to the Aral-Caspian and East Turkestan regions, and to southern Iran and the Caspian steppes. (Our own data shows that the fauna extends to the Indus River.) Wagner inadvertently proved his own point, because his findings were overlooked, and Ranck (1968) independently observed the faunal relationships, and proposed the terms 'Sahara-Sindean Faunal Region' for the area.

Biological observations on faunal relationships may receive more recognition nowadays because the data on continental drift and plate tectonics that have accumulated during the last decade have been so persuasive that some scientists who had been strongly opposed to the concepts or implications thereof now believe there may well have been faunal connections between the southern continents. Thus, in the noteworthy series on possible austral affinities among mammals which appeared in the Quarterly Review of Biology in 1968-1969, all of the authors (5, 19, 59, 68, 82-84, 147) strongly supported the concepts of northern origins and the immutability of the continents. However, by the time this series appeared in book form (1972), several of the authors (60, 85-88) accepted the idea of possible Gondwanaland links regarding at least marsupials. Among the other workers who also reversed themselves are Hoffstetter (64, 65) and Lavocat (106-108, 110, 111), regarding the possible derivation of South American caviomorph rodents from African phiomorphs ('hystrichomorphs'). Clemens in 1971 accepted an Antarctic route of entry for marsupials, in contrast to his views in 1968. Nevertheless, some of our outstanding authorities have not been convinced and prefer to still think in terms of stable continents. As Mayr (1972) has stated, 'No avian distribution is known at the present time that is more readily explained by continental drift than by land-bridges. Even where continental drift occurred, this apparently took place long before the current avian distribution pattern'. Simpson (1970, 1971) expressed similar sentiments about mammals. It must be stressed that the principles and distributions proposed by Simpson and Mayr are undoubtedly valid for a significant proportion of the world fauna, and it should also be recognized that some well-known scientists are opposed to the new global tectonics, e.g. Meyerhoff & Meyerhoff (1974a, 1974b) and most of the writers in the volume edited by Kahle (1974). Kerr (1978a) indicated that many workers in the USSR disagree with the theory of plate tectonics. While phytogeographers and paleobotanists predominantly support the concept of mobility of the continents (e.g. Raven & Axelrod, 1974, 1975), some eminent workers remain opposed to that view. For instance, van Steenis, when writing of the origins and distribution of *Nothofagus,* one of the most-cited 'examples' of southern connections, stated (1971): 'If continental drift ever occurred in the austral regions it must have taken place long before the Cretaceous'.

2.1 *Important features of the parasitological evidence*

Before citing and discussing the data on the zoogeography and affinities of fleas and lice and the pertinent implications thereof, it is necessary to consider such points as the nature and features of the parasitological evidence, commencing with some inherent drawbacks to this approach. If the hosts in question are widely separated geographically and their similarities are held to be only due to convergent evolution, as in the case of the ratite birds and the phiomorph-caviomorph controversy, then data showing that they share several groups of parasites constitute powerful arguments in favour of a common ancestry for the hosts in question. If, however, as in the case of the South American and Australian marsupials, it is generally agreed that ultimately they did arise from the same stocks, then the parasitological evidence may be denigrated because parallelism is involved – mammals (and birds) with joint ancestry are expected to share some related parasites no matter how they reached their current ranges. In this potential category, for example, are myobiid mites of the genus *Archemyobia* and certain genera of trematodes and cestodes, all of which are shared by Australian and South American marsupials. The evidence must therefore be supported by many examples, and there should be some aspects that pertain to the proposed route itself, preferably affecting other kinds of hosts.

Another factor that should be borne in mind is that the value of siphonapteran data in zoogeography has been depreciated because of the purported low degree of specificity in fleas (91), but that attitude is unwarranted, as well as misleading, for some of the apparent examples of broad host-tolerance actually contribute to our understanding of zoogeography and evolution. Thus, specificity is relative, and while in some instances the close relationship is at the species level for the host, in others it applies to the genus, or even the family or order level of host. Certain fleas, such as those of insectivores, are highly specific, as are those with special adaptations serving to enhance attachment to the hairs of the host, e.g. *Leptopsylla segnis* (209). A multitude of examples have been cited wherein chaetotaxic modifications of fleas reflect the type of pelage or habits of the host (209), and such tailored features would have little survival value if the fleas were catholic regarding their food-source. For example, Siphonaptera with hosts whose habits pose unusual hazards for a flea that falls off the host, such as birds, bats or nocturnal, arboreal mammals, tend to have hyperdeveloped combs, spines and/or bristles. Some fleas are temporarily attached to the host by means of 'crowns of thorns'. Sticktight or sessile fleas, which actually or essentially lack combs and spines, have compensated by hyperdevelopment of tibiae and tarsi, coupled with unusually stout leg and plantar bristles and a leathery integument which permits abdominal distension upon engorgement with blood, e.g. the pygiopsyllid *Lycopsylla* of wombats, and the malacopsyllids. In the pygiopsyllid *Uropsylla,* in which the larvae are highly exceptional in being parasitic on their host, the Australian *Thylacinus,* there are both a hyperdeveloped pronotal comb and such specialized tarsi as well as other modifications. Such adaptations reflect some attributes of the host and entail a conspicuous degree of specificity to that particular mammal. It is true that certain fleas, such as *Xenopsylla,* feed on a variety of theraphions that may enter their particular microhabitat, but they are exceptional. Other highly specific fleas include the vermipsyllid genus *Chaetopsylla,* which is virtually restricted to den-inhabiting Carnivora in the Holarctic region (primarily Palearctic), and the coptopsyllids, which are gerbilline parasites in the Palearctic desert region.

An important feature about specificity in fleas that has been generally overlooked is that there is a definite phylogenetic factor involved. If the specificity of a flea is not limited to the species level of the host, then the association may concern a higher taxon, but only exceptionally is it with mammals belonging to another group entirely. For example, ceratophyllid fleas are characteristically parasites of sciurids and certain cricetids (205; Traub et al., in prep.), and no members of that family infest soricid insectivores, and there is but one on talpids. Similarly, only one or two leptopsyllids are found on insectivores. Such relationships are to be expected, because the main taxa of Siphonaptera presumably have evolved simultaneously with their particular hosts and have accompanied them on their peregrinations ever since, and hence the fleas must be attuned physiologically (and often morphologically) and ecologically (including larval requirements) to those precise groups of mammals. As a corollary development, those groups of fleas may later become adapted to a newly emerging set of hosts and then evolve with them, but only rarely can they 'regress' and switch to theraphions of a lower phylogenetic class, perhaps because 1) as a result of co-evolution, they are too highly specialized re physiological requirements or adaptations to the vestiture or other features of their original hosts or 2) the more primitive hosts are already in equilibrium with an ectoparasite fauna of their own which occupies all available 'niches' associated with that group of mammals and which brooks no competition. Thus, primitive mammals tend to have primitive fleas (e.g. hystrichopsyllids on insectivores; certain pygiopsyllids on marsupials), while the youthful hosts, evolutionarily speaking, bear the most recently evolved fleas as characteristic parasites (204, 209). In other words, both host and parasite ascend the evolutionary tree together. Thus, microtine-fleas are much more commonly found on peromyscines than vice versa, while sciurid-fleas like the *Neopsylla setosa*-group have secondarily adapted to murines, but true murid-fleas (e.g. some genera of pygiopsyllids such as *Metastivalius* and *Papuapsylla*) rarely include species that are true marsupial-fleas, but it is noteworthy that when they do, the spines of the pronotal comb become secondarily modified, i.e., stiletto-like, in the typical manner seen in the more primitive fleas of these coarse-haired marsupials (204, 209, 211). Apparently, there has been sufficient contact in time and place to permit adaptation and the consequent imprint of a marsupial host upon the framework of a murine parasite. Such a regressive evolutionary trend is exceptional, and the general tendency is to either remain with the aboriginal group of hosts or else switch to a newly-emerging and higher level of host, again as a result of some co-evolution. It is these reasons, in my opinion, that account for the virtual absence of ceratophyllids and leptopsyllids on shrews occupying the same runways and burrow-systems as their regular hosts. The association between primitive fleas and primitive host applies to individuals as well as phylogenetic taxa. Shrew-fleas are only rarely found on other kinds of hosts and the same is true of mole-fleas. Moreover, such genera very seldom include species which have adapted to other hosts, and the exceptions are of special interest, e.g. *Palaeopsylla setzeri,* the only rodent-flea amongst 36 species of *Palaeopsylla* occurring on moles or shrews, and one that infests a burrowing vole in restricted parts of the Himalayas, where moles are unknown. All of these features are of definite value in evaluating zoogeographic data, whereas lack of specificity at the level of host-species may at times be irrelevant in that regard. This point of phylogeny of host and type of ectoparasite is discussed further on p. 158 after sundry pertinent data have been treated.

Another point that seems important but inadequately appreciated is that primitive taxa, both host and parasite, seem to be conservative regarding evolutionary change, particularly taxonomic ones. Of course, if they were not, there would be no primitive features for us to discern, but more is involved than that – they are slow to change at the generic and even the species level. Thus, undoubtedly ancient groups like the Australian stephanocircid helmet-fleas (206, 211) consist of just two genera and eight species in the midst of what appears to be a plethora of suitable hosts. The genus *Palaeopsylla* dates back to at least the Oligocene (36 m.y.a.) (149). *Ctenophthalmus,* by far the largest genus in the Order, and which has adapted to many different kinds of hosts (p. 143), is another ancient taxon that has been so resistant to morphological change that taxonomists have had difficulty in even designating subgenera. The Mallophaga of marsupials are other examples of non-proliferating taxa. The shrews and many kinds of marsupials seem to have changed very little through the eons, as may be expected by their success in an environment that has likewise persisted. The king crabs and sharks also demonstrate this fundamental trait of continuously maintaining a proven model through the countless millenia. If the hosts remain unchanged, there is no pressure on the parasites to become modified either. This conservatism is in marked contrast to the recently evolved, advanced and highly successful forms like the murines (and their ectoparasites like *Leptotrombidium* chiggers), where rapid change is so much the rule that their classification is the bane of systematists (206, 215).

Certain species of fleas parasitize hosts that are highly unusual for the genus concerned, but this phenomenon apparently represents evolutionary changes and not lack of specificity. Thus, some adventitious species have switched from scansorial rodents to birds, e.g. *Lentistivalius insolli* and *Bibikovana* n.sp., and these show the structural modifications (199, 202, 204) which are characteristic for bird-fleas regardless of the family or genus of Siphonaptera involved. Other examples include the exceptional species in the large genera *Papuapsylla* and *Metastivalius* which parasitize coarse-haired marsupials instead of rats, as already mentioned. There are other examples of a shift to a second host which occupies the same microhabitat. Jellison (1947) postulated that the North American genus *Opisocrostis* was originally a *Cynomys*-flea in which some species 'later spread to the closely related genus *Citellus* whose habits are in many ways similar to *Cynomys*'. However, the genus 'has not reached the limit of the range of *Citellus* to the south, west or north and has not spread to *Citellus* in Asia'. Hopkins (1957b) accepted this hypothesis. Jellison (1945) alluded to several genera of North American fleas 'of which one or more species are characteristic of tree squirrels (Sciuridae) and one or more species are characteristic of deer mice, *Peromyscus,* or wood rats, *Neotoma,* both of which are in the family Cricetidae', e.g. *Orchopeas* and *Conorhinopsylla.* Holland (1964) also noted this 'sharing' of sciurid and cricetid hosts. Some *Peromyscus* actually inhabit old nests of sciurids and *Neotoma,* or live in tree-holes, and I believe that such habits enhanced the opportunities for the divergence regarding hosts. The same principle is shown by the odd species of *Caenopsylla* which infests *Vulpes* and other carnivores instead of the customary macroscelids or gundis in xeric areas, and which has extended its range as a result. Another example of this type is *Nearctopsylla brooksi* which is found on mustelids instead of insectivorans.

There is yet another kind of transfer that has occurred in the course of evolution, and that is a switch to a new host that is evolving in the area or which has just penetrated therein. An example of the latter is *Xenopsylla cheopis* which transferred from *Arvicanthis* in northern Africa to commensal *Rattus* (207). The other type is highly pertinent to the history of the marsupials and concerns a shift to placentals, which appeared on the scene millions of years later than the metatherians. Thus, the genus *Stephanocircus* (and the entire family of helmet-fleas, Stephanocircidae) primarily parasitizes marsupials, but some Australian species are found on native rats, and the shape of their spines and bristles reflect the change (209, 211). The sole genus of pygiopsyllid which occurs in the New World, *Ctenidiosomus,* is found in South America and Central America, but little is known about its hosts. It is undoubtedly closely related to marsupial fleas of Australia, but the few extant records are mainly from complex-penis type of cricetids. Its original host may have become extinct (p. 115) as may also have been the case for the helmet-fleas on cricetids in extreme southern South America, where marsupials are unknown, at least today. The marked radiation of Pygiopsyllidae on murids in New Guinea illustrates this proposition, since 1) those fleas are clearly derived from marsupial-Siphonaptera; 2) many of the genera share critical features and hence were placed in *Stivalius* until recently, points that are indicative of evolutionary youth and a common ancestry; 3) the murids post-dated by far the marsupials in the Australian Region and 4) the older types of New Guinean murids are believed (4, 19) to have been derived from but one or two original immigrants.

3 DATA AND OBSERVATIONS

3.1 *Affinities among Siphonaptera of the Southern Hemisphere*

3.1.1 *Lack of relationships at the generic level* – In considering faunal relationships among the Siphonaptera of the Northern and Southern Hemispheres or of the austral zoogeographic regions, it is imperative to note that the affinities among the southern continents themselves (Africa, Australia and South America) are at the level of families and subfamilies, and not that of the genera. The data also clearly oppose the Matthew/Simpson doctrines of the stability of the continents and the derivation of 'southern' mammals' northern forebears.

Excluding introduced forms, there are no genera of mammal-fleas that are found on four or more continents (regarding Europe and Asia as an entity – Eurasia). There are only two genera found on three: 1) *Tunga,* the chigoe or burrowing fleas, from the Neotropical region, southern China and southwestern North America and 2) *Ctenophthalmus,* known from Eurasia, North America and Africa. For both *Tunga* and *Ctenophthalmus,* the distribution encompasses only one of the three southern continents, but even so, as indicated below, continental drift was involved. No genus of fleas occurs on all three of the austral land-masses, and only two genera are found on two of them, viz., the pulicids *Echidnophaga* and *Xenopsylla* in Africa and Australia. However, unusual circumstances may possibly account for that singular distribution. Thus, some species of *Echidnophaga* infest birds and bats, and the genus could have been introduced into Australia from Africa by those far-ranging hosts, and there secondarily adapted to marsupials. (The wealth of species on marsupials there, and the occurrence of mammal-*Echidnophaga* in Africa and Asia militate against that argument.) Precedents exist for such a role by volant hosts – a variety of species of *Parapsyllus,* a genus which is undoubtedly a derivative of Neotropical rodent-fleas (Traub, 1972b), infest sea-birds of the coasts of Africa, Australia, South America, New Zealand, etc. Also, bat-fleas of the genus *Thaumapsylla* occur in Africa, Southeast Asia and the Asiatic-Pacific islands, clear to New Guinea and Australia. As for *Xenopsylla* in Australia, at most there is only one species (*X.australiaca*) associated with native hosts (murines) and it may not be a valid taxon. The other two Australian *Xenopsylla* are fleas of commensal rats: the cosmopolitan *X.cheopis*, and *X.vexabilis*, which ranges as far as Indo-China and has been introduced into Hawaii. *Xenopsylla* in Australia may therefore represent a relatively recent phenomenon.

The picture remains relatively unchanged even if the Indian subcontinent and the Indonesian and Philippine archipelagos and the Wallacean region were considered as austral areas. Only one genus of Ceratophyllidae (*Nosopsyllus*) and one Leptopsyllidae (*Leptopsylla*) extend from the Oriental into the Ethiopian region, and in each case only a very limited number of species are found in the latter area. One leptopsyllid (*Sigmactenus*) ranges from the Oriental region into the Australian. The porcupine flea, *Pariodontis*, and *Synosternus*, both pulicids, are found in the Ethiopian region and Asia to and including the Indian subcontinent, and probably illustrate the effectiveness of the land connections between Africa and Asia, which were in operation intermittently, and as recently as 18 million years ago. Genera like *Coptopsylla* and *Caenopsylla* are not germane because, while occurring on both of these continents, in Africa they are limited to the Palearctic portion.

Moreover, there are even relatively very few genera that inhabit both the Northern and Southern Hemispheres, and in most cases only one or two species are involved in the peregrination from the homeland. For example, one species each of the Neotropical genera of pygiopsyllids and stephanocircids, and merely six species of the rhopalopsyllids, a large Neotropical family, occur in the Nearctic region. It is clear then, that if there are faunal links in the Siphonaptera of Africa, Australia and South America, they are really not at the generic level.

3.1.2 *Affinities at the family, subfamily and tribal levels* – Data on the zoogeographic distribution of the genera and species of the main families of fleas are shown in table 1, which, as noted, excludes introduced or commensal species and the fleas of volant hosts and deals with the representation of the various families in each region (often designated as 'R' in what follows, with 'SR' designating subregion). In other words, the percentages indicated add vertically to approximately 100 per cent, with discrepancies due to fractions in percentages, and with the additional important qualification that certain genera occur in more than one area. These data include new but undescribed taxa known to me, viz., at least four new genera and over 110 new species.

From table 1, it is instantly apparent that there are considerable differences in the distribution of these families. The Stephanocircidae (helmet-fleas) for example are restricted to the Australian and Neotropical regions, with the exception of a South American genus which has one species occurring in the mountains of Central America. More than three-quarters of the genera and species of the Pygiopsyllidae are found in the Australian and Wallacean regions, but elements occur in the Ethiopian and Neotropical ones as well, with the South American genus reaching Central America. All the known genera of Rhopalopsyllidae are Neotropical but two of these (and but six species) have penetrated the Nearctic.

The Hystrichopsyllidae and Pulicidae, in contrast, have representation in all the regions, although the former is firmly established in the Northern Hemisphere, where 75 per cent of the genera are found, while 68 per cent of the genera of Pulicidae occur in the Ethiopian R, and a total of 27 per cent on the other two southern continents. Although 27 per cent of the pulicid genera are in the Oriental R, five of those six genera occur elsewhere as well.

Table 1. Numbers and percent of genera and species of some major families of fleas occurring in various zoogeographic regions. (excludes introduced species and fleas of volant hosts)

Region (genera)	Pygiopsyllidae 43 gen (246 sp.)	Stephano- circidae 9 gen. (38 sp.)	Hystricho- psyllidae 45 gen. (460 sp.)	Pulicidae 22 gen. (148 sp.)	Tungidae 2 gen. (17 sp.)	Rhopalo- psyllidae 10 gen. (88 sp.)	Leptopsyllidae 24 gen. (184 sp.)	Ceratophyllidae 39 gen. (323 sp.)
Australian	29^{2*} = 67% (183) = 74%	2 = 22% (8) = 21%	2 = 4% (2) = 0.4%	2 = 9% (13) = 9%	0	0	1* = 4% (2) = 1%	0
Wallacean	4^{3*} = 9% (5) = 2%	0	0	0	0	0	1* = 4% (1) = 0.5%	0
Neotropical	1* = 2% (5) = 2%	7* = 78% (29) = 76%	5 = 11% (15) = 3%	4^{4*} = 18% (6) = 4%	2* = 100% (13) = 76%	10^{2*} = 100% (92) = 99%	0	0
Ethiopian	2* = 5% (18) = 7%	0	3* = 7% (92) = 20%	15^{6*} = 68% (83) = 56%	0	0	3* = 12.5% (10) = 5%	2* = 5% (16) = 5%
Oriental	10^{4*} = 23% (44) = 18%	0	4^{3*} = 9% (14) = 3%	6^{5*} = 27% (11) = 7%	1* = 50% (2) = 12%	0	3* = 12.5% (9) = 5%	5^{2*} = 13% (41) = 12.5%
Palaearctic but not Nearctic	1* = 2% (1) = 0.4%	0	8* = 18% (45) = 10%	8^{6*} = 36% (33) = 22%	0	0	14* = 58% (94) = 51%	10^{2*} = 26% (65) = 20%
Nearctic but not Palaearctic	1* = 2%	1* = 11% (1) = 3%	15 = 33% (63) = 14%	4^{3*} = 18% (7) = 5%	1* = 50% (2) = 12%	2^{2*} = 20% (6) = 6%	2 = 8% (3) = 2%	15 = 39% (104) = 31%
Holarctic genera	0	0	10^{2*} = 24%**	1 = 5%	0	0	4 = 17%	8 = 21%
1. Palaearctic sp.	0	0	(188) = 41%	(1) = 0.6 %	0	0	(51) = 28%	(21) = 6%
2. Nearctic sp.	0	0	(58) = 13%	(1) = 0.6%	0	0	(9) = 5%	(42) = 13%
3. Holarctic sp.	0	0	(1) = 0.2%	0	0	0	(5) = 3%	(7) = 2%

* One or more genera, not Holarctic, occur in another region besides that indicated, and are tallied in each column.
A very few species occur in more than one region.

** Includes *Ctenophthalmus* and *Neopsylla,* which occur in another region as well.

( ) Numbers of species. No ( ) Numbers of genera

The Leptopsyllidae are primarily Palearctic, viz. 75 per cent of the genera (and 51 per cent of the species) occur therein, and of these, 58 per cent are absent from the Nearctic R, while 17 per cent of the genera are truly Holarctic. The four genera that are Holarctic include 51 Palearctic species (28 per cent of the known leptopsyllid fauna). However, there are three genera (12.5 per cent) in the Oriental R, and also in the Ethiopian, and in each case, one of these is also Palearctic. The Ceratophyllidae are overwhelmingly Holarctic, with 86 per cent of the genera and 72 per cent of the species found in those northern regions. Moreover, there are no ceratophyllids (of non-avian hosts) in the Neotropical, Wallacean or Australian regions, and only five per cent of the genera are Ethiopian and 13 per cent Oriental, while some of the genera in those last two areas are also found elsewhere. (The few such ceratophyllids in South America are really in extensions of the Nearctic R in the northern montane portions.)

Table 2 presents zoogeographic data for these families in another manner, viz. the fauna in each region is tabulated regarding its generic components, expressed in percentages, according to family. Thus, 92 per cent of the 25 genera in the Papuan subregion, and 63 per cent of those in the Australian SR, belong to the Pygiopsyllidae. (Since there is only one genus found in the Wallacean R which is not found elsewhere, that area is deleted from the table, for simplification.) The Hystrichopsyllidae, Pulicidae and Stephanocircidae each have two genera (12.5 per cent) in the Australian SR. In the Neotropical R, the Rhopalopsyllidae, Stephanocircidae and Hystrichopsyllidae are the dominant families, in that sequence. In sub-Saharan Africa, well over half of the genera are pulicids, and in the Malagasy SR 50 per cent belong to that group, while the Malagasy Leptopsyllidae are surprisingly manifest (33 per cent). For the 38 genera that are Nearctic but not Holarctic, 42 per cent are Ceratophyllidae and 39 per cent Hystrichopsyllidae. The former also dominate (39 per cent) in those restricted to the Palearctic R.

Table 2 also presents data on the transitional areas between the Northern and Southern Hemispheres, thereby facilitating analysis. The intermediate zones include: the northern mountains of the Oriental R where Palearctic elements enter; the mountains of Mexico, Central America and extreme northern South America, which I have deemed Nearctic (Traub, 1972c); and the southern portion of the eastern Palearctic subregion, where there is faunal interchange with the Oriental R. The column for the Oriental R in the section dealing with the southern regions, excludes the Oriental elements in the adjacent column. In what thus may be regarded as truly Oriental, the genera of pygiopsyllids (34 per cent) outnumber the other families by at least a two to one ratio, and the next most abundant group is the Hystrichopsyllidae (14 per cent). The Oriental pygiopsyllids are fleas of murines and sciurids, with the siphonapteran roots clearly in the Papuan subregion (Traub, 1972c, 1972d). As for the transitional areas, the fauna is predominantly northern – e.g. 57 per cent of the genera in the Oriental portion are either ceratophyllids or leptopsyllids, and two-thirds of the records in the little-studied southern part of the eastern Palearctic subregion pertain to those families. It is edifying to note how the Neotropical stephanocircids and rhopalopsyllids encroach upon the southern limits of the Nearctic R, and how few genera (and species – table 1) are so involved. There are only two genera of Tungidae known, and with only 17 species, it cannot be expected that they would dominate any area. Nevertheless, the distribution of these hypodermal species (e.g. *Tunga*) is noteworthy, since indigenous representatives are found in south China, California, Utah (a new species of *Tunga*), while 76 per cent of the species are Neotropical. Further, if the anomalous *Neotunga* is a tungid, instead of a pulicid, as some workers believe, the Tungidae would

Table 2. The various zoogeographic regions and certain areas and the numbers and percent of genera of eight major families of fleas occurring in them.

Family	No. of genera & species	Southern regions						Northern regions					
		Australian		Neo-tropical	Ethiopian		Oriental	Transitional areas			Palae-arctic	Nearc-tic	Hol-arctic
		Austral. subreg.	Papuan		Malagasy	Other	Other	Oriental northern Mts.	Nearctic temperate mid. Amer.	Palaearctic E. subreg. (SO.)	(excl. Holarctic)		
Cerato-phyllidae	39 (323)	0	0	0	0	2 11%	5 17%	3 43:%	12 33%	2 33 %	13 39%	16 42%	8 33%
Leptopsyll-idae	24 (184)	0	1 4%	0	2 33%	1 5%	3 10%	1 14%	0	2 33%	4 12%	2 5%	4 17%
Hystricho-psyllidae	45 (460)	2 12.5%	0	5 17%	1 17%	3 16%	4 14%	3 43%	10 32%	0	8 24%	15 39%	10 42%
Pulicidae	22 (148)	2 12.5%	1 4%	3 10%	3 50%	11 58%	6 21%	0	3 10%	0	8 24%	4 11%	1 4%
Tungidae	2 (17)	0	0	2 7%	0	0	1 3%	0	2 6%	1 17	0	1 3%	0
Pygiopsyll-idae	43 (246)	11 69%	23 92%	1 3%	0	2 11%	10 34%	0	1 3%	1 17%	0	0	0
Rhopalo-psyllidae	10 (88)	0	0	11 38%	0	0	0	0	2 6%	0	0	0	0
Stephano-circidae	9 (38)	2 12.5%	0	7 24%	0	0	0	0	1 3%	0	0	0	0
Totals	194 (1504)	±100 %	±100%	±100%	±100%	±100%	±100%	±100%	±100 %	±100%	±100%	±100%	±100%
	No. of genera	16	25	29	6	19	29	7	31	6	33	38	24

( ) Numbers of species   No ( ) Numbers of genera

Table 3. Distribution of pygiopsyllid fleas (43 genera and 246 species) of non-volant hosts.

Geographical range of genus	Genera		Species in those genera	
	No.	%	No.	%
Australian subregion alone	6	15	10	4
New Guinean subregion alone	17	40	91	37
Limited to both Australian and New Guinean subregions	4	9	57	23
Both New Guinean subregion and Wallacea alone	1	2	5	2
Australia, New Guinea and Borneo	1	2	12	5
Malayan and Indian subregions, Ethiopian region & Japan	1	2	6	2
Wallacean region alone	1	2	1	0.4
Wallacea and Malayan subregion	1	2	2	0.8
Wallacea; Indian and Malayan subregions	1	2	7	3
Wallacea; Malayan and Burmo-Chinese subregions	1	2	27	11
Malayan and Burmo-Chinese subregions	1	2	1	0.4
Malayan subregion alone	4	10	5	2
Non-Malagasy Ethiopian	1	2	17	7
Neotropical (Andean) and montane mid-American	1	2	5	2
Truly Palaearctic	0	–	0	–
Nearctic	0	–	0	–

then occur in Africa as well, suggesting, even more strongly, connections between the southern continents (some authorities include southern China in Gondwanaland (32, 189)).

The zoogeographical features of some of the families merit further analysis and table 3 presents more information on the Pygiopsyllidae, which includes 246 species. This table lists the numbers of genera in various designated areas, as well as the number of species in those genera. (The table does not purport to show the numbers of species in Japan or India, etc., but there are two genera reported for India, with a total of five species, while there is only one genus and one species listed for Japan.) As can be seen at a glance, the Pygiopsyllidae provide an excellent example of how the distribution of fleas suggest faunal affinities between the southern continents. There are no truly Nearctic or Palearctic genera, while representatives occur in sub-Saharan Africa, South America, Australia, New Guinea and Wallacea. As also is apparent from tables 1 and 2, the heartland is obviously the Australian Region and that is where I am convinced this family arose (Traub, 1972c, 1972d). A total of 29 genera (67 per cent of the total) occur in the Australian R and/or Wallacea, and it is significant that one genus (*Bibikovana*) with 12 species, is found in Australia, New Guinea and Borneo. *Bibikovana* is primarily associated with the more primitive forms of *Rattus.* There are five genera (12 per cent) in Wallacea and eight (19 per cent) that occur in the Malayan subregion, of which four (nine per cent) are bioendemic. One genus (*Lentistivalius*) is wide-ranging, and is known from Borneo, the Philippines (a new species), and the Indian subcontinent, Japan and Central Africa. This is the only taxon that could even conceivably be cited as evidence that the family arose in the Holarctic and moved south from there, but such an argument is specious in view of 1) the broad distribution of the family in the various parts of the Southern Hemisphere and the extremely sparse representation and lack of bioendemicity in the north, and 2) the adaptive modifications seen in *Lentistivalius,* which suggest an ancient history for the genus, with time available for spread to remote areas.

The distribution of the subfamilies and tribes of the family Hystrichopsyllidae is pre-

Table 4. Distribution of genera of subfamilies and tribes of Hystrichopsyllid, Pulicid and Leptopsyllid fleas of non-volant hosts.

Taxa and number of genera		Australian							
		Austra-lian	Papuan	Neo-tropical	Ethiopian	Oriental	Palaearctic not Nearc.	Nearactic not Pal.	Hol-arctic
I. Hystrichopsyllidae	(45)								
A. Hystrichopsyllinae	(4)			25			25		50
1. *Hystrichopsyllini*	(3)						33		67
2. *Ctenoparini*	(1)			100					
B. Anomiopsyllinae	(7)						29	71	
C. Ctenophthalminae	(6)			50	17*		17*	17	17*
1. *Ctenophthalmini*	(2)				50*	25*	50		50*
2. *Agastopsyllini*	(1)			100					
3. *Carterettini*	(1)							100	
4. *Neotyphloceratini*	(2)			100					
D. Dinopsyllinae	(1)				100				
E. Doratopsyllinae	(6)	33		17			17		33
1. *Doratopsyllini*	(3)						33		67
2. *Acedestini*	(1)	100							
3. *Idillini*	(1)	100							
4. *Tritopsyllini*	(1)			100					
F. Listropsyllinae	(1)				100				
G. Neopsyllinae	(11)					18*	9	64	9*
1. *Neopsyllini*	(3)					67*		33	33*
2. *Paraneopsyllini*	(2)						100		
3. *Phalacropsyllini*	(6)							100	
H. Rhadinopsyllinae	(8)						13	74	13
1. *Rhadinopsyllini*	(5)					20	20	60	20
2. *Corypsyllini*	(2)							100	
3. *Wenzellini*	(1)							100	
I. Stenoponiinae	(1)								100
II. Pulicidae	(22)								
A. Pulicinae	(5)	20*		20*	80*		20*		20*
B. Archaeopsyllinae	(5)				20* ETH 20 MAL	40*	40*		
C. Xenopsyllinae	(7)	14*	14*		86^{4}*ETH 14 MAL	28^{2}*	43^{3}*		
D. Spilopsyllinae	(4)			50^{2}*			25	50*	25*
(Neotunga)	(1)				100				
III. Leptopsyllidae	(24)								
A. Leptopsyllinae	(6)		17*		17* ETH 33 MAL	33*	17*		17
B. Amphipsyllinae	(18)					6 *	72*	11	17

Introduced species excluded. * Genus occurs in more than one region
ETH Africa south of Sahara. MAL Malagasy subregion. Figures in percentages

sented (table 4) because of its significance in evolution and zoogeography. This is an ancient family, as evidenced by: 1) Many primitive morphological features have been retained. 2) It occurs on all continents, and endemic tribes or subfamilies exist there. 3) It is diversified at the higher levels of classification, with nine well-separated subfamilies and 17 named tribes – far more than in any other family. 4) Many taxa are specialized, e.g. the highly modified, true nest-fleas, suggesting a long history. 5) There is a tendency to infest hosts of a primitive nature, and 6) many taxa exhibit a relict, scattered or otherwise distinctive distribution (197, 205, 206, 213, Traub et al., in prep.). The oldest, clearly defined fossil fleas, are members of this family (Peus, 1968), viz. two species in Baltic amber, but belonging to the modern genus *Palaeopsylla,* which thus has a putative age of 35-50 million years.

Of the nine subfamilies, five occur on the austral continents, and an additional two approach as close as Java (Neopsyllinae) and Central America (Rhadinopsyllinae). The Ctenophthalminae and Doratopsyllinae each occur on two of the southern continents (and in the Holarctic region as well) and two of the subfamilies are bioendemic. The distribution of the 11 endemic tribes shown in table 4 is noteworthy. Four are Neotropical (of which three are in the Chilean SR) and two are Ethiopian. Of the two Nearctic monobasic tribes, Carterettini is in the Californian subregion, and Wenzellini is in the temperate Middle American subregion; Paraneopsyllini is in the southeastern Palearctic region and two tribes are in the Australian R. Thus, nine are found in the Southern Hemisphere, and two approach that area. The family is also well-developed in the Holarctic R.

As shown in table 4, the subfamilies of Pulicidae also exhibit an austral distribution. Three of the four subfamilies, and the majority of the genera (and species) occur in the Ethiopian region. One of those subfamilies, Pulicinae, is also represented in the Neotropical and Australian regions, while another, Xenopsyllini, has constituents in the Oriental and Australian regions. The Archaeopsyllinae have their center in Africa, but with one species in the Indian subregion and a monotypic genus in Java. While the Pulicinae are mainly African, the genus *Pulex* is essentially Neotropical, where one species even has adapted to a subterranean host (a pocket gopher) and is accordingly virtually eyeless (Traub, 1950, 1972b). Only the Spilopsyllinae bespeak of a possible boreal distribution and these are leporid fleas and could have arisen in northern Africa and adapted to rabbits in the Mediterranean area and gone with those hosts to eastern North America via Atlantic connections (p. 149).

3.1.3 *Summary of data on austral relationships* – Even from this brief outline, it is apparent that the zoogeography of fleas does not jibe with what one would expect if the continents were immutable (at least since the evolution of mammals) or if the bulk of the mammals and ectoparasites had boreal roots. If the Matthewsian concept of the origins of the aboriginal mammalian fauna (and their ectoparasites) were correct, then it would be expected that the following would apply regarding the southern continents: 1) there would be little or no evidence of faunal affinity between the fleas (and mammals) of the Australian, Ethiopian and Neotropical regions (excluding introduced forms). 2) If there were such elements in common in the Southern Hemisphere, they would either be: a) relatively recent invaders, or b) relicts with a somewhat ancient history, and with roots traceable to the Northern Hemisphere and with signs of their route of entry. In either case, no southern region should be the obvious center of development for such a family or subfamily. 3) The bioendemic components of the fauna should be: a) primitive forms with demonstrable affinity to northern taxa; b) derivatives thereof, with tangible northern roots, even though long isolation may have resulted in the evolution of endemic groups; or c) relatively youthful emigrants from the north, or their descendants. 4) The locally evolved fauna on the southern continents should consist of a relatively advanced type as compared to the most primitive elements of the northern fauna (unless mass extinctions have completely eliminated, without trace, the ancestral stocks in the north). 5) The degree of differentiation at the higher levels (tribes to families) should be less than in the north, due to the significantly shorter period of time available for evolution.

In the Northern Hemisphere, as per the boreal origins of mammals, it would be anticipated that: 1) primitive fleas of various diverse groups would occur, even though perhaps scarce. 2) Those taxa which are shared by two or three of the southern continents would clearly have relatives in the north (unless eliminated by 'competition' with more recently evolved forms). 3) Large numbers of distinctive tribes and subfamilies of fleas would have evolved, in view of the long history of the area. 4) Advanced, obviously youthful forms at the tribal level and higher would also occur, but would not be the predominant components of the fauna. 5) The roots of the youthful families or subfamilies would be clearly evident in the higher local levels of taxa. 6) Relict groups would be common.

Actually, the data on the zoogeography of fleas are at odds with many of those basic expectations. Thus, as shown above, the Stephanocircidae of Australia and South America is completely unrepresented north of Central America and can only be a recent emigrant across the isthmus of Panama. The Pygiopsyllidae, with members in the Australian, Wallacean, Ethiopian and Neotropical regions, surely had its origin and center of development in the first region, and is wholly unknown north of Middle America and virtually unrepresented in the Palearctic R. It is inconceivable that if the pygiopsyllids had originated in North America, or had ever even occurred there, they could have disappeared without trace, for the family is flexible enough for members to have transferred to murids as they appeared on the scene in the Australian R and then evolve with them to produce a rich fauna of genera and species. Similarly, other pygiopsyllids readily and successfully switched to callosciurines in the Oriental R (Traub, 1972a, 1972c). Certain relicts (Hystrichopsyllidae, Doratopsyllinae) occur in Australia whose nearest relatives are thousands of miles distant (p. 115 et seq.), while allied forms (Ctenophthalminae) show an austral type of distribution at the subfamily level. South of the equator there are both primitive and relatively recently evolved forms which are unrepresented in the northern land masses, e.g. the pygiopsyllids *Pygiopsylla (s. str.)* in Australia and New Guinea and *Ctenidiosomus* in South America for the former category, and the New Guinea murid fleas *Traubia* and *Metastivalius,* for the latter. The numbers of subfamilies and tribes occurring in the Gondwana areas greatly exceed those on Holarctic continents (even in the same family), viz. 11 of the 15 subfamilies in table 4 occur on the austral continents, and those southern taxa are apt to be bioendemic and are always well separated taxonomically (and geographically) from their allies. Such a high rate of endemism may be ascribed to prolonged isolation but that explanation fails regarding the austral relationships. Very significantly, the groups of fleas (and their hosts) that show austral connections are the more primitive forms, e.g. stephanocircids and hystrichopsyllids, on marsupials. This is true of the Gondwanaland-derived fauna in general, e.g. other insects (44, 89, 127) and many groups of vertebrates (Cracraft, 1975).

Two families of fleas (Ceratophyllidae and Leptopsyllidae) are overwhelmingly northern, and are the dominant members of the fauna there. Moreover, these are among the most recently evolved groups of fleas (although the latter is relatively ancient time-wise).

On the basis of comparative anatomy, morphology, habits and behaviour, phylogeny of their hosts, I am convinced that the ceratophyllids are the most lately-derived group of fleas (209, Traub et al., in prep.), and the fact that no subfamilies seemed to have developed in the family fits in with this concept. It seems logical to presume that they have not been on the scene long enough to develop subfamilies, viz. compare with the obviously ancient hystrichopsyllids, with nine subfamilies. It also seems axiomatic that the ceratophyllids and leptopsyllids were boreal in origin. On the other hand, there are elements in the fragments of Gondwanaland that are obviously northern in origin and affinity, such as *Sigmactenus* in New Guinea. These represent relatively recent migrations, and are fleas of recently evolved hosts like *Rattus* whose means of entry are clear (p. 149 et seq. below).

3.2 *The origin of the Australian marsupials*

3.2.1 *Background* – The origin of the Australian marsupials is a fascinating question that is complicated by the facts that: 1) marsupials had radiated into all sorts of ecological niches in Australia, so that there were marsupial 'moles', 'cats', 'rats', as well as herbivores, etc., eons before immigrant placentals appeared on the scene; 2) there are no fossil or living metatherians known in Africa, Madagascar or Asia (including the Pacific islands north and west of Sulawesi); 3) marsupials are well represented in the North American fossil record (dating to at least Mid Cretaceous) and in the European one as well, although much later; 4) there is a rich metatherian fauna in South America, dating back to at least the Late Cretaceous (ca. 75 m.y.a.); 5) the oldest recorded marsupial in Australia is from the Late Oligocene (ca. 26 m.y.a.); 6) no herbivorous marsupials developed in South America (12, 13, 16, 21, 24, 25, 29, 47, 74, 86, 87, 91, 120-122, 129, 131, 152, 153, 177, 190, 192, 193, 201, 205, 206, 211, 218). A variety of theories have been proposed to explain some or all of these features, including 1) origin in North America, with dispersal through Alaska or Europe to Asia and then via the Pacific islands (e.g. 13, 82, 83, 133, 146-148, 166, 172, 217); 2) entry from South America via Antarctica, island-hopping en route (e.g. 14, 15, 22, 23, 25, 29, 30, 53, 55, 61, 62, 64, 74, 87, 137, 143, 152, 153, 155, 156, 183-185, 192, 196, 201, 205, 206, 211, 223); 3) postulation of a pre-existing Gondwanaland fauna, of which the marsupials persisted in South America and Australia, with other elements becoming extinct (47, 101, 103, 161, 218); 4) a lost Pacific continent serving as the source, with one part drifting towards Australia and another towards North America in the Cretaceous (e.g. 100, 129, 130, 145). Kirsch (1977b) thought that the marsupials may have arisen in Australia and then penetrated South America via Antarctica.

Croizat (1958, 1970) averred that the current distribution of animals and plants reflects the whereabouts of their ancestors in the Cretaceous, but he did not believe in 'centers of origin' or that dispersals took place since that period. Croizat regarded the marsupials in Australia and South America as being the results of parallel evolution occurring in ancestral isolated stocks. Mani (1974), while not concerned with marsupials, expressed similar ideas about parallelism accounting for disjunct, widely separated populations. Croizat (1958) also ascribed anomalous distribution to a Pacific land mass. Martin (1977b) modified his former Pacifica hypothesis by suggesting that marsupials may have originated on islands north of New Guinea that later became incorporated with the rest of New Guinea.

In order to meet the known facts, the sundry theories about Australian marsupials postulated such proposals as: 1) the marsupials becoming extinct without trace in Asia or

Africa (22, 38, 152, 155, 156, 173), 2) ecological, physical, physiological, spatial or chronological factors, or combinations thereof, serving to bar placentals from reaching Australia, regardless of route (22-24, 28, 29, 74, 120, 167, 172, 192); 3) dispersals by hopping the volcanic arc of islands between South America and Antarctica which existed from the Early Cretaceous (ca. 120 m.y.a.) and throughout that Period, or even to the Cenozoic (65 m.y.a. et seq.), and then entering into Australia (27, 29, 31, 35, 36, 74, 86, 90, 91, 192); 4) competition from placentals leading to the extinction of the marsupials in Europe and North America as well as in Africa and Asia (17, 18, 24, 38, 121, 173, 192). Some workers objected to this concept of 'placental chauvinism' and argued that marsupials were far more adaptable and competitive than had been assumed, pointing to the success of *Didelphis* in northern USA despite its being a recent immigrant, and to that of *Trichosurus* in Australian suburbs (73, 98, 99, 187, 218). Oddly enough, some of the most ardent supporters of the theory of placental superiority also believed that the 'inferior' marsupials succeeded in getting from mainland Asia to Australia millions of years before the placentals could do so.

Until the last decade, the theory of an Asian-North American source for the Australian marsupials was generally accepted by mammalogists and paleontologists despite the reservations by the workers already mentioned and in spite of the observations on definite faunal relationships among the southern continents reported by entomologists (6, 43, 86, 152, 153, 162) and herpetologists and ichthyologists (29, 31, 43, 86). Traub (1968, 1972c, 1972d) listed evidence from the zoogeography of fleas and mammals in an attempt to refute the hypothesis of an Asia/Indonesia route for marsupials. As indicated, it was the onset of overwhelming data on plate tectonics, from diverse fields like geography, geology, paleomagnetism, sea-floor spreading, paleoclimatology, biogeography, that converted so many to the idea that continental drift must have been an important factor in the present distribution of certain mammals, and led to the ultimate realization that: 1) The dates originally proposed, Late Cretaceous (73 m.y.a.) or Early Paleocene (64 m.y.a.) were far too recent. 2) At the critical time the marsupials were supposed to have entered Australia, that continent was far south of its present position and was remote from Southeast Asia and perhaps still affixed to Antarctica, close to South America. 3) Antarctica shared a tetrapod fauna with India, Africa and South America, and 4) at the pertinent times, there was an effective marine barrier to keep European marsupials from entering Asia, and another between North America and Asia (14-16, 27, 29, 74, 86, 125, 126, 152, 153, 192). Even so, Simpson in 1970 averred that 'continental drift did not then (Late Mesozoic) have any influence on land faunas and has little, if any, bearing on the present distribution of mammals' (176).

The theory that is by far the most favoured today is that the forebears of the Australian marsupials travelled there from South America in the Cretaceous, 'hopping' the volcanic belt of islands on the western side of the latter continent to reach Antarctica and then crossing to Australia, by-passing Africa en route. On the basis of albumin immunological data on marsupials and hyaline frogs, Maxson et al. (1975) set the date for the separation of the Australian and South American faunas as Late Cretaceous for the marsupials. If such a route is accepted, the problem of the origin of the South American metatherians remains. Tedford (1974) postulated they arose on that continent, and Clemens (1977) did not exclude that possibility, while actually favouring a North American origin, as do the majority of specialists (15, 40, 58-60, 121, 125, 147, 148, 166, 173, 177, 218). The consensus is that the penetration from North America was very early, with the Late Cretaceous-Early Paleocene usually indicated, but Hoffstetter (1972) felt that the ancient mammal fauna of South America, including the marsupials, by far antedates the Late Cretaceous. Galton (1977) regarded the question of Southern versus Northern Hemisphere as academic since there were land connections between Laurasia and Gondwanaland in the Upper Jurassic (160 m.y.a.). Hoffstetter (1971) cautioned against specifically designating any of these continents, or Antarctica, as the place of origin, while McKenna (1973) and Tedford (1971) believed in an austral birth-place but apparently felt it premature to name it.

Since the question of where the insectivores arose is germane to the discussion on marsupials and their fleas (pp. 116, 137), I point out here that it is generally accepted that there were Insectivora in the Cretaceous (15, 123, 167), and that experts now believe the Insectivora arose and differentiated in Asia and then dispersed into North America (15, 120). It therefore seems that the marsupials were a New World derivation and the placentals (or at least Insectivora) were Asian. Marsupials and insectivorans have been reported as occurring together in Early Cretaceous (100 m.y.a.) beds in Texas (178, 179), but some doubt exists about whether these placentals were properly identified to Order (Lillegraven, 1974).

3.2.2 *Stephanocircid and pygiopsyllid fleas* – The zoogeography of fleas does more than indicate relationships between the marsupials of South America and Australia – the data clearly point to austral connections between the two continents, and even suggest that the route of entry of the marsupials was from South America to Australia, instead of vice versa. The helmet-fleas, Stephanocircidae, well demonstrate these points, as discussed by Traub (1972c, 1972d) and Traub & Dunnet (1973). This group of highly modified fleas is limited to Australia (*s.str.*) and, essentially, to South America (one species in Central America), as indicated above, and is intimately associated with marsupials in each area. The stephanocircids are really unique in the Order and are distinct from even the Macropsyllidae, the second family of Australian helmet-fleas. There is no evidence that stephanocircids have ever been on the other continents or have aboriginally infested other hosts. Since, in Australia and South America, some stephanocircids are associated with murids and cricetids – and bear such imprints on their vestiture as discussed elsewhere (204, 209), it would appear that the family is adaptable enough to have switched to other hosts if marsupials had been in Asia, Java, Africa, etc., and had become extinct.

As has been pointed out (211), there are too many fundamental similarities between the South American and Australian stephanocircids for them to be the result of convergence. Holland (1964) also regarded the family as monophyletic. Yet there are sufficiently great differences between the two groups to clearly show they represent wholly separate subfamilies. Distinctions at that level, coupled with basic primitive features, are in accord with a separation and isolation since the Cretaceous. It is noteworthy that specialized structures like the crown of spines and 'helmet' on the head of stephanocircids can be such very ancient evolutionary developments. This is additional evidence that the order already had a long lineage prior to Early Cretaceous.

The distribution and host relationships of the Pygiopsyllidae also strongly support the idea that there were connections between South America and Australia and that the marsupials did not reach the latter continent from Asia. Not only is the Australian region the center of development for the family, as shown above (p. 107 and table 3), but the most primitive members are marsupial fleas, and two of these are supposed to represent distinct subfamilies. Of the 11 genera of fleas of non-volant hosts occurring in Australia, nine are associated wholly with metatherians. In New Guinea, all the marsupial fleas are pygiopsyllids. In Australia, they constitute the majority. All of the Siphonaptera-bearing Australian marsupials have fleas that are more primitive than those of murines, a point that reinforces the idea that the rodents entered Australia long after the marsupials. The exceptional marsupial pygiopsyllids that represent recent switches from murine hosts and which are of a higher evolutionary type (e.g. a few *Papuapsylla, Metastivalius,* etc.), have only been encountered in New Guinea. The pygiopsyllids found in Wallacea and the Oriental region and other areas remote from the heartland in the Australian R are later derivatives of Australian/New Guinean stocks, and there is nothing in the data to suggest boreal origins or penetration from Asia (Traub, 1972b, 1972d).

The data on pygiopsyllids, then, clearly point to an original fundamental association with marsupials in Australia. It is therefore highly significant that there is one genus of pygiopsyllid (*Ctenidiosomus,* with five species) that occurs in South America, and is unknown north of Costa Rica, while the family is represented by merely two species in the Palearctic region. A northern origin is precluded for these and other reasons (204). It is theoretically possible that *Ctenidiosomus* is the sole vestige of a former widespread Gondwanaland fauna of pygiopsyllids (and marsupials), and that competition with placentals led to the extinction of the bulk of the metatherians and their fleas. This seems most unlikely, however, since: 1) there is no real evidence to support this oft-cited argument of inherent placental superiority, as we have seen (p. 112). 2) There is quite a rich fauna of marsupials extant in South America, presumably sufficient for support of pygiopsyllids. 3) The stephanocircids have thrived in South America (seven genera, 30 species). 4) *Ctenidiosomus* is obviously closely akin to some Australian marsupial fleas, so much so that a direct relationship by immigration seems more likely than that the genus is the result of parallelism and remote ancestry, particularly since no other pygiopsyllids occur in the Neotropical fauna. 5) There is no contemporary or fossil fauna of herbivorous marsupials in South America, in contrast to Australia, but if the two continents had fully shared a common fauna, both metatherian carnivores and herbivores should have existed in South America. The customary explanation for this last point is that competition from eutherian ungulates, known to be present in the earliest Neotropical fauna, precluded the development of marsupial herbivores. If so, why were there no such condylarths or notoungulates in the Australian common fauna?

It seems far more plausible that there was no such widespread Gondwanaland fauna before the breakup of the super-continent, as is supported by the evidence on lice (pp. 121, 127). Thus, Australian marsupials and some South American metatherians and rodents have amblyceran Mallophaga, but such lice are unknown on mammals in Africa, Madagascar, etc. The South American genus of pygiopsyllid therefore apparently represents an immigrant from Australia, travelling by a tenuous or filtered route, after the emigrants from South America had established themselves. It has been pointed out that groups of rats must have moved back and forth between the archipelagos of Southeast Asia and Australia (205) and there is no reason to believe the mammal dispersal between South America and Australia must have been solely, rather than predominantly, in one direction. Indeed, Kirsch (1977b) suggested that the marsupials may have actually originated in Australia because of Archer's conclusion (1976) that the Australian thylacinids (marsupial 'wolves') were more closely related to the extinct South American borhyaenids than to their compatriot dasyuroids. Kirsch reasoned that a single entry from Australia could explain the current distribution, whereas the alternative route would entail two immigrations. I suggest the possibility that such a proto-thylacinid move to South America may have occurred after radiation in Australia of the original South American marsupial stocks. Such a thylacinoid dispersal could have transported the forebears of *Ctenidiosomus.* Most of the pygiopsyllid line in South America could then have become extinct with the borhyaenids. (It will be recalled that the true hosts of *Ctenidiosomus* are not definitely known, and that most of the few available records are from cricetids.) Regardless, the existence of *Ctenidiosomus* argues strongly for faunal connections between the two continents, and discussion of the routes of dispersal will be renewed (p. 134) after consideration of further evidence provided by fleas and lice.

3.2.3 *Data from doratopsylline fleas and their hosts*

3.2.3.1 Doratopsyllines, marsupials and insectivorans – The highly anomalous distribution of Doratopsyllinae fleas (family Hystrichopsyllidae) also suggests a history of faunal connections between the southern continents (Traub, 1972c, 1972d), as indicated below by a review of their affinities, distribution and host relationships and of pertinent features concerning their marsupial and insectivoran hosts. Other fleas infesting such hosts are also briefly discussed. As shown in tables 4 and 5, the subfamily is found on all continents except Africa, and in Australia there are reportedly two tribes, each with a monotypic genus, *Idilla* and *Acedestia,* both of which parasitize dasyurid marsupials. The monobasic Latin American tribe, Tritopsyllini, with a single genus *Adoratopsylla* and two subgenera, primarily infests didelphine marsupials, and the five species are all found in the Brazilian

Table 5. Distribution and hosts of species, genera and tribes of the subfamily Doratopsyllinae (Hystrichopsyllidae).

Region	Subregion	*Doratopsyllini*			*Tritopsyllini*	*Acedestiini*	*Idillini*
		Doratopsylla 5 species	*Corrodopsylla* 4 species	*Xenodaeria* 1 species	*Adoratopsylla* 5 species	*Acedestia* 1 species	*Idilla* 1 species
		Shrews	Shrews	Shrews	Marsupials	Marsupials	Marsupials
Holarctic		*	*				
Palaearctic 3 genera 6 species		4	1	1			
Nearctic 2 genera 3 species	Temp. Mid.Amer.		2**				
	Other	1	2**				
Australian 2 genera 2 species	Australian					1	1
	New Guinea						
Neotropical 1 genus 5 species					5		
Oriental							
Ethiopian							

* Genus is Holarctic, but no species are. ** One species occurs in more than one area.

subregion, with one of these extending to the Mexican subregion. The tribe Doratopsyllini includes three genera of soricid parasites, two Holarctic, and one, a monotypic Himalayan genus, parasitizing *Crocidura.*

The absence of Doratopsyllinae on talpids is noteworthy, particularly since 1) moles occur together with shrews in certain parts of the range of the latter, and 2) other genera of shrew-fleas infest moles as well, if the hosts co-exist (p. 138).

The two Australian genera are thus separated from other members of the subfamily by many thousands of miles – i.e., there are none closer than the Asian mainland and South America. In fact, there are no other hystrichopsyllid fleas in Australia, as shown in tables 4 and 5; none at all in New Guinea, and the closest any other hystrichopsyllid comes to Australia is Java and Borneo, where dwell a very few neopsylline and ctenophthalmine fleas, as mentioned below.

Since doratopsylline fleas are restricted to marsupials and insectivores, the distribution of these mammals is pertinent. Outside the Australian region, marsupials occur only in South America and North America (with only a few migrants found in the latter area). Native shrews are not found closer to Australia than Java and Sulawesi, the Philippines and Sri Lanka. There are none in South America, save for one species of the Nearctic *Cryptotis,* restricted to the northern portion thereof. Some endemic genera of shrews occur in the Ethiopian R. There are no insectivores whatever in the Australian R, and the erinaceids (gymnurans, etc., which are not known to carry hystrichopsyllids) extend no further southeast than Java, Borneo and Mindanao, while the true moles do not range south of the mountains of Malaya. Macroscelids (elephant shrews) and chrysochlorids (golden moles) are limited to Africa. No fossil insectivorans are known from the Australian region, and no fossil marsupials have been reported from Asia, or the Pacific islands. The doratopsylline *Corrodopsylla,* which infests soricids, includes four species, of which one is Palearctic, the remainder Nearctic, with one species occurring on *Sorex* in southern Mexico. This is the southernmost record for the genus, although I believe that it may occur in Central America or even north South America in the temperate Middle American subregion of the Nearctic, on *Cryptotis.* The five known species of *Doratopsylla* parasitize soricids. Four are Palearctic and one (*D. blarinae*) is Nearctic and is essentially limited to *Blarina,* although it is apparently much more restricted geographically than is its host, since the flea is not known south of Tennessee and Georgia, while the shrew reaches Florida and the Gulf states.

The intriguing question, then, is how the doratopsyllines reached Australia. It seems highly unlikely that they accompanied shrews from Asia and then became extinct, along with their hosts, south and east of Java, leaving no trace whatsoever. Marsupials are not known west or north of Sulawesi, and the same comments above apply to extinction of such hosts and their characteristic fleas in the Asian and Malaysian regions. Discussion of this point is resumed after a background review of the other hystrichopsylid fleas of insectivorans (p. 117) and the origin of the marsupials (p. 137).

3.2.3.2 Other insectivoran fleas – Hystrichopsyllid genera which include some species characteristically infesting insectivorans are found in the subfamilies Ctenophthalminae, Dinopsyllinae and Rhadinopsyllinae, in addition to the Doratopsyllinae, just treated. Of the six genera of Ctenophthalminae, three are in two wholly Neotropical tribes; one is Californian; one (*Palaeopsylla*) is mainly Palearctic, but ranges into Indonesia, as mentioned below, and one (*Ctenophthalmus*) is Holarctic (ranging to Central America) but is also very well represented in Africa and Asia. The Ctenophthalminae thus are found in the

Southern Hemisphere to a notable degree. Among the four ctenophthalmine tribes, some marsupial-fleas occur in the Neotyphloceratini (two genera and three species; Chilean subregion), and shrew-fleas and mole-fleas are found in the Ctenophthalmini (*Ctenophthalmus* and *Palaeopsylla*). The great majority of the 36 species of *Palaeopsylla* are Palearctic (as were the two fossil species), but four occur in Nepal and three in Yunnan and northern Burma, in the Oriental R. One shrew-infesting species is known from the mountains of Malaya, and two (including a new one) are montane in Java. The Oriental species can be considered as extensions of the Palearctic R into suitable montane habitats. There are at least 142 species of *Ctenophthalmus,* and about one-third (33 per cent) of these are in the Ethiopian R and about 80 (56 per cent) are Palearctic. Only about six per cent are Nearctic and there are none known in the Australian or Neotropical regions. The genus has adapted to a variety of hosts, including microtines, other cricetines, a few murines and two families of 'mole-rats'. The subgenus *Alloctenus* (three species) which infests shrews and exhibits modifications in the genal and pronotal spines characteristic of shrew-fleas (Traub, 1972b; Traub & Barrera, 1966), is Mexican. Similar chaetotaxic variations are seen in our two new species of shrew-fleas from Ethiopia belonging to the very large subgenus *Ethioctenophthalmus.* A central African shrew-flea, the monotypic subgenus *Idioctenophthalmus,* is even more modified. There are 29 species of *Dinopsyllus* (including three new ones) and all but one are in sub-Saharan Africa; the exception is in Madagascar. At least one species is a true shrew-flea; the remainder occur primarily on rats. The rhadinopsylline insectivoran fleas are Holarctic. As for other hystrichopsyllid fleas occurring on southern continents, the subfamily Neopsyllinae is represented in Java and Malaya by two genera: *Rothschildiana* and *Neopsylla,* and the latter also occurs in Borneo. Both are restricted to rats in the foothills and mountains and usually occur on native hosts (as discussed below on p. 150). *Neopsylla,* which also parasitize rats and sciurids elsewhere, is a Holarctic taxon (vide table 7, p. 152). The monobasic Listropsyllinae is in the Ethiopian R, primarily infesting murids.

Insectivorans like macroscelids, erinaceids and chrysochlorids possess Siphonaptera belonging to completely different families, if they carry specific fleas, e.g. Chimaeropsyllidae and Leptopsyllidae for the macroscelids (which some authorities place in another order); and Pulicidae for the hedgehogs. There are no characteristic fleas recorded for the golden moles and gymnurans (except for *Hylomys,* which carries a leptopsyllid).

3.2.3.3 Affinities of the Australian doratopsyllines – This review makes it clear that although hystrichopsyllids occur on all continents, and have been in Africa, South America and Australia long enough to have established endemic subfamilies and tribes, the zoogeography of the living representatives provide no direct clues to account for the presence of *Idilla* and *Acedestia* in Australia. Their nearest relatives are too remote to indicate a route of entry. A truly detailed consideration of comparative morphology of doratopsyllines might shed light, but this has not yet been done. When Smit described the hitherto unknown male of *Idilla* in 1962, that sex of *Acedestia* had not yet been discovered. Dunnet & Mardon (1976) prepared illustrations and diagnoses of the male of *Acedestia,* but did not discuss zoogeographic or phylogenetic implications. Smit (1962:114) suggested that marsupials and Insectivora originally 'shared the same proto-doratopsylline flea . . . (and) must have retained it after the separation and evolution of Didelphoid marsupials into Didelphids in America and Dasyurids in the Australian region, the proto-doratopsylline fleas subsequently evolving into various genera, as their hosts did. The disappearance of Marsupials from Eurasia and northern America left the flea only placental insectivores as hosts – perhaps these insectivores acquired the flea from Didelphoids when the latter began to vanish in the Palaearctic area'. Smit went on to point out that the genal ctenidium of *Doratopsylla coreana* (Korea, Japan, Nepal; on soricids) and *Xenodaeria telios* (Himalayan *Crocidura*) resembles that of the Australian doratopsyllines more than that of other members of the tribe, and he thought that since Australian marsupials apparently originated in Asia, perhaps the closest living relatives of *Acedestia* and *Idilla* are those Asian shrew-fleas. However, as indicated above, the theory of an Asian route of entry for Australian marsupials has been discredited, and as shown in a series of my papers (204, 210, 213) and in the article on adaptations in this volume (209), the ctenidial modifications are regarded as examples of convergence and are often clearly adaptive. The details of their structure can hence be of little value in indicating phylogeny. Moreover, the head of *Xenodaeria* has some unique morphological features, and that genus is specialized in other ways (Traub & Evans, 1967). In the absence of Australian material for critical study, and on the basis of illustrations alone, I cannot say how much *Acedestia* and *Idilla* either resemble or differ from the tritopsyllines and doratopsyllines in fundamental ways. Nevertheless, there are cogent reasons to believe their origins lie in marsupial-fleas of the New World.

First, penetration via the Indonesian Archipelago can be excluded for the very reasons that rule out that route for the marsupials, e.g. the vast distance separating Australia and Asia in the Cretaceous and the various biological evidence outlined above and elsewhere (204, 206) regarding both marsupials and insectivorans. Secondly, derivation in Australia itself by parallel evolution from a basic autochthonous hystrichopsyllid stock of marsupial fleas (proto-doratopsylline or other) seems unlikely because: 1) A Gondwanaland fauna should include Africa, but there is no sign of doratopsylline-like fleas (or marsupials) on that continent; 2) There are so few genera and species of doratopsyllines in Australia; 3) In my opinion, both *Idilla* and *Acedestia* belong to the same tribe, despite their current classification, and hence the marked local radiation is apparent rather than real; 4) No other representatives of the family are present on the continent. Instead, importation of a single or greatly limited group of hystrichopsyllid fleas seems indicated, as has been suggested for the marsupials themselves.

Thirdly, critical study may disclose that fairly close phylogenetic relationships exist between the Australian and South American doratopsyllines. Still, it must be borne in mind that the difference in time and space between the South American and Australian representatives cannot be much different than between *Adoratopsylla* and *Corrodopsylla*/*Doratopsylla*. Then there is the intimate association of the subfamily with marsupials and Insectivora, coupled with 1) the widespread view that the former entered Australia via Antarctica and South America, and 2) the absence of any insectivorans (and their fleas) for vast distances in any direction from Australia. The absence of any known hystrichopsyllids between Australia and Java/Borneo, and the small numbers of such taxa anywhere in Southeast Asia likewise argue against Asian connections. Also, the level of differentiation between these tribes of marsupial doratopsyllines is in accord with the concept of common origin but long separation (since the Cretaceous), as in the case of the South American versus Australian stephanocircids (211).

It seems clear to me that the Australian doratopsyllines are immigrants from South America via Antarctica, even though it might be contended that the paucity of this Neotropical fauna bespeaks of recent entry into that region. In rebuttal one needs only to recall that only two genera and a total of two species are known in Australia. Moreover, the two South American subgenera may warrant status as full genera (as they had formerly), and there are five species described (all occurring in South America), with the likelihood that other undescribed taxa exist since fleas from that part of the world are inadequately known, especially those from marsupials. Inasmuch as only one species of insectivore has been recorded from South America, and that is a northern and relatively recent invader, it does not seem probable that the Neotropical doratopsyllines are a youthful group, derived from Nearctic shrew-fleas, especially since no *Corrodopsylla* have been reported south of Mexico and *Doratopsylla* is even more boreal in nature. Instead, a long history in South America is indicated, and the existence of endemic tribes of hystrichopsyllids on that continent supports this contention of ancient roots in South America.

The dearth of doratopsylline taxa in both Australia and South America may represent three phenomena: 1) extinctions of the aboriginal lines of hosts on those continents; 2) the relative inability of hystrichopsyllids to adapt to other hosts; 3) the unusual affinity of hystrichopsyllids for hosts with permanent nests. (The host versatility exhibited by *Ctenophthalmus* is exceptional, but even here – or anywhere in the family – there has been no switch to birds as hosts.) Further, in connection with the second point, it is reiterated that hystrichopsyllids tend to parasitize the more primitive of the locally available kinds of hosts (e.g. moles, shrews, microtines, etc.), and where they are associated with sciurids, for example, they are nest-fleas or else there is a dearth of locally available insectivorans. The special affinity of hystrichopsyllids for nests that are re-used by their hosts may be the evolutionary product of special requirements of the flea larvae or other stages, or else the result of the enormous length of geological time (available to their family) that is necessary for the evolution of the nest-flea habitus and mode of existence (Traub, 1972b). Strikingly, the great majority of nest-fleas belong to this family, and it seems pertinent that most marsupials lack a permanent-type of nest. Indeed, the marsupium and the concomitant style of metatherial life seems to automatically exclude a 'den' for most marsupials. In the case of the Australian metatherians, there is a potent fourth factor – the general inability of fleas to successfully parasitize vast-ranging, nest-free hosts like herbivores, such as the macropodids, for reasons already mentioned above and elsewhere (204, 209).

South America, in my opinion, is the aboriginal homeland of the entire subfamily Doratopsyllinae, and not just the marsupial-infesting components. This important point is treated below (p. 137) after discussion of the evidence on lice.

3.2.4 *Data from other fleas* – There are additional points about the zoogeography of fleas that contribute towards the view of Australian-South American marsupial relationships, although of a somewhat negative nature. If Australian marsupials were derived from Asian stocks, and the South American ones had northern roots, offhand it might be expected they would carry ceratophyllid or leptopsyllid fleas, which are the quintessence of boreal Siphonaptera. Instead, not only are those groups conspicuous for their lack of association with marsupials, but to a notable extent those families are either absent or scarce in the metatherian heartlands. Insofar as concerns fleas from non-volant hosts and non-cosmopolitan species, there are no records of ceratophyllids closer to New Guinea or Australia than Borneo or the Philippines, but undescribed species may occur on squirrels in Sulawesi (204). Even ceratophyllid bird-fleas have not been reported in New Guinea or Australia. Similarly, except for a few bird-fleas, ceratophyllids are unrepresented in over the southern half or two-thirds of South America, and what few members there are in the north are obvious spill-overs from the Nearctic (205, 224). Moreover, although marsupials occur together with rodent hosts of ceratophyllids in South America, and likewise in Central America and Mexico, where ceratophyllids abound, there are virtually no records of such fleas occurring on marsupials, even as strays. Tipton & Machado-Allison (1972) do not list any for Venezuela, and Tipton & Méndez (1966) report only a few specimens of such a rodent-flea on a marsupial in Panama. The extensive records available to me are likewise negative. As for leptopsyllids, they rarely co-exist with marsupials anywhere as natives, but where they do, e.g. *Sigmactenus* in New Guinea, they do not parasitize marsupials.

To some degree the marked dearth of records of leptopsyllids and ceratophyllids from marsupials illustrates the point made above about the scarcity of reports of fleas infesting hosts on a lower evolutionary scale, but there are obviously other factors as well. Since lagomorphs radiated in the Paleocene, and leptopsyllids are common parasites of hares and rabbits, it seems probable that these fleas were extant on various kinds of hosts in Asia at the time Simpson (1965) and others postulated that marsupials were in Asia en route to Australia. If so, a transfer to those hosts seems possible, but only one genus of leptopsyllid is found as far southeast as New Guinea, and it is obviously a relatively recent arrival with *Rattus* (Traub, 1972c, and below), providing another reason to believe there were no metatherians in Asia. The ceratophyllids probably arose in the Eocene or Oligocene with sciurids (Traub et al., in prep.) and hence appeared on the scene too late to encounter marsupials on their alleged dispersals from the Holarctic to Australia. Since neither they, nor their sciurid hosts ever got beyond Sulawesi, and their cricetid hosts apparently never even reached the southeast Asian mainland from the Palearctic, more doubt is cast upon Asian-Australian mammalian faunal connections, with the possible exception of murids (p. 150 below). Regardless, the distribution of the ceratophyllids and leptopsyllids are in accord with the concept that the marsupials either arose in the Southern Hemisphere or developed and radiated in South America and Australia without having acquired fleas of indisputable northern origins. It cannot be ruled out that ceratophyllids or leptopsyllids parasitized marsupials in Europe and North America in the Paleocene-Miocene, but if so, then they became extinct with their hosts there in the Miocene, and almost undoubtedly could not have reached Australian or South American marsupials because of the isolation of those southern continents at that time.

There is another group of Siphonaptera that is fairly common on Australian mammals, namely the pulicid sticktight flea, *Echidnophaga* (six species on marsupials or monotremes). Five other species of *Echidnophaga* occur in the Ethiopian region (including the cosmopolitan *E.gallinacea,* while five others are found in the southwestern and eastern Palearctic region. As already indicated, entry from Africa to Australia seems probable, but could have been effected by birds or bats, since native *Echidnophaga* have not been reported in South America, where its presence would be expected if dispersal had been effected by terrestrial hosts. Thus, Africa separated from South America while Australia, via Antarctica, was still attached to South America. There are no indigenous *Echidnophaga* in the Oriental, Wallacean or New Guinean regions or subregions. The overall distribution of the genus, coupled with the austral origin of the family Pulicidae (pp. 109, 142), effectively rule out an Asian path of dispersal to Australia.

Research on this subject has been handicapped by the still inadequate knowledge of marsupial fleas insofar as concerns both the Australian and Neotropical regions. It is surprising how few records there are from western Australia and from the interior. *Idilla* and *Acedestia* are still known only from a few specimens. Moreover, there are no records of specific fleas from 8 of 10 genera of kangaroos, which constitute a major component of the Australian metatherian fauna. In this case, the dearth may be due to the principle that large grazing and browsing mammals seldom have fleas, probably because of their enormous homeranges and lack of specific nest sites and the consequent reduced chance for suitable conditions for larval development or finding a host (209).

Riek (1970) reported two purported fossil fleas from Lower Cretaceous beds in Australia, and from their presence deduced that marsupials must have been contemporary. These claims are fraught with enormous significance in evolution and zoogeography and merit discussion. Smit (1972) categorically denied that one of these – a species with nematocerous type of antenna and long, thin hind-legs that were not modified for jumping – could possibly be a flea. I concur. However, the second species, which I have seen, definitely is a flea, and it seems incontrovertible that fleas were in Australia about 125 m.y.a. However, it does not automatically follow that marsupials were the host of this particular species since monotremes probably co-existed at the time and perhaps there also were proto-mammals of some other type that subsequently became extinct. Nevertheless, I too believe that there were metatherians in Australia in the Lower Cretaceous, if not earlier, based upon 1) the distribution and systematics of stephanocircid and other fleas, as outlined above and elsewhere (205, 206, 211); 2) the Late Jurassic-Early Cretaceous dating of the onset of rifting between east Antarctica and Australia (29, 74, 86, 154); and 3) the known occurrence of Cretaceous marsupials in South America.

3.2.5 *Evidence from lice* – The evidence from lice (Anoplura and Mallophaga) is of greater consequence to zoogeography than seems to be indicated in the literature and supports the conclusions herein presented regarding austral connections and other effects of continental drift. Nevertheless, there are inherent limitations in the data that must be borne in mind, namely 1) certain groups of mammals definitely have not yet been properly searched for lice. 2) lacunae exist in the information available so that it is often impossible to determine whether the lack of reports of lice from certain hosts means that the hosts are truly free of infestation by lice or whether they have been inadequately examined, especially by modern techniques involving maceration of the pelt, etc.; and 3) a definitive, widely accepted classification of all the lice, especially at the higher level, has not yet been presented (but see Addendum, p. 163).

Background data on the historical biogeography of the major hosts of pertinent Anoplura, e.g. insectivores and rodents, are presented in this section because conclusions based thereon are relevant to the origins and dispersals of the marsupials. The terms primary infestation and secondary infestation are used as per Hopkins (1949). In primary infestation, the condition is ancestral, dating back to primordial divergence of parental stocks and hence most or all of the related members of the host-group are parasitized by lice that are clearly kin. In secondary infestation, the lice on a host were acquired subsequent to the divergence of the host from the aboriginal stock, and the lice are obviously derived from a main branch found on other kinds of hosts.

Among the excellent articles on lice that contain data pertinent to the discussion herein but which do not mention zoogeography because they dealt with other subjects are those of Hopkins (1949, 1957a) and Clay (1957), dealing with host relationships, and those of Clay (1970, 1971, 1976), Ferris (1951), von Kéler (1971) and Kim & Ludwig (1978) on taxonomy. The notable example of the case regarding the ratite birds has been cited above, but analysis of the data on lice offers additional support for the theme in this article, as indicated below, following some prefatory generalizations about the Mallophaga (chewing-lice) and Anoplura (sucking-lice) of mammals. (All groups of birds have Mallophaga but none have Anoplura and birds will not be discussed here.) These remarks will concern only the taxa of mammals relevant to our purpose. The lice of carnivores and hoofed animals, etc., will be mentioned only in passing, since major taxa of those hosts occur on several different continents and/or have done so in the past, and there are inadequate data about their complex evolutionary history and geography. Also, in general, these comments apply only to those infestations which are clearly primary, not those that are secondarily derived and recent.

3.2.5.1 The distribution of some mammal-Mallophaga – With these qualifications, it can be stated that the following taxa apparently lack Mallophaga: Insectivora; Lagomorpha; all rodents in the Old World; all sciurid, murid, cricetid, heteromyid, dipodid or gliroid rodents. This stark observation seems both highly expressive and pertinent, since I am convinced that, with the possible exception of the murids, all of these mammals arose in the Northern Hemisphere. Two families of the primitive mallophagan 'suborder' Amblycera, are associated with marsupials: 1) the Boopidae, exclusively in Australia, where they infest solely marsupials, with the two exceptions of a recent transfer to dogs (Hopkins, 1949) and the newly discovered species found on the cassowary (Clay, 1971), and 2) the Trimenoponidae, wholly in South America or Central America, where they commonly infest marsupials. There are Amblycera on rodents, but they are restricted to South America, and there infest only caviomorph ('hystrichomorph') rodents, i.e. trimenoponids on Caviidae, Chinchillidae and Echimyidae; and Gyropidae on the same three, as well as on Dasyproctidae. Although the origin of these rodents is a matter of controversy, as shown on p. 138, there is unanimity of opinion that these are the oldest of the South American rodents, having entered in the Eocene or Oligocene. Of the rich fauna of cricetid rodents in South America occupying the same geographical and ecological niches, only one, *Scapteromys,* has Mallophaga and that genus of gyropid is considered to have recently transferred to this host. While members of the family Trichodectidae (suborder Ischnocera) are legion on fissiped Carnivora, Perissodactyla, Artiodactyla and Hyracoidea, they are found on only two groups of rodents, viz. an isolated branch occurring on the caviomorph Erethizontidae (New World porcupines) and a second distinctive taxon on the geomyids (pocket gophers). Clay (1970) believes that the suborder Ischnocera is more closely allied to the Anoplura (sucking-lice) than to the Amblycera.

There are few remaining mammalian Mallophaga and some of these are noteworthy and pertinent. The Xenarthra (edentates), together with the marsupials, constitute the most ancient of the extant inhabitants of South America, but no lice are known from edentates like anteaters and armadillos, and the sole recorded louse (a trichodectid ischnoceran) from sloths bears 'a marked likeness' to those on Hyracoidea in Africa (Hopkins, 1957a). Ischnoceran genera that are related occur on a Southeast Asian lorisid and on ceboid monkeys of the New World (70). An amblyceran gyropid occurs on ceboid monkeys in South America. The ceboids thus have two kinds of Mallophaga and it is of interest that some authorities believe that, like the caviomorphs, these monkeys have African ancestors (p. 138 below). However, the Anoplura (except for *Pediculus*) and Mallophaga found on South American monkeys do not occur on the African non-human primates. Peccaries (with living representatives essentially Neotropical and recently penetrating southern North America) have a gyropid louse (and an anopluran). (Fossil peccaries are known from the Old World: Asia

(Miocene); Europe (Miocene and Oligocene); and, fide Hendey (1976), South Africa (Pliocene).) There is a peculiar ischnoceran genus (Philopteridae) infesting Madagascar lemurs, but all other members of that family parasitize birds.

As indicated in this summary: 1) the mammal-Amblycera are basically entirely Neotropical or Australian, and 2) the only rodents with Amblycera are also Neotropical, and caviomorphs at that. Excluding the Mallophaga of wide-ranging hosts like feloids and the hoofed animals, and with but two interesting exceptions, all the remaining representatives of mammalian chewing-lice are restricted to the Southern Hemisphere, especially South America, as indicated above. The two exceptions are trichodectids (Ischnocera) of rodents and are either of South American origin, i.e. Erethizontidae, or have their center of radiation in Central America (Geomyidae).

3.2.5.2 Origins of the marsupial Mallophaga – Harrison (1916), in the same article in which he pointed out the monophyly of the rhea and the ostrich, predicted that Mallophaga would be found on South American marsupials and that they would be related to those found on Australian metatherians. He then described such a louse in 1922 and was satisfied that its relationships were with the Australian Mallophaga (52, 54) and Hopkins (1949, 1957a) concurred. Harrison (50, 52, 53) felt that the Australian family (Boopidae) was most closely related to the South American family Gyropidae, which are lice of rodents, rather than to the Neotropical Trimenoponidae, to which the South American marsupial lice belong (all are amblycerans). Harrison was convinced that all the mammal Amblycera had a common origin and were originally marsupial lice. Hopkins (69, 70) regarded the trimenoponids as close to the boopids, not to the gyropids, and presumably likewise considered these families as monophyletic, and felt that both the boopids and trimenoponids were primary parasites of marsupials. Hopkins (69, 70) indicated that the South American rodent Amblycera apparently evolved from trimenoponid stocks infesting marsupials. Vanzolini & Guimarães (1955a, 1955b) also believed that the closest relatives of the trimenoponids were the boopids, and considered the association of the trimenoponids with South American marsupials to be very ancient and primary. They stated they had reached their conclusions about affinities on purely morphological criteria, independent of host-data and inferences.

Clay (1970) felt otherwise, and reported that 1) in her opinion, five of the seven known boopid genera do not differ materially from one another and 2) the three main groups (superfamilies) of Australian marsupials all include members which are parasitized by boopids. Clay reasoned that if indeed the boopids were primary parasites of marsupials, then the extant boopids must have descended from a common ancestor infesting the progenitor of the three host superfamilies, probably in the Paleocene. If that were true, then, fide Clay (1970:94), 'it would be expected that the present Boopidae would show greater diversity and that each superfamily of the marsupials would have a specific genus or genera of parasite'. For similar reasons, Clay pointed out that there is too great a similarity between the Boopidae and the Menoponidae (avian lice), and too much dissimilarity between the former and the Trimenoponidae for the trimenoponids to be closer to the boopids than to any other family. She obviously strongly disagreed with Harrison (1915) who placed *Menopon* etc. in a separate family from the marsupial lice. Clay (1970:94-96) therefore hypothecated that the 'infestation of the Australian marsupials was comparatively late and that it arose from an avian menoponid stock . . . The Trimenoponidae and Gyropidae might have descended from a common amblyceran stock, probably avian-infesting, which became established first on the South American marsupials giving rise to the Trimenoponidae. Perhaps part of this stock, before great divergence had taken place, became established on the New World Hystrichomorph rodents, giving the Gyropidae; secondary infestation by Trimenoponidae on this group of rodents would also have taken place. It can be postulated that the Hystrichomorphs arrived in S. America without any amblyceran parasites. This hypothesis would explain the presence of both Trimenoponidae and Gyropidae on the New World Hystrichomorphs and the absence of members of both families on any of the Old World mammals. The Gyropidae were also able to establish themselves on other mammals which entered S. America at a later date.' In 1971, when describing a new genus of Boopidae, the first from a bird (the Australian cassowary), Clay mentioned the possibility that the genus could have originated on a marsupial (either recently or in the past) or else could have been derived from the ancestral avian boopid she believes gave rise to the Australian marsupial Amblycera.

Dr Clay is undoubtedly one of the World's leading authorities on lice and her views on their systematics and affinities merit and receive the utmost respect. At the risk of being presumptuous, but emboldened because Dr Clay did not mention zoogeography at all in the 1970 and 1971 articles, I make some comments about the theories cited above, but first stress that Murray & Calaby (1971) agreed with Clay and stated that the Boopidae and Trimenoponidae 'almost certainly had different origins'.

It is entirely within the realm of possibility that the Boopidae and Trimenoponidae represent two separate transfers from birds to marsupials. After all, rodent fleas of six families have effected a transfer to birds, and the combed representatives of these bird fleas share at least seven or eight features which are not seen on related forms (199, 202, 204, 209). It is not clear whether Clay (1970) was thinking in terms of static continents when suggesting a polyphyletic origin for the marsupial Mallophaga, but her theories would admirably explain the observed circumstances in an unchanging world. They also would fit if continental drift had occurred, but if one accepts the idea of the breakup of Gondwanaland and subsequent separation and drift of Australia and South America, then it would be far simpler to think in terms of a possible common ancestry for the marsupial lice of each of these continents (even if the ancestral form was a bird louse). Monophyly seems equally indicated for other marsupial ectoparasites, like the fleas and mites, and for the metatherians themselves. For example, how could one account for the existence of the two subfamilies of helmet-fleas except in terms of monophyly, regardless of how the marsupials reached Australia? There are far too many fundamental similarities throughout the body of these fleas for one to espouse convergence (211). If, however, the boopids and trimenoponids (regardless of the origin of the gyropids) had a common forebear, then the ancestral form must date back to at least Early Cretaceous, the time of divergence of these two groups of marsupials. This, then, would appear to even strengthen Clay's arguments about the relatively small degree of differentiation shown by lice of such purported ancient lineage. The crux is the inherent implication that geological age and the extent of morphological variations are correlated in members of a taxon, and this is a debatable point.

True, it has often been contended that ectoparasites provide admirable clues to the affinities of their hosts because the rate of evolution in lice, fleas, etc. is much slower than in the mammals and birds they infest. It was this postulation that led to the perceptive proposition that the ratite birds must be monophyletic, for example. However, there are examples where this is clearly not the case. Thus, *Lorentzimys nouhuysii* is infested by at least five distinctive species of *Astivalius* and by three specialized *Metastivalius* in the mountains of New Guinea, and not even subspecies of the host have been described. *Calomyscus bailwardi* (southwestern Asia) is associated with five or more species of fleas of *Phaenopsylla.* In contrast, some fleas do seem to have a much slower rate of evolution than the host, e.g. *Corrodopsylla curvata* may be collected from several genera and species of soricid shrews in the absence of other shrew fleas. As previously mentioned, the flea *Palaeopsylla* has not changed at the generic level for 50 million years (149, 206). With Anoplura, either some taxa are slow to differentiate into new genera or else they are highly adaptable in accommodating themselves to new kinds of hosts. Thus, the genus *Hoplopleura* is found on mammals of two orders, viz. lagomorphs (*Ochotona*) and rodents, such as sciurids and muroids (cricetids, including microtines and gerbillines; murids, including murines, hydromyines, etc.).

On this basis, it seems logical to suggest that the Mallophaga may have been equally slow to change morphologically while infesting marsupials for eons in both Australia and South America. It has been pointed out above (p. 101) that primitive groups are conservative and tend to form new genera and species at a low rate. It seems typical that the entire known subfamily Stephanocircinae (Australian helmet-fleas) consists of only two genera and eight species, and that one of these genera was first recognized as such in 1973 (211). Moreover, some of these are murine fleas, not parasites of marsupials. If it is argued that the subfamily must have been richer in the past and is relatively impoverished due to extinctions, then the same applies to the Amblycera. Actually, however, the existence of seven genera and 35 species, as in the Boopidae (93) suggests a fair degree of radiation. Moreover, the group, while mainly infesting macropodids, also parasitizes vombatids, dasyurids and peramelids (93, 144). The absence of boopids on the arboreal possum-like families appears significant but cannot be evaluated without more data. I do point out, however, that 1) many of those hosts are densely woolly and lice would not be expected in such pelage and 2) the koala lacks fleas and presumably other ectoparasites as well, and this may be due to the physiological side-effects of a diet of eucalyptus leaves.

The widely disjunct distribution within South America of the sole trimenoponid marsupial genus *Cummingsia* (three species), namely 1) Peruvian Andes, 2) a small focus in eastern Brazil, and 3) northeastern Brazil (219, 220) does not seem in accord with a relatively recent conversion from a bird louse. The host distribution likewise is anomalous for such an origin, i.e. two species on didelphids and one on a coenolestid. To me this disjunct host and geographical distribution, coupled with a limited fauna, suggest a slowly-evolving group with marked extinctions.

It is unfortunate that von Kéler died before his paper (1971) was in final form and that his ideas on the affinities between the Boopidae and Trimenoponidae were not presented. The abstract does refer to the Trimenoponidae as being the family most closely related to the Boopidae. Also, some intermediate features are cited in the descriptions in the text, and these may suggest possible links, but not necessarily so. Nothing in the article seems to rule out monophyly.

The above data and observations on lice lead me to agree with Harrison (50-52, 54), Hopkins (69, 70) and Vanzolini & Guimarães (219, 220) that the evidence indicates that the lice of South American and Australian marsupials have a common origin. These authors also state or imply that the marsupials are monophyletic as well, and I concur. K.C.Emerson (in litt.) also doubts that the boopids arose from avian menoponid lines, and feels that the two groups of marsupial lice are monophyletic. It seems to me that the taxonomic divergence displayed by the boopids and trimenoponids parallels the case for the two subfamilies of stephanocircid fleas and are in accord with complete isolation of the Australian and South American marsupials (and their ectoparasites) since the Cretaceous (regardless of how the metatherians reached Australia).

If mammal Mallophaga are indeed polyphyletic and are derived from avian lice, then the divergence(s) must have occurred in extremely ancient times, as witness 1) no genus infests both birds and mammals, and 2) with but two notable exceptions, even at the family level, the Mallophaga are restricted to either infesting wholly birds, or only mammals. One unconformity involves the single lemur-infesting species of philopterid in Madagascar, and this is so distinctive that some regard it worthy of family rank. The Philopteridae is a large family and is otherwise limited to birds. The second exception is the unique, monotypic genus of boopid found on cassowaries, and here the regular hosts (marsupials), like this ratite host, date back to at least the Cretaceous in Australia.

3.2.5.3 Conclusions from the data on Mallophaga – These observations strongly suggest that the Australian marsupials had their roots in South America and could not have dispersed from Asia. The data likewise support the concept of common origin of the two families of marsupial lice. The main evidence may be summarized as follows: 1) mammalian Amblycera are absent in North America, Europe, Asia, the entire Oriental region and Wallacea, which are areas astride the routes that would have been taken by marsupials dispersing to Asia from North America; 2) Amblycera can adapt to non-marsupial hosts, e.g. rodents and a peccary in South America; a cassowary in Australia; 3) there were barriers along the route at the time the movement to Australia was supposed to be in progress – i.e., the broad seas separating Europe from Asia from the Jurassic to Mid Eocene, and the Beringian bridge between Alaska and Siberia was not established until Late Cretaceous (126, 231); 4) the only mammal-infesting Amblycera that occur on hosts which are not marsupials are all in South America (e.g. on ceboid monkeys and caviomorph rodents) and are ancient inhabitants at that.

Even concerning Ischnocera from pertinent hosts, the data point towards the Southern Hemisphere, viz.: 1) reportedly related lice occurring on a Southeast Asian lorisid and a South American ceboid; 2) the alleged resemblance between a hyracoid (African) louse and one from a sloth (South American) and 3) the South American or Central American roots of the only North American rodents with Mallophaga. However, there are problems in interpreting such distributional patterns because of factors like parasitism on birds and the consequent dispersal over many parts of the world and possible transfer to mammalian hosts. Nevertheless, the absence of Mallophaga on groups of mammals which are clearly of northern origin, such as insectivores, sciurids, dipodids, gliroids, leporids, cricetids (except for the single recent transfer, in South America), etc., supports the concept of austral origin for all the chewing-lice of mammals. Thus, it is logical to deduce that the reason these hosts lack Mallophaga is because they arose in the Northern Hemisphere, where such lice were absent. Murids, a group which apparently originated in the islands of Southeast Asia from Palearctic forebears (as indicated by their Anoplura, p. 150 below), lack Mallophaga, and this reinforces the argument regarding South America as the homeland for mammalian Amblycera, since the rats have penetrated Africa, New Guinea and even Australia without having acquired chewing-lice anywhere. Murids never reached the New World, except as commensals in historical times, and hence their lack of Mallophaga is to be expected, under this hypothesis. The absence of primary Mallophaga on the rich fauna of cricetids in South America can thus be ascribed to the relatively late entry of the forebears of those hosts from the Nearctic region, and to the supply of Anoplura and other ectoparasites transported with them en route. In short, the presumed primary infestation of marsupials by Amblycera; their radiation on those hosts; and the restriction of mammal Amblycera to those parts of the world where the metatherians occur, strongly support the hypothesis of austral origin and dispersal of the marsupials.

3.2.5.4 Data from Anoplura – background – The place and date of origin of the Anoplura and its various components are highly pertinent to such points as where the marsupials and insectivores, etc. arose. Although it may be possible to supply answers as to when and where certain groups of Anoplura originated – and even here there are opinions differing by over 60 million years, for example – there are insufficient data to definitely resolve the critical question as to the nativity of the order itself. One of the problems is that too little is known about the classification of the precursors of the contemporary orders of mammals, e.g. the placentals which co-existed with the North American Cretaceous marsupials, and it is useless to speculate if they were infested with lice. Another difficulty is that if an order or family lacks primary infestations with Anoplura, it is often unclear as to whether this is due to an ancestral condition or to a subsequent development, such as the loss of sucking-lice due to competition with Mallophaga, which Hopkins (1949, 1957a) believed is the reason for the absence of Anoplura on virtually all Carnivora. However, I believe that the louse-free group of hosts may have arisen in a part of the world where Anoplura were unknown, and hence may already have had a 'balanced' ectoparasite fauna of its own upon first coming in contact with Anoplura-bearing immigrants. This may be why the Australian marsupials did not acquire Anoplura from the murids when they first appeared on the scene in the Oligocene or Miocene. Another possible reason is that the marsupials are physiologically or otherwise unsuitable for the murid-Anoplura, especially since they evolved millions of years earlier than the rodents (p. 129). Special physical features, such as an exceptionally tough hide or lack of hair have been cited (70) as reasons why the rhinoceros and hippopotamus and allies lack lice, as have deliterious habits, such as being amphibious (muskrats, beaver, hippopotamus, etc.), although the highly modified echinophthirids exist on seals and one otter despite their aquatic habits. As per the material that follows, it seems safe to conclude that the Anoplura arose in North America and that there were lice on rodents in the Paleocene and that there may have been Anoplura earlier (p. 131).

If I am correct about the provenance of the order and its existence by at least the Paleocene, then the following would be expected in consequence: 1) mammals that arose in other parts of the world would be expected to lack primary infestations with lice; 2) hosts that evolved before that era would also tend to lack Anoplura (p. 158); 3) mammals that arose in North America and were contemporary with, or subsequent to, the initial radiation of the Anoplura, and which were in contact with hosts infested with sucking-lice, would come to harbour such lice, e.g. later groups of rodents; 4) the derivative of such hosts, dispersed elsewhere (e.g. Xerini in Africa, Funambulini in south Asia and Africa, South American myomorphs), carried their Anoplura with them; 5) secondary infestations would occur on other kinds of mammals in contact with rodents and other louse-bearing hosts, but only to a limited degree; 6) transfer to a more primitive host would occur, but such would be uncommon. The observed facts seem to fit these deductions, so much so that the associations and correlations seem fundamental. Thus, groups of mammals that apparently originated in the Southern Hemisphere completely lack Anoplura, viz., monotremes, marsupials and all of the many taxa of edentales. Other taxa of mammals which are currently restricted to the Southern Hemisphere but which may have roots ultimately extending to the north prior to the origin of the Anoplura, also apparently lack sucking-lice, e.g. the insectivorans like solenodontids, tenrecids, potomogalids, and chrysochlorids. Perhaps Carnivora like viverrids and ferret-badgers are in this category. Other denizens of the Southern Hemisphere for which Anoplura have not been reported include the tapiroids, the rhinoceratoids, and hippopotamids. There are some indigenous austral hosts with sucking-lice but either these rodents have an ultimate boreal ancestry, or the Anoplura fauna is very limited and clearly secondary. These lice are all polyplacids (but not *Polyplax*) viz. one genus on lorises, one on lemurs and two on tupaiads, and bioendemic genera on 'hystrichomorph' rodents from Africa or South America (p. 131 and 156). Kim & Ludwig (1978) regarded this group as having 'coevolved with rodents', and the infestation of primates and tupaiads as having 'occurred in recent evolutionary times'. The austral polyplacids thus have boreal roots. The primitive status of their hosts (especially among rodents) and the disjunct distribution of both lice and mammals lead me to believe these polyplacids must be among the oldest extant Anoplura. Another instructive case is provided by the distinctive monotypic family and genus of Anoplura on the flying-lemur (Dermoptera), whose austral distribution, i.e., restricted to Southeast Asia, the Philippines, etc., appears anomalous for a mammal with Anoplura. Actually, Dermoptera existed in the Paleocene and Eocene in North America and hence the extant louse may be a relict of a former broad distribution.

3.2.5.5 Anoplura and Insectivora, Carnivora and Lagomorpha – The Insectivora, with presumed origin in Asia in the Cretaceous (15, 16, 120, 123) well antedate the rodents, which are Paleocene (229). The oceanic barrier which separated Asia and North America was terminated about 70 m.y.a. when a continental collision made possible the Beringian bridge (McKenna, 1975) which had formerly (recapitulated by Simpson, 1965) been regarded as the route for intercontinental faunal exchange since the dawn of the era of large tetrapods. Such a Late Cretaceous date for the portal could possibly account for Insectivora of Asian descent in North America in the Cretaceous as well. Significantly, Insectivora are only sporadically infested with Anoplura, i.e., macroscelids, on some Old World soricids and New World talpid moles (Scalopinae). The macroscelid lice are secondary infestations (97). Only two of the 30 *Polyplax* listed by Ferris (1951) infest shrews, and such soricid lice are known from Europe, Southern Asia and central Africa (70). The species are related to murid lice, whose hosts abound locally. Another genus (two species) is known from Old World shrews, and is in the Hoplopleuridae, a family whose evolution 'centered in Rodentia' (97), and also includes a few representatives that also switched secondarily to lagomorphs and prosimian primates. The absence of lice from New World shrews is striking (although it conceivably may represent inadequate collecting) and fits the concept that the order and perhaps the family as well, arose in Asia at a time when no Anoplura were present. Singularly, there is a genus of Hoplopleuridae which is found on Nearctic moles (Scalopinae), but has not been reported from the Palearctic. This may represent a transfer from the ubiquitous Nearctic rodents, due to long association in a common habitat. (The Scalopinae apparently have had a very different history than the Talpinae moles, because 1) the former are limited to North America and eastern Asia, with a probable Nearctic center of origin, and the latter to Europe and Asia and 2) the rich ctenophthalmine genus *Palaeopsylla* (essentially restricted to shrews and moles) is limited to the Palearctic, and the records of this genus of flea on scalopines in Japan may be due to the occurrence of talpines in the area.)

According to published data, the fissipede Carnivora lack Anoplura, save for three exceptions: one species on the dog and arctic fox; one on the South American wild dog, and one on a North American river otter (45, 97). It has been suggested that this marked dearth is a secondary condition – all the Fissipeda having had lice originally (69, 97) but I find it hard to believe that so large and varied a group as the civets, mongooses, felids, other canids, ursids, procyonids, remaining mustelids, etc., found in so many diverse areas and habitats, could all have lost their Anoplura, whether because of competition with Mallophaga or not. This skepticism is encouraged by a fourth exception, kindly provided by K.C.Emerson (in litt.), namely, the occurrence of a species of *Linognathus* on the fox, *Vulpes ruppelli* in Iran and Mauritania, which is clearly derived from a louse of a hoofed animal. If competition between the two kinds of lice were the answer, why should 1) some ostensibly suitable hosts like tapirs lack both kinds of lice (assuming the absence of records is a valid observation) and 2) certain Artiodactyla like bovids and cervoids have both Anoplura and Mallophaga? It is easier to believe that the Carnivora never had any until the four exceptions above acquired them secondarily, just as it is generally accepted that the boopid mallophagan occurring widely on dogs represents a transfer from Australian marsupials. Why should these carnivorans be the only ones to have retained their original infestations, particularly when the otter louse is placed in the same family as the seal lice and has the same adaptations for an amphibious existence that they do? Also, the Carnivora are an ancient group, dating back to the Paleocene, and representatives of modern families like the mustelids, viverrids, canids and felids were in Eurasia or North America in the Late Eocene (102), whereas the majority of contemporary rodent families like Cricetidae, Heteromyidae, etc. reportedly did not appear until the Oligocene, and the remainder in the Miocene (229). The places of origin for the families of Carnivora are apparently unresolved but their current distribution suggests that most, if not all, the major taxa were originally Eurasian rather than North American. If so, that could be the main reason for their lack of Anoplura. The possible significance of the date of origin in this regard seems to be illustrated by the Aplodontidae. The contemporary representative, the mountain beaver (North American) has no lice, which is highly unusual among Nearctic rodents, but it presumably arose in the Eocene, long before other existing groups of rodents. Of course there can be no carnivores in the absence of prey, and in the provenance of the Carnivora there obviously must have been some mammalian food-supply that was more numerous than the predators. Perhaps they were insectivorans, proto-rodents or some sort of herbivores, and at least those in North America may have harboured lice, but at present there is no way of knowing.

The lagomorphs' homeland is in the Old World, presumably in the Paleocene (146). They have few primary lice. *Ochotona* have *Hoplopleura,* apparently derived from rodent members of the genus. Only one genus of louse, a polyplacid, is found on leporids. The lice on Brazilian hares are secondary and recently established from rodents (70). The dearth of Anoplura and the affinities with rodent lice, fit with what is expected of a taxon that did not arise in North America.

3.2.5.6 Anoplura and Rodentia – In contrast to the Insectivora, Carnivora and Lagomorpha, the Rodentia in general have a rich fauna of Anoplura, although a few taxa lack sucking-lice. The history of the rodents is here briefly reviewed as a prerequisite for discussion. Wood (1977a) suggested that the order arose in North America in the Paleocene, probably in the southeastern part of the continent. The earliest known fossils are all paramyids, from western USA, in the Late Paleocene (ca. 59 m.y.a.), and by that time the basic characters of the order were already established. In the Early Eocene (ca. 54 m.y.a.) western North America and Europe had an almost identical paramyid fauna at the generic level, accomplished by facile terrestrial transatlantic connections (104, 125, 126, 231). This land bridge was severed about 49 m.y.a., in Mid Eocene, and has remained inoperable since. Another possible transatlantic route, the 'Thulen Bridge', which may also have functioned in Early Tertiary, also failed by this time (126). The Beringian bridge was already operable at the time the rodents arose and Wood (1977a) felt that 'some migration must have taken place, presumably by the Middle Eocene, between the two continents'. However, it has only recently become appreciated that there were vast marine barriers (the huge Obik Seaway and the Turgai Straits) between Europe and Asia from the Jurassic to Mid Eocene, and hence no faunal exchange seems to have occurred between those two continents until the seas dried in the Oligocene (126, 231). Rodents rapidly exploited these Europe-Asia and Siberia-Alaska routes.

In the Eocene, then, there was traffic of rodents between North America and Europe, but it was soon to cease, while the dispersals between North America and eastern Asia continued. The paramyids, now extinct, were sciuromorph rodents (231, 232) and were represented in the earliest Asian rodents (Eocene), but the known Asian rodent fauna consisted primarily of ctenodactylids (232), a paramyid-derivative that today is known only in Africa. The rodents radiated rapidly in North America, and the first recorded non-paramyid was another sciuromorph group now extinct, the Sciuravidae, appearing in Late Eocene. Extensive dispersal into Asia of North American rodent stocks took place at the end of the Eocene, e.g. the cricetids, since by that period the Alaskan-Siberian connections were better established than formerly. Among Oligocene emigrants to Asia were additional cricetid lines, and presumably, zapodids. 'The Bering migration route also explains the Holarctic distribution of several Late Tertiary and Pleistocene microtine genera' (231). Paramyid rodents reached the Indian subcontinent 'presumably from northern Asia, by the Mid-Eocene' (231).

The original European rodents (paramyids) were immigrants from North America and were already present in Europe by earliest Eocene. Evolution there in isolation in limited areas, followed the disruption of the land connections with North America in Mid Eocene (ca. 45 m.y.a.) but, except for the glirids, it is apparently not clear which of the modern European taxa descended from the original paramyids or from the paramyid-derived families that arose in Europe. When connections were first established between Europe and Asia in the Early Oligocene (ca. 38 m.y.a.), faunal exchange became possible. Asian emigrants to Europe included the Cricetidae, the Eomyidae, and presumably the Castoridae and Sciuridae, all of which had apparently arisen in North America (Wood, 1977a).

Further analysis of the history and movements of the rodents is not feasible or really germane to the questions under consideration in this paper, but three remaining points are stressed: 1) Although the ancestral homeland of the murids is unknown, the rats and their allies are an evolutionarily youthful group (probably Miocene, *fide* Simpson, 1965) that has succeeded in penetrating and radiating in all parts of the Old World except for areas of extreme aridity and low temperature (and even in such biotopes some unusual murids may occur). 2) With the exception of the caviomorphs, there were no rodents in South America prior to the opening of archipelagic or land-connections with Central America in the Late Miocene (ca. 12 m.y.a.) or Pliocene (ca. 5 m.y.a.) (p. 139). Thereafter there was a large influx of Nearctic rodents and other mammals (59, 60, 147, 148, 173,

175). 3) The African rodents are supposed to have Asian roots, but no African records exist before the Oligocene (19, 20, 231, 232). The question of possible African origin of the caviomorphs is treated elsewhere (p. 138).

The Anoplura associated with rodents may be briefly outlined as follows. The Sciuromorpha are primarily infested with enderleinellids, and certain of these genera of lice are confined to specific groups of hosts and localities, e.g. some restricted to funambulines or petauristines. Sciurids at times have polyplacid or hoplopleurid lice. The heteromyids (geomyoid sciuromorphs) have a characteristic polyplacid genus. The Cricetinae are largely parasitized by *Polyplax* (Polyplacidae), *Hoplopleura* (Hoplopleuridae) and by the polyplacid *Neohaematopinus.* The first two of those genera regularly infest the Gerbillinae, a generally xerophilic subfamily of Cricetidae. The microtines are also largely parasitized by those genera, as are the Muridae throughout their range, even the native genera in New Guinea, Australia and Africa. The glirids (dormice) have a distinctive hoplopleurid genus. Lice are unreported from anomaluroids save for the pedetids, which carry a polyplacid. As for the 'hystrichomorphs' (i.e., the South American caviomorphs and African phiomorphs), among the Caviomorpha at least three families bear polyplacids, and two, hoplopleurids. At least two of the Phiomorpha have polyplacid lice. The polyplacids thus have a striking and highly pertinent geographic and host distribution. *Polyplax* infests rodents (and occasionally other hosts as well) in most parts of the world, but the other genera in this family are often bioendemic and restricted to one host, infesting such theraphions as phiomorphs like bathyergids and thryonomids; caviomorphs like caviids, chinchillids and echimyids; dipodids; pedetids; the only two true cricetids in the Ethiopian subregion; other cricetids elsewhere; xerines and other sciurids; heteromyids; loricids; lemurids; tupaiads and leporids. The majority of these associations are surely primary. The geographic areas represented include various parts of Africa, Madagascar, South America, Asia, Europe and North America, with the majority of species in bioendemic genera occurring in the austral regions. The remarkable range of the polyplacid *Eulinognathus* is discussed below (p. 156).

The distribution of hosts and localities for the Polyplacidae, coupled with marked endemism, immediately suggests antiquity and continental drift, and hence I question the chart of Kim & Ludwig (1978) which shows this family as dating only from the Mid or Late Oligocene. The primary, ancestral hosts of these lice must have been in Madagascar, Africa and South America in the Eocene, as per, for example, the extent of the radiation exhibited by the caviomorph Oligocene fossils. The Enderleinellidae, depicted as Oligocene, with possible roots in the Late Eocene (97) also appear to be more ancient than indicated, since these are characteristic sciurid-lice, with representation on xerines, flying-squirrels, etc., and the sciuromorphs are the oldest of rodents. Similarly, the indicated age of Oligocene for the *Hoplopleura* seems too youthful.

3.2.5.7 The place and time of origin of Anoplura – If the Rodentia arose in the Paleocene (231), and the proposed Late Eocene-Oligocene dates of origin for these lice (97) were correct, then rodents existed free of lice for about 20 million years. If the rodent lice are derived from the Anoplura of Perissodactyla and Artiodactyla, as Kim & Ludwig (1978) suggest in their figure, and the lice of the hoofed mammals originated in the Eocene as depicted therein, then the interval before the appearance of the rodent Anoplura is still a minimum of 10 million years. Neither alternative appears plausible without more evidence. Even if the lice did arise on the hoofed mammals, the host and geographic data summarized above indicate that such lice probably were present on rodents in North America in the Paleocene. (Regardless, it seems probable that the hoofed animals arose in North America, since, as shown by Simpson (1945) the vast majority of the earliest fossils of the group (Lower Eocene for Perissodactyla) were found there, rather than in Asia or Europe.) Further, if the distinctive glirid Anoplura is primary, then that genus (or its forebears) may date back to the Eocene, because the dormice presumably arose from paramyid stocks isolated in Europe prior to the opening of the portals from Asia, about 70 m.y.a. (231). That would also automatically mean that the paramyids emigrating to Europe from North America carried lice with them and that the order Anoplura dates back at least 60 million years.

The distribution and ubiquity of the rodent lice and the history of the rodents, etc., outlined above suggest that the primordial rodents arose in North America, were or became infested with Anoplura and transported that group of lice with them on their dispersals. These observations are in marked contrast to those reported above for the rodent Mallophaga. The sucking-lice occur on virtually all groups of rodents, wherever their hosts are found, with such exceptions as 1) the geomyids and New World porcupines, which have Mallophaga instead; 2) Old World porcupines; 3) aquatic hosts like beaver and muskrats *(Ondatra),* etc. (It therefore seems that the attributes of the host have an important effect on infestation by Anoplura, whose tarsi are adapted to clinging to hairs, not spines, for example, and whose spiracles and bristles ordinarily cannot cope with prolonged immersion. Thus, this may be one reason why spiny animals like echidna, tenricids and hedgehogs, etc. lack Anoplura, and obviously why the only sucking-lice (echinophthirids) found on amphibious hosts like seals and otters have special adaptations re chaetotaxy and spiracles, etc.)

Except for the caviomorphs, some of which have Mallophaga and, occasionally, Anoplura as well, rodents have either Mallophaga or Anoplura, but not both. Hopkins (1949, 1957a) credited this to 'competition' and Patterson (1957) ascribed the striking absence of Mallophaga on Insectivora to that cause but, as we have seen, the insectivorans generally lack Anoplura as well. In contrast, Jellison (1942) favoured the influence of geography and believed that the reason erethizontids and geomyids are unique among North American rodents in having Mallophaga instead of Anoplura is because both groups of hosts are of austral origin. He further postulated that rodents of Holarctic roots have Anoplura. Jellison (1942) also indicated that the data on heteromyids supported this theory because they are of Sonoran origin, and hence intermediate geographically and because he was led to believe (mistakenly) that these rodents have both Anoplura and Mallophaga. Today we know that the heteromyids have only Anoplura, and that the ancestry of the geomyids is probably Nearctic although pocket gophers do occur as far south as the Nearctic portions of Panama. The New World porcupines definitely are Neotropical in origin, even if their roots may be African (p. 138). Jellison's conclusions about the geographical aspects of these data are sound and of profound significance, as indicated by the observations reported above regarding the Nearctic origins of the rodent Anoplura. Jellison (1942) also stated that the lagomorphs likewise demonstrate how the geographic origins influence parasitism with lice, pointing out that the two families (Leporidae and Ochotonidae) not only are Holarctic in distribution, with some leporids extending into the Neotropical region, but that North American leporids and Asian ochotonids have Anoplura. Vanzolini & Guimarães (1955a, 1955b) reached some noteworthy conclusions concerning South American Anoplura, viz.: 1) There are no primary infestations of parasites whatever on any of the

Neotropical mammals classified as 'ancient immigrants' or 'old island hoppers' by Simpson (1950) (and by Patterson & Pascual, 1968, 1972), i.e., marsupials and edentates in the first category, and ceboid monkeys and caviomorph ('hystrichomorph') rodents in the second. (It is important to note that Vanzolini & Guimarães regarded the 'hystrichomorph' Anoplura as secondary and relatively recently acquired parasites, in contrast to Hopkins (1949, 1957a), who viewed them as primary and ancient (p. 136 below). 2) The late immigrant mammals that are alleged to have entered after the Panama bridge opened in the Pliocene (146-148, 170, 175), such as the cricetids, sciurids, carnivores, llamas, peccaries, deer, etc., do have Anoplura, having brought these with them from North America. (Some of the other conclusions reached by Vanzolini & Guimarães (1955a, 1955b) in my opinion are controversial and are mentioned on p. 141 below.) Vanzolini & Guimarães thus agree with Jellison (1942) on the northern affinities of the Anoplura, although they do not mention his paper. My own ideas on the importance of the geographical factor go further since I believe the Amblycera of mammals are Neotropical in origin (p. 126), and the Anoplura, Nearctic. Since the Insectivora apparently are not primary hosts of sucking-lice, those hosts either antedate the Anoplura or there were no such lice in Asia, where the Insectivora presumably arose. Since, as shown above, the rodents, and apparently the hoofed animals are of North American extraction, and both groups are infested with Anoplura throughout their ranges, North America seems to be the most probable aboriginal homeland for these lice, especially in view of their absence as primary parasites on truly austral hosts, etc.

As for the nativity and age of the Anoplura, it seems justifiable to state that: 1) There were Anoplura on rodents in the Paleocene. 2) It cannot be excluded that the Anoplura arose earlier on other hosts, e.g. on ungulates, or even on multituberculates in the Cretaceous. Ungulate emigrants from North America were supposed to have dispersed to South America in the Paleocene (148), but if so, they either lacked Anoplura when starting, or the lice became extinct with their hosts in South America. Still, Hopkins (1949) may very well have been correct in suggesting that the Anoplura arose in the Cretaceous, if not even earlier.

3.2.5.8 Anoplura and the origin of the Australian marsupials – The data in this review on the history of Anoplura and their pertinent hosts corroborate the view that the marsupials could not have dispersed to Australia via North America and Asia. Since the marsupials everywhere lack Anoplura, it is easy to conclude that they had arisen where and/or when sucking-lice were absent. If Hopkins (69, 70) was correct about the Anoplura having originated in the Jurassic or Cretaceous, then the marsupials should have been in contact with infested placentals in North America or Asia, and acquired Anoplura secondarily. The Australian marsupials would then have Anoplura – but they lack them. Even though the Insectivora may not have been infested with any kind of lice, especially in Asia in the Cretaceous, there may very well have been other anopluran-infested placentals extant en route to infest the transients with lice, e.g. the mammals doubtfully reported as insectivorans in the Albian of Texas 100 m.y.a. (121, 178, 179). Even if the Anoplura did not evolve until the Paleocene, then it would still seem likely that peregrinating marsupials would encounter infested hosts, for Simpson (1965) named that period as a possibility for Nearctic metatherians entering Australia. There are data on South American Anoplura that indicate austral relationships, namely, the affinities between the sucking-lice of Neotropical caviomorphs and those of African phiomorphs (p. 140).

3.2.5.9 South America as the provenance of the Australian marsupials – Since the data on the zoogeography of fleas and lice indicate that the marsupials of Australia and South America are monophyletic, as are the ectoparasites in question, we can now enquire whether the Australian metatherians are derived from the South American, or vice versa. Nowadays the consensus is that the route was from South America to Australia via Antarctica, and the above data on lice and fleas support this contention.

In comparing the Siphonaptera of Neotropical versus Australian marsupials, it is apparent that South America has the richer fauna despite the much greater diversification of these hosts that has taken place in Australia, where, for example, Archer & Kirsch (1977) recognize 13 families of living metatherians, and where Keast (1972c) lists 122 species of marsupials, and Ride (1970) cites 50 genera. In South America there are only two extant families, with a total of 14 genera and 60 living species (Hershkovitz, 1972). The richer marsupial fauna in Australia, like the case of the Mallophaga and the presence of macropsyllids there, probably reflects the greater geographical, geological and zoological isolation of Australia as compared to South America. Even so, in Australia there are only two genera and eight species of stephanocircids; in South America and Central America, seven genera and 29 species. Even more impressive is the case for the two doratopsylline genera in Australia, for, as indicated above (p. 117), all the signs point to an origin in South America. The data on the pygiopsyllids obviously suggest that there was a movement of marsupials from Australia to South America, e.g. one genus (*Ctenidiosomus*) and five species in the latter area versus 11 genera found in Australia, with a total of 78 species of those genera occurring in the Australian region. However, this passage almost undoubtedly was a later event that the original faunal link between those continents and must have been undertaken at a time when the connections were highly tenuous, because if this pygiopsyllid dispersal represented the main flow of traffic, a far richer fauna of that family would be expected in South America.

If one assumes that 1) rodent trimenoponids are descended from marsupial lice, and 2) that they have a common origin with the boopid marsupial lice of Australia, it can be concluded that the evidence on Mallophaga likewise suggests South American origin for the Australian marsupials. (Whether the Gyropidae arose from the Trimenoponidae or from avian lice is irrelevant in this regard.) There is a much richer fauna of marsupial lice in Australia (e.g. seven genera and 35 species) than in South America, but this may reflect the far greater variety of extant metatherians which are available as hosts on that continent.

3.3 *The place of origin of the marsupials*

If South America was the source of the Australian marsupials, there still remains the intriguing riddle of the origin of the South American metatherians themselves. The classical explanation (146-148, 170, 171, 173), and the still-favoured one (16, 58, 60, 121, 125), is that they were immigrants from North America, probably in the Lower Cretaceous (ca. 120 m.y.a.). The main reasons for this opinion are: 1) at least until recently, the known marsupial fossils from North America (Mid Cretaceous, ca. 95 m.y.a.) (13, 74, 122, 192) by far out-dated those from South America (Paleocene, ca. 60 m.y.a.) and 2) there was a good fossil record from Europe as well, suggesting an ancient Holarctic fauna. However, there now is a Late Cretaceous fossil marsupial record from South America (23, 85, 91), which Lillegraven (1974) claimed was already a 'remarkably advanced' form, implying it had ancient local roots. Moreover, the obvious broad and distinctive radiation exhibited in the Paleocene fauna suggest that the marsupials must have had a very long history in South America. Also, it is now recognized that faunal exchanges were occurring via transatlantic land connections as late as Mid Eocene (50 m.y.a.) (124-126) and hence, once having penetrated North America, the marsupials could easily reach Europe. The data on ectoparasites outlined above, and other points, indicate to me that Tedford (1974) was correct in proposing a South American origin for the marsupials, as was Savage (1974) in espousing at least that route of dispersal.

This contention is based upon the following premises documented above: 1) The order Siphonaptera arose only once (206). 2) There are austral affinities in the order at the levels of subfamily, etc. 3) The stephanocircids are an ancient monophyletic family with many primitive features and arose in either South America or Australia (211), probably the former, and have an aboriginal association with marsupials. 4) The pygiopsyllids originated in the Australian region (205, 206). 5) All marsupials lack Anoplura (69, 70, 219, 220), and this represents the primary situation. 6) The marsupial lice (Amblycera) are monophyletic. 7) Insectivora do not have Mallophaga (69, 70, 219, 220) and this absence is fundamental. 8) Insectivora lack primary infestations with Anoplura. 9) The Anoplura of rodents probably arose in the Paleocene. 10) There were no Anoplura in South America until well after the radiation of marsupials (69, 70, 219, 220). 11) The marsupials arose in the New World (e.g. 16, 74, 192). 12) The Insectivora have Asian roots (15, 120) although they co-existed with the marsupials in time, and both groups probably had a common ancestor. 13) The Beringian bridge did not become operable until Late Cretaceous (McKenna, 1975). 14) The soricids and talpids evolved long after the marsupials, even though (Patterson, 1957) these families had clearly become established along extant lines by the time of the first known fossils (Oligocene).

Prerequisites requiring emphasis are: 1) There must have been Siphonaptera in existence on Pangea, prior to the formation of Gondwanaland and Laurasia, and where the fleas parasitized the forebears of the orders and families of mammals extant. There seems to be no other way that the presence of a specialized (albeit primitive) family of fleas like the stephanocircids in the Early Cretaceous could be explained. Similarly, one could not otherwise readily account for the hystrichopsyllid subfamilies associated with Insectivora in the Holarctic, particularly when endemic tribes or subfamilies likewise developed in the Southern Hemisphere. The only alternative would be to postulate independent evolution of fleas in Laurasia and Gondwanaland, but monophyly of the order, universally accepted, would rule this out. 2) The living marsupials lack Anoplura today because they never had them, and not because they originally had both chewing-lice and sucking-lice and lost the latter. There were no Anoplura where and when the marsupials arose. 3) Similarly, there are no Mallophaga on insectivorans because the two groups have not come into contact except in a limited geographical way, and this was subsequent to establishment of an equilibrium between various ectoparasites and their shrew and mole hosts.

3.3.1 *Data from lice* – If the marsupials had entered South America from the north, it seems likely they would have brought Anoplura with them and then it would be expected that today they would be infested with such sucking-lice instead of bearing amblyceran Mallophaga throughout their range. If Hopkins (1949, 1957a) was correct in dating the Anoplura as Cretaceous, or even if the emigration was Paleocene, the time factor jibes. Marsupials dispersing from North America should then have encountered Anoplura on the earliest rodents (Paleocene) if not on other Nearctic placentals (Cretaceous). The only primary Anoplura in South America are on 'hystrichomorphs' (caviomorphs) and (perhaps) ceboids, both of which entered South America eons after the marsupials were ensconced on the scene; and on the undoubted late arrivals (Late Miocene-Pleistocene, p. 139). Hopkins (1957a) logically reasoned that since the highly specialized echinophthirid lice are ubiquitous on seals, the anoplurans must have parasitized ancestral seals before they became marine, and hence the sucking-lice must at the very least antedate the Eocene. Relationships of the seal lice with those of camels and with other families of lice led him to espouse origin in the Cretaceous for the Anoplura. Vanzolini & Guimarães (1955a, 1955b) surprisingly, did not consider this type of evidence when estimating the age of Anoplura and when trying to account for the anomalous situation in South America, where Anoplura existed on 'hystrichomorphs' whose ancestors purportedly arrived from North America in the Oligocene, but where the older inhabitants all lacked primary Anoplura. Their solution was to regard the caviomorph lice as obviously secondary and relatively recent, i.e. a derivation from cricetid immigrants allegedly reaching South America in the very late Pliocene. Vanzolini & Guimarães (1955a, 1955b) implicitly accepted the concepts of Simpson (e.g. 1951) regarding stability of continents, Nearctic origins of the entire Neotropical fauna, and his proposed dates of entry. Accordingly, they did not consider the possibility of African origin of the caviomorphs and, instead, argued that if Anoplura had been in existence at the time the 'hystrichomorphs' emigrated to South America from the north, the rodents would have brought Anoplura. Since, as per their theory, the 'hystrichomorph' lice are secondary and youthful parasites on cricetids, the former could not have been infested with sucking-lice at the time of emigration. Although they stated (219) the Anoplura must have risen in the 'Early Tertiary'; as Hopkins (1957a) pointed out, they implied elsewhere (220) that the date was closer to the period of the Pliocene migration. This entire argument appears flawed to me because: 1) African origin of the caviomorphs would resolve the discrepancies. 2) There is evidence that the caviomorph lice are indeed related to African lice of pertinent hosts (p. 140 below). 3) None of the other older residents have Anoplura. 4) It has been contended that the dates of entry proposed by some (148, 149, 171) for the Nearctic emigrants, were far too late (58, 60, 206) and 5) Vanzolini & Guimarães (1955a, 1955b) ignored the fact that the caviomorphs also have amblyceran lice, and such an infestation suggests a long history in South America, especially since the other rodents lack such lice as primary parasites. For example, Emerson & Price (1976) described a new genus of Amblycera, representing a distinctive family, from the rat-chinchilla *Abrocoma* in Chile, geographically (and taxonomically) remote from its closest relative, in the middle of the continent. These points likewise are indicative of great age. Also, spiny rats (Echimyidae) are infested with an anopluran whose lateral margins are uniquely modified to form scale-like projections, in what clearly looks like an adaptation to the spinose fur of their hosts. Development of such modifications, unknown in boreal lice, must take eons. Another echimyid anopluran is remarkable in possessing spiniform bristles. 6) Caviomorph fleas are of bioendemic, undoubtedly ancient, stocks, in contrast to the fleas on the late invaders. This too indicates a very long lineage in that area. 7) The Anoplura seem to be an older group than that implied by Vanzolini & Guimarães (1955a, 1955b) (p. 131).

If I am correct, the South American marsupials bear amblyceran Mallophaga instead of Anoplura because the first two evolved together in South America, in the absence of sucking-lice. Such a primary infestation would then more or less effectively exclude infestation by Anoplura when they did appear eons later on the ancestral caviomorphs, and still later (Late Pliocene, according to some), on cricetid immigrants, which also transported other new faunal elements such as ceratophyllid fleas. Some Amblycera switched to caviomorph rodents as these mammals arrived, and the Gyropidae, derived from trimenoponid stocks, or else originating from avian lice, evolved on such rodents.

If the mammal Mallophaga in general arose in South America, then it is not surprising that hosts moving north would carry Mallophaga with them, e.g. the sole Nearctic rodents (porcupines and geomyids) with chewing-lice, and which have Neotropical or Central American roots respectively.

3.3.2 *Data from fleas* – South American origin of the marsupials would also account for the antiquity of the stephanocircids and for their unusual distribution, including the slight spill-over into Panama and Costa Rica, and their complete absence north of Central America and from Eurasia and Africa. Inasmuch as the northwards dispersal must have been in the Early Cretaceous to fit with the fossil record, that early date may have precluded infestation of those particular taxa by stephanocircids. Another possibility is that the helmet-fleas in the north became extinct with their hosts.

The zoogeography of the doratopsylline fleas can apparently only be adequately accounted for on the basis of a Neotropical metatherian homeland. Reasons were advanced above for postulating that *Idilla* and *Acedestia* were of South American extraction. At first thought, it might seem that those Neotropical roots could be easily explained by doratopsylline fleas accompanying marsupials as they moved south from North America after having encountered Insectivora in the latter area, and after some doratopsyllines had switched to marsupials from their original insectivoran hosts. This theory is also attractive in that it also fits with the distribution of the tribe Doratopsyllini on shrews in the Holarctic region. There are serious drawbacks to such a hypothesis, however, viz.: 1) If Insectivora arose in Asia, as is now believed (15, 16) (p. 113) and as the data on Anoplura suggest, they would have had to have been in North America in the very Early Cretaceous (if not Jurassic, 140 m.y.a.) to permit contact with marsupials, transfer of doratopsyllines, and dispersal to South America and Australia so that the ancestors of *Acedestia* and *Idilla* would be in Australia in the Early Cretaceous. The available data do not indicate such an ancient history for the Insectivora. 2) The Beringian corridor between Asia and North America was apparently not functioning before the Cretaceous/Paleocene boundary (126) and this is another reason why emigrant Insectivora might have been unable to meet the marsupials in time to effect any transfer of fleas. 3) If shrews (or other Insectivora) were in sufficiently close contact for such a transfer (and evolution) of fleas, then it would be expected that the insectivorans would have accompanied the 'inferior' marsupials on the southward dispersal, along with the edentates and other of the South American 'ancient inhabitant' that are supposed to have emigrated from North America in the Cretaceous. There is no definitive evidence that there ever were Insectivora in South America prior to the relatively recent opening of the Panama bridge or the Late Miocene insular links. (The Lower Eocene necrolestids, once considered as questionable insectivorans (167) are regarded as marsupials by some (148) and of unknown affinity by others (91).) *Cryptotis* (one species) seems to be the only immigrant soricid, and is restricted to the northern part of the continent. The soricids are also poorly represented in Central America. 4) The absence of other shrew-fleas (and Corypsyllini) in the Neotropical region also suggests that no Insectivora made the journey. 5) It is not certain that Insectivora and marsupials actually co-existed at the critical time and place. Slaughter (1968, 1971) reported that both groups were represented in Albian beds in Texas (about 100 m.y.a.), but there is some doubt (Lillegraven, 1974) as to whether the placentals were truly Insectivora. 6) The Doratopsyllini are parasites of soricids, but the earliest records for the host-family are Oligocene (36 m.y.a.) (146). For these reasons, it seems more logical to postulate that the Doratopsyllinae are autochthonous to South America, on marsupials, and moved into North America with their hosts, and there had descendants that adapted to, evolved with, and dispersed with, soricids. (The subfamily Doratopsyllinae therefore apparently antedates the Soricidae.) This theory also implies that the North American Insectivora had no fleas at that time to pass to the marsupials – and this in turn supports the idea that these hosts were immigrants from Asia. The Beringian bridge could account for the occurrence of shrew-doratopsyllines in Asia-northern Europe.

In my opinion, the surprising absence of Doratopsyllinae on talpids lies in the proposed Neotropical origin of the subfamily and their move northwards with their hosts, as suggested by the current distribution of the moles in North America. There the talpids are restricted to the Pacific coastal sections and, roughly, to the eastern half of the USA and southern Canada. Thus, they are completely absent from the Rocky Mountain areas, and even as far east as the western borders of Nebraska and Kansas, etc. This distribution cannot be ascribed solely to climate or to limitations like soil suitable for burrowing, for, hundreds of miles of arable territory east and west of the Rockies are uninhabited by moles, and hence presumably can validly be contrasted with that of soricids, which covers most of North America, including the mountains of Central America. It seems likely that there were no talpids (or their forebears) along the main northward route of the marsupials carrying doratopsyllines, but that proto-soricids were present. Thus, in the Cretaceous, marsupials were better represented in diversity and numbers in western North America than in the east, judging by the fossil record (14, 15, 192), even after compensating for the fact that more and better fossil beds are known in the west than in the east. At this period the two parts of the continent were essentially islands, separated by a vast intercontinental sea extending north and south. Europe, which to a certain extent shared a marsupial fauna with eastern North America, as it did regarding dinosaurs in prior times, may not have had marsupials until the Mid Eocene (Tedford, 1974). Between 70-75 m.y.a., marsupials and multituberculates were dominant in the western USA fauna but they nearly became extinct following an increase in the Eutherian fauna 65 m.y.a. (192).

3.4 *African-South American faunal connections*

The vast majority of mammalogists and zoogeographers have strenuously opposed the idea that any South American mammals could have African roots (102, 103, 146-148, 170, 175, 228, 230, 234). The resemblance between the Old World 'hystrichomorphs' (phiomorphs) and those of the New World (caviomorphs), as well as those of the monkeys of the two continents were ascribed to convergence, not parallelism. Recently, however, some of the most ardent antagonists of such southern relationships (64, 109-111) have now come to definitely believe that the 'hystrichomorphs' are monophyletic and likewise that the South American platyrrhine (ceboids) and the African catarrhine monkeys have a common ancestry, and opine that entry into South America was by rafting in the Eocene. Wood (1977a) writing about rodents, agreed that such emigration was possible, at the end of the Eocene, but still preferred to regard North America as the probable source of the caviomorphs. It is now generally accepted that the two continents were much closer together in the Paleocene and Eocene than at present. At about 65 m.y.a., 'only 600 km probably separated Africa and South America at their closest points' (152, 153) and numerous islands were interspersed. By the end of the Eocene, they were ca. 1 400 km apart. As Washburn (1978) pointed out, about 35-40 m.y.a., Africa was just as close to South America as was North America, and hence mammal dispersal by rafting was just as likely from one continent as the other. (The prevailing currents probably were a factor, but data are lacking.) Hershkovitz had also come to believe that continental drift affected the distribution of mammals, and in 1972, suggested that some of the South American myomorph rodents, as well as the 'hystrichomorphs' have African affinities, and specifically mentioned ancestral sigmodontines (Cricetinae) as probably having penetrated South America after rafting from Africa in the Lower Tertiary. He also made the same assertion about the background of the West Indian Insectivora. Cracraft (1975) pointed out that South America and Africa share some critical elements of the fish, amphibian, reptilian and avian faunas, and he concurred with the idea of African roots for the platyrrhines and caviomorphs and the concept of entry by rafting.

3.4.1 *History of the South American mammals* – According to the traditional concept of the Nearctic origin of the South American fauna, the caviomorphs and platyrrhines appeared in the Upper Eocene or Lower Oligocene while the 'ancient inhabitants' (e.g. Marsupialia, Edentata, Condylarthra, etc., but nothing like rodents or Insectivora) arrived in the Cretaceous-Palaeocene border (146-148, 170, 175). In each instance, the ancestors were supposed to be waif immigrants, arriving by sea from North America. Those sources also aver that when the Panama landbridge was first established in the Pliocene-Pleistocene, 16 families of Nearctic mammals, hitherto unrepresented in South America, crossed into that continent and radiated rapidly there. The cricetid, heteromyid and sciurid rodents; Soricidae (*Cryptotis*); leporids and procyanids were supposed to be included with those 'late immigrants'. Patterson & Pascual (1972) categorically stated that 'The South American fauna does not support any hypothesis of Cenozoic or late Cretaceous connections with either of the other southern continents, whether by land-bridge or continental drift'. Simpson (1950, 1969) of course held similar views. Savage (1974) mentioned evidence indicating that 1) intermittent island links existed between South America and North America in the Late Miocene and that 2) from the earliest Pliocene, the Panamanian connection was continuous. As mentioned above and elsewhere (205, 206, 211), the data on fleas and lice have long convinced me not only regarding faunal affinities of the austral continents but that the fleas and their hosts are far older than generally believed. The tremendously distinctive morphological and taxonomic features of some of the bioendemic families of South American fleas associated with rodents, to say nothing of the subfamilies and tribes, cannot conceivably have developed in the geologically short time-span postulated by the traditionalists for the existence of the isthmian bridge, especially when those Siphonaptera are compared to taxa such as the ceratophyllid *Pleochaetis,* of peromyscines, which are clearly recent immigrants. Even if the bioendemic taxa are ascribed to entry with the caviomorphs, the Simpson/Patterson dating of Late Eocene/Oligocene seems too late, aside from the fact that there are no palpable Nearctic roots for some of the major autochthonous groups, e.g. the Rhopalopsyllidae, which has even been considered a superfamily (Johnson, 1957). The ancestral rhopalopsyllids may have been contemporary with the stephanocircids as primeval autochthonous inhabitants of South America, perhaps as

Xenarthra-parasites. The enigmatic necrolestids (Lower Miocene), still of unknown affinity, suggest that other potential hosts of fleas existed earlier and await discovery. There are no obvious rhopalopsyllid links with the African fauna (save for the constituent *Parapsyllus* which emigrated to South Africa and the sub-Antarctic islands on marine birds). If the primordial rhopalopsyllids came from Africa, or arose on ancestral caviomorphs within South America, it must have been no later than the Paleocene or Early Eocene, judging from the unique characters of the family or superfamily.

Wenzel & Tipton (1966) and Hershkovitz (1966, 1969) independently also reached similar conclusions concerning the impossibility of fitting so late a date as a Pliocene-Pleistocene entry with their concepts of the zoogeography and taxonomy of various Neotropical ectoparasites or mammals. However, they believed in the ultimate Nearctic origins of the Latin American fauna and did not consider faunal connections with Australia and Africa, as pointed out by Traub (1968). Savage (1974) provided excellent reasons for believing the sigmodontines must have reached South America (by an over-water route) well prior to the uplift of the Panamanian connection, but, unlike Hershkovitz (1972) believed they came from North America, and did not regard Africa as a possibility. Croizat (1958) also objected vehemently to the time-frame imposed by Simpson (1950) and the others, arguing that Pleistocene immigrants could not achieve such a level of development in South America in so short a geological span. Kurtén (1971) maintained that the Paleocene fossil marsupials, notoungulates, edentates and condylarths in South America were derived from a former 'World Continent', thereby indicating a common stem for the forebears of some extinct and contemporary mammals in the southern continents. The island links proposed by Savage (1974) for faunal exchange between North America and South America in Late Miocene seem much more plausible than a Pliocene/Pleistocene bridge as the sole means of entry.

3.4.2 *Data from lice* – The zoogeography of lice (and fleas) support the concept of African/South American faunal connections by rafting, probably occurring in the Eocene, although the data seem to have been overlooked, because the various experts were not concerned with such factors as continental drift and did not discuss the possible implications. Hopkins (1957a) and Ferris (1951) pointed out that the 'hystrichomorphs' of South America (caviomorphs) have Anoplura that are related to the lice of the African 'hystrichomorphs' (phiomorphs), and that most of the genera found on those hosts are restricted to those particular groups of mammals. Moreover, as pointed out above (p. 122), the caviomorphs are exceptional among the rodents on that continent in that 1) they carry Mallophaga (Amblycera) and 2) they are the only members of the Neotropical fauna that was supposed to be *in situ* prior to the opening of the Panamanian bridge and yet which are parasitized with Anoplura. Hopkins (69, 70) believed that originally many kinds of mammals had infestations of both Mallophaga and Anoplura, and that in most cases one or the other type of lice was eliminated by 'competition'. This would account for the disjunct distribution of related forms. He admitted the plausibility of a zoogeographic basis for the presence or absence of Mallophaga, but reiterated (1957a) his theory of competition as the reason, and stated in defense that the Anoplura of caviomorphs 'constitute the test case', maintaining that these sporadic polyplacine lice 'are so near to genera found on African rodents that have been referred to the hystrichomorphs . . .' that they scarcely are distinguishable. This argument, in my opinion, renders strong support to precisely the opposite (geographic) theory because of the presumed African origin of the caviomorphs.

The lice on these rodents, according to my hypothesis represent: 1) descendants of anopluran lice imported on the ancestral African immigrants and 2) transfer of trimenoponid Mallophaga from marsupials to the newly evolving group of caviomorphs, and 3) the development of the Gyropidae on caviomorphs. There were no other groups of rodents available for infestation by Mallophaga or African-derived Anoplura at the time, and the later arrivals from North America carried their own kind of Anoplura. In support of my contentions may be cited the allusion of Johnson (1960) to some unnamed as well as described species of Anoplura from some South American rodents (known or presumed to be caviomorphs), which were related to the polyplacid *Eulinognathus.* At the time, Johnson was discussing the taxonomy of a *Eulinognathus* from the African cricetid *Lophiomys.* As shown on p. 156, *Eulinognathus* has an interesting austral transcontinental distribution.

Vanzolini and Guimarães (1955a, 1955b) denied any such relationship between the African and South American 'hystrichomorph' lice, thinking in terms of Simpson's concepts (1951, etc.) of the stability of continents and sole North American origins. They stated that their conclusions support Wood's claim (1950) that the similarity of the African and South American 'hystrichomorphs' represent pure convergence. Apparent deficiencies in the argument of Vanzolini & Guimarães, e.g. the consequent unacceptable youth of the Anoplura, have been discussed above (p. 136), and another is that they ignored the evidence provided by the lice of the ceboid monkeys. If the platyrrhine monkeys had been derived from Nearctic stocks, it would be expected that they would be infested with aboriginal Anoplura, like the cricetids and carnivores emigrating from the North. Instead they carry amblyceran and ischnoceran Mallophaga, which is in accord with the theory of a long history in South America, following introduction from Africa. However, there also is a species of *Pediculus* occurring on the ceboid monkeys in the wild in various parts of northern South America. Hopkins (1957a) regarded this as a probable derivative of the human louse following contact with man in the geologically recent past. Since the African chimpanzee has its own species of *Pediculus* (95), it appears to me that it is possible that the Neotropical monkey-*Pediculus* actually represents a primary infestation, dating back to the African homeland of the platyrrhines. Other species of *Pediculus* may perhaps exist on the ceboids but there is inadequate information available to permit a decision. The New World monkey-*Pediculus* could have arisen by parallel evolution and need not necessarily imply that *Pediculus* per se dates back to the Eocene. Such independent evolution, from common primordial roots, of a taxon which meets the definition of a genus occurring elsewhere, fits the concepts of Croizat (1958, 1970) and Mani (1974) regarding parallel evolution. It is true that the African catarrhine monkeys have *Pedicinus* lice (Anoplura), but such parasites may have been lost after the ancestral hosts reached South America. Parasitology is replete with such examples. The absence of Mallophaga on African monkeys (and rodents, etc.) (like the absence of marsupials) suggest that the route of dispersal was from one-way Africa to South America, and of a filter type, as by rafting by a definitely prevailing current.

There can be little question that there were Anoplura in Africa at the time of the suggested emigration of ancestral 'hystrichomorphs'. The marked bioendemicity and individuality of certain anopluran lice indicate that not only do some groups of mammals like the phiomorphs have an ancient background in Africa, but that some major taxa originated there. Examples of such lice include the utterly distinctive genus on the equally enigmatic aardvarks; another unusual taxon, *Scipio,* on ancient groups like *Thyronomys* and *Petromus* (both phiomorphs); and a peculiar genus on macroscelid insectivores. The autochthonous and specialized nature of some subfamilies and genera of African fleas supports this contention. However, Wood (1977a) suggested that phiomyids, the presumed ancestral family of the phiomorphs, only reached Africa, from southwest Asia, in the Late Eocene. Travel of ancestral phiomorphs from North America to Europe and thence to Africa seems to be excluded by geological evidence, i.e., dating (as well as the available fossil data). Overland dispersal between Europe and Africa was possible at various times until Early Paleocene (ca. 63 m.y.a.), but the links were severed until about 18 m.y.a. (20, 153) and hence were absent during the periods in question.

3.4.3 *Data from fleas* – Among the evidence provided by Siphonaptera for African/South American relationships is the distribution of the Pulicidae (p. 109 and table 4). Four of the five genera of Pulicinae occur in the Ethiopian region. The fifth, *Pulex,* is found in South America, although with a center of development and probable origin in Central America. (The question immediately arises as to how *Pulex* reached the Old World and became a cosmopolitan flea of man. The authentic record of *Pulex irritans* as a tenth century 'Viking flea' (Rothschild, 1973) rules out introduction by ship from the New World in the post-Columbus era. However, it is now generally accepted that the Vikings reached North America by the tenth century and may even have entered the interior of the USA. Also, archaeological evidence clearly points to cultural contact between Japan and Ecuador 5 000 years ago (meggers et al., 1965). Either or both exposures could have resulted in the acquisition of *Pulex* from American Indians and its subsequent emigration.) Of the 22 genera of Pulicidae, 14 (64 per cent) are found in Africa. The family Hystrichopsyllidae is global, but there are four bioendemic tribes in South America and two such subfamilies in the Ethiopian region, bespeaking of great antiquity in each (as well as the Nearctic region, and Palearctic, table 4, p. 108). The Ethiopian Dinopsyllinae and Listropsyllinae have no near relatives, but the affinities of the autochthonous Neotropical hystrichopsyllid tribes are clear and significant. The marsupial-infesting Tritopsyllini (Doratopsyllinae) have already been discussed regarding Australia. The Ctenoparini are found in the extreme southern part of South America but the closest any relative occurs is Mexico. Derivation from a primeval hystrichopsyllid introduced from Africa is conceivable. The Ctenophthalminae, represented by four tribes and six genera in the New World, constitute the group of greatest interest re possible African connections.

3.4.3.1 Ctenophthalminae and *Ctenophthalmus* – Two tribes of Ctenophthalminae occur in South America and both are bioendemic. In the monotypic Agastopsyllini, *Agastopsylla* is essentially a *Ctenophthalmus* which has become modified along the lines of a nest-flea (198, 204). Despite its bizarre genitalic modifications, *Neotyphloceras* (Neotyphloceratini), has been regarded as a truly representative ctenophthalmine (78), while the other member of the tribe, *Chiliopsylla,* resident in the southern areas, is a typical member of the subfamily.

The most germane taxon is *Ctenophthalmus,* by far the largest genus in the order. Jordan (1929) pointed out that the single species in the USA was unusual in being confined to the eastern part of the country. Today the situation seems even more anomalous, since we know of a total of eight Nearctic species, seven of which are found in Mexico or Central America, and one, *C.(Nearctoctenophthalmus) pseudagyrtes,* is limited in the USA to the area east of the Great Plains and south of southern Canada, but extending at least as far as the mountains of southern Mexico. The range of its main host, *Microtus,* greatly exceeds that of *C.(N.)pseudagyrtes.* There are four Mexican or Central American *C.(Nearctoctenophthalmus),* on *Microtus* primarily, and three *C.(Alloctenus)* on montane Mexican *Cryptotis* shrews. Restriction to an area east of the Rockies is highly unusual among North American fleas, since virtually all genera of fleas in the eastern part of the USA and Canada are not only also found in the West, but generally are much better represented there regarding numbers of species. The only eastern genera absent in the Rockies are the monotypic *Tamiophila* and *Conorhinopsylla* (two species) and both of these are specialized nest-fleas. Moreover, many western genera are not found east of the Mississippi River. Another very unusual feature about *Ctenophthalmus* in North America is its absence in the northeastern part of the continent, to say nothing of the northwestern section. All other genera of fleas occurring in New York State, for example, extend well into Canada, frequently to the subarctic or Arctic. Also, it is unprecedented for a species of fleas from the northeastern part of the USA to be found as far south as southern Mexico. *Ctenophthalmus* is unique as a taxon that is far better represented in Mexico and Central America than in the northeastern USA. There is only one other species of flea that is found in the latter region and in central Mexico, namely *Corrodopsylla curvata* and significantly, that is a shrew-parasite. (The rhopalopsyllid *Polygenis gwyni,* an opossum-flea, is a recent, Neotropical-derived invader into the southeastern USA, along with its host.)

How can this strange distribution of North American *Ctenophthalmus* be explained? First we must consider its antiquity, for *Ctenophthalmus* must be an ancient taxon. Thus, the closely related *Palaeopsylla* (Palearctic) has apparently remained unchanged at the generic level for 50 million years (Peus, 1968). Moreover, *Palaeopsylla* seems to be a derivation of *Ctenophthalmus* in which the genal comb moved to an upright position and developed greatly modified spines whose design depends upon whether the host is a mole or shrew (209, 213). Parallel evolution of such a comb has occurred in another ctenophthalmine in southern South America, and the similarities are striking. *Ctenophthalmus* therefore appears to be the more primitive and ancestral form. As mentioned above (p. 101), *Ctenophthalmus* is a prime example of a primordial form that changes slowly at the generic level, but there are sufficient differences to have permitted Hopkins & Rothschild (1966) to designate subgenera. More important, *Ctenophthalmus* has adapted to a surprising gamut of hosts and often is quite specific thereto, and its morphology reflects such changes, viz. the characteristic spines in the combs of its shrew-fleas, sciurid-fleas; and of the types of combs and other modifications seen in the *Ctenophthalmus* of the polyphyletic 'mole-rats'; the variations in the size of the eye and in type of chaetotaxy, depending upon the habits of the host (204, 209, 210, 213). The evolution of such features must surely reflect a long association with the particular hosts. There are also distinctive and consistent modifications in the aedeagal apodeme of the various groups of *Ctenophthalmus,* and basic differences in that organ are seldom seen in congeneric fleas. The kind of hosts are also suggestive of antiquity – soricid and talpid hosts are of ancient lineage even though they are not the oldest of the Insectivora – and primitive fleas tend to parasitize primitive hosts. Microtines, frequently infested by members of this genus, are among the oldest living rodents. While *Ctenophthalmus* in the Ethiopian region are often associated with murines, which are a relatively youthful group, those fleas are among the most specialized members of the genus (indicating they were evolved more recently than the others), and their hosts are the autochthonous, more primitive forms of rats.

Most suggestive of all regarding the age of *Ctenophthalmus* is its zoogeographic distribution, and that of other members of the subfamily. The Ctenophthalminae is the only subfamily besides the Pygiopsyllinae to occur on all three of the southern continents (and on the Indian subcontinent as well). *Ctenophthalmus* is the only genus of mammal flea with native species in the Ethiopian, Palearctic and Nearctic regions. In each of these regions, some *Ctenophthalmus* infest Insectivora. A long history seems indicated for the subfamily in South America, and although *Ctenophthalmus* itself is unknown there, its derivative *Agastopsylla* is well represented. Another telling point is that *Ctenophthalmus* tends to be restricted to certain hosts or biotopes, whereas relatively recently-evolved forms are generally rapidly speciating, highly successful forms that can promptly accommodate themselves to a spectrum of environments, e.g. the genus *Rattus* and its various subgenera in the Malayan and Papuan subregions, and fleas like *Sigmactenus.* In contrast, *Ctenophthalmus* frequently shows a relict distribution, viz. survival in circumscribed favourable habitats in the midst of palpably unsuitable terrain, as in the case of the 'Siberian' *C.golovi* in the Pakistan Himalayas (213). Moreover, the subgenera of *Ctenophthalmus,* which were established on morphological grounds, generally fall into geographic patterns, as indicated by prefixes like *Ethio-*, *Nearcto-*, *Sino-* and *Palaeo-* attached to the stem *Ctenophthalmus.* The taxonomy of the genus, then, also bespeaks of long occupancy of certain areas.

The putative view about the distribution of *Ctenophthalmus* seems to be that the genus was originally extremely widespread (and entered North America via the Bering Strait connection), but has been extirpated in many parts of its range during periods of glaciation or other environmental changes such as development of aridity in mesic foci, etc. One major drawback to such a theory is that *Ctenophthalmus* is remarkably poorly represented in eastern Asia as compared to western Asia, Europe and Africa. There are flaws in a generalization based upon mass extinction because at least some species of *Ctenophthalmus* are visibly highly adaptable, e.g. *C.golovi* in subarctic terrain (213), and a rich fauna in the high mountains of Ethiopia under conditions of aridity which last for months; and species of the subgenus *Paractenophthalmus* infesting desert rodents in Central Asia. Why should the North American glaciation have eradicated *Ctenophthalmus* in Alaska and the western USA and northern Canada, when intercontinental (Holarctic) hystrichopsyllid genera like *Catallagia* and *Corrodopsylla* survived there (and in the eastern USA as well)? Holland (1958, 1963) excellently discussed intercontinental faunal connections and cited many significant examples of Holarctic fleas, but did not mention *Ctenophthalmus,* since it is absent from northern North America. Other questions that have not been asked hitherto and require answers are: Why should only one of the northern species of *Ctenophthalmus* survive, and why should it be essentially restricted to the eastern USA? The answer seems to be that *Ctenophthalmus* had a very different history than the doratopsyllines and the bulk of hystrichopsyllids and the ceratophyllids in North America.

The zoogeographic peculiarities of *Ctenophthalmus* in North America could be most readily explained by African roots for the subfamily, with entry into South America, or Middle America, by rafting of the hosts in the Eocene. The center of development and radiation for the genus seems to have been sub-Saharan Africa, regardless of whether the group arose there or in the Palearctic. It therefore seems likely that a *Ctenophthalmus* or ctenophthalmine could have accompanied the African 'hystrichomorphs' on the journey. If Hershkovitz's suggestion (1972) is correct about the African background of the South American sigmodontines and West Indian Insectivora, then those primeval hosts may have transported the ctenophthalmine stocks en route. A *Ctenophthalmus* or a proto-*Ctenophthalmus* entering South America in the Paleocene or Eocene could very well have given rise to *Agastopsylla,* which is closely akin to *Ctenophthalmus.* If the immigrant were a proto-*Ctenophthalmus,* the New World *Ctenophthalmus* may have arisen by parallel evolution, à la the concepts of Croizat (1958, 1970) and Mani (1974). If this hypothesis is correct, entry of *Ctenophthalmus* or the prototype into Middle America could have been effected by island-hopping across an archipelago linking the two Americas. Tedford (1974) and Cracraft (1974a) mentioned such links as existing through much of the Cenozoic. An archipelago involving the Antillean plate was postulated by Wood (1977a) on the basis of data on Caribbean plate tectonics provided by Ladd (1976) but Wood was suggesting that the caviomorphs dispersed from the USA to South America in that manner.

It is possible that elemental ctenophthalmines could have given rise to *Ctenophthalmus* after coming into contact with *Cryptotis* in northern South America in Late Miocene or Early Pliocene as the shrews moved back and forth over isthmian connections, and then developed into the two subgenera of *Ctenophthalmus* (one on *Cryptotis* and one on cricetids) in Central America and Mexico. There are so very few records of *Cryptotis* being checked for fleas in South America that it cannot be stated with certainty whether or not *Ctenophthalmus* occurs there. A dispersal from Mexico into southeastern USA is quite feasible, judging from botanical, avian and mammalian evidence beyond the scope of this paper, and as indicated by an isolated population of a species of the Neotropical rhopalopsyllid *Polygenis* on *Peromyscus* in Florida. Immigrant rodent stocks from Africa could also have brought with them other proto-hystrichopsyllids to account for the endemic tribe Ctenoparini in South America.

An alternate route of entry from Africa could have been, via Insectivora, to the West Indies. Thus, Croizat (1958) proposed a 'track' for Tenrecoidea as follows: Central Africa (and Madagascar) across the Atlantic to the Caribbean and thence to the southern USA. From the Caribbean islands, entry to Central America and South America would have been easy since (60, 161) many kinds of Neotropical mammals are supposed to have made the reverse trip. Hershkovitz (1972) also suggested an African-West Indies route, by rafting, for Insectivora.

An important question is why there had not been western and northwestern dispersal of *Ctenophthalmus* in the USA following entry from the Caribbean or Mexico. The answer may lie in the formidable barrier posed by a remnant of the Western Interior Sea (Cretaceous) or Cannonball Sea (Paleocene) which rent the continent, originally from arctic seas to the Gulf of Mexico (16, 121, 158, 192).

3.4.3.2 *Stenoponia* – The distribution of *Stenoponia* in North America somewhat suggests that of *Ctenophthalmus.* Two species are known in the New World: *S.ponera,* in the mountains of Durango, Mexico and in New Mexico, and *S.americana,* ranging south from Quebec to Alabama and west across the Great Plains to the foothills of the Rocky Mountains in Alberta. The genus is thus unrecorded from the Rockies proper and further west or north, including Alaska. Therefore, a Beringian dispersal seems unlikely. The subfamily Stenoponiinae includes only the one genus, and is well represented in the Mediterranean subregion, the Caucasus and the Far East. It is conspicuously absent in Western Europe, which militates against a transatlantic route of dispersal, as does *S.americana*'s absence in northeastern Canada. Since *S.americana* and other members of the genus are 'winter' fleas, it appears doubtful that low temperatures alone have excluded it from the northern Nearctic areas. If *Stenoponia* had existed south of the Sahara (where its temperature requirements may have led to subsequent extinction), its distribution would make sense, for even the

Far Eastern species could easily have been derived from the relatively rich group in North Africa and the Middle East, and the forebears of the Nearctic species could have traversed the route proposed for *Ctenophthalmus* via the South Atlantic.

3.4.3.3 Origins and dispersals of the phiomorphs – On the basis of the information extant, 'hystrichomorphs' cannot be termed major hosts for ctenophthalmines in either Africa or South America, although nothing is really known about the fleas of the phiomorph genera *Thyronomys* and *Petromus (= Petromys)*. There are some records of ctenophthalmine fleas from caviomorphs, but they generally are infested with fleas of bioendemic Neotropical families or with helmet-fleas. (The latter point suggests the hosts have been on the scene since antiquity or otherwise the late-entering ceratophyllids and Nearctic-type anoplurans would be parasitizing these hosts.)

It must be reiterated that there is no fossil evidence to support the idea of African roots for the South American 'hystrichomorphs', sigmodontines, etc. The caviomorphs suddenly appear in the Oligocene fossil records, with six of the living families recognizable (147, 148), and that is one reason why immigration seems so logical a solution as to their origin. However, there seems to be no reason to exclude Africa as the source, especially since there apparently is no paleontological material to strongly support the idea of Nearctic derivation either. Similarly, there are no fossil data to indicate that the sigmodontines were in North America at the appropriate time (161).

The data on ectoparasites suggests that the opposing views about the origin of the South American caviomorphs are really not as far apart as the literature appears to indicate. It is now generally accepted that the phiomorphs and caviomorphs have a common ancestry. The remaining and critical point of issue is not so much how the latter reached South America but whether their forebears originated in Africa or in North America. In discussing these questions, two important statements by Wood (1977a) should be borne in mind: 1) '. . . Middle America is the only area in the world, except for Western United States, where hystrichognathous rodents, potentially ancestral to the caviomorphs, are known to have lived during later Eocene times . . .' 2) '. . . there are no described Old World Eocene rodents that are demonstrably related to either the African Phiomorpha or the South American caviomorphs'. Wood (1977a) also suggested that the extinct phiomorph family Phiomyidae reached Africa from southwest Asia about the end of the Eocene but believed that the group actually originated in North America. Inasmuch as both the caviomorphs and the phiomorphs have Anoplura (and related taxa at that) and the Anoplura apparently arose in North America, it is logical to conclude that the 'hystrichomorphs' did originate in North America. The only reasonable alternative would be that these rodents had secondarily acquired sucking-lice from other hosts in South America or Africa. As we have seen, the caviomorphs could not have secondarily obtained their anopluran fauna in South America, for there was no source available. If the phiomorph Anoplura were secondary infestations from contemporary unknown rodents or perhaps anomaluroids or pedetids or others of doubtful affinity, then it could be argued that those hosts should be the ones that are related to the caviomorphs. No such candidates are readily apparent as substitutes for the phiomorphs. Moreover, their presence in Africa would have to be explained and the Anoplura would still point to North America.

The conflict would be readily resolved but for the dating of the events. By Early Oligocene, the South American rodents (caviomorphs) were quite diversified, more so than the known African fossils (231, 233), and if the former had Asian-African roots, then in the 3-5 million years' interval between the end of the Eocene and Early Oligocene, the proto-phiomorphs had to become established in Africa, raft to South America and, fide the chart by Patterson & Pascual (1972), radiate to form eight families. An entry into Africa in the Early Eocene would seem to be the minimum prerequisite for such African-South American connections (by rafting in Mid or Late Eocene), but it would fit with the concepts of monophyly of caviomorphs and phiomorphs; origin of the ancestral group in North America; infestation with Anoplura; the relationships of these Ethiopian and Neotropical lice; and the data on Siphonaptera. The alternate theory of caviomorph-phiomorph ancestry in Middle America and rafting or island-hopping to South America (231) is in somewhat direct opposition to the data on Siphonaptera and is inconsistent with the observations on lice, since there are no Middle American or US Anoplura that suggest close kinship to the 'hystrichomorph' lice.

Wood (1977a) wrote that the 'African Early Oligocene phiomyids are clearly related at the subordinal level to the caviomorphs, but could not have given rise to them, although, morphologically, the reverse could have occurred'. It would be unfortunate if a reader inferred from this that the African phiomorphs were derived from the South American rodents. Travel in the direction South America-Africa seems to be completely excluded by the data on ectoparasites. For example, although pygiopsyllid fleas occur in both South America (*Ctenidiosomus*) and Africa (*Afristivalius* and one species of *Lentistivalius*), the evidence regarding this group, and that on lice, contravene a hypothesis of eastward transatlantic dispersal, viz.: 1) *Ctenidiosomus* is a primitive taxon of the type associated with marsupials. The African genera are more advanced and with different affinities, and metatherians are unknown in Africa, living or fossil. 2) The species of *Afristivalius* are essentially murid fleas and there is no sign that these hosts (or their fleas) were ever in South America. 3) The murids are evolutionarily a youthful group and probably entered Africa long after the ancestral 'hystrichomorphs' emigrated. 4) The forebears of *Afristivalius* probably entered Africa, with the murines, later than the putative Early Oligocene date for phiomorphs in Africa. 5) Regardless of whence or how *Lentistivalius* reached Africa, there is no evidence to link it with *Ctenidiosomus* or South America. 6) If a pygiopsyllid had accompanied the emigrating 'hystrichomorph' stocks to Africa, a rich fauna of pygiopsyllids would be expected to have developed on the phiomorphs which had radiated so extensively. 7) The evidence suggests *Ctenidiosomus* originated in the Australian region. 8) The phiomorphs would be expected to have some Mallophaga, acquired by the caviomorphs during their sojourn in South America. 9) Anoplura related to those on Phiomorpha and Caviomorpha should occur in reasonable numbers in Central America but they do not.

The data on pygiopsyllids (1-7 above) also militate against the idea of faunal links between Africa and South America subsequent to the epoch of phiomorph emigration. Of course we are not treating introductions effected by man, e.g. the importation of *Tunga penetrans,* the chigoe, into much of Africa from South America as a result of the slave-trade.

3.4.4 *Evidence from viruses and endoparasites* – One possible source of evidence for Africa/South American faunal connections that does not seem to have been investigated is that of the virus of yellow fever. This infection is limited to South America and Africa concerning natural reservoirs, though of course outbreaks of human disease have occurred in many parts of the world due to ship-borne *Aedes aegypti* mosquitoes and infected personnel. On both continents the reservoir of the yellow fever virus lies in various species of arboreal primates (including galogos and monkeys in Africa), with canopy-mosquitoes as the vectors. It seems possible that the association predates the Eocene entry of the primates into South America by rafting from Africa. Mattingly (1960) in discussing mosquito-borne viruses, claimed that the 'picture of primate evolution during the early Tertiary presented by the fossil record suggests an origin during the Eocene in Europe and North America'. This probably would be too recent a date to account for an African origin of yellow fever virus in South American ceboid monkeys. However, in recent years the fossil record on primates has been vastly improved, and even advanced hominids date back to 4 m.y.a. in Africa (Washburn, 1978), while Cooke (1972) lists Eocene primates and believe their history in Africa may go back to the Paleocene. An Eocene date for the origin of arthropod viruses may hence be too recent. In 1969 Mattingly (136) wrote that viruses closely related to yellow fever occur only in Africa, and for that reason 'coupled with the generally greater (laboratory) tolerance on the part of African primates, seems strongly to suggest a post-Columbian introduction of yellow fever into South America by man'. This argument has lost much of its force because 'silent' foci of the infection area are now known in large areas of South America and Central America where the monkeys are obviously resistant to the virus.

The oft-posed question, 'Why is the yellow fever virus absent in Asia?' seems answerable by this hypothesis of the African origin of the agent. Thus another member of the Flavivirus Yellow Fever – Dengue Complex of Group B viruses occurs in Southeast Asia, namely the Dengue-Series itself. Like the yellow fever virus, these agents: 1) can cause a hemorrhagic disease; 2) are vectored by the same complex of *Aedes* mosquitoes as the African agent and 3) naturally infect primates in sylvan habitats. In short, the analogy with yellow fever is complete and I believe that the dengue and yellow fever groups of viruses are siblings, i.e., they had an ancestor in common, and that the former is the Asian derivative of that joint stock. It is premature to state whether the dengue virus infection known to occur in forest monkeys in Nigeria is the aboriginal condition or an introduction. Certainly, dengue (and the vector mosquitoes) has been imported into parts of the southern U.S.A. and various Pacific islands and into the Caribbean region, but no native virus in this group is known in North America, which, according to those who deny African-South American connections, was the homeland of the Neotropical monkeys.

Protozoological evidence superficially resembles the virological. The monkeys of Asia, Africa and South America have enzootic *Plasmodium* (malarial) parasites. However, in our present state of knowledge, it cannot be ruled out that this represents either an ancestral condition, regardless of the origin of the South American monkeys, or else local transfer of *Plasmodium* from other hosts. There are other infections whose distributions clearly suggest austral origins of the vectors but not necessarily African-South American connections at a date that would involve mammals, e.g. leishmaniasis and psychodid sandflies; *Onchocerca* filarial worms and simuliid blackflies, especially since the simuliids of the austral continents are supposed to be related. Primates lack specific fleas and hence there are no siphonapteran data concerning possible African origin of the ceboid monkeys.

3.5 *Transatlantic connections*

There seems to be only one other hypothesis that could explain some of the odd features of the distribution of *Ctenophthalmus* in North America, and that is to ascribe the original entry to transatlantic connections, from Europe via Greenland, etc., a route that is now believed to have been operable until about 50 m.y.a. (Mid Eocene) (124, 126, 231), even though the very idea of such faunal links was considered unacceptable to most mammalogists and paleontologists just a few years ago. Simpson (1947), on the basis of the fossil evidence, concluded that mammals had not dispersed via transatlantic routes, and offered evidence that the Palearctic-Nearctic exchanges were solely effected via Beringian connections. These points have been well reviewed by McKenna (1975), and Kurtén (1971) also mentioned an Atlantic landbridge in the middle of the Eocene. Wood (1977a) wrote of 'easy direct terrestrial migration' for rodents between Europe and North America in the Early Eocene. Until recently the case against Atlantic links was so strong that Holland (1958) naturally did not consider such a possibility when discussing the obviously close relationship between *Atyphloceras bishopi* of the eastern USA and *A.nuperus* of central and southern Europe, and the absence of this group in all areas in between (western North America and Asia, etc.). Holland logically concluded that the *A.bishopi* group must have originally been spread across the breadth of North America and Asia and then became extinct everywhere but where it exists today. With the new data on continental drift, faunal exchange via Greenland and other Atlantic links offers a better explanation, as is also suggested by the zoogeography of several other species of fleas. The leporid flea *Odontopsyllus* is known only from two species in North America and one (two subspecies) in Spain and France. Also, the Nearctic (and Neotropical) genus of rabbit-flea, *Cediopsylla,* is closely related to the European rabbit-flea, *Spilopsyllus.* How could a Beringian connection account for such a huge gap (Asia and eastern Europe) in the distribution of these fleas? The pattern, nevertheless, is not fully what one would expect regarding fleas that entered North America by the alternate route, from the northeast. Thus, *A.bishopi* and the western species of *Cediopsylla* barely enter southern Canada, and there are no published records of eastern *Odontopsyllus* and eastern *Cediopsylla* in that country. Even so, glacial extinctions may account for the observed distribution, and Atlantic connections seem to offer the best explanation overall. That solution does not apply well for *Ctenophthalmus,* however, since the genus is absent in northern North America, and there is but one species known north of Mexico, in contrast to the eight in that country or Central America. While there are a few genera of fleas that are more populous south of the US border than north of it (e.g. *Kohlsia, Jellisonia,* and *Pleochaetis*), unlike *Ctenophthalmus,* those are ceratophyllids and are clearly of Nearctic origin. The unusual distribution of *Stenoponia* has already been discussed (p. 145).

The fact that the Ceratophyllidae provide no examples suggesting transatlantic connections actually supports the concept of such links. The family is a youthful one and probably did not arise until after the Atlantic routes were closed by marine barriers. There are instances of trans-Beringial exchange in the ceratophyllid fauna, however (66, 67), but that bridge was intermittently operable until Mid Pleistocene (66, 126).

3.6 *The zoogeography of* Rattus *and their fleas*

One of the most important and intriguing riddles in biology is the source of the murid rodents, but limitations of time and space, as well as on data on the fleas of primitive murines from the Philippines, Sulawesi, etc., preclude discussion at this time. Previously (205, 206), I outlined reasons why I felt that murids could not have arisen in India or mainland Asia, thereby disagreeing with some of the basic tenets of Simpson (1961), par-

Table 6. Distribution and major hosts of *Leptopsylla* and allies (Leptopsyllidae, Leptopsyllinae)

Taxa		*Leptopsylla (Leptopsylla)*			*Leptopsylla (Pectinoctenus)*		*Peromyscopsylla*		*Sigmactenus*	
No. of species		9			6		18		6	(Total)
Major hosts		Rats or *Apodemus*	*Mus*	Other	*Apodemus*	*Cricetines*	*Apodemus* or *Rattus*	Other	*Rattus* etc.	
No. of species thereon		5	1**	3	5	1	1	17	6	39
Palaearctic region (23)		4		3	5	1	1*	9		23
	Indian						1*			1*
Oriental region (3)	Malayan A. Borneo								1	1
	B. Philippines						1*		1	2*
Australian region (2)	New Guinean subregion								2	2
Wallacean region (2)									2	2
Ethiopian region (1)		1***	1***							1
Nearctic region (6)								6		6
Holarctic region (2)								2		2

* One species occurs in more than one area. ** *L.segnis*, cosmopolitan species. *** Species found on several hosts.
Holarctic species are *not* tallied in Nearctic or Palaearctic columns.

ticularly since I also regarded his proposed Miocene date as too recent to permit the radiation and specialization seen in the murine ectoparasites or to fit the data on their zoogeography. None of the available data contradict the idea that the murids *per se,* or at least the murines, could have arisen in the Wallacean subregion or in the nearby Malfilindo Archipelago. However, the evidence on Anoplura clearly suggests that the ancestral proto-murids arose in the Palearctic, and at a time subsequent to the origin of the sciurids, cricetids, phiomorphs, etc., viz. 1) the ubiquitous nature of the Anoplura infestations on murids (e.g. on hydromyines, otomyines, dendromyines and Murinae, wherever they occur); 2) only two genera, *Polyplax* and *Hoplopleura,* are involved; 3) a wealth of species of murid lice are nevertheless represented in these two genera; 4) there is virtually a complete lack of the relatively primitive, non-nominate polyplacids on murids; 5) there are no native murids in the New World; 6) murids lack Mallophaga, which would seem to rule out Australia as a homeland, particularly since the bioendemic murids on that continent have the same genera of Anoplura as occur in the Palearctic; 7) murids lack the genera (and some families) of lice that are typical of rodents like sciurids, dormice, geomyids, etc.; 8) the Anoplura arose in North America, not in the tropics; and 9) the Asian rodents (and their lice) are Nearctic in origin. If, for example, the Anoplura of the murids in Africa had been derived from local sources, it would be expected that they would have species of the polyplacid *Eulinognathus* as do the phiomorphs, and *Pedetes* and dipodids do in the area (p. 156), but none of the authorities (45, 69, 70, 79) cite such reports. Except for one or two obvious strays, the only records of a non-*Polyplax* polyplacid from a 'murid' pertain to *Cricetomys,* a primitive African rodent that has been considered a cricetid and is still regarded so by some specialists. (That *Cricetomys* may have had an exceptional history and ancestry is indicated by that it is the only taxon of rodent truly infested by undoubted parasitic earwigs, the genus *Hemimerus.*) *Hoplopleura* and *Polyplax* infest rodents in many parts of the world where native murids do not occur, e.g. the New World and extreme northern Asia. Species of *Hoplopleura* occur on squirrels in various parts of the world, and *Polyplax* infests microtines and other cricetids, indicating that these genera could have arisen on non-murid hosts. All in all, it seems likely that the protomurids acquired their Anoplura in the Palearctic region and carried their lice with them before they evolved into true murids.

It is important to note that the data on fleas continue to support Simpson's view (1961) that *Rattus* entered the Australian region 'through the East Indies' and did so relatively recently, geologically speaking (although the Simpson date of Pleistocene may not be early enough). For example, new data on fleas extend the evidence previously reported (205, 206) and which is based upon the distribution of *Sigmactenus, Neopsylla, Rothschildiana* and *Stivalius (s.str.).* Thus, among the Leptopsyllidae, which are almost undoubtedly of Palearctic origin and are primarily so in their current geographic range, *Sigmactenus* is found on rats in Borneo, the Philippines, the Wallacean region (Timor and Sulawesi) and New Guinea, as shown in table 6. (The Wallacean species are new to Science, and were collected by the US Naval Medical Research Unit No. 2). The related genera, *Leptopsylla* and *Peromyscopsylla* are mainly Palearctic but are found as far afield as Ethiopia *(s.str.)* for the former, and the Oriental region (Indian Subcontinent and the Philippines) as well as the Nearctic region, for the latter (table 7). (The Philippines record, which was in conjunction with the finding of another leptopsyllid, *Frontopsylla* sp., may conceivably represent an introduction and hence was not listed in tables 1 and 2.)

The distribution of the neopsyllines *Neopsylla* and *Rothschildiana* (table 7) is in

Table 7. Distribution and major hosts of *Neopsylla* and allies (Hystrichopsyllidae, Neopsyllinae)

Taxa		*Neopsylla***		*Rothschildiana*	Nine other genera
		setosa-group	*stevensi*-group		
Major hosts		Cricetinae & Sciurids	Murines	Murines	Rodents, etc.
Number of species (88)		14	15	2	57
Distribution of species					
Holarctic (1)					1
Palaearctic (26)		14*	7		5
	Indo-China		3*		
	South China	1*	3*		
Oriental (11)	Malaya		1*	1	
	Borneo		1		
	Java		2	1	
Nearctic (52)		1			51

* One species occurs in more than one area. ** Excludes three Chinese species with inadequate data.

accord with these concepts. In the subfamily there are 88 species for which we have adequate data for analysis, and 77 (88 per cent) of these are found in the Palearctic or Nearctic region. Eighty-two per cent of the genera are restricted to northern areas, and these taxa include over 60 per cent of the species. The genus *Neopsylla* includes 29 species (33 per cent of the total), and the two taxonomic groups into which these were divided on purely morphological grounds, have turned out to apply to be equally applicable regarding the hosts (205). Thus, 52 per cent of the *Neopsylla* are in the *stevensi*-group and these all parasitize murines. It is predominantly this group that occurs in the Oriental region, viz. eight of nine (89 per cent) of such species of *Neopsylla.* Notably, although squirrels occur in the region, including the Malfilindo Archipelago as far as Sulawesi of the Wallacean region, primarily these are callosciurines, a southern derivative. The *setosa*-group, and other squirrel-infesting Neopsyllinae, are associated with sciurids of Holarctic location or affinity. Oriental *Neopsylla* constitute 38 per cent of the total, and three species are endemic in the Indomalayan Insular Areas (including an undescribed species from Java). The closely allied *Rothschildiana* is known from two *Rattus*-infesting species in Malaya and Java respectively. As in the case of *Sigmactenus,* the Asian mainland origin and association with *Rattus* are obvious for these neopsyllines.

The genus *Stivalius s. str.* is also intimately associated with *Rattus,* particularly the subgenus *Rattus* (203). This specialized taxon has its roots in the Papuan subregion and apparently encountered the later waves of the rats (including *Rattus rattus*) as they moved back and forth across the Malfilindo Archipelago and accompanied *R. rattus* to Sri Lanka, the Indian subcontinent, Thailand, Indo-China, southern China and the Philippines, speciating en route (203, 205). To this list may now be added a new species from Timor and another from Sulawesi, recently collected by the Navy Unit, and several new subspecies from the Philippines, gleaned from other sources.

Reasons for regarding all these taxa and associations with *Rattus* as relatively recent as compared to the more primitive rats of the Papuan subregion, the Wallacean region and Indo-Malayan subregion, and their respective fleas include: 1) the unusual aedeagal modifications seen in *Stivalius,* 2) the ornate chaetotaxic developments of *Sigmactenus* as compared to its mainland allies, and 3) all the familiar taxonomic problems seen in recently evolved, rapidly-changing taxa, analagous to those that have made the systematics of *Rattus* a scourge to mammalogists (203-206, 209). Subsequent observations support these contentions, as well as the view that while these fleas and their hosts have evolved more recently than the bulk of the pygiopsyllid fleas and their more primitive murine hosts, the actual entry of *Rattus* into New Guinea probably was earlier than the Pleistocene date indicated by Simpson (1961). Thus, dealing with the latter point first, *Sigmactenus* has been in New Guinea long enough for the evolution of a species (*S.cavifrons*) which is specific to *Pogonomelomys,* a rat of ancient lineage, in contrast to *S.toxopeusi,* which infests *Rattus* and *Melomys.* Moreover, the latter species of flea has apparently developed subspecies characteristic of *Rattus* or *Melomys.* The *Sigmactenus* of the Wallacean region and the Indo-Malayan subregion often seem to be parasites of native subgenera of *Rattus,* rather than *R.(Rattus),* indicating entry on an earlier wave of invasion than that of the latter, and subspeciation of *Sigmactenus* has occurred on some of the isolated mountains of Mindanao, if not on some of the various islands of the Philippines. As for the comparative youthfulness of the *Rattus,* the fleas in question have specialized features suggesting recent evolution. *Rothschildiana* is obviously a derivative of *Neopsylla* which has heavier tanning and other taxonomic characteristics superimposed on a *Neopsylla* framework. The *Neopsylla* of murids have supernumerary spines and bristles, which are adaptive features so often seen in fleas of scansorial nocturnal hosts (Traub, 1979).

Another telling point is that the murids in the Insular Indo-Malayan Area are not infested with ceratophyllid fleas, even though members of that family are present in the local fauna, infesting sciurids. This suggests that contact between squirrels and rats in that region has not been in existence long enough to permit transfer of ceratophyllids from sciurids to rats, a procedure that would be complicated by the already existing infestation of those murines by pygiopsyllids, leptopsyllids and neopsyllines. The sciurids in Borneo, Mindanao and Indonesia are of northern origin, as are their ceratophyllids (Traub, 1972c). Notably, murids have very few ceratophyllid parasites anywhere in their range. There is only one such genus (*Nosopsyllus*) in Africa, and it is present on murines, gerbillines and sciurids in the Middle East and the Oriental region as well. Some species of the aberrant ceratophyllid *Paraceras* infest rats in the southern Palearctic region and mountains of Malaya, etc., but others occur on badgers and other carnivores, and these may represent the original group of hosts. A few *Callopsylla* are found on murines on the Indian subcontinent, but again other types of hosts are also involved. The sparsity of the ceratophyllid rat fauna indicates limited contact in time and space, and that is an argument for the concept of relatively recent evolution of the rats, as well as for my contention that the murids, and especially the rats, did not originate on the Indian subcontinent.

There is another group of *Rattus*-fleas that may indicate comparatively recent entry of those rats into the Australian region via Indonesia, and that is the *Xenopsylla vexabilis* complex (Traub, 1972c; Traub & Elisberg, 1972). *X.vexabilis* is found in Thailand and Vietnam, where it infests *Rattus (Bullimus) berdmorei* and occasionally *R.(Rattus).* It has been reported from the Philippines, Java, New Guinea and Australia, always in relatively dry climes or areas, as in savannah, and generally on the subgenus *Rattus,* and the US Navy Team has recently collected it on Timor on that host. It has been introduced into

Hawaii on commensal rats, and no doubt occurs in parts elsewhere in the Pacific, and probably on native rats in unreported foci as well. A sibling species has been reported in Australia, parasitizing *Rattus* and bioendemic murids, namely *X.australiaca,* which may be of doubtful validity. A related species, *X.nesiotes* occurs on *Rattus* on Christmas Island, south of Java. These data may suggest that *X.vexabilis* originally entered the Australian Region via a path from the Palearctic through Indo-China and Thailand and then the Australo-Asian archipelagos, infesting *Rattus* en route. The situation is far more complex, however. There are no bioendemic *Xenopsylla* known in Malaya, Borneo, Sumatra, etc., but this may be because the current conditions are constantly far too mesic for *Xenopsylla.* Even the cosmopolitan *X.cheopis* is much more abundant in the dry season there than in the wet (Traub, 1972e). More significant is the fact that highly modified, bioendemic species exist in New Guinea (*X.papuensis*) and the mountains of Luzon (*Xenopsylla* n.sp.), infesting native rats. Since these fleas are eyeless, are poor jumpers, etc., it seems that their evolution would have been a long process – one that perhaps would antedate invasion by *R.(Rattus),* if not *Rattus.* This suggests the possibility that the Pacific *Xenopsylla* are derived from Australian roots, which in turn, would have had their origin in the Ethiopian region, where *Xenopsylla* is extremely well developed, constituting about 62 per cent of the known species. However, if such were the case, one would expect a rich fauna of *Xenopsylla* in Australia, but at most there are only two known species that are conceivably bioendemic. More data are needed before the questions about the origin of the Pacific *Xenopsylla* can be resolved.

It should not be inferred from these observations that, since the sciurids are an older group than the murines, the squirrels appeared on islands like Sulawesi, Borneo, Mindanao, etc. prior to the rats. On the contrary, the evidence suggests that the murines were there first, viz.: 1) the squirrel-flea genus *Medwayella,* common on sciurids in these archipelagos is a derivative of the rat-pygiopsyllids, with its ultimate roots in the Australian region (203). The same is true for *Farhangia.* 2) The local murines are considerably more differentiated than the squirrels, more so than can apparently be accounted for on grounds of a more rapid rate of evolution. 3) Bioendemic genera of fleas occur on some of the rats on these islands and in general the murine flea fauna is more diversified than the sciurid Siphonaptera. 4) Giant tree-rats exist only on islands where squirrels do not occur (e.g. Luzon, Flores, New Guinea), indicating that on the other islands these tree-rats could not compete with the newly invading squirrels (205). These points also indicate a longer period in situ for the rats than the squirrels. If the murines entered in the Pleistocene, as has been alleged, when did the sciurids do so?

The youthfulness and other aspects of the geological history of the New Guinean highlands has made it possible to estimate the age of the pygiopsyllid *Tiflovia* and that of the two component species (208) and that in turn has some bearing on the age of murines. Thus, it was suggested that this genus at the earliest is a Pliocene development. If, however, this alpine taxon turns out to inhabit the mountains of West Irian with peaks of eternal snow, then it may very well be more ancient. If not, then in three million years there has evolved a specialized genus of murid flea with some of the marked modifications seen in nest-fleas. Regardless, to the maximum of Pliocene age must be added the time required for the ancestor of *Tiflovia* and the forebears of its apparent host, the rather primitive murid *Pogonomelomys,* to reach New Guinea and to evolve to their current level. At present there are insufficient data to enable one to estimate that interval of time. It appears, however, that the two species of *Tiflovia* became differentiated since the

Pleistocene (208). While the rate of speciation of *Rattus*-fleas cannot be extrapolated from this observation, it seems reasonable to believe that such a dating of 1-3 million years could account for the formation of the various species of *Sigmactenus* and other *Rattus*-fleas. However, surely a far greater period of time would have been required for the development of *Sigmactenus* from murine-leptopsyllid stocks on or near mainland Asia, especially since the head comb and tibial combs are both well-developed in this taxon (209).

The data on fleas clearly suggest that several waves of *Rattus* moved between mainland Asia and the Australian/Asian archipelagos, regardless of where the murids, or murines, themselves originally arose. The wave of *Rattus* was predominantly in the direction of mainland Asia towards Australia, as indicated by their Palearctic fleas. One major dispersal, effected later, involved *R.(Rattus)* and was largely in the direction away from the Australian region, as denoted by *Stivalius (s.str.)* fleas with Australian roots, on *R.(R.) rattus* and allies, in India, Sri Lanka, Java and the Philippines, etc. (203, 205, 206).

3.7 *Madagascar connections*

The original position of Madagascar on Gondwanaland and the movements of the island after the breakup of the supercontinent are highly controversial subjects. Its present position and major components of the fauna and flora immediately suggest an original fit with Africa, and most authorities accept such a geological relationship, as outlined by Cracraft (1973a). Precisely where Madagascar was positioned there is a matter of dispute, and a site against Mozambique has been proposed (e.g. 189), while others have suggested southern Africa. Some have opted for a position opposite Kenya, Tanganyika or even Somalia (182, 196). Some workers have doubted that Madagascar was ever joined to Africa or drifted at all (166, 190). Fooden (1972) used the extraordinary mammalian fauna of Madagascar as a basis for interpreting Madagascar's past history, and considered such points as: 1) The four groups of extant mammals consist of 38 genera, of which no less than 34 are bioendemic, viz. 10 tenricid insectivorans; 10 lemuroids; seven viverrids; and seven genera of rodents, all in the cricetid subfamily Nesomyinae, which is restricted to Madagascar. 2) Six of the seven orders which are missing on the island occur in Africa. 3) The aardvarks (Tubulidentata), today found only in the Ethiopian subregion, are represented by a Malagasy Pleistocene fossil. 4) Certain families present in Africa are missing in Madagascar, i.e., four families of Insectivora, three of primates and 11 of rodents. 5) Native murids are unknown on Madagascar, but the Nesomyinae, an older group, have radiated there. 6) Marsupials and monotremes were absent. Fooden (1972) hypothesized that 1) Madagascar detached from Gondwanaland after Australia and South America had already done so, and 2) its current fauna (eutherian only) reflects the stage of mammalian development at the time of fragmentation, i.e., the monotremes and metatherians had already become extinct. McKenna (1973) and Lillegraven (1975) and others, objected to this theory, which automatically assumes marsupials had existed in Africa, etc. A simpler explanation for the unusual contemporary fauna is suggested by the data on fleas, namely that prior to drifting to its present position, Madagascar lay close to Somalia. In that way it could have received a filtered Palearctic fauna, e.g. cricetids and leptopsyllid fleas, from Southwest Asia, and, later, selected elements of mammals and ectoparasites of the Ethiopian subregion. For example, although Madagascar is a typically impoverished insular area with a total of only five genera of fleas (of non-volant hosts) which contain indigenous species, two of these genera are leptopsyllid fleas (with a total of nine species), primarily on autochthonous cricetid hosts. In contrast, in all of Africa, with its vast size and varied habitats, and immensely rich fauna of rodent hosts, there are only two genera of leptopsyllids, both Palearctic, and with a total of four species in Africa, all North African or in Ethiopia. The Leptopsyllidae, as outlined above (p. 105) are a northern family, predominantly Palearctic and no doubt aboriginally so. The case of *Caenopsylla* is illustrative. This African and Middle Eastern genus includes a specific parasite of the gundi, *Ctenodactylus,* a North African rodent. The family Ctenodactylidae arose in central Asia and, in the Miocene, was represented on the Indian subcontinent, North America and Sardinia (Wood, 1977b). It is not surprising that a rodent with such a history carries a leptopsyllid. Interestingly enough, some workers had considered this group as 'hystrichomorph' but Wood (1977b) regarded the resemblances as due to convergence. The data on fleas support that opinion since the phiomorphs lack leptopsyllids. No lice have been reported on gundis, so the absence of phiomorph Anoplura cannot be evaluated.

Other aspects of the siphonapteran fauna of Madagascar reflect its present geographical position, e.g. a *Dinopsyllus* rodent-flea; a monotypic pulicid genus adapted to a tenricid insectivore and another bioendemic pulicid genus that has radiated somewhat on the cricetids – all endemic species with roots in the Ethiopian SR. The African background, however, is clearly not as dominant in the rodent/flea fauna as is the Palearctic, a point strikingly demonstrated by the absence of murids. The long isolation of Madagascar is insufficient per se to account for the characteristics of the fauna.

The data on rodent lice are in general accord with these points. Thus, the few known Anoplura from the native Malagasy cricetids include lice that are polyplacids and belong to groups of cricetid lice from other areas, e.g. the *Polyplax jonesi* complex (96) although one member thereof infests the murid *Saccostomus* in the Ethiopian subregion. There are only two cricetids in the Ethiopian SR, and one of these, *Mystromys,* is parasitized by a member of the *P.jonesi*-group. Another Madagascar cricetid-louse is in the polyplacid genus *Eulinognathus,* a taxon whose distribution is edifying and pertinent to austral connections, viz. one species each in: 1) South America (on *Ctenomys,* an echimyid caviomorph); 2) Shensi, China (on a dipodid); 3) Tunis (ex a dipodid); 4) East Africa (on the second of the two cricetids in the Ethiopian region). Two species occur in South Africa – one on *Bathyergus* (the Cape mole, which some have regarded as a 'hystrichomorph') and one from *Pedetes,* an obscure anomaluroid (96). *Eulinognathus* thus suggests African/South American connections as well as Palearctic elements (cricetids) in Africa and Madagascar. Parts of southern China are believed by some authorities (32, 143, 189) to have been part of Gondwanaland, but *Eulinognathus* is probably too youthful to reflect such connections directly, and the association re Shensi probably is one via dipodid rodents and southwest Africa and North Africa.

Instead of the above hypothesis regarding an aboriginal northern position, it might be argued that the reason why Madagascar has cricetids and cricetid-fleas is because the ancestral fauna was derived from Africa at a time when the rodents on that continent were predominantly cricetids, prior to their virtual extinction there through competition with newly-invading murids. According to that theory, the cricetids and their fleas survived in Madagascar because of its isolation and the absence of murids. Thus, Darlington (1963) believed that murids 'replaced most cricetids'. The basis for his claim that cricetids had once been 'dominant' in Africa was that the group is represented on Madagascar, presumably since in a world of stable continents, no other explanation could readily account for such Malagasy rodents. If this were so, it would follow that the Palearctic-aligned fleas of Madagascar were descended from such hypothetical African cricetid-fleas. However, there is nothing in the fossil fauna to indicate that cricetids had been abundant in Africa. Thus, only murid fossils were mentioned by Cooke (1972) for the Ethiopian region for the Pliocene and earlier, not cricetids. There is nothing in the siphonapteran data to suggest that cricetids were ever abundant in Africa, even in the Palearctic portions thereof. With the possible exception of *Ctenophthalmus,* none of the Siphonaptera in the Ethiopian R could be considered as cricetid-fleas, and if *Ctenophthalmus* did not arise on an insectivore or phiomorph in Africa, it could have entered from Eurasia on such a host. Finally, it is difficult to envisage how the cricetids could manage to reach an immobile Madagascar, but not the eminently successful murids (except for the recent introduction of commensal rats).

3.8 *Origins of the major groups of Siphonaptera*

On the basis of the material presented herein, the history and origins of some of the major taxa of Siphonaptera appear to have been as follows. (Families like the Ischnopsyllidae, Vermipsyllidae, Tungidae, Coptopsyllidae, Malacopsyllidae, etc. are not being considered.) Fleas as such were in existence on Pangea and were represented on both Laurasia and Gondwanaland before their fragmentation into the continents and subcontinents. Laurasia included ancestral hystrichopsyllids which gave rise to the Hystrichopsyllinae, Neopsyllinae, Rhadinopsyllinae and Anomiopsyllinae. The evolution of the Leptopsyllidae, in Eurasia, was a much later event, and that of the Ceratophyllidae, probably in North America, as indicated by a basic association with sciurids, was even later than that. The Gondwanaland Siphonaptera apparently differentiated into families or subfamilies after the continents were split off, and it seems likely that on the supercontinents fleas were spottily distributed, since some of the subsequent components presumably lacked Siphonaptera originally (or else they became extinct before the islands, e.g. India and Madagascar, drifted to their present position). South America included the precursors of the Stephanocircidae and Macropsyllidae (helmet-fleas); the Doratopsyllinae (of the Hystrichopsyllidae) and the Rhopalopsyllidae, although it is barely conceivable that the roots of the last lay in hystrichopsyllids brought in from Africa in the Eocene. The Ctenoparini were either derived from autochthonal hystrichopsyllids or else their ancestors entered as African emigrants. Africa was the homeland of the Pulicidae, at least for the Archaeopsyllini, Xenopsyllinae and Pulicinae; and the Spilopsyllinae arose either in North Africa or elsewhere in the southwest Palearctic region. Hystrichopsyllid roots gave rise to the Ctenophthalminae, Listropsyllinae and Dinopsyllinae in the Ethiopian R and perhaps the Stenoponiinae originated there as well. Elements of the Ethiopian Ctenophthalminae and Stenoponiinae and perhaps of other hystrichopsyllid stock and Pulicidae, accompanied the primogenitors of the caviomorphs and/or other rodents across the south Atlantic to South America. The Chimaeropsyllidae, which have not been discussed, are a purely Ethiopian SR development. The Pygiopsyllidae arose in Australia.

These conclusions are based in part upon the premise that the most primitive fleas extant are associated with the more primeval hosts, as indicated above (p. 100 et seq.) and elsewhere (e.g. 204). It can be no coincidence that such fleas are generally unusually large and show hyperdevelopment of combs and/or bristles, e.g. the macropsyllids *Stephano-*

psylla and *Macropsylla;* the pygiopsyllids *Ctenidiosomus* and *Uropsylla;* certain stephanocircids of marsupials; and the hystrichopsyllid *Hystrichopsylla.* This suggests that primordial fleas may have been large and multi-combed (but not necessarily the truly ancestral one, which probably could not yet have become so well adapted re ctenidia etc.). However, it also seems likely that the development of these chaetotaxic features also represents the intrinsic factor of the size of the flea itself, for ectoparasites that are inordinately large must be especially susceptible to damage by the teeth and claws of the host, and the combs and bristles presumably enhance their chances of escape or remaining in situ (209). Significantly, those exceptional giant fleas which are not very hairy, and which have no combs (e.g. certain vermipsyllids) infest hosts which have very dense fur and/or are hoofed animals, and hence the fleas are not at special risk because of their size.

3.9 *The ectoparasitic fauna and the time of origin of the host*

On p. 100, mention was made of the principle that ectoparasites rarely infest hosts which are at a lower phylogenetic and evolutionary level than their original ones. This point is not restricted to the fleas and lice of mammals and appears to have a broad application. There are no Siphonaptera, Anoplura or Mallophaga on amphibians and reptiles and this cannot be due solely to the lack of hair or feathers, because starving fleas will feed on lizards, for example. Poikilothermy or aquatic habits are obviously important factors here, but the pattern of infestation transcends such traits, since mosquitoes feed on reptiles and amphibians, and those hosts do not have characteristic insect ectoparasites. In fact, there is a marked correlation between the position of the host on the tree of life and the component fauna of ectoparasites, but as we have seen, *where* the hosts originated superimposes a picture on that regarding *when* they did so. Considering just the factors of time and primary infestations, the situation can be summarized as follows for ancient major groups of hosts. The Amphibia are the oldest (Paleozoic) and they are parasitized only by mites. The reptiles, also Paleozoic but much later, have mites and ticks. The birds apparently date back to the Jurassic and have mites, ticks, certain parasitic Diptera, Mallophaga on all groups, and secondarily, and much later, scattered infestations of fleas. The marsupials, possibly Late Jurassic, have mites, ticks, Amblycera, stephanocircids, doratopsyllines and primitive pygiopsyllids. The Insectivora probably are close to the age of the marsupials and are infested with mites, ticks, doratopsyllines, other primitive hystrichopsyllids and, occasionally, specialized leptopsyllids or pygiopsyllids in Africa. They presumably lack primary Anoplura. The bats (Chiroptera) may date back to the Cretaceous (146) and are infested with mites, ticks, Diptera and hemipteran and other unusual ectoparasites, as well as a characteristic siphonapteran family (Ischnopsyllidae) related to the Leptopsyllidae, and a few pulicids. The lagomorphs probably arose in the Early Paleocene and possess mites, ticks, leptopsyllids, pulicids, and occasionally a ceratophyllid, while the very few Anoplura and Mallophaga present are presumed to be secondary. The rodents originated in the Paleocene and have mites, ticks, Anoplura, Mallophaga and various groups of fleas which again are correlated with the age of the family of host involved. Thus, hosts arising before the Paleocene lack Anoplura, but all have mites and ticks, etc. The times of origin of host and ectoparasite exhibit analogous interrelationships with various pathogens of mammals (e.g. 215), but discussion of these points must await another occasion.

4 CONCLUDING REMARKS

In general, then, the fleas and lice offer many clues of prime significance regarding the historical geography and evolution of their hosts. This faunal approach can also contribute to other branches of biology. For example, now that data are finally being obtained on lice from certain wild mammals instead of merely specimens in zoos, it can be argued that evidence on Anoplura support the concept of links between man and the great apes. Pubic lice of the distinctive genus *Pthirus* (often misspelled *Phthirus*) occur on both man and the gorilla, and the two species involved are instantly separable (Kim & Emerson, 1968), so there is no question of misidentification or accidental transfer. Chimpanzees, like man, have a *Pediculus* (p. 141) and here again the two species cannot be confused (95). The obvious inference is that the *Pediculus* and *Pthirus* stocks antedate the divergence of the hominoid branch from that of the anthropoids, which Washburn (1978) places as 5-10 m.y.a. It therefore appears that 'Adam had 'em' is not only the world's shortest poem, but is accurate as well. Still, it would help if we knew whether orang-utans and gibbons had lice and if so, what their affinities were, but such information is unavailable. Regrettably, such gaps in our knowledge of ectoparasites are common. There are no records of fleas or lice in some of the most critical areas regarding zoogeography or from hosts like the phiomorphs, the bioendemic murines of the Philippines, the giant tree-rats of Indonesia, some tenrecs, etc. The situation is even worse regarding lice, because of the absence of records from many kinds of hosts which may be expected to bear such ectoparasites, such as the grizzly and polar bears, moose, Old World moles, and perhaps North American shrews, to say nothing about exotic mammals (and birds). As K.C.Emerson points out (in litt.), data on lice may resolve the controversy as to whether the English settlers introduced the red fox into the USA (for purposes of hunting). The undoubtedly native, grey fox has lice which are distinctive from those of the European red fox, but lice from the US red fox have not been collected (nor, apparently, even sought). Another problem is that frequently it is impossible to say that a dearth of records means lack of infestation of a particular taxon, since the real reason may be a complete lack of data – or inadequate methodology or sampling. It is hoped that this article may encourage the collection of data on parasites and their use as evidence in zoogeography and evolution.

4.1 *Conclusions*

Data on the distribution, taxonomy, phylogeny and evolution of fleas and lice can contribute materially to studies on the zoogeography, history and relationships of their hosts.

The alleged low degree of specificity of fleas does not constitute valid grounds for depreciating their significance in zoogeography. Thus, there still may be important clues in their generic relationships and in those at the higher levels. There are also noteworthy phylogenetic associations, e.g. primitive hosts have primitive fleas (and lice) and the most advanced mammals tend to have what are clearly the most evolutionarily youthful fleas. The 'exceptions' in specificity generally also provide evidence that is pertinent.

Fleas tend to parasitize the group of hosts with which they co-evolved or else infest hosts which developed later and to which they could become adapted. In general, they do not infest mammals lower on the evolutionary scale, presumably because such hosts either are already in equilibrium with their own fauna of ectoparasites, or because the fleas are so specialized for life on the hosts with which they co-evolved that they cannot transfer to others in the absence of prolonged association. In the few instances of a switch to a more primitive host, there is evidence of such sustained contact, e.g. the development of pointed pronotal spines in rat-fleas adjusting to coarse-haired marsupials. Primitive hosts, and their fleas and lice, tend to be conservative and change very slowly at the generic or even species level, especially as compared with taxa that have developed relatively recently such as murids and their parasites. The close correlation between the kind of ectoparasite occurring on tetrapods and the geological epoch when the hosts arose extends to class and order, with only mites occurring as true parasites of amphibia; both mites and ticks on reptiles, etc.; extending to a gamut of ectoparasites on rodents. Thus, hosts arising before the Paleocene lack Anoplura, etc.

The zoogeography of fleas does not fit the pattern expected if they had boreal roots, e.g. some families in the Southern Hemisphere show little or no evidence of either origin or occurrence in the northern region. In contrast, several groups exhibit a distribution demonstrating faunal relationships among some of the southern continents, e.g. stephanocircids, doratopsyllines, pygiopsyllids, pulicids, etc. Some primitive forms occur on the austral continents, and there are advanced ones as well. Significantly, the groups of fleas and their hosts which demonstrate signs of austral affinities and links are the more primitive forms. When there are northern roots, they are clearly discernable and pertain to recent invaders. The features of the siphonapteran fauna in the Palearctic and Nearctic regions do not jibe with the anticipated pattern either. Thus, there are no signs of roots of some of the characteristic austral families. The main families (Ceratophyllidae and Leptopsyllidae) there are overwhelmingly consistently boreal and are clearly evolutionarily younger than those indicating austral relationships. There are few relict or distinctive tribes or subfamilies as compared to families found in Australia, South America or Africa.

The siphonapteran relationships exhibited among the southern continents are at the level of subfamily or family level and not at the generic.

The marsupials arose in South America, in the Late Jurassic or Early Cretaceous, and dispersed to Antarctica by island-hopping in the archipelago between the west coast of South America and Antarctica, and then penetrated Australia by at least Early Cretaceous. Other elements of the fauna moved north into North America at about the same time, and descendents later dispersed into Europe by transatlantic connections. The geological and parasitological evidence completely exclude travel to Australia by the route North America-Asia-Indonesia-New Guinea. The marsupials emigrating to Australia carried stephanocircid and proto-doratopsylline fleas and amblyceran Mallophaga. It is possible that later an ancestral thylacinid made the reverse trip to South America (carrying the forebears of the pygiopsyllid *Ctenidiosomus*) and gave rise to the borrhyaenids, which subsequently became extinct.

Fossil and other evidence indicate that fleas were present in Australia in the very Early Cretaceous, if not earlier. The apparent primary infestation of marsupials by Amblycera in both Australia and South America and the absence of mammal Amblycera in areas where the marsupials are absent support the concept of austral origin of the marsupials.

The current impression of mammalogists and paleontologists that the Insectivora arose in Asia is corroborated by the parasitological evidence. They lack amblyceran Mallophaga, which presumably originated in South America (at least regarding mammal-lice), and likewise have no primary infestations with Anoplura, which have North America as a provenance. The data on fleas are in accord with the new concept that the Insectivora entered North America without fleas and acquired doratopsyllines from marsupials.

The order Siphonaptera is monophyletic and must date back to the Jurassic, and fleas, as such, were in existence on Pangea. Laurasia included ancestral hystrichopsyllids which gave rise to subfamilies like Hystrichopsyllinae, Neopsyllinae, Rhadinopsyllinae and Anomiopsyllinae. The Leptopsyllidae arose in Eurasia at a much later date, and the Ceratophyllidae, probably North American in origin, is the most youthful group of all. South America was the provenance of the Stephanocircidae, Malacopsyllidae, the Doratopsyllinae (Hystrichopsyllidae) and the Rhopalopsyllidae. Africa was the homeland of the Pulicidae, but *Pulex* was native to Central America and South America. The Ctenophthalminae, Lystropsyllinae and Dinopsyllinae arose in the Ethiopian subregion, and perhaps Stenoponiinae did too. The Pygiopsyllidae are Australian in origin, as are the Macropsyllidae.

The Anoplura arose in North America and probably were already infesting rodents in the Paleocene. In all likelihood, they were parasitizing other hosts (but not marsupials or Insectivora) prior to the rodents, and thus may date back to Early Paleocene. Since the amblyceran Mallophaga infest both Australian and South American marsupials, they probably arose in the Early Cretaceous or Late Jurassic. These hosts, as well as the Amblycera, and their stephanocircid and doratopsylline fleas, all appear to be monophyletic.

There were African/South American faunal connections, by rafting, in the Early Eocene, involving such mammals as the 'hystrichomorph' rodents and ancestral ceboid monkeys and their ectoparasites. The Anoplura of the caviomorphs (South American) have such African (Phiomorpha) roots, as do the Neotropical ctenophthalmine fleas (and, ultimately, the Nearctic members of that subfamily). The phiomorphs have Anoplura that could only have been ultimately derived from North America, via Asia. Wood (1977a) proposed such an origin and route of dispersal for the ancestral phiomorphs. African extraction of the Nearctic *Ctenophthalmus* would explain the unique distribution of the genus in the New World, viz. only one species in the USA, and that restricted to the east, coupled with a comparative wealth of species in Mexico and Central America. Related taxa occur in South America. The Nearctic *Stenoponia* may also have a similar history. The Pulicidae likewise are of African origin but one branch apparently rafted with its host to South America and gave rise to *Pulex*.

The lice of the ceboid monkeys, like those of the caviomorphs (both Anoplura and Mallophaga), bespeak of a long history in South America for the hosts, probably dating back to the Eocene. It is possible that the *Pediculus* of South American monkeys are derived from the *Pediculus* of African primates (e.g. the chimpanzee), perhaps by parallel evolution, rather than representing secondary infestations from humans in the New World. Virological evidence may also support this idea of African-South American relationships, i.e., the occurrence of infection with the virus of yellow fever in wild monkeys on both continents. The distribution of malaria, with enzootic infections in monkeys and other primates, is also suggestive. The observations on ectoparasites in general clearly indicate that the connections were from Africa to South America instead of vice versa. Since, like man, chimpanzees have a *Pediculus* and gorillas a pubic louse (*Pthirus*), it seems that those primate lice antedate the division of the hominoid-anthropoid stocks.

There were land connections between North America and Europe in the Eocene, but not thereafter, as suggested by the distribution of *Odontopsyllus* and the *Atyphloceras bishopi* group of fleas, coupled with the absence of such evidence for ceratophyllids, a more youthful group. Cricetid rodents and some other theraphions were in South America at a significantly earlier date (Miocene) than the Pliocene-Pleistocene dates usually quoted. Among the rodents that arose in the Northern Hemisphere are sciurids, dipodids, cricetids and gliroids. The leporids also are of boreal origin.

Regardless of where the Muridae and Murinae actually arose, their roots go back to the Asian mainland, for there is no other place they could have acquired their Anoplura. The distribution of fleas like *Sigmactenus* and *Neopsylla,* which are undoubtedly of Palearctic origin, indicates that *Rattus* moved from Southeast Asia towards Australia, although their progenitors probably arose in Wallacea or other Asiatic-Pacific islands. That dispersal was in both directions is also indicated by the pygiopsyllid *Stivalius (s.str.)* fleas of *Rattus (Rattus),* for the provenance of the pygiopsyllids is in the Australian region. The absence of murid-ceratophyllids, coupled with the presence of fleas of that family on squirrels on some of the same islands, support the concept of the relative youthfulness of the murids. The zoogeography of these fleas suggests that *Rattus* entered New Guinea in the Pliocene instead of as recently as the Pleistocene. Sciurids penetrated islands like Sulawesi, Borneo, Mindanao, etc. subsequent to the advent of the murids, e.g. giant tree-rats.

Madagascar at one time was much further north in position than it is at present and probably was opposite Somalia, as suggested by: 1) the bioendemic cricetids have Anoplura and leptopsyllid fleas which are Palearctic in affinity and 2) the relative dearth of taxa of African affinity. It seems unlikely that murids replaced cricetids in Africa but rather that the latter were only poorly represented originally on that continent.

There is an urgent need for more data about the fleas and lice of many hosts, even some common ones, and for information about the ectoparasites from many parts of the world that are critical regarding zoogeography.

5 ACKNOWLEDGEMENTS

Preparation of this article and the bulk of the attendant research were made possible by Grant AI-04242 of the National Institutes of Health, Bethesda, Maryland, USA, with the Department of Microbiology, University of Maryland School of Medicine, Baltimore, Maryland, USA, and undertaken with the partial support of Contract N00014-76-C-0393 with the Office of Naval Research, Arlington, Virginia, USA and the Naval Medical Research and Development Command, Bethesda, Maryland, USA. Much is owed to the Honourable Miriam Rothschild because of her encouragement of studies on austral affinities, and her own investigations on the subject, both several decades ago, when thoughts on continental drift were deemed a nigh-anathema. Once again I am indebted to Helle Starcke, of our Department, for considerable editorial assistance and for aid in obtaining loan of references. Mrs Jytte Pedersen likewise helped regarding bibliographic material, as did J.F.Marquardt of the Smithsonian Institution Libraries. Some of the specimens yielding significant data were obtained through the courtesy of Capt P.F.D.Van Peenen, USN, then Commanding Officer of the US Naval Medical Research Unit No. 2 (Taipeh) or of the field-teams thereof operating in Indonesia. Other important specimens were received because of our joint investigations in New Guinea and Asiatic-Pacific islands undertaken with the Wau Ecology Institute, Wau, Papua New Guinea, and B.P.Bishop Museum, Honolulu, Hawaii, with the co-operation of Dr J.L.Gressitt and Dr F.J.Radovsky. Dr G.P.Holland kindly verified the distribution of some eastern Nearctic fleas. This analysis on the zoogeography of lice could not have been accomplished without the foundation provided by the late G.H.E.Hopkins (1957) on host relationships. Dr Theresa Clay kindly and fully answered many questions raised in correspondence regarding lice and posed stimulating and edifying ones of her own. Dr R.A.Ward obligingly read the manuscript and made some very useful suggestions. Dr K.C.Emerson not only critically reviewed the sections on lice but, eruditely and helpfully discussed the available data. To all these friends and colleagues I am much indebted and express my thanks.

The opinions and assertions herein expressed are the author's and are not to be construed as necessarily reflecting the views of those who read various parts of the manuscript.

ADDENDUM

Subsequent to the submission of this manuscript, an important paper by K.C.Kim and H.W.Ludwig appeared, entitled 'Phylogenetic relationships of parasitic Psocoidea and taxonomic position of the Anoplura' (Ann. Ent. Soc. Amer. 71(6):910-922, 1978). In this notable cladistic analysis, the authors have modified their former views (97) about the age of the Anoplura (p. 131 herein) and state '. . . the Anoplura must have arisen not later than the Eocene' and that 'it is reasonable to speculate that the Anoplura were fairly well established on the archaic mammals in the Paleocene or early Eocene'. They also apparently believe that the Amblycera arose in the Late Cretaceous from 'Protoamblycera' which had roots in the Jurassic, and that the ancestral Ischnocera diverged in the Jurassic, but later than the forebears of the Amblycera. These new conclusions are in accord with mine except perhaps that I feel the Amblycera were already distinct in the Early Cretaceous, or else they could not infest marsupials in both Australia and South America. Kim & Ludwig also concurred that the Anoplura and Mallophaga are distinct at the ordinal level. They opined that the mammalian Mallophaga were derived from avian lice and that the switch of proto-Mallophaga from birds to mammals took place more than once, but expressed no opinion as to whether the Amblycera of marsupials are monophyletic. Their chart does show diversification in the suborder occurring in the Mid Eocene, but this seems so late that the authors may not have been considering marsupial Amblycera in that respect. Kim & Ludwig believe that the Anoplura arose from 'Protoanoplura' and not from an ischnoceran ancestor. The division in the psocoid stock, with one branch leading to the Mallophaga and Anoplura and another to Psocoptera, is dated Mid Triassic. In this noteworthy paper, the authors also present new material on the higher classification of the lice.

REFERENCES

1. Archer, M.A. 1976, The dasyurid dentition and its relationships to that of didelphids, thylacinids, borrhyaenids (Marsupicarnivora) and peramelids (Peramelina, Marsupialia). Aust. J. Zool. Suppl. Ser. no. 39, 1-34.
2. Archer, M.A. & J.A.Q.Kirsch 1977, The case for the Thylacomyidae and Myrmecobiidae, Gill, 1872, or why are marsupial families so extended? Proc. Linn. Soc. N.S.Wales 102(1):18-25.
3. Baker, A.W. 1931, Insect parasites of vertebrates and host phylogeny. Canad. Field Nat. 45:189-191.
4. Baverstock, P.R., C.H.S.Watts & J.T. Hogarth 1977, Chromosome evolution in Australian rodents. I. The Pseudomyinae, the Hydromyinae and the *Uromys/Melomys* group. Chromosoma (Berlin) 61:95-125.
5. Bigalke, R.C. 1968, Evolution of mammals on southern continents. III. The contemporary mammal fauna of Africa. Quart. Rev. Biol. 43(3):265-300.
6. Brundin, L. 1966, Transantarctic relationships and their significance, as evidenced by chironomid midges. Kungl. Svenska Vetenskapsakad. Handl. Ser. 4, Band 11, No. 1, 472 p.
7. Cameron, T.W.M. 1960, Southern inter-

continental connections and the origin of the southern mammals. p. 79-89. In: Cameron, T.W.M. (ed.), Evolution: its science and doctrine. Symp. Roy. Soc. Canad. 1959. Univ. Toronto Press. 232 p.
8. Clay, T. 1957, The Mallophaga of birds. In: First symposium on host specificity among parasites of vertebrates. Inst. Zool., Univ. Neuchatel. :120-158.
9. Clay, T. 1970, The Amblycera (Phthiraptera: Insecta). Bull. Brit. Mus. (Nat. Hist.) 25(3):73-98.
10. Clay, T. 1971, A new genus and two new species of Boopidae (Phthiraptera: Amblycera). Pacif. Insects 13(3-4):519-529.
11. Clay, T. 1976, Geographical distribution of the avian lice (Phthiraptera): a review. J. Bombay Nat. Hist. Soc. 71(3):536-547.
12. Clemens, W.A. 1966, Fossil mammals of the type Lance Formation, Wyoming. Part II. Marsupialia. Univ. Calif. Publ. Geol. Sci. 62:1-122.
13. Clemems, W.A. 1968, Origin and early evolution of marsupials. Evolution 22:1-18.
14. Clemens, W.A. 1970, Mesozoic mammalian evolution. Ann. Rev. Ecol. Syst. 1: 357-390.
15. Clemens, W.A. 1971, Mammalian evolution in the Cretaceous. J. Linn. Soc. London Zool. 50, Suppl. 1, 165-180.
16. Clemens, W.A. 1977, Phylogeny of the marsupials. p. 51-68. In: Stonehouse, B. & D.Gilmore (eds.), The biology of marsupials. Baltimore-London-Tokyo, Univ. Park Press. 486 p.
17. Colbert, E.H. 1973a, Continental drift and the distribution of fossil reptiles. In: Tarling, D.H. & S.K.Runcorn (eds.), Implications of continental drift to the earth sciences. 1:395-412. NATO Advan. Study, Inst. New York, Academic Press.
18. Colbert, E.H. 1973b, Wandering lands and animals. New York, E.P.Dutton & Co., Inc. 323 p.
19. Cooke, H.B.S. 1968, Evolution of mammals on southern continents. II. The fossil mammal fauna of Africa. Quart. Rev. Biol. 43(3):234-264.
20. Cooke, H.B.S. 1972, IX. The fossil mammal fauna of Africa. p. 89-140. In: Keast, A., F.C.Erk & B.Glass (eds.), Evolution, mammals, and southern continents. Albany, N.Y., State Univ. N.Y. Press. 543 p.
21. del Corro, G. 1977, Parasitos, marsupiales y deriva continental. Paleontol. 2(3):35-67.
22. Cox, C.B. 1970, Migrating marsupials and drifting continents. Nature 226:767-770.
23. Cox, C.B. 1973, Triassic tetrapods. p. 213-223. In: Hallam, A. (ed.), Atlas of Palaeobiogeography. London, Elsevier Publ. Co.
24. Cox, C.B., I.N.Healey & P.D.Moore 1973, Biogeography. An ecological and evolutionary approach. London, Blackwell Scientific Publ. 184 p.
25. Cracraft, J. 1972a, Mesozoic dispersal of terrestrial faunas around the southern end of the world. Biogeographie et Liaisons Inter-Continentales au cours du Mesozoïque, Theme 1, 35 p. 17 Int. Congr. Zool. Monaco.
26. Cracraft, J. 1972b, Continental drift and Australian avian biogeography. Emu 72: 171-174.
27. Cracraft, J. 1973a, Continental drift, paleoclimatology, and evolution and biogeography of birds. J. Zool. London 169: 455-545.
28. Cracraft, J. 1973b, Vertebrate evolution and biogeography in the Old World tropics: Implications of continental drift and palaeoclimatology. In: Tarling, D.H. & S.K.Runcorn (eds.), Implications of continental drift to the earth sciences. 1: 373-393. New York, Academic Press.
29. Cracraft, J. 1974a, Continental drift and vertebrate distribution. Ann. Rev. Ecol. Syst. 5:215-261.
30. Cracraft, J. 1974b, Phylogeny and evolution of the ratite birds. Ibis, J. Brit. Ornithol. Union 116(4):494-521.
31. Cracraft, J. 1975, Historical biogeography and earth history: perspective for a future synthesis. Ann. Missouri Bot. Gard. 62: 227-250.
32. Crawford, A.R. 1974, A greater Gondwanaland. Science 184:1179-1181.
33. Croizat, L. 1958, Panbiogeography. Vol. I, 1018 p.; Vols. IIa & IIb, 1778 p., Caracas, Venezuela.
34. Croizat, L. 1970, A selection of notes on the broad trends of dispersal mostly of old world Avifauna. Nat. Hist. Bull. Siam Soc., Deignan Mem. Issue 23(3):255-324.
35. Dalziel, I.W.D. & D.H.Elliott 1971, Evolution of the Scotia Arc. Nature 233: 246-251.
36. Dalziel, I.W.D. & D.H.Elliott 1973, The Scotia Arc and Antarctic margin. In: Nairn, A.E.M. & F.G.Stehli (eds.), The

ocean basins and margins. 1:171-246.
37. Darlington, P.J., Jr. 1959, Area, climate and evolution. Evolution 13(4):488-510.
38. Darlington, P.J., Jr. 1963, Zoogeography: The geographicl distribution of animals. 675 p. New York, John Wiley & Sons, Inc. (2nd edition of 1957 book)
39. Darlington, P.J., Jr. 1965, Biogeography of the southern end of the world. Cambridge, Harvard Univ. Press. 236 p.
40. Dawson, T.J. 1977, Kangaroos. Sci. Amer. 237(2):78-89.
41. Dunnet, G.M. & D.K.Mardon 1976, A monograph of Australian fleas (Siphonaptera). Austral. J. Zool., Suppl. Ser. no. 30:1-273.
42. Emerson, K.C. & R.D.Price 1976, Abrocomophagidae (Mallophaga: Amblycera), a new family from Chile. Florida Ent. 59(4):425-428.
43. Evans, J.W. 1958, Insect distribution and continental drift. Symposium on Continental Drift, Hobart, 1956, 375 p., :134-161.
44. Evans, J.W. 1959, The zoogeography of some Australian insects. In: Bodenheimer, F.S. & W.W.Weisbach (eds.), Monographiae Biologiae. 8:150-163.
45. Ferris, G.F. 1951, The sucking lice. Mem. Pacific Coast Ent. Soc. 1:1-320.
46. Fittkau, E.J. 1969, The fauna of South America. In: Biogeography and ecology in South America. 2:624-655. The Hague, W.Junk N.V. Publisher.
47. Fooden, J. 1972, Breakup of Pangaea and isolation of relict mammals in Australia, South America, and Madagascar. Science 175:894-898.
48. Galton, P.M. 1977, The ornithopod dinosaur *Dryosaurus* and a Laurasia-Gondwanaland connection in the Upper Jurassic. Nature 268:230-232.
49. Harrison, L. 1914, The Mallophaga as a possible clue to bird phylogeny. Austral. Zool. 1:7-11.
50. Harrison, L. 1915, The respiratory system of Mallophaga. Parasitol. 8:101-127.
51. Harrison, L. 1916, Bird parasites and bird-phylogeny. Ibis (10)4:254-263.
52. Harrison, L. 1922, On the mallophagan family Trimenoponidae, with the description of a new genus and species from an American marsupial. Austral. Zool. 2: 154-158.
53. Harrison, L. 1924, The migration route of the Australian marsupial fauna. Austral. Zool. 3:247-263.
54. Harrison, L. 1926, Crucial evidence for Antarctic radiation. Amer. Nat. 60(669): 374-383.
55. Harrison, L. 1928a, Host and parasite. Presidential Address. Proc. Linn. Soc. N.S. Wales 53:IX-XXXI.
56. Harrison, L. 1928b, The composition and origins of the Australian fauna, with special reference to the Wegener hypothesis. p. 332-396. In: Maitland, A.G. (ed.), Report of 18th Mtg. Austral. Assoc. Advanc. Sci. (Australia and New Zealand). Perth, F.W.Simpson, Gov't. Printer. 913 p.
57. Hendey, Q.B. 1976, Fossil peccary from the Pliocene of South Africa. Science 192: 787-789.
58. Hershkovitz, P. 1966, Mice, land bridges, and Latin American faunal interchange. In: Wenzel, R.L. & V.J.Tipton (eds.), Ectoparasites of Panama. Chicago Field Mus. Nat. Hist. p. 725-751.
59. Hershkovitz, P. 1969, The evolution of mammals on southern continents. VI. The recent mammals of the Neotropical region: A zoogeographic and ecological review. Quart. Rev. Biol. 44(1):1-70.
60. Hershkovitz, P. 1972, VII. The recent mammals of the Neotropical region: A zoogeographic and ecological review. p. 311-432. In: Keast, A., F.C.Erk & B. Glass (eds.), Evolution, mammals and southern continents. Albany, N.Y., State Univ. N.Y. Press. 543 p.
61. Hoffstetter, R. 1970a, L'Histoire biogéographique des marsupiaux et la dichotomie marsupiaux-placentaires. C.R. Acad. Sci. 271:388-391.
62. Hoffstetter, R. 1970b, Radiation initiale des mammifères placentaires et biogéographie. C.R. Acad. Sci. 270:3027-3030.
63. Hoffstetter, R. 1971, Le peuplement mammalien de l'Amérique du Sud. Rôle des continents austraux comme centre d'origine, de diversification et de dispersion pour certains groupes mammaliens. An. Acad. Brasil. Cienc. 43 (Suppl.):125-144.
64. Hoffstetter, R. 1972, Relationships, origins and history of the Ceboid monkeys and Caviomorph rodents: a modern reinterpretation. Evol. Biol. 6:323-347.
65. Hoffstetter, R. & R.Lavocat 1970, Découverte dans le Déséadien de Bolivie de genres pentalophodontes appuyant les affinités africaines des Rongeurs Cavio-

morphes. C.R. Acad. Sci., Ser. D 271: 172-175.
66. Holland, G.P. 1958, Distribution patterns of northern fleas (Siphonaptera). Proc. 10. Int. Congr. Ent. (1956) 1:645-658.
67. Holland, G.P. 1963, Faunal affinities of the fleas (Siphonaptera) of Alaska, with an annotated list of species. In: Pacific Basin Biogeography, a symposium. Bishop Mus. Press:45-63. Symp. 10. Pacific Sci. Congr., 1961.
68. Holland, G.P. 1964, Evolution, classification, and host relationships of Siphonaptera. Ann. Rev. Ent. 9:123-146.
69. Hopkins, G.H.E. 1949, The host-associations of the lice of mammals. Proc. Zool. Soc. Lond. 119(2):387-604.
70. Hopkins, G.H.E. 1957a, The distribution of Phthiraptera on mammals. In: First symposium on host specificity among parasites of vertebrates. Inst. Zool., Univ. Neuchatel:88-119.
71. Hopkins, G.H.E. 1957b, Host-associations of Siphonaptera. In: First symposium on host specificity among parasites of vertebrates. Inst. Zool., Univ. Neuchatel:64-87.
72. Hopkins, G.H.E. & M.Rothschild 1966, An illustrated catalogue of the Rothschild collection of fleas (Siphonaptera) in the British Museum. Vol. 4, 549 p. London, Brit. Mus. (Nat. Hist.).
73. Hunsaker, D. (ed.) 1977, The biology of marsupials. New York, Acad. Press. 537 p.
74. Jardine, N. & D.McKenzie 1972, Continental drift and the dispersal and evolution of organisms. Nature 235(5332):20-24.
75. Jellison, W.L. 1942, Host distribution of lice on native American rodents north of Mexico. J. Mammal. 23(3):245-250.
76. Jellison, W.L. 1945, Siphonaptera: A new species of *Conorhinopsylla* from Kansas. J. Kansas Ent. Soc. 18(3):109-111.
77. Jellison, W.L. 1947, Siphonaptera: Host distribution of the genus *Opisocrostis* Jordan. Trans. Amer. Miscrosc. Soc. 66 (1):64-69.
78. Johnson, P.T. 1957, A classification of the Siphonaptera of South America with descriptions of new species. Mem. Ent. Soc. Wash. No. 5, 299 p.
79. Johnson, P.T. 1960, The Anoplura of African rodents and insectivores. U.S. Dept. Agric. Tech. Bull. No. 1211, 116 p.
80. Jordan, K. 1929, On some problems of distribution, variability and variation in North American Siphonaptera. 4. Int. Congr. Ent., Ithaca, N.Y., 1928, 2:489-499.
81. Kahle, C.F. 1974, Plate tectonics – assessments and reassessments. Memoir 23. Amer. Assoc. Petroleum Geologists, Tulsa, Oklahoma. 514 p.
82. Keast, A. 1968a, Evolution of mammals on southern continents. I. Introduction: The southern continents as backgrounds for mammalian evolution. Quart. Rev. Biol. 43(3):225-233.
83. Keast, A. 1968b, Evolution of mammals on southern continents. IV. Australian mammals: Zoògeography and evolution. Quart. Rev. Biol. 43(4):373-408.
84. Keast, A. 1969, Evolution of mammals on southern continents. VII. Comparisons of the contemporary mammalian fauna of the southern continents. Quart. Rev. Biol. 44(2):121-167.
85. Keast, A. 1972a, I. Introduction: The southern continents as backgrounds for mammalian evolution. p. 19-22. In: Keast, A., F.C.Erk & B.Glass (eds.), Evolution, mammals, and southern continents. Albany, N.Y., State Univ. N.Y. Press. 543 p.
86. Keast, A. 1972b, II. Continental drift and the evolution of the biota on southern continents. p. 23-87. In: Keast, A., F.C. Erk & B.Glass (eds.), Evolution, mammals, and southern continents. Albany, N.Y., State Univ. N.Y. Press. 543 p.
87. Keast, A. 1972c, V. Australian mammals: Zoogeography and evolution. p. 195-246. In: Keast, A., F.C.Erk & B.Glass (eds.), Evolution, mammals, and southern continents. Albany, N.Y., State Univ. N.Y. Press. 543 p.
88. Keast, A. 1972d, VIII. Comparisons of contemporary mammal faunas of southern continents. In: Keast, A., F.C.Erk & B.Glass (eds.), Evolution, mammals, and southern continents. Albany, N.Y., State Univ. N.Y. Press. 543 p.
89. Keast, A. 1973, Contemporary biotas and the separation sequence of the southern continents. p. 309-343. In: Tarling, D.H. & S.K.Runcorn (eds.), Implications of continental drift to the earth sciences. New York, Academic Press.
90. Keast, A. 1977a, Zoogeography and phylogeny: The theoretical background and methodology to the analysis of mammal and bird fauna. p. 249-312. In: Hecht, N.K., P.C.Goody & B.M.Hecht (eds.),

Major patterns in vertebrate evolution. New York, Plenum Press. 908 p.

91. Keast, A. 1977b, Historical biogeography of the marsupials. p. 69-95. In: Stonehouse, B. & D.Gilmore (eds.), The biology of marsupials. Baltimore-London-Tokyo, Univ. Park Press. 486 p.

92. Keast, A., F.C.Erk & B.Glass 1972, Evolution, mammals, and southern continents. Albany, N.Y., State Univ. N.Y. Press. 543 p.

93. von Kéler, S. 1971, A revision of the Australasian Boopiidae (Insecta: Phthiraptera), with notes on the Trimenoponidae. Austral. J. Zool., Suppl. Ser. No. 6:1-126.

94. Kerr, R.A. 1978, Precambrian tectonics: Is the present the key to the past? Science 199:282-285, 330.

95. Kim, K.C. & K.C.Emerson 1968, Descriptions of two species of Pediculidae (Anoplura) from great apes (Primates, Pongidae). J. Parasit. 54(4):690-695.

96. Kim, K.C. & K.C.Emerson 1974, A new Polyplax and records of sucking lice (Anoplura) from Madagascar. J. Med. Ent. 11(1):107-111.

97. Kim, K.C. & H.W.Ludwig 1978, The family classification of the Anoplura. Syst. Ent. 3:249-284.

98. Kirsch, J.A.W. 1977a, The classification of marsupials, with special reference to karyotypes and serum proteins. p. 1-50. In: Hunsaker, D. (ed.), The biology of marsupials. New York, Acad. Press. 537 p.

99. Kirsch, J.A.W. 1977b, The six-percent solution: second thoughts on the adaptedness of the Marsupialia. Amer. Sci. 65:276-288.

100. Kraglievich, L. 1931, Cuatro notas paleontológicas. (Sobre *'Octomylodon aversus'* Amegh., *'Argyrolagus palmeri'* Amegh., *'Tetrastylus montanus'* Amegh. y *'Munizia paranensis'* N.Gen., N.Sp.). Physis 10:242-266.

101. Kurtén, B. 1967, Continental drift and the paleogeography of reptiles and mammals. Commentationes Biologicae, Soc. Sci. Fennica 31:1-8.

102. Kurtén, B. 1971, The age of mammals. London, Weidenfeld & Nicolson. 250 p.

103. Kurtén, B. 1972, The age of mammals. New York, Columbia Univ. Press. 250 p.

104. Kurtén, B. 1973, Early tertiary land mammals. p. 437-442. In: Hallam, A. (ed.), Atlas of palaeobiogeography. London, Elsevier Sci. Publ. Co.

105. Ladd, J.W. 1976, Relative motion of South America with respect to North America and Caribbean tectonics. Bull. Geol. Soc. Amer. 87:969-976.

106. Lavocat, R. 1955, Quelques progrès récents dans la connaissance des rongeurs fossiles et leur conséquence sur divers problèmes de systématique de peuplement et d'évolution. Colloq. Intern. Cent. Nat. Rech. Sci. Paris 60:77-85.

107. Lavocat, R. 1969, La systématique des rongeurs hystrichomorphes et la dérive des continents. C.R. Acad. Sci. 269: 1496-1497.

108. Lavocat, R. 1971, Affinités systématiques des Caviomorphes et des Phiomorphes et origine africaine des Caviomorphes. An. Acad. Brasil. Cienc. 43, Suppl. :515-522.

109. Lavocat, R. 1973, Les rongeurs du Miocène d'Afrique orientale. 1. Miocène inférieur. Ecole Pratique des Hautes Etudes, Inst. des Montpellier, Mém. 1, 284 p.

110. Lavocat, R. 1974a, What is an hystrichomorph? Symp. Zool. Soc. Lond. 34:7-20.

111. Lavocat, R. 1974b, The evolution of the old world and new world hystrichomorphs. Symp. Zool. Soc. Lond. 34:21-60.

112. Lavocat, R. 1976, Rongeurs Caviomorphes de l'Oligocène de Bolivie. II. Rongeurs du Bassin Deséadien de Salle-Luribay. Palaeovertebrata 7:15-90.

113. Lewis, R.E. 1972, Notes on the geographical distribution and host preferences in the order Siphonaptera. Part 1. Pulicidae. J. Med. Ent. 9(6):511-520.

114. Lewis, R.E. 1973, Notes on the geographical distribution and host preferences in the order Siphonaptera. Part 2. Rhopalopsyllidae, Malacopsyllidae and Vermipsyllidae. J. Med. Ent. 10(3):255-260.

115. Lewis, R.E. 1974a, Notes on the geographical distribution and host preferences in the order Siphonaptera. Part 3. Hystrichopsyllidae. J. Med. Ent. 11(2):147-167.

116. Lewis, R.E. 1974b, Notes on the geographical distribution and host preferences in the order Siphonaptera. Part 4. Coptopsyllidae, Pygiopsyllidae, Stephanocircidae and Xiphiopsyllidae. J. Med. Ent. 11(4): 403-413.

117. Lewis, R.E. 1974c, Notes on the geographical distribution and host preferences in the order Siphonaptera. Part 5. Ancis-

tropsyllidae, Chimaeropsyllidae, Ischnopsyllidae, Leptopsyllidae and Macropsyllidae. J. Med. Ent. 11(5):525-540.

118. Lewis, R.E. 1975, Notes on the geographical distribution and host preferences in the order Siphonaptera. Part 6. Ceratophyllidae. J. Med. Ent. 11(6):658-676.

119. Lidicker, W.Z. & A.C.Ziegler 1968, Report on a collection of mammals from eastern New Guinea, including species keys for fourteen genera. Univ. Calif. Publ. Zool. 87:64 p.

120. Lillegraven, J.A. 1969, Latest Cretaceous mammals from the upper part of the Edmonton formation of Alberta, Canada, and review of marsupial-placental dichotomy in mammalian evolution. Univ. Kansas Publ., Paleont. Contr. Art. 50 (Vertebrate 12) p. 122.

121. Lillegraven, J.A. 1974, Biogeographical considerations of the marsupial-placental dichotomy. Ann. Rev. Ecol. & System. 5:263-283.

122. Lillegraven, J.A. 1975, Biological considerations of the marsupial-placental dichotomy. Evolution 29:707-722.

123. McKenna, M.C. 1970, The origin and early differentiation of therian mammals. Ann. N.Y. Acad. Sci. 167:217-240.

124. McKenna, M.C. 1972, Was Europe connected directly to North America prior to the Middle Eocene? Evol. Biol. 6:179-189.

125. McKenna, M.C. 1973, Sweepstakes, filters, corridors, Noah's arks, and beached Viking funeral ships in palaeogeography. p. 295-308. In: Tarling, D.H. & S.K.Runcorn (eds.), Implications of continental drift to the earth sciences. New York, Academic Press. 622 p.

126. McKenna, M.C. 1975, Fossil mammals and Early Eocene North Atlantic land continuity. Ann. Missouri Bot. Gard. 62: 335-353.

127. Mackerras, I.M. 1970, Composition and distribution of the fauna. In: The insects of Australia. Melbourne Univ. Press. p. 182-203.

128. Mani, M.S. 1974, Biogeographical evolution in India. XXIV. p. 698-724. In: Mani, M.S. (ed.), Ecology and biogeography in India. Ser. Monog. Biol. 23, 773 p. The Hague, W.Junk Publisher.

129. Martin, P.G. 1970, The Darwin Rise hypothesis of the biogeographical dispersal of marsupials. Nature 225:197-198.

130. Martin, P.G. 1975, Marsupial biogeography in relation to continental drift. Mem. Mus. Nat. Hist. Nat. Ser. (A) 88:216-237.

131. Martin, P.G. 1977, Marsupial biogeography and plate tectonics. p. 97-115. In: Stonehouse, B. & D.Gilmore (eds.), The biology of marsupials. Baltimore-London-Tokyo, Univ. Park Press. 486 p.

132. Marvin, U.B. 1973, Continental drift. The evolution of a concept. Smithsn. Inst. Press, 239 p.

133. Matthew, W.D. 1915, Climate and evolution. Spec. Publ. N.Y. Acad. Sci. 1:1-223. (Republished 1939).

134. Matthew, W.D. 1928, Outline and general principles of the history of life. Univ. Calif. Syllabus Ser. No. 213:1-253.

135. Mattingly, P.F. 1960, II. Ecological aspects of the evolution of mosquito-borne virus diseases. Trans. Roy. Soc. Trop. Med. Hyg. 54(2):97-112.

136. Mattingly, P.F. 1969, The biology of mosquito-borne disease. London, George Allen & Unwin, Ltd. 184 p.

137. Maxson, L.R., V.M.Sarich & A.C.Wilson, 1975, Continental drift and the use of albumin as an evolutionary clock. Nature 255:397-400.

138. Mayr, E. 1957, Evolutionary aspects of host specificity among parasites of vertebrates. In: First symposium on host specificity among parasites of vertebrates. Inst. Zool., Univ. Neuchatel :7-14 (In discussion of article by T.Clay).

139. Mayr, E. 1972, Continental drift and the history of the Australian bird fauna. Emu 72(1):26-28.

140. Meggers, B.J., C.Evans & E.Estrada 1965, Early formative period of coastal Ecuador. The Valdivia and Machalilla phases. Smithsn. Contrib. Anthrop. 1, 234 p.

141. Meyerhoff, A.A. & H.A.Meyerhoff 1974a, Tests of plate tectonics. p. 43-145. In: Kahle, C.F. (ed.), Plate tectonics – assessments and reassessments. Memoir 23. Amer. Assoc. Petroleum Geologists, Tulsa, Oklahoma. 514 p.

142. Meyerhoff, A,A, & H.A.Meyerhoff 1974b, Ocean magnetic anomalies and their relations to continents. p. 411-422. In: Kahle, C.F. (ed.), Plate tectonics – assessments and reassessments. Memoir 23. Amer. Assoc. Petroleum Geologists, Tulsa, Oklahoma. 514 p.

143. Molnar, P. & P.Tapponier 1977, The collision between India and Eurasia. Sci. Amer. 236(4):30-41, 148.

144. Murray, M.D. & J.H.Calaby 1971, The host relations of the Boopiidae. Appendix II. In: von Kéler, S. (ed.), A revision of the Australasian Boopiidae (Insecta: Phthiraptera), with notes on the Trimenoponidae. Austral J. Zool., Suppl. Ser. No. 6:81-82.
145. Nur, A. & Z.Ben-Avraham 1977, Lost Pacifica continent. Nature 270:41-43.
146. Patterson, B. 1957, Mammalian phylogeny. In: First symposium on host specificity among parasites of vertebrates. Inst. Zool., Univ. Neuchatel:15-49.
147. Patterson, B. & R.Pascual 1968, Evolution of mammals on southern continents. V. The fossil mammal fauna of South America. Quart. Rev. Biol. 43(4):409-451.
148. Patterson, B. & R.Pascual 1972, VI. The fossil mammal fauna of South America. p. 247-310. In: Keast, A., F.C.Erk & B.Glass (eds.), Evolution, mammals, and southern continents. Albany, N.Y., State Univ. N.Y. Press. 543 p.
149. Peus, F. 1968, Über die beiden Bernstein-Flöhe (Insecta, Siphonaptera). Paläontol. Zt. 42(1/2):62-72.
150. Ranck, G.L. 1968, The rodents of Libya. Taxonomy, ecology, and zoogeographical relationships. U.S. Natl. Mus. Bull. 275, 264 p.
151. Raven, P.H. 1975, Summary of the Biogeography Symposium. Ann. Missouri Bot. Gard. 62:380-385.
152. Raven, P.H. & D.I.Axelrod 1974, Angiosperm biogeography and past continental movements. Ann. Missouri Bot. Gard. 61 (3):539-673.
153. Raven, P.H. & D.I.Axelrod 1975, History of the flora and fauna of Latin America. Amer. Sci. 63:420-429.
154. Rich, P.V. 1975, Antarctic dispersal routes, wandering continents, and the origin of Australia's non-passeriform avifauna. Mem. Natl. Mus. Victoria, No. 36: 63-125.
155. Ride, W.D.L. 1964, A review of Australian fossil marsupials. J. Proc. Roy. Soc. West. Austral. 47:97-131.
156. Ride, W.D.L. 1970, A guide to the native mammals of Australia. Melbourne, Oxford Univ. Press. 249 p.
157. Riek, E.F. 1970, Lower Cretaceous fleas. Nature 227:746-747.
158. Rosen, D.E. 1978, Vicariant patterns and historical explanation in biogeography. Syst. Zool. 27(2):159-188.
159. Rothschild, M. & T.Clay 1952, Fleas, flukes and cuckoos, a study of bird parasites. London, New Naturalist. 304 p.
160. Rothschild, M. 1973, Note given at meeting on *Pulex irritans* found in Viking pit excavations. Proc. Roy. Ent. Soc. Lond. 38(7):29.
161. Savage, J.M. 1974, The Isthmian link and the evolution of Neotropical mammals. Contrib. Sci. Nat. Hist. Mus. Los Angeles County 260:1-51.
162. Schlinger, E.I. 1974, Continental drift, *Nothofagus*, and some ecologically associated insects. Ann. Rev. Ent. 19:323-343.
163. Schwarz, E. 1924, On the evolution and radiation of mammalian faunae. Acta Zool. 5:393-423.
164. Serventy, D.L. 1972, Causal ornithogeography of Australia. Proc. 15. Int. Ornithol. Congr., The Hague. 1970. p. 574-584.
165. Simpson, G.G. 1940a, Antarctica as a faunal migration route. Proc. 6. Pacific Sci. Congr. p. 755-768.
166. Simpson, G.G. 1940b, Mammals and land bridges. J. Wash. Acad. Sci. 30(4):137-163.
167. Simpson, G.G. 1945, The principles of classification and a classification of mammals. Bull. Amer. Mus. Nat. Hist. 85: 350 p.
168. Simpson, G.G. 1947a, Holarctic mammalian faunas and continental relationships during the Cenozoic. Bull. Geol. Soc. Amer. 58:619-688.
169. Simpson, G.G. 1947b, Evolution, interchange, resemblance of the North American and Eurasian Cenozoic mammalian faunas. Evolution 1:218-220.
170. Simpson, G.G. 1950, History of the fauna of Latin America. Amer. Sci. 38:361-389.
171. Simpson, G.G. 1951, History of the fauna of Latin America. p. 369-408. In: Gaitsell, G.A. (ed.), Science in Progress. New Haven, Conn., (7th Ser.), Yale Univ. Press.
172. Simpson, G.G. 1961, Historical zoogeography of Australian mammals. Evolution 15:431-446.
173. Simpson, G.G. 1965, The geography of evolution. Collected essays. Philadelphia, Chilton Books. 249 p.
174. Simpson, G.G. 1966, Mammalian evolution on the southern continents. Neues Jahrb. Min. Geol. Palaeont. Abh. 125:1-18.
175. Simpson, G.G. 1969, South American mammals. In: Fittkau, E.J. et al. (eds.), Biogeography and ecology in South

America. Monogr. Biol. 19:879-909.

176. Simpson, G.G. 1970, The Argyrolagidae, extinct South American marsupials. Bull. Mus. Comp. Zool. 139(1):1-69.

177. Simpson, G.G. 1971, The evolution of marsupials in South America. An. Acad. Brasil. Ciênc. 43:103-118.

178. Slaughter, B.H. 1968, Earliest known marsupials. Science 162:254-255.

179. Slaughter, B.H. 1971, Mid-Cretaceous (Albian) therians of the Butler Farm local fauna, Texas. In: Kermack, D.M. & K.A. Kermack (eds.), Early mammals. J. Linn. Soc. (Zool.) Suppl. 50:131-143.

180. Smit, F.G.A.M. 1962, A description of the male of *Idilla caelebs,* with a discussion of the relationships of this species (Siphonaptera: Hystrichopsyllidae). Proc. Roy. Ent. Soc. London, Ser B. Taxonomy 31(9-10):109-114.

181. Smit, F.G.A.M. 1972, On some adaptive structures in Siphonaptera. Folia Parasit. (Praha) 19:5-17.

182. Smith, A.G. & A.Hallam 1970, The fit of the southern continents. Nature 225:139-144.

183. Smith, G. 1909, A naturalist in Tasmania. Oxford, Clarendon Press. 147 p.

184. Smith, J.M.B. 1974, Southern biogeography on the basis of continental drift. A review. Austral. Mammalogy 1(2):213-229.

185. Spencer, B. 1896, Summary of the zoological, botanical, and geological results of the expedition. p. 139-199. In: Spencer, B. (ed.), Report on the work of the Horn Scientific Expedition to Central Australia. London and Melbourne. 220 p,

186. van Steenis, C.G.G.J. 1971, *Nothofagus,* key genus of plant geography in time and space, living and fossil, ecology and phylogeny. Blumea 19:65-98.

187. Stonehouse, B. 1977, Introduction: The marsupials. p. 1-8. In: Stonehouse, B. & D.Gilmore (eds.), The biology of marsupials. Univ. Park Press, 486 p.

188. Sullivan, W. 1974, Continents in motion. The new Earth debate. New York, McGraw-Hill Book Co. 399 p.

189. Tarling, D. 1972, Another Gondwanaland. Nature 238(5359):92-93.

190. Tarling, D. & M.Tarling 1971, Continental drift. A study of the Earth's moving surface. New York, Doubleday & Co. 140 p.

191. Tedford, R.H. 1971, Marsupials and global tectonics. Abstr. Geol. Soc. Amer. 3(7): 730-731.

192. Tedford, R.H. 1974, Marsupials and the new paleogeography. Soc. Econ. Paleo. & Mineral., Spec. Publ. No. 21:109-126.

193. Tedford, R.H., M.R.Banks, N.R.Kemp, I.McDougall & F.L.Sutherland 1975, Recognition of the oldest known fossil marsupials from Australia. Nature 255: 141-142.

194. Tipton, V.J. & C.E.Machado-Allison 1972, Fleas of Venezuela. Bull. Brigham Young Univ. Sci., Biol. Ser. 17(6):1-115.

195. Tipton, V.J. & E.Méndez 1966, The fleas (Siphonaptera) of Panama. p. 289-338. In: Wenzel, R.L. & V.J.Tipton (eds.), Ectoparasites of Panama. Chicago Field Mus. Nat. Hist.

196. du Toit, A.L. 1937, Our wandering continents: an hypothesis of continental drift. Edinburgh, Oliver & Boyd. 366 p.

197. Traub, R. 1950, Siphonaptera from Central America and Mexico. A morphological study of the aedeagus, with descriptions of new genera and species. Fieldiana, Zool. Mem. 1:1-127.

198. Traub, R. 1952, Records and descriptions of fleas from Peru (Siphonaptera). Proc. Ent. Soc. Wash. 54(1):1-22.

199. Traub, R. 1966, Some examples of convergent evolution in Siphonaptera. (Abstract of paper presented at meeting of Society, Dec. 1966: Proc. Roy. Ent. Soc. Lond. (C) 31(7):37-38.) Reported in *Ibid.* 31(8):46-47 (1966-1967) with errata in *Ibid.* 31(11):79 (1966-1967).

200. Traub, R. 1968, *Smitella thambetosa,* n. gen. and n.sp., a remarkable 'helmeted' flea from New Guinea (Siphonaptera, Pygiopsyllidae) with notes on convergent evolution. J. Med. Ent. 5(3):375-404.

201. Traub, R. 1968, Book Review. 'Ectoparasites of Panama' by R.L.Wenzel & V.J. Tipton (eds.), (Chicago, Ill.: Field Museum of Natural History, 861 p., 1966). Bull. Ent. Soc. Amer. 14(2):143-145.

202. Traub, R. 1969, *Muesebeckella,* a new genus of flea from New Guinea, with notes on convergent evolution (Siphonaptera: Pygiopsyllidae). Proc. Ent. Soc. Wash. 71(3):374-396.

203. Traub, R. 1972a, Notes on zoogeography, convergent evolution and taxonomy of fleas (Siphonaptera), based on collections from Gunong Benom and elsewhere in South-east Asia. I. New Taxa (Pygiopsyllidae, Pygiopsyllinae). Bull. Br. Mus. Nat. Hist. (Zool.) 23(9):201-305.

204. Traub, R. 1972b, Notes on zoogeography, convergent evolution and taxonomy of fleas (Siphonaptera), based on collections from Gunong Benom and elsewhere in South-east Asia. II. Convergent evolution. Bull. Br. Mus. Nat. Hist. (Zool.) 23(10): 307-387.
205. Traub, R. 1972c, Notes on zoogeography, convergent evolution and taxonomy of fleas (Siphonaptera), based on collections from Gunong Benom and elsewhere in South-east Asia. III. Zoogeography. Bull. Br. Mus. Nat. Hist. (Zool.) 23(11):389-450.
206. Traub, R. 1972d, The colloquium on the zoogeography and ecology of ectoparasites, their hosts and related infections at the Second International Congress of Parasitology, Washington, D.C., 1970. 2. The zoogeography of fleas (Siphonaptera) as supporting the theory of continental drift. J. Med. Ent. 9(6):584-589.
207. Traub, R. 1972e, The colloquium on the zoogeography and ecology of ectoparasites, their hosts and related infections at the Second International Congress of Parasitology, Washington, D.C., 1970. 27. Notes on fleas and the ecology of plague. J. Med. Ent. 9(6):603.
208. Traub, R. 1977, *Tiflovia,* a new genus of pygiopsyllid fleas from New Guinea, with notes on convergent evolution and zoogeography (Siphonaptera). J. Med. Ent. 13(6):653-685.
209. Traub, R. 1980, Some adaptive modifications in fleas. (In press) In: Proceedings, International Conference on Fleas, Ashton, England, June 1977.
210. Traub, R. & A.Barrera 1966, New species of *Ctenophthalmus* from Mexico, with notes on the ctenidia of shrew-fleas (Siphonaptera) as examples of convergent evolution. J. Med. Ent. 3(2):127-145.
211. Traub, R. & G.M.Dunnet 1973, Revision of the siphonapteran genus *Stephanocircus* Skuse, 1893 (Stephanocircidae). Austral. J. Zool., Suppl. Ser., 1973, No. 20: 41-128.
212. Traub, R. & B.L.Elisberg (eds.) 1972, The colloquium on the zoogeography and ecology of ectoparasites, their hosts and related infections at the Second International Congress of Parasitology, Washington, D.C., 1970. Nos. 1-32. J. Med. Ent. 9(6):583-605.
213. Traub, R. & T.M.Evans 1967, Description of new species of hystrichopsyllid fleas, with notes on arched pronotal combs, convergent evolution and zoogeography (Siphonaptera). Pacif. Insects 9(4):603-677.
214. Traub, R. & C.L.Wisseman, Jr. 1968, Ecological considerations in scrub typhus. 1. Emerging concepts. Bull. Wld. Hlth. Org. 39(2):209-218.
215. Traub, R. & C.L.Wisseman, Jr. 1974, The ecology of chigger-borne rickettsiosis (scrub typhus). J. Med. Ent. 11(3):237-303.
216. Traub, R., C.L.Wisseman, Jr. & A.Farhang-Azad 1978, The ecology of murine typhus – A critical review. Trop. Dis. Bull. 75(4): 237-317.
217. Troughton, E. le G. 1959, The marsupial fauna: its origin and radiation. p. 70-88. In: Keast, A., R.L.Crocker & C.S.Christian (eds.), Biogeography and ecology in Australia. The Hague, W.Junk.
218. Tyndale-Biscoe, H. 1973, Life of marsupials. New York, Elsevier Publ. Co., Inc. 254 p.
219. Vanzolini, P.E. & L.R.Guimarães 1955a, Lice and the history of South American land mammals. Rev. Brasil. Ent. 3:13-46.
220. Vanzolini, P.E. & L.R.Guimarães 1955b, Notes and comments. South American land mammals and their lice. Evolution 9(3):345-347.
221. Wagner, J. 1932, Die Bedeutung der Flöhe für die Frage nach der Genesis der Säugetierfauna. Zoogeographica 1(2):263-268.
222. Washburn, S.L. 1978, The evolution of man. Sci. Amer. 239(3):194-208, 242.
223. Wegener, A. 1912, Die Entstehung der Kontinente. Petermanns Mitteilungen, p. 185-195, 253-256, 305-309. (English Translation by Biram, J., 1966, 'The origin of continents and oceans' of 1929 edition 'Die Entstehung der Kontinente und Ozeane'. New York, Dover Publications, Inc. 246 p.)
224. Wenzel, R.L. & V.J.Tipton 1966, Ectoparasites of Panama. Chicago Field Mus. Nat. Hist. 861 p.
225. Wolfson, A. 1948, Bird migration and the concept of continental drift. Science 108: 23-30.
226. Wolfson, A. 1955, Origin of the North American bird fauna: Critique and reinterpretation from the standpoint of continental drift. Amer. Midl. Nat. 53(2):353-380.
227. Wood, A.E. 1950, Porcupines, paleogeo-

graphy and parallelism. Evolution 4:87-98.

228. Wood, A.E. 1973, Eocene rodents, Pruett Formation, southwest Texas; their pertinence to the origin of the South American Caviomorpha. Texas Mem. Mus., Pearce-Sellards Ser. (20):1-40.
229. Wood, A.E. 1974, Rodentia. Encyclopaedia Britannica 15:969-980.
230. Wood, A.E. 1975, The problem of the hystricognathous rodents. Studies on Cenozoic Paleontology and Stratigraphy in Honor of Claude W.Hibbard, Univ. Mich. Pap. Paleont. 12:75-80.
231. Wood, A.E. 1977a, Paleontology and plate tectonics with special reference to the history of the Atlantic Ocean. West, R.M. (ed.), Proc. Symp. N. Amer. Paleont. Conv. II, Lawrence, Kansas. Milwaukee Publ. Mus. Spec. Publ. 2:95-109.
232. Wood, A.E. 1977b, The evolution of the rodent family Ctenodactylidae. J. Palaeont. Soc. India 20:120-137.
233. Wood, A.E. & B.Patterson 1959, The rodents of the Deseadan Oligocene of Patagonia and the beginnings of South American rodent evolution. Bull. Mus. Comp. Zool. 120:279-428.
234. Wood, A.E. & B.Patterson 1970, Relationships among hystricognathous and hystricomorphous rodents. Extr. de Mammalia 3(4):628-639.

R. L. C. PILGRIM
University of Canterbury, Christchurch, New Zealand

THE NEW ZEALAND FLEA FAUNA

ABSTRACT

The New Zealand zoogeographic subregion includes the Kermadec Islands, mainland New Zealand, several sub-Antarctic islands, and a sector of the Antarctic continent. The bird fauna is unusual, and there are very few native mammals. This is reflected in the flea fauna of the area, in which the endemic bird fleas outnumber the endemic mammal fleas by 19:1.

Native land mammals are represented by only two species of bats (Chiroptera), with one flea, *Porribius pacificus* (Ischnopsyllidae). The other mammal fleas are nine on introduced hosts.

The bird flea species and subspecies are in the Ceratophyllidae (1), Pygiopsyllidae (7) and Rhopalopsyllidae (11); there is one introduced bird flea, *Ceratophyllus gallinae,* which presumably arrived with domestic poultry. The native species are almost entirely confined to ground-nesting or flightless or poor-flying birds. The relationships of their fleas are with Australia (Pygiopsyllidae) and South America (Rhopalopsyllidae).

1 INTRODUCTION

The New Zealand zoogeographic subregion of the Australasian region includes, as the term is here used: the Kermadec Islands, mainland New Zealand (North, South and Stewart Islands and their immediate offshore islands), the Chatham Islands, the several groups of sub-Antarctic Islands between longitude 150°E and 150°W (Bounty, Antipodes, Snares, Auckland, Campbell, Macquarie, Balleny) and the corresponding sector of the Antarctic continent.

The subregion contains very few native mammals (table 1), and a bird fauna which is strikingly different in its taxonomic composition from the bird faunas of Europe (table 3), and even of Australia. The unusual composition of the New Zealand mammal and bird fauna is reflected in the flea fauna, in which the bird fleas very greatly outnumber those on mammals (table 6). The situation is thus strikingly different from the flea-host relationship for the order as a whole, in which approximately 95 per cent of the world's known fleas occur on mammals.

Comprehensive distribution records for the subregion have been published by Smit (1965), with additional information for the Kermadec Islands by Watt (1971). In the

Table 1. New Zealand mammal fauna and associated fleas

	Fleas
Native mammals:	
Carnivora-Pinnipedia – 7 spp.	–
Chiroptera – 2 spp.	+
Introduced mammals:	
Carnivora-Fissipedia – 5 spp.	+
Artiodactyla – 13 spp.	+
Perissodactyla – 2 spp.	–
Rodentia – 5 spp.	+
Insectivora – 1 sp.	–
Lagomorpha – 2 spp.	–
Primates – 1 sp.	+
Marsupialia – 6 spp.	–

Table 2. New Zealand flea fauna – mammal hosts

Flea species	Principal hosts in New Zealand
ISCHNOPSYLLIDAE	
Porribius pacificus	bat *(Chalinolobus tuberculatus)*
PULICIDAE	
**Ctenocephalides canis*	*dog *(Canis familiaris)*
**Ct.felis*	*cat *(Felis catus)*
**Pulex irritans*	*man *(Homo sapiens)*
**Xenopsylla cheopis*	*rats *(Rattus rattus, R.norvegicus)*
**X.vexabilis*	*rat *(R.exulans)*
LEPTOPSYLLIDAE	
**Leptopsylla segnis*	*mouse *(Mus musculus)*
CERATOPHYLLIDAE	
**Nosopsyllus fasciatus*	*rats *(R.rattus, R.norvegicus)* *mouse *(Mus musculus)*
**N.londiniensis*	*rats *(R.rattus, R.norvegicus)* *mouse *(Mus musculus)*
PYGIOPSYLLIDAE	
**Pygiopsylla hoplia*	*rat *(R.exulans)*

* = introduced

following account, these are supplemented from other publications and from unpublished records of the writer and colleagues; only the more regular host records are given, since in some cases there are isolated records from suspiciously 'incorrect' hosts. These are regarded as accidentals, contaminants, or as having arisen from inadequacies in collecting or curating.

2 MAMMALS AND MAMMAL FLEAS (tables 1 and 2)

Native mammals of the subregion are confined to several species of Carnivora-Pinnipedia which do not carry fleas, and two species of Chiroptera. Of the latter, *Chalinolobus tuber-*

culatus (Vespertilionidae) is host to *Porribius pacificus* (Ischnopsyllidae), the only native mammal flea. *Chalinolobus* contains five further species in Australia-Tasmania-Norfolk Island-New Caledonia, and some ten species in a separate subgenus in southern Africa. *Porribius* has three further species in Australia-New Guinea on several genera of Vespertilionidae and on one of Molossidae (Dunnet & Mardon, 1974). *P.pacificus* has clearly been derived during earlier connections of New Zealand with Australia and northwards.

The other native bat *Mystacina tuberculata* is the sole member of the family Mystacinidae, of obscure affinities. It has not been collected as frequently in the past; *P.pacificus* has been recorded from it once only but the record appears questionable. Recent work by Holloway (1976) revealed no sign of fleas on more than 150 live *M.tuberculata* examined in her studies, or in the vicinity of the roost containing over 500 bats. Daniel (pers. comm.) informs me that he has handled several hundreds of *M.tuberculata* in the past few years and has seen no sign of any fleas associated with this bat. It seems very likely that *M.tuberculata* is not a normal host of *P.pacificus.*

A number of mammals have been successfully introduced, either deliberately or accidentally, into New Zealand and are established as domestic or feral faunal elements. Several of these bear fleas which must therefore be regarded as components of the flea fauna (tables 1 and 2).

2.1 *Pulicidae*

Ctenocephalides canis is widespread and common, particularly on farm dogs, but has also been taken occasionally on cat and man.

Ct.felis is widespread on cats, from which host it is perhaps the most common household flea. There are occasional records also from rat (*Rattus rattus*) and hedgehog (*Erinaceus europaeus;* nests).

Pulex irritans was apparently very common in earlier times, but later declined in abundance. Recently, there are signs of increasing numbers associated with the establishment of close-packed living 'communes' and their poor standards of hygiene. It has been taken from dogs, rats (*R.norvegicus* and *R.rattus*) but not from pig in New Zealand.

Xenopsylla cheopis was associated with the occurrence of plague in this country during the first decade or so of this century (MacLean, 1955). Although the flea has been occasionally reported since then (see Smit, 1965 for references), plague has not recurred. The most recent specimens (3) I have seen were obtained from a food-processing factory in Auckland in 1971 (Gibson, 1972); the Department of Health (in litt.) informs me that the species was recovered from a vessel in port, in 1975.

There are no positive records of murine typhus in New Zealand.

X.vexabilis is normally a parasite, in New Zealand, of the Pacific rat *R.exulans.* This host was once widely distributed throughout mainland New Zealand where it was brought in the early Maori immigrations. It is now confined almost entirely to a few offshore islands following the later introduction by European settlers of *R.rattus* and *R.norvegicus,* which have largely displaced *R.exulans.* Despite earlier overlap of the ranges of these three rats, *X.vexabilis* has not been unequivocably recorded from *R.rattus* or *R.norvegicus* except in Kermadec Is. (Watt, 1971). *X.vexabilis* is also found in southeast Asia, New Guinea, Hawaii and Australia, on a wide range of rats.

2.2 *Leptopsyllidae*

Leptopsylla segnis is abundant on the mouse (*Mus musculus*) throughout mainland New

Zealand; it has also been taken from all three *Rattus* species, and occasionally from stoat (*Mustela erminea*).

2.3 *Ceratophyllidae*

Nosopsyllus fasciatus is common on *Rattus rattus* and *R.norvegicus* throughout the country. It has been found on mouse, hedgehog (nests), ferret (*Mustela putorius*), weasel (*M.nivalis*), stoat, dog and several birds. Smit (1965) and Gibson (1972) recorded the species on Cuvier I., Hen I., and Little Barrier I. (all lying northeast of Hauraki Gulf, North Island). These islands are reported to be free from *R.rattus* and *R.norvegicus*, so that transfer of the flea to *R.exulans* probably occurred while *R.exulans* was a mainland inhabitant. The possibility that *R.rattus* and/or *R.norvegicus* inhabited the islands in the past is, however, not easily dismissed; if their former presence can be demonstrated, an alternative opportunity for flea transfer can be postulated.

N.londiniensis was reported by Smit (1965) from mouse and rat (*R.rattus*) from the North Island only. I have since had numerous records of specimens from mouse in the vicinity of Christchurch, South Island.

2.4 *Pygiopsyllidae*

Pygiopsylla hoplia is widespread on *R.exulans* in the latter's offshore island habitats. It has been taken from both *R.exulans* and *R.norvegicus* on Raoul I. (Kermadec Is.). Gibson (1972) recorded it from *R.rattus* and *R.norvegicus* on islands off mainland Coromandel Peninsula where *R.exulans* is not known to occur; similar considerations to those suggested for *X.vexabilis* may apply to this situation also.

Pygiopsylla is known only from Australia-New Guinea-North Borneo and is found on rodents, both marsupial and eutherian. *P.hoplia* has a very wide host distribution in Australia (see Dunnet & Mardon, 1974), including monotremes, marsupials and true rodents.

2.5 *Discussion on mammal fleas*

There are some unexpected absences from the New Zealand mammal flea fauna. The lack of native rodents is surely the reason for the absence of the large family Hystrichopsyllidae and other characteristic rodent fleas. More surprising is the absence of fleas typical of several of the introduced mammals: *Archaeopsylla e.erinacei* from hedgehog, and *Spilopsyllus cuniculi** from rabbit (*Oryctolagus cuniculus*), since these fleas are common on those hosts in Britain whence they were introduced to New Zealand. There have been several successful introductions of marsupials (five species of Macropodidae, one of Phalangeridae) from Australia; none of these has been found to carry a flea typical of the host in its original home.

3 BIRDS AND BIRD FLEAS (tables 3 to 5)

The New Zealand bird fauna is characterised, inter alia, by a very large proportion of the world's species of Sphenisciformes, Procellariiformes and Phalacrocoracidae (Pelecani-

* *S.cuniculi* was released on Macquarie Island in 1968 as a possible vector for myxomatosis (Sobey et al., 1973). Since this was a deliberate introduction, it has not been included in tables 1, 2 and 6.

Table 3. New Zealand bird fauna – some striking features

NEW ZEALAND (species and subspecies)		BRITAIN (species and subspecies)
14	Sphenisciformes (Penguins)	–
72	Procellariiformes (Albatrosses, Petrels, Shearwaters, Fulmars)	14
	Pelecaniformes:	
14	Phalacrocoracidae (Shags, Cormorants)	2
13 (+3*)	Psittaciformes (Parrots, Parakeets)	–
	Gruiformes:	
13	Rallidae (Rails)	7
53 (+17*)	Passeriformes (Song-birds)	188
4	Falconiformes (Diurnal birds of prey)	24
1 (+7*)	Galliformes (Pheasants, Partridges, Quail)	10
	Charadriiformes:	
3	Laridae (Gulls)	15

* = introduced

formes). In comparison with Britain, the subregion also contains considerable numbers of Psittaciformes but is deficient in Passeriformes, Galliformes, Falconiformes and some smaller taxonomic groups. The bird flea fauna reflects these differences.

3.1 *Ceratophyllidae* (table 4)

Ceratophyllus gallinae is widespread and common, particularly on starling (*Sturnus vulgaris*) from whose nests it invades dwellings. With *Ctenocephalides felis,* it is among the most frequently found attacking man. Other birds infested, less intensively and extensively, are blackbird (*Turdus merula*), song thrush (*T.philomelos*), house sparrow (*Passer domesticus*) and silvereye (*Zosterops lateralis*). Occasional records are known from a further range of birds (see Smit, 1965). Domestic chickens (*Gallus domesticus*) appear to be less commonly infested than in the past, following the introduction of 'battery-type' poultry-keeping now in commercial practice.

Glaciopsyllus antarcticus is the sole endemic representative of this family in the New Zealand subregion. It was first described (Smit & Dunnet, 1962) from islands off the Antarctic coast between 77°E and 110°E (Australian Territory). It was obtained from nests of two birds, *Fulmarus glacialoides* and *Pagodroma nivea.* Since then it has been taken from nest material of *P.nivea* on the Antarctic mainland at Cape Christie, ca. 72°35′S 170°E, in the New Zealand Territory (unpublished records). The flea evidently has a wide distribution range, related to that of the hosts. Interest in this species lies not only in the fact that it represents a predominantly Holarctic family, but in the extremely rigorous conditions which the larval stages must undergo during the absence of the hosts. Larvae and pupae have been collected from the nests of both host species.

Table 4. New Zealand flea fauna – bird hosts of Ceratophyllidae and Pygiopsyllidae

Flea species	Host genera
CERATOPHYLLIDAE	
**Ceratophyllus gallinae*	Galliformes: **Gallus*
	Passeriformes: **Sturnus, *Turdus*
Glaciopsyllus antarcticus	Procellariiformes: *Fulmarus, Pagodroma*
PYGIOPSYLLIDAE	
Notiopsylla k.kerguelensis	Procellariiformes: *Diomedea, Pterodroma, Pachyptila, Puffinus, Pelecanoides*
	Charadriiformes: *Larus*
N.kerguelensis ssp. A	Procellariiformes: *Procellaria*
	Psittaciformes: *Cyanoramphus*
N.e.enciari	Procellariiformes: *Pterodroma, Pachyptila, Procellaria, Puffinus*
	Psittaciformes: *Cyanoramphus*
N.enciari ssp. C	Procellariiformes: *Pterodroma, Pachyptila, Puffinus*
	Gruiformes: *Gallirallus*
Notiopsylla sp. B	Procellariiformes: *Diomedea, Puffinus*
	Passeriformes: *Bowdleria*
Notiopsylla sp. D	Procellariiformes: *Puffinus*
Stivalius galliralli	Procellariiformes: *Pterodroma*
	Gruiformes: *Gallirallus*
	Charadriiformes: *Coenocorypha*
	Passeriformes: *Bowdleria, Petroica, Zosterops*

* = introduced

3.2 *Pygiopsyllidae* (table 4)

This family is largely associated with rodent hosts in Australia, SE Asia, S.America and Africa. The sole mammal flea from this family in New Zealand is *Pygiopsylla hoplia* mentioned above. Bird fleas dominate the New Zealand flea fauna in the family, and fall into two genera.

Notiopsylla contains large fleas found primarily on sea birds in the high southern latitudes; it is, however, unknown from penguins.

Notiopsylla k.kerguelensis is associated with a number of birds, mainly Procellariiformes – *Diomedea melanophris impavida, D.chrysostoma, Pterodroma lessoni, Pachyptila desolata, Puffinus griseus, Pelecanoides urinatrix,* and a few others outside the subregion (Smit, 1964). All the records are from sub-Antarctic islands, and many are derived from nests, some of which were unoccupied at the time of collecting. The species is also recorded from *Larus dominicanus* on Macquarie I.

Notiopsylla e.enciari is similarly reported from a variety of *Procellariiformes – Pterodroma lessoni, Pt.inexpectata, Pachyptila turtur, Procellaria cinerea, Puffinus griseus,* as well as two *Psittaciformes – Cyanoramphus n.novaezelandiae,* and *C.unicolor.* All records are from sub-Antarctic islands and the species appears to be restricted to the New Zealand subregion.

When described (Smit, 1957), the species was obtained from the parakeet, *C.unicolor,* and it was suggested that the flea would be found associated with petrels or gulls of the sub-Antarctic. Dunnet (1964) postulated that parakeets were the original hosts but admitted that more data were required. Numerous records now show the species widespread on procellariiform birds, some of which are from islands (Snares, Macquarie) without parakeets.

Several further species and subspecies awaiting description (Smit, pers. comm.) occur on mainly Procellariiformes:

Notiopsylla kerguelensis ssp.A on *Procellaria cinerea* and *Cyanoramphus unicolor* is found on Antipodes I. along with *N.e.enciari* on both of these hosts.

Notiopsylla enciari ssp.C is found on *Pterodroma inexpectata, Pachyptila turtur, Puffinus griseus* and *Gallirallus australis scotti* but all records are confined to Stewart I.

Notiopsylla sp.B is recorded only from Snares I. on *Diomedea bulleri, Puffinus griseus,* and *Bowdleria punctata caudata.*

Notiopsylla sp.D is known only from *Puffinus huttoni,* on the New Zealand mainland.

Stivalius galliralli was described by Smit (1965) from a single female collected on *Gallirallus australis* (Gruiformes: Rallidae; South Island) but Smit expressed doubt that this was the true host. Since then many speciments have been taken from *Bowdleria punctata caudata, Petroica macrocephala dannefaerdi* (both Passeriformes: Muscicapidae; Snares I.), *Coenocorypha aucklandica meinertzhagenae* (Charadriiformes: Scolopacidae; Antipodes I.) and *Pterodroma inexpectata* (Procellariiformes: Procellariidae; Snares I.). Thus the hosts from which the bulk of the specimens are now known are from three orders very distinct from that of the original record. Further, the subsequent material has all been collected on sub-Antarctic islands, which may prove to be the prime habitat. Large numbers of larvae were obtained in the nests of some of the hosts on Snares I., but they cannot yet be placed uncategorically as those of *S.galliralli,* since other flea adults were also present.

The genus *Stivalius* contains mainly marsupial or rodent parasites, with a few species infesting birds. Holland (1969) discussed the geographical and host distribution of this genus and concluded that the original concept of *Stivalius* was too broad. He erected several new genera to accommodate species in the New Guinea fauna, and some of these are also present in Australia (Holland, 1969; Dunnet & Mardon, 1974). The New Zealand *galliralli* does not, in fact, belong to *Stivalius* s.str., but requires a new genus to contain it (Smit, pers. comm.).

3.3 *Rhopalopsyllidae* (table 5)

This family is chiefly associated with rodents in the Neotropical Region. *Parapsyllus,* the sole representative in the New Zealand subregion, is exceptional in parasitising birds and in having a circumpolar distribution in high southern latitudes. The first species described were obtained from penguins and it was thought (Hopkins, 1957:76) that the genus was confined to Sphenisciformes. However, subsequent discoveries showed that species are widespread on other sea birds and several occur on land birds (see Smit, 1965:42-3; Dunnet & Mardon, 1974:45-51, for discussion). As with *Notiopsylla,* there are, apart from those given in Smit (1965), several species and subspecies awaiting description (Smit, pers. comm.).

Parapsyllus cardinis is recorded from mainly the sub-Antarctic islands (Macquarie, Auckland, Antipodes) on six genera of procellariiform birds: *Macronectes giganteus, Pterodroma lessoni, Pachyptila desolata, Procellaria cinerea, P.aequinoctialis steadi, Pelagodroma marina maoriana, Pelecanoides urinatrix exsul.* It has also been taken, on Antipodes I. only, from *Coenocorypha aucklandica meinertzhagenae, Cyanoramphus unicolor* and *C.novaezelandiae hochstetteri.*

P.n.nestoris is confined to land birds. The chief host appears clearly to be *Nestor notabilis* but there are occasional records from *Strigops habroptilus, Gallirallus a.australis* and

Table 5. New Zealand flea fauna – bird hosts of Rhopalopsyllidae

Flea species	Host genera
RHOPALOPSYLLIDAE	
Parapsyllus cardinis	Procellariiformes: *Macronectes, Pterodroma, Pachyptila, Procellaria, Pelagodroma, Pelecanoides*
	Charadriiformes: *Coenocorypha*
	Psittaciformes: *Cyanoramphus*
P.n.nestoris	Psittaciformes: *Nestor, Strigops*
	Gruiformes: *Gallirallus*
	Charadriiformes: *Larus*
P.nestoris ssp. C	Procellariiformes: *Pachyptila*
	Psittaciformes: *Cyanoramphus*
P.jacksoni	Sphenisciformes: *Eudyptula*
	Procellariiformes: *Pterodroma, Pachyptila, Puffinus, Pelecanoides, Pelagodroma*
P.l.lynnae	Procellariiformes: *Diomedea, Pterodroma*
	Passeriformes: *Bowdleria*
P.lynnae ssp. A	Procellariiformes: *Pachyptila*
	Charadriiformes: *Coenocorypha*
P.lynnae ssp. E	Procellariiformes: *Puffinus*
P.magellanicus	Sphenisciformes: *Eudyptes*
	Procellariiformes: *Diomedea, Pterodroma, Pachyptila, Pelecanoides*
	Pelecaniformes: *Leucocarbo*
	Charadriiformes: *Larus*
	Psittaciformes: *Cyanoramphus*
P.longicornis	Sphenisciformes: *Eudyptula*
	Procellariiformes: *Diomedea, Phoebetria, Macronectes, Daption, Pachyptila*
	Pelecaniformes: *Sticticarbo, Phalacrocorax*
	Charadriiformes: *Coenocorypha, Larus*
	Psittaciformes: *Cyanoramphus*
Parapsyllus sp. B	Psittaciformes: *Cyanoramphus*
Parapsyllus sp. D	Gruiformes: *Gallirallus*

Larus dominicanus, all on mainland New Zealand. These hosts are all ground-nesting birds. Smit (1965:12, 42-3) provided a reasoned argument for the appearance of fleas of this genus on such land birds.

P.nestoris ssp.C, closely related to *P.n.nestoris,* has been taken from *Cyanoramphus* spp. on Antipodes Is., and from *Pachyptila vittata* on Chatham Is. If the true host is the ground-nesting parakeet(s), it may be expected to have acquired the flea in a fashion similar to that postulated for *P.n.nestoris* by *Nestor notabilis* etc. (Smit, 1965:42).

P.jacksoni is included in the *cardinis* species group with the foregoing, and is closely related to *P.n.nestoris.* First described mainly from *Puffinus griseus* and *Pachyptila turtur* (Smit, 1965), it has now been taken from a further range of procellariiform hosts (*Pterodroma macroptera gouldi, Puffinus carneipes hullianus, P.pacificus pacificus, P.g.gavia, Pelagodroma marina maoriana, Pelecanoides u.urinatrix*), as well as from penguins (*Eudyptula minor* sspp.). Most records are from mainland New Zealand.

P.l.lynnae, described from a male and female (Smit, 1965) from an unknown host on Solomon's I. (off Stewart I.) has since been taken from *Diomedea bulleri, Pterodroma inexpectata* and *Bowdleria punctata caudata,* all at Snares I.

P.lynnae ssp.E is known only from *Puffinus huttoni,* a mainland bird breeding in the South I.

P.lynnae ssp.A is known only from Chatham Is., taken on *Pachyptila vittata* and *Coenocorypha aucklandica pusilla.*

P.magellanicus has a circumpolar distribution in the sub-Antarctic. Its hosts are '. . . penguins, petrels, cormorants and skuas . . .' (Dunnet & Mardon, 1974:50). In the New Zealand subregion, it has been taken from *Eudyptes crestatus, Diomedea exulans, Pterodroma lessoni, Pt.inexpectata, Pachyptila turtur, Pelecanoides urinatrix exsul* and *Leucocarbo carunculatus chalconotus,* with occasional records from *Larus novaehollandiae scopulinus* and *Cyanoramphus unicolor.* All records are from sub-Antarctic islands. In the present account, *P.magellanicus heardi* is not separately recognised (see Smit, 1964:333).

P.longicornis is also a widely distributed flea with numerous host records in the temperate and circumpolar high latitude zones. On the New Zealand mainland it is the commonest flea on the smaller penguins (*Eudyptula minor* sspp.); on the sub-Antarctic islands, it is found mainly on procellariiform hosts (*Diomedea epomophora, D.melanophris impavida, D.chrysostoma, Phoebetria palpebrata, Macronectes halli, Daption capense, Pachyptila vittata*). It has also been taken from *Phalacrocorax carbo novaehollandiae, Sticticarbo p. punctatus,* and *Larus novaehollandiae scopulinus* (all on the mainland), *Coenocorypha aucklandica pusilla* (Chatham Is.) and *Cyanoramphus unicolor* (Antipodes I.).

Two further species awaiting description are known from limited localities and hosts: *Parapsyllus* sp.B from *Cyanoramphus novaezelandiae vhathamensis* on Chatham Is., and *Parapsyllus* sp.D from *Gallirallus a.australis* on the mainland.

3.4 *Discussion on bird fleas*

Several interesting aspects of zoogeographical relationships are illustrated by the bird fleas in the New Zealand subregion. Firstly, there is the intriguing occurrence of *Glaciopsyllus antarcticus,* the sole endemic member of the Ceratophyllidae, a family which has a '. . . predominantly Holarctic distribution, with relatively few representatives in the Neotropical, Ethiopian and Oriental Regions' (Smit & Dunnet, 1962. See also Holland, 1964:139, for a discussion of relationships).

Secondly, the Pygiopsyllidae with its distribution predominantly in Australia-New Guinea-southeast Asia, indicates links between those areas and New Zealand. The family '. . . has an obvious connection with Marsupialia and Rodentia . . .' (Hopkins, 1957:68) and in Australia itself '. . . has radiated dramatically to parasitize monotremes, most of the Australian families of marsupials, terrestrial and arboreal, terrestrial and aquatic rodents and land- and seabirds' (Dunnet & Mardon, 1974:51). In the absence of suitable endemic mammal hosts, the family is represented in New Zealand by a single mammal flea, *Pygiopsylla hoplia* on the early introduced *R.exulans,* and by seven species on birds. Of the latter, *Notiopsylla* with six taxa is confined to birds of the New Zealand and higher latitude zones. *N.k.kerguelensis* is apparently circumpolar while the other species have much more closely circumscribed distributions. *N.e.enciari* was first recorded from parakeets, the types and numerous subsequent specimens being taken from *Cyanoramphus unicolor* at Antipodes Is. Further collections, including those on other islands – some now without parakeets – have broadened the host list. Nevertheless it is apparent (Smit, 1965:23, 42; Dunnet, 1964:234; Dunnet & Mardon, 1974:78) that an origin from parakeets is postulated. However, the genus is predominantly found on procellariiform birds which are probably the general hosts for most species. Since the parakeets nest in close proximity with the ground, exchange of fleas with the ground-nesting procellariiform birds is readily envisaged.

Thirdly, the Rhopalopsyllidae is by contrast essentially a family associated with rodents of South America. *Parapsyllus* is widespread on sea birds (with some records of a reversion (?) to rodents in the absence of birds; Dunnet & Mardon, 1974:46), but it also contains several species and subspecies on strictly land birds. Thus, its hosts include Procellariiformes, Sphenisciformes, Pelecaniformes and Charadriiformes (Laridae) all of which are marine; and Charadriiformes (Scolopacidae), Gruiformes, Psittaciformes, Apterygiformes and Passeriformes all of which are terrestrial. The apparent anomaly here may be resolved from the habit, common to almost all these hosts, of nesting on or near the ground or even of burrowing into it. Further, it must be admitted that many records, quoted as 'taken from' some birds, are more strictly derived from sampling of nests. In some cases, this may have introduced misleading host-parasite records, since several of the hosts are known to nest in close approximation, either in space or in time, and the 'true' host(s) of some fleas may be more restricted than the foregoing account indicates. To be sure, some species of *Parapsyllus* appear, from the relative abundance of records, to be fairly host-promiscuous; nevertheless, critical re-examination of the data and meticulous attention to collecting conditions in the future are highly desirable. It may well be that some of the more glaring anomalies in the host-parasite data will prove to be due to 'collecting errors' s.l.

Another aspect to which attention should be paid is the tendency of some flea taxa to be confined to a locality rather than to a single host or even a closely related group of host birds. Thus, *Notiopsylla kerguelensis* ssp.A has been taken only from Antipodes Is., *Notiopsylla enciari* ssp.C only from Stewart I., *Notiopsylla* sp.B and *Parapsyllus l. lynnae* only from Snares I., and *P. lynnae* ssp.A only from Chatham Is. Particular attention to nesting habits of the recorded hosts in these localities may resolve the apparent host promiscuity of these fleas.

The taxonomy of the fleas in the New Zealand subregion, though fraught with its own idiosyncracies, should be reasonably stabilised when the remaining known material is worked up and the new species and subspecies described. As a good deal of collecting has been carried out over the past few decades, especially in the critical sub-Antarctic islands, it is not expected that many new taxa remain undiscovered. The very long freedom (geologically speaking) of the mainland bird fauna from endemic rodents seems to have resulted in a native terrestrial bird fauna which is relatively free from fleas; the marine birds, with their more recent and even continuing communication with other continents, have an abundance of fleas. (Both marine *and* terrestrial birds of the endemic fauna, of course, carry many Mallophaga.)

In view of the approaching satisfactory state of the taxonomy of the adult fleas, and bearing in mind the relatively small size of the flea fauna, it is opportune to outline what could be regarded as the next stages in New Zealand Siphonapterology. Apart from checking the validity of present records, as mentioned above, there should be an endeavour to discover the 'normal' host range of each species (as opposed to records of 'stragglers'); many species of most orders of birds have not been examined for fleas, though a few unpublished negative records are known to the writer. Little progress has been made on ecological studies such as temperature or humidity tolerances and their normal limits in nests. Above all, life history observations and the determination of larvae, with a view to

Table 6. Species and subspecies of fleas in New Zealand – Summary

	on native hosts	on introduced hosts	total
on mammals	1	9	10
on birds	19*	1	20

* includes undescribed species and subspecies from tables 4 and 5.

working up the taxonomy of the immature stages, is a worthy project.

Larvae are known and have been described for most of the introduced fleas of the subregion, with the probable exception of *Xenopsylla vexabilis* and *Pygiopsylla hoplia.* With regard to the endemic fleas, Brown (1964) has briefly described and illustrated the larva of *Parapsyllus magellanicus heardi* (see note regarding this subspecies; Smit, 1964), but no comparative treatment was possible at that time. Given the relative paucity of taxa in the subregion, the taxonomy of the flea larvae is not an impracticable task; much collecting has been done by field collectors and the writer holds most of the resultant larvae in his collections. Larval taxonomy must of course be preceded by definitive adult taxonomic publications, but once completed it will open the way for more meaningful studies of nest ecology as a whole. This in turn could lead to a better understanding of host-flea distributions within the New Zealand bird fauna.

Table 6 summarises the host-flea distribution in the subregion. The great preponderance of bird fleas is one of the most striking features of the endemic flea fauna. Bearing in mind the concept that bird fleas probably arose from mammal fleas, the absence of native rodents in the subregion probably accounts for the lack of fleas evolved onto those endemic birds which nest in the tree canopy (e.g. most Passeriformes). The bird fleas which do occur are associated with ground-nesting Procellariiformes, Charadriiformes, Gruiformes, Psittaciformes; even among the Passeriformes, *Bowdleria* nests in low tussock, effectively in contact with the ground.

4 ACKNOWLEDGEMENTS

I am indebted to numerous colleagues for permission to quote from their records of flea collecting data, and to Dr J.Warham for advice about the nesting habits of the birds mentioned. I am especially grateful to Mr F.G.A.M.Smit for making available his unpublished data, including the distribution of the undescribed taxa. Grants in aid of travel expenses are gratefully acknowledged from the University of Canterbury, and the British Council.

REFERENCES

Brown, K.G. 1964, The insects of Heard Island. Australian National Antarctic Research Expeditions, Reports. Series B, Vol. 1, Zoology, Siphonaptera, p. 30-33.

Dunnet, G.M. 1964, Distribution and host relationships of fleas in the Antarctic and Subantarctic. Proc. 1st SCAR Symposium on Antarctic Biology, p. 223-238.

Dunnet, G.M. & D.K.Mardon 1974, A monograph of Australian fleas (Siphonaptera). Austral. J. Zool. Suppl. Ser. 30:1-273.

Gibson, R.N. 1972, Metazoan parasites of rodents in New Zealand. Unpublished thesis. Christchurch, N.Z., University of Canterbury.

Holland, G.P. 1964, Evolution, classification, and host relationships of Siphonaptera. Ann. Rev. Ent. 9:123-146.

Holland, G.P. 1969, Contribution towards a

monograph of the fleas of New Guinea. Mem. Ent. Soc. Canada, 61:1-77.

Holloway, B.A. 1976, A new bat-fly from New Zealand (Diptera: Mystacinobiidae). New Zealand J. Zool. 3:279-301.

Hopkins, G.H.E. 1957, Host-associations of Siphonaptera. First symposium on host specificity among parasites of vertebrates. Université de Neuchâtel, Institut de Zoologie. p. 64-87.

MacLean, F.S. 1955, The history of plague in New Zealand. New Zealand Med. J. 54(300): 131-143.

Smit, F.G.A.M. 1957, Siphonaptera from the Isles of Amsterdam and Kerguelen, collected by Patrice Paulian, and from Antipodes Island. Proc. R. Ent. Soc. London (B) 26(11-12):189-196.

Smit, F.G.A.M. 1964, Insects of Campbell Island. Siphonaptera. Pacif. Insects Monogr. 7:330-334.

Smit, F.G.A.M. 1965, Siphonaptera of New Zealand. Trans. R. Soc. New Zealand, Zool. 7(1):1-50.

Smit, F.G.A.M. & G,M.Dunnet 1962, A new genus of flea from Antarctica. Pacif. Insects 4(4):895-903.

Sobey, W.R., K.M.Adams, G.C.Johnston, L.R. Gould, K.N.G.Simpson & K.Keith 1973, Macquarie Island: the introduction of the European rabbit flea *Spilopsyllus cuniculi* (Dale) as a possible vector of myxomatosis. J. Hyg., Cambridge 71(2):299-308.

Watt, J.C. 1971, Ectoparasitic insects on birds and mammals of the Kermadec Islands. Notornis 18(4):227-244.

CLUFF E. HOPLA
University of Oklahoma, Norman, Okla., USA

A STUDY OF THE HOST ASSOCIATIONS AND ZOOGEOGRAPHY OF *PULEX*

ABSTRACT

Six species are assigned to *Pulex;* all but one are known from the New World with the exception of *Pulex simulans* in Hawaii. The host associations for all species are reviewed; however, emphasis is placed on *Pulex irritans* and *Pulex simulans. P.simulans* has a much broader host distribution in the southern part of its range, being confined mostly to burrowing rodents in the extreme northern edge. *P.irritans* has adapted to *Speotyto* sp. in the Pacific Northwest of the United States and in the adjacent area in Canada. Central and South America are thought to be the center of origin for the genus. *Pulex sinoculus* demonstrates a unique adaptation along with the host.

1 INTRODUCTION

The original design of this study dealt with (1) the incidence, geographic distribution, and host associations of *Pulex simulans* Baker, 1895, in the southwestern United States, and (2) a review of publications dealing with *Pulex irritans* Linnaeus, 1758, in order to determine which of the two species was involved in a particular citation dealing with the North American fauna. However, in order to gain a clear image of the host relationships of the two closely related species, *P.irritans* and *P.simulans,* the scope ultimately was broadened to include research into all known history and distribution of *P.irritans* and *P.simulans* in the New World. This was accomplished by reviewing past records of *P.irritans* and attempting to obtain the specimens upon which these citations were based. This last mentioned procedure was not only important to clarify the species of flea reported, but has significant bearing on past studies regarding plague in the western United States. Where considerable material was available, a visit to certain laboratories and museums to study the specimens was undertaken, after which arrangements were made for borrowing the *P.simulans* for detailed investigation.

It is my hope that the following discussion will clearly indicate that, while a great deal of time has been put into this study, considerable more information is needed, especially from Mexico, Central America and South America. Additional effort must be expended in order to obtain a thorough understanding of the genus *Pulex.* No longer can we consider *Pulex* in a simple concept. I believe this genus to be a complex one, consisting of several species and numerous subspecific forms.

1.1 *Historical review*

Until relatively recently, the genus *Pulex* was not clearly defined. While some confusion still exists, much has happened in the previous decade that makes a definition of the genus tenable. A history of the genus proves interesting and informative, especially if limited to the New World.

Pulex irritans Linnaeus, 1758, has been considered a cosmopolitan species and essentially the only member of the genus until approximately 28 years ago. Since its original description, the name *P.irritans* has remained unsullied by the vagaries of taxonomic opinion.

In 1895, Baker described *Pulex simulans* as a species closely related to *P.irritans,* collected from 'the opossum, *Didelphis virginiana*', along Devil's River, Texas (Smit, 1958: 525). Yet in 1904 Baker reduced it to a variety of the latter. However, Jordan & Rothschild (1908) synonymized *P.simulans* with *P.irritans* without comment, where it remained until redefined by Smit in 1958.

In 1899, Baker described *Pulex irritans dugesii* collected from *Spermophilus macrourus* in Guana Juato, Mexico. Later, Baker (1904) elevated this form to a full species. Subsequently, Jordan & Rothschild (1908) placed *dugesii* as a synonym of *irritans,* where it remained until 1943 when Ewing & Fox restored it as a subspecies of *irritans* again. Hopkins & Rothschild (1953) agreed with Jordan & Rothschild (1908) thus placing *dugesii* in synonymy once more. However, Barrera (1955) re-established *dugesii* as a subspecies of *irritans.* He felt that *P.i.dugesii* differed from the nominate form by having longer palpi, the constant presence of a genal tooth, and a slightly smaller size in female specimens. Later Smit (1958) placed *dugesii* as a synonym of *Pulex simulans* Baker, 1895. Smit's concept is unquestionably a valid one and is accepted by both Barrera (personal communication) and myself.

In 1923 Jordan & Rothschild described *Pulex porcinus;* however, Wagner in 1933 transferred *porcinus* to a new genus, *Juxtapulex,* at which time he also described *echidnophagoides.* Hopkins & Rothschild (1953) returned *porcinus* to *Pulex,* retaining *echidnophagoides* in *Juxtapulex.* Barrera (1955) did not agree with Hopkins & Rothschild and proposed a new system of classification. His scheme placed *Juxtapulex* as a subgenus within *Pulex,* represented by the following species: *echidnophagoides, porcinus,* and a new species, *alvarezi.* According to Barrera's concept, the subgenus *Pulex* contained *irritans, i.dugesii,* and *sinoculus.* The last mentioned species was described by Traub (1950) from specimens taken on *Orthogeomys grandis* and *Sciurus g.griseoflavus* in Tecpan, Guatemala. The only record of *P.sinoculus* north of Guatemala is that of Barrera (1955) from *Canis familiaris,* 20 km north of Tapachula, Chiapas, Mexico.

The concept of the genus *Pulex* being divided into two subgenera is acceptable to me. In summary, the subgenus *Pulex* now contains *irritans, simulans,* and *sinoculus.* Including the species listed above for the subgenus *Juxtapulex,* the genus now contains six species.

With the exception of the records of *P.simulans* in Hawaii (Haas & Wilson, 1967), all other species of the subgenus *Pulex* but *irritans* are confined to the New World. Insofar as I have been able to ascertain, *P.irritans* (as currently defined) is a remarkable morphologically homogeneous species, particularly when one considers its distribution in both the Old and the New World. Noticeable biological differences apparently exist in various parts of the world, but considerable study is needed to elucidate this concept.

Table 1. Host associations of *Pulex irritans* and *simulans* in Colorado, Montana and Wyoming

Host	*irritans*		*simulans*	
	#Coll.*	#Specimens**	#Coll.	#Specimens
Canis familiaris	14	270		
Canis latrans	35	1 137		
Cynomys gunnisoni			01	03
Cynomys leucurus			12	27
Cynomys ludovicianus	01	05	18	60
Cynomys sp.			07	32
Homo sapiens	02	02		
Lepus californicus	02	02		
Lepus townsendii	03	10	01	01
Lepus sp.	04	07		
Marmota flaviventris			01	56
Neotoma cinerea			01	01
Mustela campestris	01	01		
Spermophilus columbianus			01	14
Spermophilus richardsonii	01	01		
Spermophilus sp.			01	01
Sylvilagus audubonii	01	01		
Sylvilagus nutallii			01	01
Sylvilagus sp.	01	01		
Taxidea taxus	01	02		
Vulpes fulva	01	07		

* #Coll. refers to the total number of positive records from the area.
** #Specimens refers to the total number of specimens taken.

1.2 *Review of past records*

Jellison & Kohls (1936) reported one of the earliest records of *P. irritans* from the black-tailed prairie dog (*Cynomys ludovicianus ludovicianus*). They collected 140 fleas from ten hosts in widely separated points within a colony in Montana. One hundred and twenty-four of these belonged to *Pulex.* I have been able to locate 38 specimens of this collection; 26 were at the Rocky Mountain Laboratory, and 12 at the United States National Museum. All are *P. simulans,* and it is reasonable to assume that most, if not all, of the remaining material was of the same species. This seems especially true when one considers that all of the other known records of *Pulex* material from the burrowing Sciuridae in that portion of the country pertain to this species. Table 1, in part, reflects the review of this area.

A few years later, Eskey & Haas (1940) reported the presence of *P. irritans* on Gunnison's prairie dog (*Cynomys gunnisoni*) in Apache County, Arizona, but did not find it in adjacent Catron County, New Mexico. Specimens received from the Plague Laboratory, San Francisco, California, indicate that these were *P. simulans.* It is interesting to note that Ewing & Fox (1943), without comment, state that the prairie dog was a true host of *'irritans'.* This is the first report, insofar as I can ascertain, where a rodent was given such a classification in regard to this flea.

The two texts by Fox (1940) and Hubbard (1947) cite only the most general of references. However, many of the specimens recorded as *P. irritans* associated with *Didelphis, Spermophilus, Mephitis,* and *Cynomys* are without doubt *P. simulans.*

Early studies on the ectoparasites of the wild Canidae are not numerous. Randolph &

Eads (1946) reported *P.irritans* collected from wolves (*Canis niger rufus*) in Lavaca County, Texas. No specimens have been recovered from this study, but on the basic information compiled from this host in Oklahoma and Texas, undoubtedly most of the *Pulex* must have been *simulans.* Eads (1948) reported on the ectoparasites removed from 90 coyotes from various regions in Texas. Of the Siphonaptera, Eads shows that *P.irritans* was, by an overwhelming majority, the most abundant flea. From data in Eads' table 1, I find that 2,024 out of 2,047 fleas (99 per cent) were *Pulex.* Undoubtedly, many of these fleas were *P. simulans,* although, unfortunately, I have not been able to examine any of these specimens. The same applies for the small number of *Canis niger rufus* reported in the same paper. I have received part of the specimens used by Gates (1945) in her study of fleas in Nebraska. Both species are represented nearly equally in this material. Ellis (1955) reported *P.irritans* from wild Canidae at the Wichita Refuge taken by Messrs McMurray & Rouse. All of these fleas were *P.simulans.* Records for the complete collection made by these two gentlemen amounted to a total of 2,575 fleas. Benton (1960) reported on a collection of *P.simulans* from northern Louisiana; the only known host reported was a dog. Table 2

Table 2. Host associations of *Pulex irritans* and *simulans* in Oklahoma and Texas

Host	*irritans*		*simulans*	
	#Coll.	#Specimens	#Coll.	#Specimens
Bos tarus	01	01		
Canis familiaris	03	01	23	223
Canis latrans			32	584
Canis niger			03	37
Dog-Coyote cross			01	03
Cynomys ludovicianus			09	29
Cynomys sp.			02	03
Dama sp.	01	18	01	05
Didelphis marsupialis			06	22
Didelphis mesmericana			02	11
Didelphis sp.				
Homo sapiens	01	02	05	09
Lepus californicus			01	02
Lepus sp.			02	06
Mephitis mephitis			06	18
Mephitis sp.			06	17
Orchids	01	01		
Peromyscus leucopus			01	01
Peromyscus truei			01	01
Peromyscus sp.			01	01
Procyon lotor	01	01	03	05
Rattus norvegicus			01	16
Spermophilus mexicanus			01	02
Spermophilus spilosoma			01	01
Spermophilus variegatus			01	02
Spermophilus sp.				
Sylvilagus sp.	02	02	01	01
Tayassu tajacu	03	08		
Urocyon cinereoargenteus			06	41
Vulpes fulva			02	08
Vulpes velox	07	76	07	129

lists the host associations of *P.irritans* and *P.simulans* in Oklahoma and Texas, based upon the literature and personal data.

Ecke & Johnson (1952) reported that *P.irritans* constituted a minor population on the black-tailed prairie dog in Colorado. A small number of specimens, believed to have been among the lot under discussion, have been received from the Plague Laboratory, San Francisco, California. They are *P.simulans.* Essentially the same is true for the records of Miles, Wilcomb & Irons (1952). Personal communication with Dr Richard B.Eads, formerly of the Texas State Department of Health, indicated that approximately 60 of the specimens collected by Miles & Wilcomb are in the permanent collection of the institution and that all of these are *P.simulans.* Since this correspondence with Eads, I have received these specimens through the courtesy of Dr Jack Wiseman, Texas State Department of Health, and have been able to confirm that all *Pulex* removed from prairie dogs and/or burrows by Miles & Wilcomb were *P.simulans.* In contrast to Ecke & Johnson (1952), Miles & Wilcomb found *P.simulans* to be one of the most abundant fleas in the burrows of the black-tailed prairie dog at the time they worked in western Texas, Benton (1960).

2 HOST ASSOCIATIONS

2.1 *Materials and methods*

With regard to the host relationships of *P.simulans* in the southwestern United States, various animals such as the prairie dog, opossum, coyote, etc., were collected, placed in cloth bags, and 'chloroformed' to recover the ectoparasites. This was a standard operational procedure in this study and has proved successful in similar studies over the years. The animals were 'brushed' over a white enamelled pan and the ectoparasites collected, labelled, and numbered the same as the host. A record was maintained of all host animals taken, regardless of whether they were infested with fleas or were free from such insects. This type of information is especially valuable, and, as more data are accumulated in the future, a valid picture of the host associations of many species of fleas (of some species little or nothing is known at the present time) can be established.

Animal burrows were checked with a 'swab' which consists of a square yard of flannel secured to the end of a flexible cable. By pushing this into the burrow as far as possible and slowly pulling it out, fleas and ticks become entangled on the surface of the cloth and can then be removed with the aid of forceps. Frequently this method produces more fleas than collecting from the host animals. This is especially true when working with rodents such as *Cynomys.*

Whenever possible, nest materials of mammals were placed in bags and emptied into modified Berlese funnels to secure the fleas. The latter technique is one of the most valuable aids in obtaining host records for certain groups of these insects. While it is a relatively simple operation, it is one method that has been largely overlooked in the past by many workers.

Figure 1 illustrates the area in the Southwest where *P.irritans* and *P.simulans* were studied by me. While this area at first glance appears to be rather homogeneous, eight Biotic Provinces (Dice, 1943) are found within its limits. The variation in climate, particularly precipitation, is significantly contrasting; for example, the Kansan and Navahonian as compared to the Texan and Austroriparian. The terrain encompassed was large enough

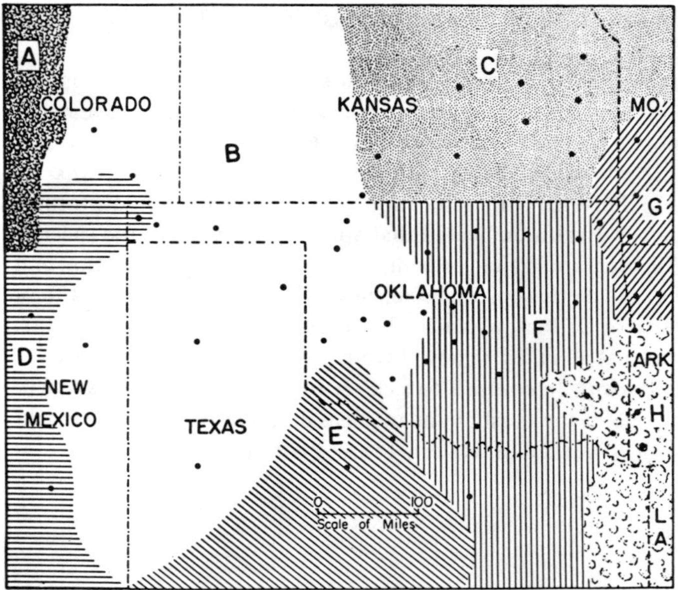

Figure 1. The Biotic Provinces in the study area (Dice, 1943). A. Coloradan, B. Kansan, C. Navahonian, E. Comanchian, F. Texan, G. Carolinian, H. Austroriparian.

to give adequate sampling within these Biotic Provinces, thereby avoiding the possibility of mere overlapping or 'fringing' from one province to another.

Once the fleas were obtained, they were processed in 10 per cent KOH, rinsed at least twice in distilled water, transferred through a series of alcohols, cleared in either oil of cloves or methyl salicylate and mounted on slides with Harleco Synthetic Resin or Neutral Canada Balsam as the mounting medium. A large number of slides which had been mounted in 'permount' medium earlier in the study had to be remounted in Harleco because of a cloudiness that developed after about six months. Preliminary experiments indicated that the Harleco medium was satisfactory and did not yellow with age (slides aged for two years were available at the beginning of this study). One of the advantages of a long term study lies in the fact that one observes his own misjudgements! In the years since 1959, the Harleco Resin was found to yellow within three to four years, and a change in refractory index apparently takes place. (I now use balsam exclusively, as do most other serious students of the Siphonaptera.)

Circular 15 mm, No. 2 cover slips were used in an effort to secure uniformity and to standardize, insofar as possible, distortions due to mounting. This procedure is important when statistical methods are applied to the specimens at a later date.

Several hundred alcoholic specimens were obtained from Dr W.L.Jellison of the Rocky Mountain Laboratory, Hamilton, Montana. These have been mounted on slides as described above and are extremely valuable not only from the aspect of geographic distribution but, because of the large numbers involved, they probably represent the most extensive collection of *Pulex* from a given area in North America.

In addition to these alcoholic specimens, Dr Jellison had a large series of this group on slides from various parts of the world. All *Pulex* specimens from the New World were studied at the US National Museum, Field Museum (Chicago, Ill.), Traub Collection, Canadian National Collection, and British Museum (Natural History) at Tring, England. Arrangements were made for a loan of all available *Pulex* material except that of the British Museum. In addition, all available *Pulex* were secured from the CDC Field Station (Harold Stark), California Vector Control Service (Richard Peters), Snow Entomological Museum of the University of Kansas (George Byers), Communicable Disease Center (Harry Pratt), and the Texas Department of Health (Jack Wiseman). Alfredo Barrera sent a list of the material contained in the Mexico City Museum of Natural History.

Scientific names of the hosts follow Hall & Kelson (1959) whenever possible. However, in the appendices, I have used the names as they occur on the slides. Stringent efforts have been made to avoid errors in 'the translation'.

2.2 *Results*

The host associations of *P.simulans* are extremely varied from one geographic location to another, but consistent within a given area. For example, in the northernmost part of its distribution (table 1), it is confined almost entirely to the burrowing mammals such as *Cynomys, Spermophilus* and *Marmota.* In the same area, *P.irritans* is found almost entirely on various Canidae and *Speotyto* (burrowing owl). The last mentioned is a most interesting association and more will be said of this later. On the other hand, Canidae taken in Nebraska are just as likely to have both species as they are to have either one; known material from Kansas shows a greater abundance of *P.simulans.* Progressing further south, by the time one reaches Oklahoma, *P.simulans* is certainly the most abundant of the two on coyotes, and certain collections from dogs in the Woodward area consist entirely of *P. simulans.* A large series of specimens obtained from coyotes by personnel of the Wichita Mountains Wildlife Refuge in Comanche County, Oklahoma, is almost entirely *P.simulans.*

On the basis of material collected to date, *Pulex* sp. is virtually absent from *Cynomys ludovicianus* in the eastward limits of that mammal's range in the southwestern United States. For example, prairie dog colonies in Comanche (Wichita Refuge), McClain and Cleveland counties in Oklahoma had densely populated prairie dog colonies of approximately 50 acres each during the primary phase of this study. Despite repeated checking of the animals and their burrows, no *Pulex* specimens were recovered. On the other hand, in the drier areas to the west (Arizona, Colorado, New Mexico, and Texas) if *Pulex* is found either on the colonial sciurids or in their burrows, almost invariably it is *P.simulans,* not *P.irritans.* Concepts regarding host associations in California are not so easily interpreted, in part because of its diverse climate and fauna. General impressions indicate that the *P. simulans* females are larger in size in the eastern part of its distribution, but statistical methods have not yet been applied sufficiently to demonstrate the degree of this difference. Dr Allen H.Benton, State University of New York (personal communication) has observed this in certain of his material from Louisiana when compared with specimens taken in Arizona.

Figure 2 reveals the known distribution of the two species in North and South America. Collections of these two fleas from mammals in Nevada and eastern Washington are only rarely encountered. Most of the data for that state and eastern Washington come from unpublished data in my collection.

In the southwestern United States, several common genera of mammals act as hosts of *P.simulans.* Principal among these are the opossum (*Didelphis marsupialis virginiana*) from

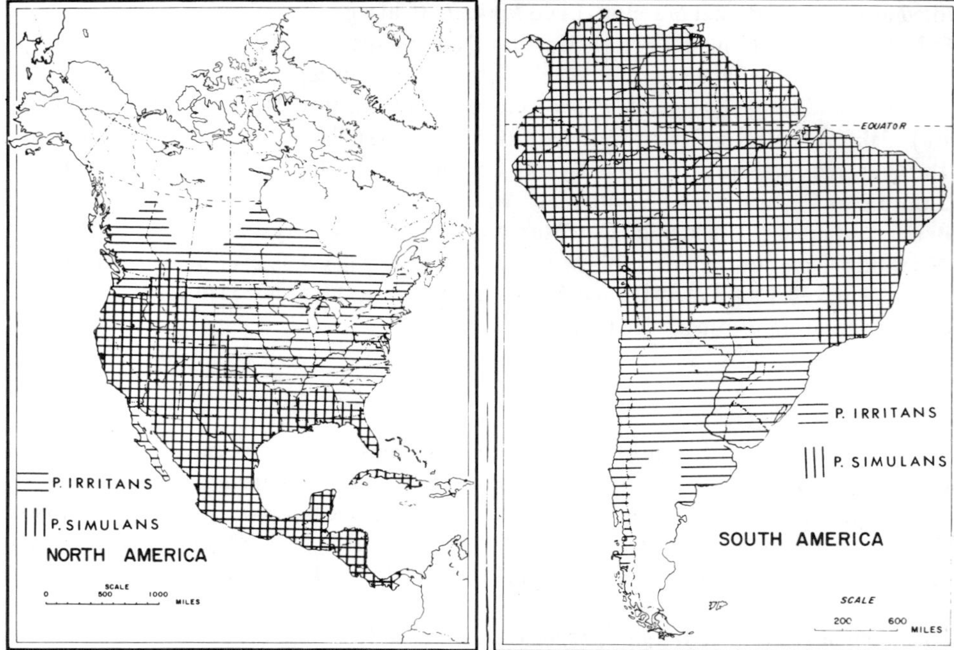

Figure 2. Distribution of *Pulex irritans* and *Pulex simulans* in North America and South America.

which Webester (1904) reported *P. irritans* and several species of the genera *Mephitis, Canis* and *Vulpes.* This observation is supported by Mead (1963), and by Wilson (1966), albeit the latter investigation was cursory. It is difficult to establish host relationships insofar as the opossum is concerned, because it lives in such a variety of habitats within a given location. Mohr & Morlan (1959) indicate that the opossum has no true fleas, but I think that in the Southwest, under sylvatic conditions, it is a true host of *P. simulans.* If the opossum is taken near urban areas, it will have a variety of 'promiscuous' fleas (Hopkins, 1957), such as *Ctenocephalides felis* (Curtis, 1826), *Nosopsyllus fasciatus* (Bosc, 1801) or *Orchopeas howardi* (Baker, 1895). Be that as it may, records to date from approximately 50 opossums live-trapped in central Oklahoma and taken in localities other than the edificarian habitat, have *P. simulans* almost to the exclusion of any other species. Essentially the same picture holds true for the various species of *Mephitis.*

The host associations of *P. simulans* in Mexico do not, on the basis of available records, differ largely from those of the south-western United States. Burrowing rodents (such as *Spermophilus variegatus* and *Cynomys mexicana*), *Didelphis marsupialis,* and several species of *Mephitis* seem to be the principal hosts. Barrera (1956, 1968) has substantiated host records with his collections. Tipton & Mendez (1968) believed the prairie dog *Cynomys* to be the optimal host and that Mexico and Central America are the geographic areas best suited for *P. simulans.* Tipton & Mendez (1968) reported on host parasite relationships of the fleas in Cerro Potosi. Barrera (1968) considered Central America as the center of radiation of *Pulex.* Here man enters the picture as a host, although not for the first time, inasmuch as I now have one record each from California and Texas. In reality, I have not seen a sufficient number of records from Mexico. Considerable information

Table 3. Host associations of *Pulex irritans* and *simulans* in Mexico

Host	*irritans*		*simulans*	
	#Coll.	#Specimens	#Coll.	#Specimens
Bassariscus astutus			03	27
Canis latrans			03	10
Cynomys mexicanus			01	04
Didelphis aurita			01	02
Didelphis marsupialis			04	72
Equus caballus	02	03		
Fox	01	?		
Homo sapiens	02	09	04	07
Mephitis mephitis			01	02
Peromyscus mexicanus			01	01
Peromyscus sp.			01	04
Spermophilus variegatus			10	43
Spermophilus sp.			01	14
Spilogale augustifrons			03	09
Tapirus bairdii			01	04
Taxidea taxus			01	02
Tayassu sp.	01	01		
Urocyon cinereoargenteus			03	04
Vulpes macrotis	01	01		

needs to be gained from this part of the Nearctic Region. Table 3 summarizes data I have from Mexico.

Records from Central America with regard to *P. simulans* are again meager; however, they do indicate a broad host range. Man regularly appears to be a host, as do certain species of 'wild rodents'. In the Republic of Panama, where records are especially sparse, I have found only four collections, which are in the Traub collection. Three of these are from man and one from a dog. Tipton & Mendez (1966) completed an extensive survey of the Siphonaptera of Panama and reviewed the previous publications. Their results were disappointing insofar as *P. irritans* and *P. simulans* were concerned. However, out of a total of 32 males and 57 females of *P. simulans,* 31 males and 53 females were from man. See table 4 for other data.

In South America adequate information is not to be had. Johnson (1957) reports a broad distribution, but the scope of the publication does not lend itself to an understand-

Table 4. Host associations of *Pulex irritans* and *simulans* in Central America

Host	*irritans*		*simulans*	
	#Coll.	#Specimens	#Coll.	#Specimens
Canis familiaris			07	11
Didelphis sp.			01	01
Homo sapiens			07	16
Mephitis sp.			01	02
Mustela frenata			01	01
Orthogeomys grandis			01	02
Sciurus griseoflavus	01	01		
Urocyon cinereoargenteus			02	02

Table 5. Host associations of *Pulex irritans* and *simulans* in South America

Host	*irritans*		*simulans*	
	#Coll.	#Specimens	#Coll.	#Specimens
Canis familiaris	06	26		
Cavia porcellus			01	01
Homo sapiens	03	12	03	10
Mustela sp.			01	05
Nasua sp.	01	01		
Rodents			01	03
Speothos sp.			01	03
Sus scrofa	01	02		
Tamandua tetradactyla			01	02

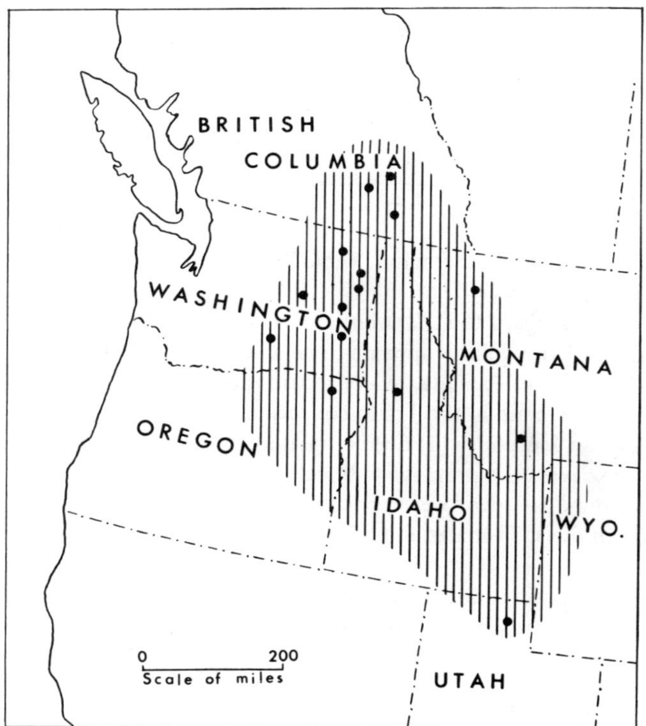

Figure 3. Distribution of *Pulex irritans* from burrowing owls in North America

ing of host associations except in the most general way. Tipton & Machado-Allison (1972) in Venezuela found *P. simulans* most commonly associated with an anteater, *Tamandua longicaudate,* followed by the crab-eating fox, *Cerdocyon thous,* and a skunk, *Conepatus semistriatus.* Tipton & Machado-Allison (1972) did not collect *P. irritans.* Other records available, though cursory, indicate host associations as far south as Bolivia are broad. Man is indicated frequently as a host. The true nature of the picture between *P. irritans* and *P.*

simulans in this particular respect is not known. On the basis of records available to me, *P.simulans* appears almost as frequently on man as does *P.irritans.*

Correspondence with Mr Frans Smit (1960, 1966) of the British Museum indicates that speciation within the genus is greater than one might suspect. He has a large series of specimens from South American rodents which undoubtedly represent new species. If this proves to be correct, it is perhaps well to indicate that they are closer to *P.simulans* than to any other species within the genus, indicating that a *P.simulans* complex may exist (table 5).

In reviewing all the host records as one proceeds south from Canada, it seems certain that *P.simulans* is a highly promiscuous flea, apparently able to adapt to many species of mammals, even more so than *P.irritans.* It has been interesting to note that all fleas taken from the domestic pig have always been *P.irritans,* which is a distinction that even man cannot share!

Ordinarily when dealing with the genus *Pulex* one does not expect to be involved in a host association with Aves. However, in the Pacific Northwest, this unique phenomenon is encountered. Figure 3 shows the distribution of a form of *P.irritans* associated with the burrowing owl (*Speotyto cunicularia*). This is indeed a surprising relationship, as indicated by Holland (1949). Before examining the material from this host, it was speculated by others and myself that likely the *Pulex* fleas reported from this avian host and its burrows would be *P.simulans.* This seemed reasonable, especially in view of the fact that this owl does not dig a burrow of its own, but modifies one from the burrowing colonial sciurids such as *Cynomys* or *Spermophilus.* Secondly, these owls prey upon cricetid rodents, insects, and amphibians, which are not likely sources of either *P.irritans* or *P. simulans* in this particular area of the United States. To me, this indicates more than ever that the burrows of the colonial sciurids should furnish the *Pulex* involved. However, after checking some 300 fleas removed from the owls and their burrows, no *P.simulans* were found. All were of the *P.irritans* type! Table 6 summarizes the data for *Speotyto.*

Whatever the processes involved in such a unique adaptation, they are truly remarkable ones. Biological differences in habitat for this form of *P.irritans* in contrast to those taken on Canidae in the same general area must be considerable. The higher humidity found in the burrow, the higher host temperature, and the feeding on nucleated erythrocytes is not to be taken lightly. Preliminary observations of this material have not provided any means of separation of the specimens from typical *P.irritans.* The longer, narrower bristles associated with fleas on avian hosts are not evident here, indicating that the transfer of this flea to the burrowing owl is a recent one. Needless to say, this is one phase of the *Pulex* concept that must be investigated further by utilizing experimental laboratory techniques.

Table 6. Associations of *Speotyto* sp. (Burrowing Owl) with *Pulex irritans* and *simulans* in North America

Locality	*irritans*		*simulans*	
	#Coll.	#Specimens	#Coll.	#Specimens
British Columbia	03	13		
Idaho	01	03		
Montana	02	02	01	01
Oregon	01	50		
Utah	02	08		
Washington	18	276		

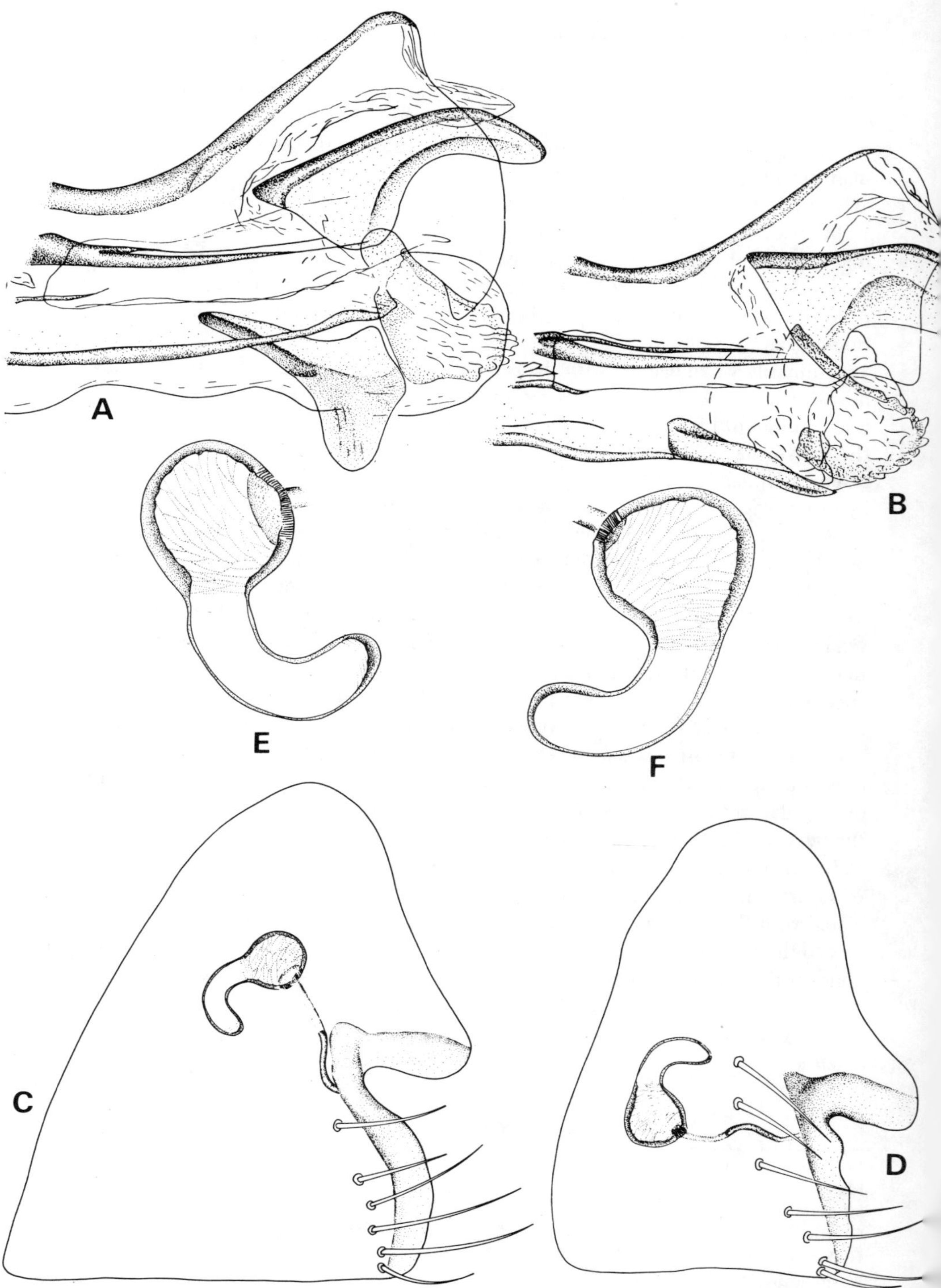

Figure 4. Crochets and dorsal aedeagal sclerites of *Pulex irritans* and *Pulex simulans* (A & B), the seventh sternites (C & D) and spermathecae (E & F).

3 TAXONOMY

It is perhaps well to recall again that Smit (1958) illustrated the pertinent male characteristics of *P. simulans* and *P. irritans;* these have been found to be remarkably consistent for the specimens (Figure 4) collected throughout my study area. The females, however, were entirely another matter and could not be separated with certainty. No combination of characteristics has yet been found that will separate the females of the two species of fleas in question, especially when specimens are from widely scattered areas. Figure 4C and D illustrate the seventh sternites for these two species. Typically, *P. irritans* has fewer bristles along the posterior margin than *P. simulans,* and usually is larger. No one characteristic is consistent. In the absence of male fleas from a particular collection record, the females are usually listed as *Pulex* sp. In a few areas one can state, with relative safety, that a female flea is one or the other species of *Pulex,* depending on the host and previous knowledge of the host associations of the area in question.

In dealing with two such closely related species as *P. irritans* and *P. simulans,* it is interesting to speculate as to what would happen if breeding populations of the two were found on the same host. Figure 5A and B illustrates the variations found under just such a circumstance. One hundred and fifteen male specimens were removed from a specimen of swift fox (*Vulpes velox velox*) taken in Cimarron County, Oklahoma, and as noted in the figure, the dorsal aedaegal sclerite and crochet, while variable, are remarkably consistent as to the general form for each species. The variations are as great as those noted in comparing several hundred specimens from many parts of the New World.

A somewhat different variation of *P. irritans* was encountered in specimens received from California, which had been removed from a mule deer some years ago (1944). Figure 5C presents the variations encountered, and at first glance it does not seem to conform to *P. irritans* as I have known it in the midwest. Certain of these may appear to some to be a hybrid of *P. irritans* and *P. simulans,* but judging from the shape of the crochet (2, 3) in these specimens, I decided it was within the normal limits of variation encountered in *P. irritans* elsewhere. Had the crochet of C-2 been associated with the dorsal aedaegal sclerite of C-3, finding an explanation would have been more difficult. To be sure, there is room for question, and some readers may take exception to my conclusion.

The question arises as to whether the specimens found on the swift fox in the above instance actually were 'promiscuous', as discussed by Hopkins (1957). However, due to the fact that the adults were of various ages (some teneral specimens were obtained), and the fact that many of the females had eggs within the abdominal cavity, it would appear to me to support the theory that the fleas were truly breeding-populations. Also the large numbers taken on Canidae in the Southwest lends additional support to the concept that *P. simulans* is promiscuous. The last point seems especially important because, paradoxically, virtually none of the small rodents (Cricetidae) within my own study area (figure 1) are known to harbour either of the species of fleas in question.

Large numbers of *P. simulans* have been removed from coyote dens, particularly when young pups were present. Such fleas are known to lay considerable quantities of eggs. One series of 19 *P. simulans* females averaged 17 eggs per day for an observation period of two weeks when placed, thrice daily, upon a pup.

Certain investigators have questioned the validity of *P. simulans,* for example Tipton & Mendez (1966, 1968), Tipton & Machado-Allison (1972). Others such as Stark (1958) have implied that likely more than two species are involved. I do not know the right an-

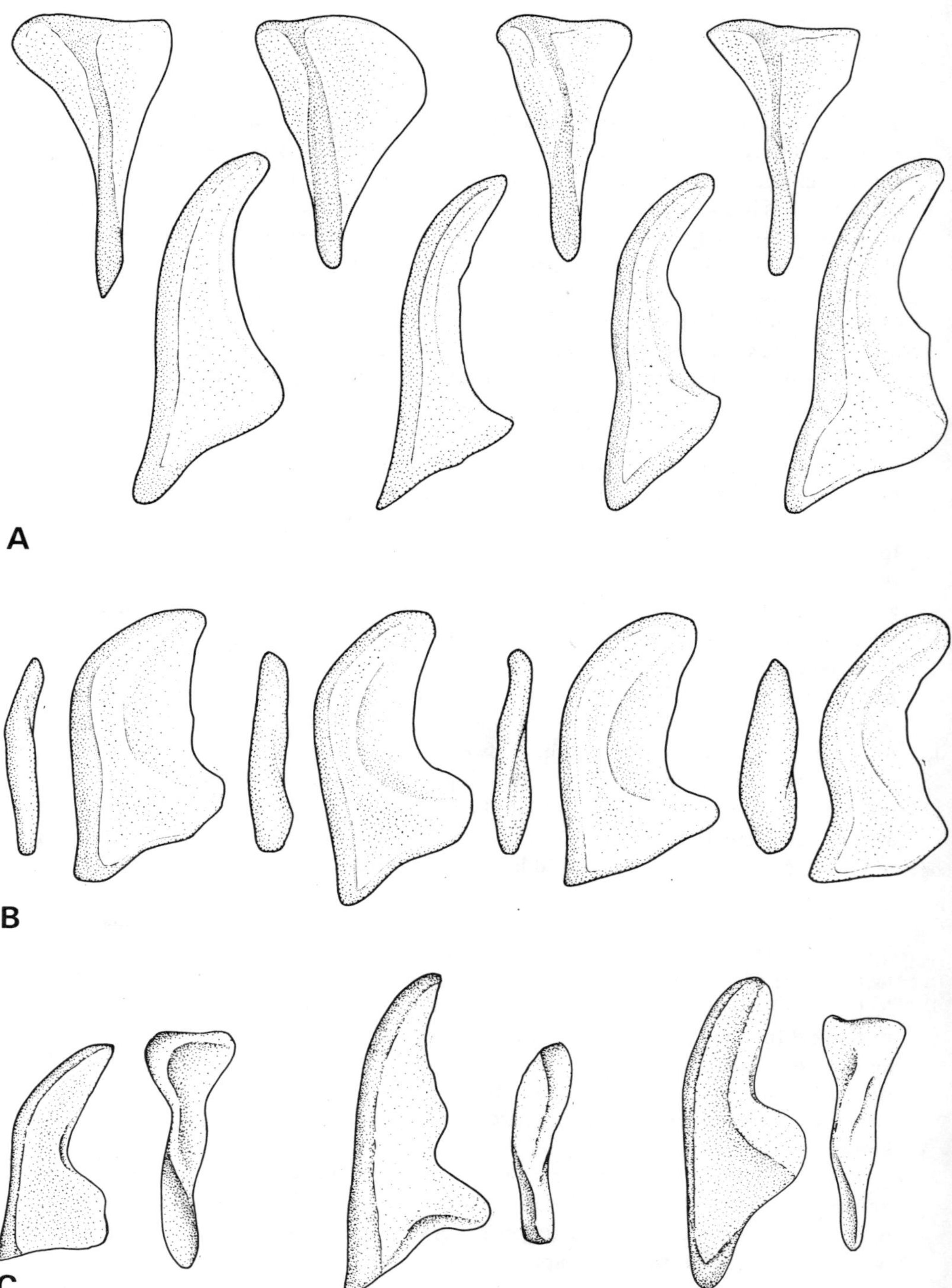

Figure 5. Variation of the crochet and dorsal aedeagal sclerite of: A. *Pulex irritans* from coyote in Oklahoma, B. *Pulex simulans* from coyote in Oklahoma, and C. *Pulex irritans* from deer in California.

swer to all of these various questions, but I am definitely of the opinion that *P. irritans* and *P. simulans* are distinct species. To me, the examples shown in figures 4 and 5 clearly indicate this is so. If these were merely races or subspecies, intergrades between the two should have been observed. This is but one example among many that could have been used with equal clarity. As to the significance of the variations in maxillary palps, the laciniae and the hypopharynx, as reported by Stark (1958) and Smit (1958), these are not fully understood by me because I have found them highly variable. I could cite innumerable other variations in other local populations. At one time, I thought certain characteristics would be valid in separating the females of the two species under discussion but soon was forced to abandon this concept as the study broadened in scope. Had I worked with only this localized population, even though the series was extensive, I would have erred in my conclusion.

In general, *P. irritans* is a somewhat larger specimen than *P. simulans.* These two species appear to vary in size in certain locations. The largest specimens of *P. irritans* that I have studied were collected in Nova Scotia.

A planned 'second part' of this study contemplates a statistical analysis of these variations. But first adequate samples must be procured from different parts of the world for which specimens are not now available. Such a study must also involve the establishment of laboratory colonies of known origin. Cross matings (if possible) should prove highly interesting. Hudson & Prince (1958) have shown that these two species can be reared in the laboratory.

The variety of habitats and different hosts of *P. irritans* in the Old World are not well understood. The Mediterranean and Siberian subregions (Hopkins & Rothschild, 1953) reveal several species of mammals from which this flea was collected. However, there is not a sufficient series on any one species other than man, domestic dog and cat, to indicate if it is of more than accidental occurrence. One is left with a feeling that it is far more complex than the data presently indicate.

4 MEDICAL IMPORTANCE

The medical importance of '*P. irritans*' has been much debated. Without doubt, a certain amount of confusion has stemmed from the poorly understood taxonomy of the *P. irritans* complex. Furthermore, the vector efficiency studies attempted to date probably are not truly indicative of the ability of a particular species of flea to transmit plague simply because so little information is known concerning the biology of the flea in question.

Generally speaking, the majority of the investigators of plague have downgraded the importance of *P. irritans.* Jordan's (1956) comment that *P. irritans* was 'known as a carrier of bubonic plague and suspected to be a transmitter of various other diseases; intermediate host of some Nematodes and Platyhelminthes' is therefore all the more interesting.

Ramos Diaz (1938), working the province of Lambayeque, Peru, was one of the first to support the concept that *P. irritans* was of importance in transmitting plague organisms from man to man and possibly from domestic animals to man. According to Ramos Diaz (1938), an epizootic among *Rattus rattus* was responsible for the appearance of plague in domestic guinea pigs (*Cavia porcellus*) and that *P. irritans* was responsible for conveying the infection from these animals to the initial human case and later to a second family in which clothes from the first victim were utilized. Ramos Diaz also reported he isolated

Pasteurella pestis from *P.irritans* found in the clothes of the second family. However, Pollitzer et al. (1965) do not consider the techniques used as fully reliable.

Macchiavello (1957, 1958) was of the opinion that *P.irritans* was of no importance in the transmission of plague, agreeing whole-heartedly with the concepts established by the Plague Commission in 1929. Yet Pollitzer et al. (1965) quote Macchiavello as stating in an unpublished article that 'if *P.irritans* were a vector of plague, the human population would have disappeared from the Andean territory'.

The abundance of *P.irritans* in parts of Bolivia, Ecuador and Peru in almost incredible numbers in clothes, on earthen floors, and intra-domestic animals (various species of *Cavia*) is of considerable importance. Highly significant is the fact that *P.irritans* frequently infested *C.porcellus* to a higher degree than the 'specific' flea *Tiamastus cavicola* (Weyenbergh, 1881). De la Barrera (1957) reported that of the eight species of fleas infesting *C.porcellus* in Loja Province, Ecuador, *P.irritans* was by far the most numerous, consisting of 1 804 specimens out of a total of 2 493. Only one specimen of *Xenopsylla cheopis* (Rothschild, 1903) was reported. Jervis Alarcon (1958) reported that *P.irritans* comprised 67 per cent of the fleas removed from this same host in the same province.

The only experimental study dealing with the vector efficiency of *Pulex* and plague organisms was attempted by Burroughs (1947). In this study, Burroughs assumed (and logically so) that he was working with *P.irritans* that had been raised for 'innumerable generations' upon deer. The data revealed that this strain of *Pluex* was not a good vector of plague in individual transmission experiments, although with en masse experiments he obtained successful transmission. At this writing, it has been impossible to secure specimens of the strain of fleas used by Burroughs (previously I had hoped that some specimens had been preserved in alcohol). Inasmuch as records cited in the current paper show that both *P.irritans* and *P.simulans* occur on deer in California, the results of Burroughs' study cannot be applied with any degree of certainty to either species. Nonetheless, Burroughs (1947) did show that only one specimen out of 57 became blocked but the flea died the following day, thus the vector efficiency remains unknown.

Blanc (1956) developed the concept reported by Blanc & Baltazard (1941, 1945) more fully. Briefly, in Morocco, *P.irritans* is stated to be a vector of plague and particularly important in transmitting plague organisms from man to man. Swellengrebel (1953) in reviewing the differences in plague transmission between Java and Morocco, supports Blanc and Baltazard, as does Meyer (1963). These authors generally agree that in those areas where *P.irritans* occurs in almost overwhelming numbers this flea can make up by sheer numbers for what it lacks in vector efficiency. The report by Beasley (1965) concerning plague in Bolivia follows this concept.

Baltazard et al. (1960) in reviewing plague in Kurdistan reports that *P.irritans* is responsible for the 'inter-human' transmission of plague in the small villages where there were no domestic rodents or 'liaison rodents' in the rural villages. This then became a familial plague.

Macchiavello (1958) indicated there was good evidence the buccal infection among some South American natives occurred from the habit of killing fleas by biting them. In such instances, there was an involvement of the cervical lymph nodes. Such a habit of crushing ectoparasites is not uncommon among aboriginal peoples.

In a particularly significant observation, Barnes (personal communication, 1968) found *P.irritans*, removed from dogs in New Mexico, infected with plague organisms.

Varela & Vasquez (1954) reported naturally infected prairie dogs (*Cynomys mexicanus*) near Gomez Farias in southeastern Coahuila. Barrera (1956) working in the same area of

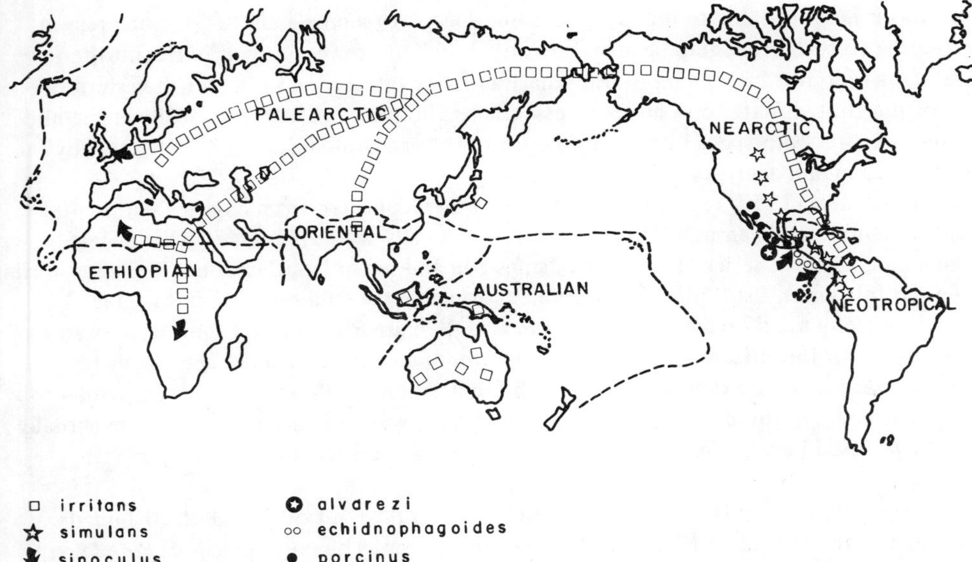

Figure 6. Origin and dispersal of *Pulex*. Central and northern South America are thought to be the center of origin for this genus.

Coahuila found the prairie dogs infested with a few specimens of *Opisocrostis* sp., and a large number of *P.simulans* (*P.i.dugesii*). Insofar as I know, this is the strongest implication that *P.simulans* possibly is of importance as a vector of plague.

P.irritans is an effective experimental vector of *Rickettsia mooseri*. As Traub, Wisseman & Farhang-Azad (1978) indicate, its role in the ecology of murine typhus has never been properly evaluated.

5 ZOOGEOGRAPHY

As indicated in figure 6, I postulate that the genus *Pulex* arose in the New World. This conclusion is supported by the fact that all but one species of the genus is limited to the New World and Smit (personal communication, 1961) has other undescribed *Pulex* from South America in his possession. Another factor not to be overlooked is the broad host distribution of this genus in South America (Johnson, 1957; Tipton & Machado-Allison, 1972) as compared to that known in the Old World, especially the Palearctic (Hopkins & Rothschild, 1953). The studies of De la Barrera (1957) and Jervis Alarcon (1958) are interesting when compared to those of Blanc (1956). Blanc was of the opinion that *P.irritans* in Morocco did not feed readily upon guinea pigs whereas Jervis Alarcon and De la Barrera found it to be the most abundant flea occurring on this animal.

6 DISCUSSION

In reflecting back on the material presented, it is apparent that the host distribution of *P. simulans* is broader than anticipated. Also, starting from the northern limits of its range and moving in a southerly direction, the host spectrum becomes increasingly varied. In part, this is no doubt due to the fact that some species of animals, limited for one reason or another to certain geographic areas, are not abundant or even present in the northern areas. However, it is interesting to note that the *Pulex* infestation in the northwestern portion of the United States on Canidae is essentially 100 per cent *P. irritans,* while in the midwestern portion both species occur. In the southwestern United States, *P. simulans* is by far the most abundant species on these animals.

Gier & Ameel (1959) reported that of 2 197 *Pulex* collected from coyotes within 40 miles of Manhattan, Kansas, all those examined critically were *P. simulans.* This differs from a much smaller series taken by Mr James Bee and me in Douglas County, Kansas, during 1948. Of this series, I still have 127 specimens available for study pooled from three animals. Among the 57 males, 31 are *P. irritans* and 26 are *P. simulans.* I cannot offer an explanation for this difference, nor do I know what percentage of those specimens reported by Gier & Ameel were examined critically. Be that as it may, these authors were wrong to state 'that most, if not all, of the fleas identified from wild animals as *P. irritans* are actually *P. simulans*'. As I have indicated previously, it depends upon the area in which one is working.

Stanley (1963) recorded that four foxes (*Vulpes vulpes*) out of a total of 20 animals examined were infested with fleas in northeastern Kansas. The only species of *Pulex* was *simulans,* which was also the most abundant flea encountered. On the other hand, Kilgore (1969) found *P. irritans* commoner than *P. simulans* on the swift fox. Kilgore's data are not compatible with mine; but neither of us have sufficient data pertaining to this host.

Tipton & Mendez (1966) point out that *P. irritans* was found at the higher elevations, whereas *P. simulans* occurred from sea level up to heights equalling that of *P. irritans.* Records supplied to me by Alfredo Barrera (personal communication, 1964) from Mexico support this concept most vividly. The highest known altitude for the collection of *P. simulans* in the United States were those taken by me from *Marmota flaviventris,* Mt. Timpanagos, Utah County, Utah, at approximately 8 300 feet. The habitat was a talus slope cohabited by *Ochotona princeps* and *Spermophilus lateralis.* On the other hand, I have no records of *P. irritans* from the intermountain west that would anywhere near approach this altitude.

My observations in the temperate region do not agree with those of Tipton & Mendez (1968) or Barrera's studies in Mexico. Haas & Wilson (1967) and Haas et al. (1972) found *P. simulans* to have a much broader geographic range in Hawaii than *P. irritans.* The latter species was more abundant in the cooler higher altitudes, in association with *P. simulans.*

Data for *P. (J.) alvarezi* are simply too limited to make any evaluation beyond the original collection cited previously. Records for *P. (J.) echidnophagoides* are not extensive but I have good reason to believe that it is now more common than the few published records indicate (Barrera, 1953). As armadillos of the genus *Dasypsus* are now considered important biological models for the study of leprosy and thus far depend upon wild caught specimens, collections of *P. (J.) echidnophagoides* should increase markedly in the near future. Recently, I have received two specimens collected from the nine-banded armadillo (*Dasypsus novemcinctus*) in Louisiana. To be sure, they have not moved northward as rapidly as the host. Layne (1972) examined 29 nine-banded armadillos with no fleas reported. The origin of the nine-banded armadillos in Florida is unclear. Cahalane (1947) states that the current population is the result of escaped pets. If Cahalane is correct, this could explain the negative records reported by Layne (1972). On the other hand, if the current population is a relict one, *P. (J.) echidnophagoides* should be present. Tipton & Mendez (1966) collected 859 *P.(J.)echidnophagoides* from 23 animals; 70 per cent of them were from seven *D. novemcinctus* with an average of 85 fleas per armadillo. *P.(J.)porcinus* is currently the most widely distributed species of the subgenus *Juxtapulex.* Although a small number of hosts have been recorded in the literature (Hopkins & Rothschild, 1953) peccaries (*Tayassu*) are the primary hosts.

Of the two species, I am forced to conclude that *P.simulans* is more adaptable than *P. irritans* in regards both to host associations and variety of habitats in North America. On the other hand, *P.irritans* is represented both farther south and north than *P.simulans* in the New World. The significance of this last point will be discussed later.

The situation in South America is poorly understood at the present because sufficient records are not available (Johnson, 1957). One can only speculate as to how many species may be ultimately involved and to the distribution of those species that are now known. Based on present information it seems *P.irritans* adjusts to a wider range of temperatures than *P.simulans.* Too, the adaptation of *P.irritans* to burrowing owls in the Western Central area of North America is remarkable. *P.sinoculus,* appearing to be a 'blind flea' undoubtedly is restricted to a subterranean habitat and likely will prove to be restricted to a small geographic area.

The extremely broad host range of *P.simulans* in Mexico, Central and South America, while coming from a relatively small number of records (when one considers the vastness of the area concerned and the number of animals for which records are not currently available), nonetheless suggests that Central or South America is probably the original or true home of this species. It is somewhat surprising that Beasley (1965) makes no mention of *P.simulans* in Bolivia. Only by adapting to various burrowing colonial Sciuridae has it been possible for this species to reach the extreme northern limit of its distribution in North America. The possibility of new species to be described by Smit in the near future indicates that the genus is far more complex than most workers had originally supposed. It appears that *P.simulans* is better adapted to the warm moist lower elevations of these two areas of the New World than *P.irritans.* One of the most intriguing aspects of the study has been the reference to wild rodents, other than the burrowing sciurids. Likely, most of these are accidental records in North America, but not so for South America. The types of hosts reported by Traub (1950) for *P.sinoculus* are most interesting and are not accidental. From the few records available, it seems likely that *P.sinoculus* is truly a 'rodent flea' and a burrowing rodent at that. The record of *P.sinoculus* previously cited by Barrera (1955) was removed from a dog that had been digging up a pocket gopher nest (Barrera, personal communication, 1964). When adequate collecting has been done, no doubt *Pulex* sp. will have been found and known to occur regularly on a wide variety of small mammals in Central and South America.

P.sinoculus, a 'blind flea', is associated primarily with the gopher *Orthogeomys grandis,* which is also virtually blind. Surely this is an adaptation to a particular type of habit, and is similar to that of fleas of *Dactylopsylla* and the gophers of *Cratogeomys* and *Geomys* in North America. This parallel with pocket gopher fleas was pointed out by Traub (1950, 1972).

The finding of *P.simulans* in Hawaii by Haas & Wilson (1967) is highly significant. Insofar as I know, this is the first indication that this flea has been introduced into a new area in historical times. It appears self-evident that the domestic dog was the 'carrier' for both species of *Pulex,* for man and pigs (even the 'feral' Suidae), in Hawaii are practically free from infestation. Haas et al. (1972) were not necessarily correct in thinking *P.simulans* gained entrance to the Hawaiian Islands by dogs brought from Central or South America. *P. simulans* could have gained entrance just as readily from a number of areas of the United States, such as the southwestern portion.

At least as interesting as the Hawaiian situation, is the possibility that *P. simulans* has undoubtedly been introduced into the Great Basin Region by domestic dogs on numerous occasions, as undoubtedly has *Ctenocephalides canis* (Curtis, 1826) and/or *felis* (Bouche, 1835). However, I have yet to find a domestic dog or cat infested with these last two mentioned fleas in that particular area. Another facet of this interesting puzzle lies in the fact that various species of wild Canidae are infested with *P. irritans,* yet records are still unknown from the domestic dog. Comparing the records from Hawaii with mine of the Great Basin Region (250 farm dogs in Utah), it is evident that more than introduction and a favourable host are necessary for the establishment of a flea. This last mentioned problem has been almost totally ignored in the study of Siphonaptera.

The data presented make me question whether *P. irritans* was introduced from the Old World to the New World by man as indicated by Jordan & Rothschild (1908). Holland (1949) reports an incident, written in 1792 by George G. Hewett, assistant surgeon on board HMS Discovery (Captain George Vancouver). According to Holland, Hewett stated that during the investigation of a certain abandoned Indian village, Vancouver's men were driven out of the area by myriads of fleas which attacked them to such an extent that they rushed out into the water to rid themselves of the fleas! Rothschild's (1973) report of a *P. irritans* from a viking burial site is most interesting. The Vikings may have introduced *P. irritans* to the New World for the first time or simply re-introduced it. If our concepts concerning the migration of mammals across the Bering Land Bridge during the Pleistocene or earlier be valid (Hopla, 1965), it seems entirely possible that *P. irritans* could have been introduced just as easily from the New to the Old World. In my own mind, I am convinced that this happened, despite the fact that 1) hundreds of native dogs, and several coyote and wolf skins in Alaska have always been found free of *Pulex* fleas and 2) M. Rothschild (1973) reported a specimen of *P. irritans* from a tenth century Viking pit.

Interesting as certain of these speculations are, they are difficult, if not impossible, to prove. It is known, however, that considerable biological variation of *P. irritans* occurs, although to date it is considered simply a cosmopolitan species. There are populations of '*P. irritans*' occurring on the foxes of Egypt and neighbouring countries which Dr Harry Hoogstraal (personal communication) believes to have a different biology from the ones found on man. Similar information such as this has been gathered from various individuals in diverse parts of the world. Recalling the burrowing owl-*P. irritans* relationship discussed earlier, the information implies that *P. irritans* must be a complex of species or subspecies and must ultimately be investigated from this point of view.

Taxonomy is not the major concern in this study, but one cannot work with two species within a genus and totally ignore the others. Differences of opinion will always exist, but new techniques available for taxonomic studies should lend themselves well to the study of Siphonaptera. On the basis of background gained in this study, a thorough study of *Pulex* and closely related genera should be undertaken. In part, the confusion during the past has resulted because we are fundamentally ignorant of the homologies of the parts making up the modified genitalia, especially that of the males. However, considerable progress has been made since Snodgrass (1946) published his observations. Traub (1950), Hopkins & Rothschild (1953, 1956, 1962), Peus (1956), Gunther (1961), Rothschild & Traub (1971), are largely responsible for new advances in our understanding of homologies. It is hoped someone can be encouraged to undertake this revision in the near future.

It would seem that many of the unresolved problems concerning this genus could be approached, in part, in the laboratory by rearing fleas of different stocks obtained from selected hosts and regions of the world. However, it is likely to prove an extremely difficult task, because while Bacot (1914) and Hudson & Prince (1958) were successful in rearing *P. irritans* under strict laboratory conditions, innumerable other attempts by other investigators were complete failures. Hudson & Prince (1958) are the only ones known to have reared *P. simulans*. The results of the latter study further support the concept that varous physiological differences exist and that additional taxonomic investigations are warranted.

ACKNOWLEDGEMENTS

This study was supported in part by Grant No. E-3052 (A) from the National Institute of Health and by the Faculty Research Council, The University of Oklahoma. I am deeply indebted to several individuals, who especially at the inception of this study were particularly helpful. They are W.L.Jellison, G.H.E.Hopkins, C.F.Muesebeck, F.G.A.M.Smit, H.Stark, and R.Traub. The technical help of Joyce Markman and Robin Young is acknowledged.

REFERENCES

Bacot, A.W. 1914, A study on the bionomics of the common rat fleas and other species associated with human habitations, with special reference to the influence of temperature and humidity at various periods of the life history of the insects. J. Hyg. 13, Plague Supp. 3: 447-654.

Baker, C.F. 1895, Preliminary studies in Siphonaptera, II. Canad. Ent. 27:63-67.

Baker, C.F. 1899, On two new and one previously known flea. Ent. News 10:37-38.

Baker, C.F. 1904, A revision of American Siphonaptera, or fleas, together with a complete list and bibliography of the group. Proc. U.S. Nat. Mus. 27:365-469.

Baltazard, M., M.Bahmanyar, M.Mostachfi, M. Eftekhari & C.Mofidi 1960, Recherches sur la peste en Iran. Bull. Wld. Hlth. Org. 23(1): 152-155.

Barrera, A. 1953, Sinopsis de los Sifonapteros de la Cuenca de Mexico. An. Esc. Nac. Cien. Biol. 7(1-4):155-245.

Barrera, A. 1955, Las especies mexicanas del genero *Pulex* Linnaeus. An. Esc. Nac. Cien. Biol. 8:219-236.

Barrera, A. 1956, Nota preliminar sobre Sifonapteros de *Cynomys* de la zona de énzootia pestosa del sureste de Coahuila, Mexico. Acta zool. Mex. 1(12):1-4.

Barrera, A. 1968, Distribución biogeográfica. An. Inst. Biol. Univ. Nat. Auton. Mexico 39. Ser Zool. (1):35-100.

Beasley, P. 1965, Human fleas (*Pulex irritans*) incriminated as vectors of plague in Bolivia. Vector Control Briefs 15:5-6.

Benton, A.H. 1960, Occurrence of *Pulex simulans* Baker in Louisiana. Southwest. Natur. 5(2):104.

Blanc, G. 1956, Une opinion non conformiste sur le mode de transmission de la peste. Rev. Hyg. Med. Soc. 4(6):535-562.

Blanc, G. & M.Baltazard 1941, Recherches expérimentales sur la peste. C.R. Acad. Sci. 213:813-816.

Blanc, G. & M.Baltazard 1945, Recherches sur le mode de transmission naturelle de la peste bubonique et septicémique. Arch. Inst. Pasteur Maroc. 3:173-349.

Burroughs, A.L. 1947, Sylvatic plague studies. The vector efficiency of nine species of fleas compared with *Xenopsylla cheopis*. J. Hyg. 43(3):371-396.

Cahalane, V.H. 1947, Mammals of North America. New York, Macmillan Co., 682 p.

De la Barrera, J.M. 1957, Informe presentado por la Oficina Sanitaria Panamericana, Officina Regional de la Organizacion Mundial de la Salud, al Gobierno del Ecuador. (Unpublished typescript) Cited in Pollitzer, et al., 1965. Plague in the Americas. Sci. Pub. 115, Panamer. Hlth. Org.

Dice, L.R. 1943, The biotic provinces of North America. Ann Arbor, Mich., University of Michigan Press.

Eads, R.B. 1948, Ectoparasites from a series of Texas coyotes. J. Mammal. 29:268-271.

Ecke, H. & W.Johnson 1952, Plague in Colorado. Publ. Hlth. Monogr. No. 6:1-37, 3-21.

Eskey, C.R. & V.H.Haas 1940, Plague in the western part of the United States. U.S. Publ. Hlth. Ser. Bull. No. 254:1-83, p. 63.

Ellis, L.L., Jr. 1955, A survey of the ectoparasites of certain mammals in Oklahoma. Ecology 36:12-18.

Ewing, H.E. & I.Fox 1943, The fleas of North America. U.S. Dept. Agric., Misc. Publ. No. 500:1-142, p. 117-118.

Fox, I. 1940, Fleas of Eastern United States. Collegiate Press, Inc., Iowa State College, 191 p., p. 19-21.

Gates, D.B. 1945, Notes of fleas (Siphonaptera) in Nebraska. Ent. News 1945:10-13.

Gier, H.T. & D.J.Ameel 1959, Parasites and diseases of Kansas coyotes. Kans. S. Univ. Agric. & Appl. Sci., Tech. Bull. 91, 34 p.

Gunther, K.K. 1961, Funktionell-anatomische Untersuchung des männlichen Kopulations apparates der Flöhe unter besonderer Berücksichtigung seiner postembryonalen Entwicklung. Dtsch. Ent. Zt. 8(3/4):258-349.

Haas, G.E. & N.Wilson 1967, *Pulex simulans* and *P.irritans* on dogs in Hawaii (Siphonaptera: Pulicidae). J. Med. Ent. 4(1):25-30.

Haas, G.E., H.Wilson & P.Q.Tomich 1972, Ectoparasites of the Hawaiian Islands. I. Siphonaptera. Contrib. Amer. Ent. Inst. 8(5):1-76.

Hall, E.R. & K.R.Kelson 1959, The mammals of North America. New York, Ronald Press Co., 1083 p.

Holland, G.P. 1949, The Siphonaptera of Canada. Dominion of Canada, Dept. Agric., Tech. Bull. 70, 306 p., p. 64, 66-67.

Hopkins, G.H.E. 1957, Host-association of Siphonaptera. First symposium on host specificity among parasites of vertebrates. Inst. de Zool., Universite de Neuchatel. 324 p., p. 64-77.

Hopkins, G.H.E. & M.Rothschild 1953, An illustrated catalogue of the Rothschild collection of fleas (Siphonaptera) in the British Museum (Natural History). Vol. 1, Tungidae and Pulicidae, British Museum (Natural History). 361 p., p. 105-118.

Hopkins, G.H.E. & M.Rothschild 1956, An illustrated catalogue of the Rothschild collection of fleas (Siphonaptera) in the British Museum (Natural History). Vol. 2, Vermipsyllidae to Xiphiopsyllidae. 445 p.

Hopkins, G.H.E. & M.Rothschild 1962, An illustrated catalogue of the Rothschild collection of fleas (Siphonaptera) in the British Museum (Natural History). Vol. 3, Hystrichopsyllidae. 560 p.

Hopla, C.E. 1965, Alaskan hematophagous insects, their feeding habits and potential as vectors of pathogenic organisms. I: The Siphonaptera of Alaska. Arctic Aeromed. Tech. Rpt., 64-12, Vol. 1:267 p.

Hubbard, C.A. 1947, Fleas of Western North America. Their relation to public health. Ames, Iowa State College Press, 553 p., p. 57-59.

Hudson, B.W. & F.M.Prince 1958, Culture methods for the fleas *Pulex irritans* (L.) and *Pulex simulans* Baker. Bull. Wld. Hlth. Org. 19:1129-1133.

Jellison, W.L. & G.M.Kohls 1936, Distribution and hosts of the human flea, *Pulex irritans* L., in Montana and other western states. U.S. Pub. Hlth. Rpts. 51:842-844.

Jervis Alarcon, O. 1958, La pest bubonica: Problem de urgente resolucion. Rev. Ecuad. de Hig. y Med. Trop. 15(3):105-137.

Johnson, P.T. 1957, A classification of the Siphonaptera of South America. Mem. Ent. Soc. Wash. 5:299 p., p. 231.

Jordan, K. 1956, Suctoria. In: J.Smart (ed.), Insects of medical importance. Smart, Jarrold & Sons, Ltd., p. 202-223.

Jordan, K. & N.C.Rothschild 1908, Revision of the non-combed, eyed Siphonaptera. Parasit. 1:1-100.

Jordan, K. & N.C.Rothschild 1923, New American Siphonaptera. Ectoparasites 1(5):309-319.

Kilgore, D.L. 1969, An ecological study of the swift fox (*Vulpes velox*) in the Oklahoma Panhandle. Amer. Midl. Nat. 8:512-534.

Layne, J.N. 1971, Fleas (Siphonaptera) of Florida. Florida Ent. 54(1):35-51.

Macchiavello, A. 1957, Estudios sobre peste selvatica en America del Sur. I. Peste en la Provincia de Loja, Ecuador. Bol. Ofic. Sanit. Panamer. 43(1):19-41.

Macchiavello, A. 1958, Estudios sobre peste selvatica en America del Sur. III. Peste selvatica en la Cordillera de Huancabamba, Peru. Bol.

Ofic. Sanit. Panamer. 44(6):484-512.

Mead, R.A. 1963, Some aspects of parasitism in skunks of the Sacramento Valley of California. Amer. Midl. Nat. 70:164-167.

Meyer, K.F. 1963, Plague. In: T.G.Hull (ed.), Diseases Transmitted from Animals to Man. Springfield, Ill. Charles C.Thomas. p. 527-587.

Miles, V.I., M.J.Wilcomb & J.V.Irons 1952, Rodent plague in the Texas South Plains, 1947-49. U.S. Publ. Hlth. Monogr. No. 6:41-53, p. 46-47.

Mohr, C.O. & H.B.Morlan 1959, The nature of parasitism of the opossum by fleas in southwestern Georgia. J. Parasit. 45(2):233-237.

Peus, F. 1956, Siphonaptera. Morphology of genitalia. Taxonomist Glossary. Genitalia in Insects, S.G.Tuxen (ed.), p. 122-131.

Pollitzer, R., K.F.Meyer, A.N.Bica & E.C.Chamberlayne 1965, Plague in the Americas. Sci. Pub. 115, Panamer. Hlth. Org.

Ramos Diaz, A. 1938, Epidemiologia de la peste bubonica en la sierre del Departamento de Lambayeque. Bol. Ofic. Sanit. Panamer. 17 (9):776-780.

Randolph, N.M. & R.B.Eads 1946, An ectoparasite survey of mammals from Lavaca County, Texas. Ann. Ent. Soc. Amer. 39:597-601.

Rothschild, M. 1973, Report of a female *Pulex irritans* in a tenth century viking pit. Proc. Roy. Ent. Soc. London 38(7):29.

Rothschild, M. & R.Traub 1971, A revised glossary of terms used in the taxonomy and morphology of fleas. G.H.E.Hopkins & M.Rothschild (eds.), An illustrated catalogue of the Rothschild collection of fleas (Siphonaptera) in the British Museum (Natural History). V. pp. 8-85.

Smit, F.G.A.M. 1958, A preliminary note on the occurrence of *Pulex irritans* Linnaeus and *Pulex simulans* Baker in North America. J. Parasit. 44(5):523-526.

Snodgrass, R.E. 1946, The skeletal anatomy of fleas (Siphonaptera). Smithsonian Misc. Colls. 104(18), pp. 89.

Stanley, W.C. 1963, Habits of the red fox in northeastern Kansas. Univ. Kans. Mus. Nat. Hist., Misc. Pub. No. 34, 31 pp., p. 25.

Stark, H. 1958, Siphonaptera of Utah. U.S. Publ. Hlth. Serv., 239 pp.

Swellengrebel, N.H. 1953, Researches on ectoparasites of man in the vicinity of Marrakech (Morocco). Doc. Med. Geogr. Trop. 5(2):151-156.

Tipton, V.J. & C.E.Machado-Allison 1972, Fleas of Venezuela. Brigham Young Univ. Sci. Bull. Biol. Ser. 17(6):1-115.

Tipton, V.J. & E.Mendez 1966, Ectoparasites of Panama. Wenzel, R.L. & V.J.Tipton (eds.), Field Mus. Nat. Hist., p. 289-338, 47 p.

Tipton, V.J. & E.Mendez 1968, New species of fleas (Siphonaptera) from Cerro Potosi, Mexico, with notes on ecology and host parasite relationships. Pacific Insects, 10(1): 177-214.

Traub, R. 1950, Siphonaptera from Central America and Mexico. A morphological study of the aedaegus, with descriptions of new genera and species. Fieldiana: Zool. Mem. 1: 1-127.

Traub, R. 1972, Notes on zoogeography, convergent evolution and taxonomy of fleas (Siphonaptera), based on collections from Gunong Benom and elsewhere in South-east Asia. II. Convergent Evolution. Bull. Br. Mus. Nat. Hist. (Zool.) 23(10):309-387.

Traub, R., C.L.Wisseman, Jr. & A.Farhang-Azad 1978, The Ecology of Murine Typhus – A Critical Review. Trop. Dis. Bull. 75(4):237-317.

Varela, G. & A.Vasquez 1954, Hallazgo de la peste selvatica en la Republica Mexicana. Infeccion natural del *Cynomys mexicanus* (Perros Llaneros) con *Pasteurella pestis.* Rev. Inst. Salubr. Enferm. Trop. 14:219-223.

Wagner, J. 1933, Aphanipteren-Material aus des Sammlung des Zoologischen Museum der Berlinger Universität. Mitt. Zool. Mus. Berlin 18(3):338-362, p. 341.

Webster, F.M. 1904, The so-called human flea, *Pulex irritans,* infesting the opossum, *Didelphis virginiana.* Canad. Ent. 36:244.

Wilson, N. 1966, A new host and range extension for *Pulex simulans* Baker with a summary of published records (Siphonaptera: Pulicidae). Amer. Midl. Nat. 75(1):245-248.

JAMES R. BUSVINE
Emeritus Professor of Entomology, London University, UK

FLEAS, FABLES, FOLKLORE AND FANTASIES

It will soon be apparent that this little paper is entirely different from those of my learned co-contributors. It could, perhaps, be regarded as providing a measure of light relief; or it might be possible to justify its inclusion in the Medical Section, as illustrating the psychological impact of fleas on man. Ectoparasites generally tend to arouse strong feelings of aversion in civilized people; and as any advisory entomologist will confirm, even the mere idea of them can cause a psychosis in neurotic people. This is known as delusory parasitosis. The sufferers are unshakably convinced that their bodies are plagued by imaginary parasites and they make persistent and pitiful attempts to cleanse themselves.

In my experience, however, fleas are not so prone to excite aversion as other ectoparasites or, indeed, quite harmless insects or spiders. In fact, they may elicit a degree of admiration. Kirby & Spence's 1815 Textbook of Entomology includes a story of a lively old lady visiting a sick friend, who complained of fleas. 'Dear Miss' she exclaimed 'Don't you like fleas? I think they are the prettiest little merry things in the world. I never saw a dull flea in my life!' Another learned textbook, the Tierleben of von Brehm (1892) includes the following little poem:

Glücklich drum preis'ich den lockeren Gesellen
Pulex, den Turner im braunen Trikot
Wenn er in Sprüngen, vergangen, schnellen
Himmelhoch jauchzet, frisch, fromm, frei und froh!

That raffish fellow I gladly praise
Pulex, the gymnast in sleek brown tights
Spontaneous springs will readily raise
His joyful jolly jumps to heavenly heights!

Fleas, we must admit, are irritating. The host of Chaucer's Canterbury Pilgrims thus admonishes a sleepy member of the band:

What eyeleth thee to slepe by the morwe?
Hastow had fleen al night, or artow dronke
Or hastow with some quene al night y-swonke
So that thou mayst nat holden up thyn heed?

And, in a rather morbid mood, the 16th century George Gascoyne complains:

The huntry fleas that frisk so fresh
To worms I can compare
Which greedily shall gnaw my flesh
And leave the bones bare.

In historical times, when standards of housekeeping left much to be desired, fleas must have been so common that only the more fastidious complained. Not only the aristocracy, but even royalty seem to have been afflicted, according to various accounts. One of my favourite stories is that of Louis XI (1423-1483). It appears that one day a louse crawled out of his finery in full view of the courtiers. One of his retinue, noticing the creature, quietly picked it off; but the King noticed his action and demanded to know what he had done. The man nervously confessed that he had found a louse on His Majesty. Instead of rebuking him, the King cheerfully improved the occasion by remarking that a louse is an excellent reminder to royalty that they are but human. Whereupon, a flatterer seized the opportunity to copy the servitor's example and pretended to find a flea on the King. But the latter (probably tiring of the subject) exclaimed, 'What! Do you take me for a dog, that I should be running with fleas? Get out of my sight!'

In polite company, one was trained not to search for fleas. Thus, about 1450, John Russell, a steward of the Duke of Gloucester, wrote a book of instruction for the superior servants of the nobility. It contains the verse:

Your hed no bak ye claw, a fleigh as though you sought
Ne your heer ye stryke, ne pyke, to pralle for a flesche mought.

A similar code of manners was compiled at the Jesuit Seminary of La Fleche, published in 1595. Rule 22 runs:

Gardez vous bien de vous arrester a tuer une puce, ou quelque sale bestiole de cette espece, en presence de ce que puisse . . . etc.

But if one could not actively try to catch fleas, there were diverse flea traps that could be worn under the clothing, some of which are illustrated in my figure 1. The little gadget hanging on a string round the lady's neck, has a sticky peg inside, to catch fleas that venture through the holes in the case. It is hard to believe in its efficacy; but the flea-fur might well have exploited the tendency of fleas to burrow into such material. The artificial head should be noted. I have myself seen a very similar head, of rock crystal set with gold and gems, which was sold at Christie's in London in 1973, for 8 500 guineas.

It is noteworthy that all these devices seem to have been used by ladies. From the late Middle Ages, fleas have been supposed to be more troublesome to women. Their tribulations are described in Der Weiber Flöh Scharmützel a comic German poem of the 16th century, by Johann Fischart, who also wrote Floh Hatz, Weiber Kratz. This concerns a verbal duel between women and fleas, at which Jupiter gives judgement that fleas may bite them, but only on their tongues (by which they belabour their menfolk), at puffed out ruffs and sleeves and at the dance. A variant of this story comes into Simplicissimus, by Grimmelshausen, in the next century.

Our own Thomas Mouffet wrote 'Fleas are a vexation to all men, but as the wanton poet hath it, especially to young Maidens . . .' The wanton poet mentioned was an anonymous writer of a mildly erotic poem in the style of Ovid, in the 11th or 12th century. It begins:

Parve pulex sed amara lues, inimica puellis.

I have made a rather free translation, but it is too long to give here. Briefly, the poet imagines himself a flea, wandering over the delectable landscape of a maiden's body; then, at a suitable moment, he changes back to human form to take advantage of the situation. This vein of erotic fancy was maintained throughout the centuries. Most people will remember John Donne's verse:

Mark but this flea, and mark in this
How little which thou deny'st me is.
It sucked me first, and now sucks thee
And in this flea our two bloods mingled be.

A more ribald allusion is this German epitaph for a flea from the early 18th century:

Nachdem ich lange Zeit die weisse Brust bewacht
So kahm ich an den Ort wo selbst die Anmuth lacht
Allein den Vorwitz must ich mit dem Leben büssen
Denn allda wil man nichts als grosse Stachel wissen.

On a snow white breast I tarried a while
Then came to the spot where pleasures smile,
For rashness now my life is ceded
Since here much bigger pricks are needed.

Later on in the same century, a more refined note was sounded by the poems celebrating the appearance of a flea on the bosom of Madame des Roches, whose salon in Poitiers was crowded with eminent Frenchmen of the time. These gallants vied with each other in composing a series of verses on the subject. Further examples can be culled from the 19th century. But, as Lehane remarks in his pleasant little book, *The Compleat Flea,* the subject begins to pall; and in any case, the 20th century needs wryer stimulants.

Perhaps I should turn to more edifying aspects of fleas; for example, their use to point moral precepts. Aesop refers to them in two fables. In one, a flea steals a ride on a heavily laden camel, but later jumps down to relieve it of the burden. The camel thanks him for nothing, the moral being (in the words of Caxton's 1482 version) 'of hym which may neyther helpe ne lette man, nede not make grete estymacion of'. The other tale concerns a man who caught a flea, asking 'Why bytest thou me?' The flea pleads that it is his nature to do so; but the man kills him anyway. The moral is: 'For men ought not to leve none evyll unpunyshed how be but nat greate'.

La Fontayne brings a flea into a moral-pointing verse in which an exasperated man calls on Heaven to relieve him of flea bites, noting:

Our importunate clamour doth weary high Heaven
Often for trifles unworthy of man.

John Gay, of Beggar's Opera fame, wrote a poem on the theme of human vanity; it recounts how man regards all Nature as being created for his especial needs. 'Not so' says a flea feeding on his nose:

For thee made only for our need
that more important fleas might feed.

In the 18th century, certain poets were inclined to deride scientific curiosity. Pope, for example, in the Rape of the Lock, speaks scornfully of

Cages for gnats and chains to yoke a flea
Dried butterflies and tomes of casuistry

and classes them with such suspect items as

The courtier's promises and sick men's prayers
The harlot's smile and the tears of heirs.

William Cooper, in a long poem on Charity, implies that science is only acceptable if pursued in a God-fearing spirit; otherwise the scientist, whether he

Weigh sunbeams, carve a fly or split a flea
The solemn trifler, with his boasted skill
Toils much and is a solemn trifler still.

Again, Peter Pindar taunts Sir Joseph Banks, President of the Royal Society, in a poem in which Banks is supposed to be investigating the suitability of fleas for food, on account of their resemblance to lobsters. The experiment fails and

'How!' roar'd the President, and backwards fell;
'There goes then my hypothesis to hell!'

In these examples, I have tried to illustrate some particular themes; but many other examples of passing references to fleas as comic, exasperating, or trivial, could be found in literature, from Aristophanes onwards. Shakespeare mentions them at least half a dozen times.

Space, however, is limited, and I will conclude on a very different subject. The interest in fleas and other ectoparasites shown by early naturalists, could serve as a microcosm of the growth of biological science. To give one illustration, since they were so common in past times, they were one of the first objects to be examined under the earliest lenses and microscopes. It is very interesting to note the different degrees of accuracy with which they were drawn from such observations, which had little to do with the scientific stature of the investigation. Thus, the well-known diagram published by Robert Hooke in 1665 is remarkably accurate, though his textual description is very superficial. However, the engraving contrasts well with the highly imaginative picture of the Italian Pietro Lucatelli, published at about the same time (figure 2). This seems to be covered with fur and has a wicked gleam in its eye! Another Italian, Mario Cestoni, produced some passable illustrations of different stages in the life history, in 1699; but de Geer's pictures of some 50 years later are very crude, despite his scientific eminence. They compare sadly with the fine coloured illustrations of von Rosenhof, published about the same time; but the German was a miniature painter, who happened to be interested in entomology.

In this short article, I have only been able to hint at the remarkable curiosities to be encountered in the relations between man and fleas. I have said nothing about 'performing' fleas, for example, which would require several pages. Possibly, however, enough has been written to arouse curiosity. If so, it is no secret (as the journalists say) that I have recently published a book on the subject (*Insects Hygiene and History,* Athlone Press, London, 1976, 262pp. £6.95).

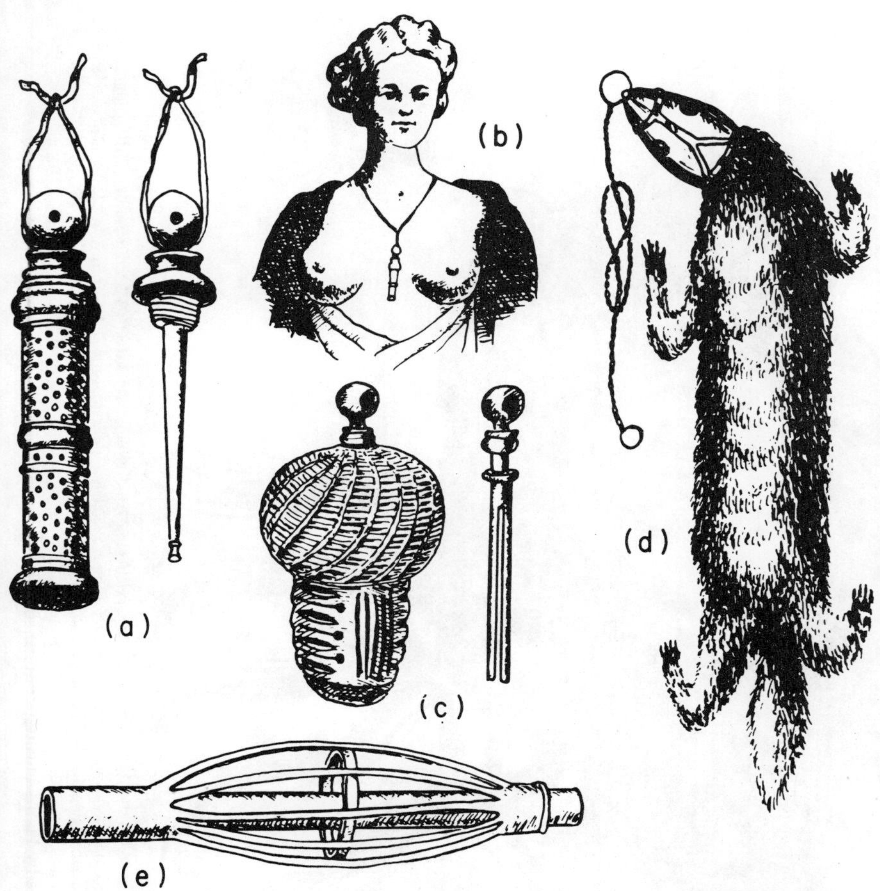

Figure 1. Diverse flea traps. a, b, as described by Brückmann, F.E. (1729) Die neuerfundene curiose Flöh-Falle. Wolfenbuttel; c, a similar type described by Madel, W. (1955) Des Flohes Strauss mit der Laus. Werkztg. Firm C.H.Boehringer, No. 3 Ingelheim; d, a flea fur with a crystal head; e, a Chinese flea trap as described by Eysell, A. (1913) Handbuch der Tropenkrankheiten, 1, 87.

Figure 2. Picture of a flea by Pietro Lucatelli, published in Rome about 1686. By courtesy of Dr M. Colluzzi.

MEDICAL AND VETERINARY

NORMAN G. GRATZ
World Health Organization, Geneva, Switzerland

PROBLEMS AND DEVELOPMENTS IN THE CONTROL OF FLEA VECTORS OF DISEASE

ABSTRACT

Attempts have been made to control pest fleas for a long time. However, it was only when the role of fleas as vectors of disease, first of plague, and later of murine typhus, that efforts were made to develop materials which would halt transmission of disease by controlling flea populations. Only with the appearance of the chlorinated hydrocarbon insecticides and later of other chemical groups of synthetic organic insecticides, has it been possible to control flea populations over large geographical areas. The use of insecticide dusts applied as tracking powders or insufflated into burrows has proven highly effective in achieving rapid control of rodent flea ectoparasites. The application of DDT and other insecticides in tropical areas has also sharply reduced rat flea indices in the treated zones and, consequently, rodent-borne disease. This type of insecticide application has been one of the major causes of the selection of rat and other flea populations for insecticide resistance. In many geographical areas, this resistance has reached levels so high as to preclude further use of DDT or other compounds to which resistance has developed. It is essential, therefore, to encourage the development and field trials of alternative compounds, and the paper describes the laboratory and field trials that have been carried out already to overcome the problem.

1 INTRODUCTION

Aside from the interest of the siphonapterologist in their biology, physiology or classification, the main importance of fleas to man is in their role as important vectors of disease as well as pests of man and his domestic animals. As with all other insects, it was only following the discovery of the part that fleas played in transmitting a disease – plague, did concern with their control assume a greater importance than one of only that of how to control an annoying pest.

The role of fleas as vectors of disease was only suspected for the first time 80 years ago by Ogata (1897) on epidemiological grounds when working on plague in Formosa; at that time Ogata wrote (as quoted in Pollitzer, 1954) 'one should pay attention to insects like fleas for as the rat becomes cold after death, they leave their host and may transmit the plague virus direct to man'. Ogata studied this by injecting a suspension of crushed fleas which he had collected from plague infected rats into mice and found that the latter developed plague. Shortly thereafter, Simond (1898), working in India, noted that fleas from rats would bite man and that rats sick with plague often carried many fleas. He examined fleas taken from a plague-infected rat and found bacilli similar to those of plague in their alimentary canal; these bacilli were absent from fleas taken from healthy rats. Simond further succeeded in transmitting plague from a sick rat to a healthy one by placing the healthy rat into the same jar as a plague-infected one but separated from it by a wire grill which was only big enough for fleas to pass through from one side to the other. He repeated the experiment but first freed both rats from all ectoparasites and the plague-free rat remained healthy. Simond considered these experiments and suggested that if his views on the importance of fleas in the transmission of plague were proven to be correct, disinfection of houses against fleas should be carried out before the disinfection with bacteriocides as was the common practice at that time. This viewpoint and approach is, of course, quite correct and is, in fact, the foundation of any flea-borne disease control programme. In his paper, Simond further advised that a rat or other animal which had died of plague, should never be removed before its ectoparasites had been killed by pouring boiling water, or some other substance which would kill the ectoparasites, upon the body; if a house is suspected of harbouring plague-infected rats, it should be disinfected immediately, preferably if it could be well closed, by sulfur fumes. Otherwise boiling water should be poured on the floors to rid them of parasites.

Though the evidence seemed convincing, there was heated opposition to Simond's views on the transmission of plague by a number of the plague experts of the time, including the eminent parasitologist, G.F.H.Nuttall and the controversy is well described in Hirst's book (1953) on the history of plague. Very soon, however, a succession of workers produced still further evidence to support Simond's hypothesis, among them Gauthier and Raybaud (1903), Tidswell and Thompson (1903), Verjbitski (1908), culminating in the more extensive studies of the Indian Plague Commission. The Commission, whose work began in Bombay in 1905, conclusively demonstrated the role of the rat flea in the transmission of plague from rat to rat and from rat to man. Shortly thereafter, McCoy (1908) showed that rodents other than rats could be infected by plague in his study in the Western United States.

2 THE EARLY DEVELOPMENT OF FLEA CONTROL METHODS

Flea control was a matter of concern even to the ancient Egyptians. The Ebers papyrus which dates from between 1555-1335 BC and was translated by Ebbell 1937, gives several prescriptions of how to expel fleas from a house, for example sprinkling the walls with Natron water, which is apparently a solution of sodium carbonate.

Once it had been confirmed that rats, and other rodents, were the reservoirs of plague and that fleas were the vector, a more rational control of plague began to be developed. While most of the measures taken against outbreaks of the disease were still aimed primarily against rats, the common practice of fumigating vessels, houses and particularly grain-stores with HCN gas or burning sulfur killed not only the rats but also their flea ectoparasites. Ironically, even from the time of the Black Death onwards, the use of sulfur or sulfur fumes frequently appears among the numerous recipes which it was advised would hopefully stop the spread of plague. Unfortunately, no one up to the last century carried out such a sulfur fumigation in a systematic enough manner to realize or understand its true potential for controlling the spread of the plague. Probably because rat control was usually the first measure to be taken during a plague outbreak in the early part of the century, it may well be that many epidemics instead of being brought to a halt were, in fact, spread or prolonged by the many plague-infested fleas that left the bodies of the rats that had been killed by clubbing, trapping or poisoning.

As more experience and understanding was gained, it became generally accepted that in the case of outbreaks of bubonic plague, the chief danger lay in the presence of plague-infected rats and the fleas that infested them, whereas in outbreaks of pneumonic plague it was the patient himself who spread the infection. By 1922, Heiser in his book on the practice of medicine in the tropics emphasized that control of outbreaks of bubonic plague depended upon preventing rat fleas that had bitten plague-infected rats, or patients ill with plague, from biting other people. Several authors at that time proposed various types of protective clothing to prevent such flea bites while others drew attention to the relative immunity from plague enjoyed by certain groups in South-east Asia whose habit it was to rub their bodies with coconut oil, as well as workers in oil stores whose skin was constantly soaked with paraffin oil which apparently discouraged flea bites.

The carrying out of effective flea control programmes in the early part of the century was severely hampered – as was insect control in general – by the relatively few active insecticides known and the general lack of persistence of those that were available; the short action of fumigants, pyrethrums, naphthalene, etc., often meant that areas or individual houses which it was wished to protect from fleas, had to be frequently re-treated if a high level of control was sought. As a result, only a few more favoured or critical areas were so protected and a good deal of effort was expended to containing plague outbreaks and limiting them to the area in which they had occurred.

During the 1920's and 1930's, there was a good deal of concern about the possible transfer of bubonic plague from one geographical area to another by means of imported goods carrying either plague-infected rats or plague-infected fleas. As early as 1913, Swellengrebel had observed that in East Java 'the only merchandise, the maximum import of which coincides more or less closely with new conquests of plague, is the rice from British India and Southern China' (quoted in Hirst, 1931). Hirst himself (ibid.) emphasized that grain was 'the most important vehicle for plague infection since it is the favourite food of rats and its debris forms a suitable diet for the larvae of the plague carrying fleas'; he observed that plague never persisted for any length of time in Colombo, Ceylon except in the vicinity of grain stores infested with *Xenopsylla cheopis.* He thought that both rats and fleas could be responsible for the transfer of bubonic plague from one place to another but that in certain circumstances, such as in the transfer of goods from vessels off shore to the port on lighters, fleas were more important. He then reviewed in considerable detail the desirability of fumigating cargoes considered as dangerous and, after analysing the different fumigants available, recommended the use of liquid hydrocyanic acid impregnated into discs or discoids of wood pulp or Fuller's earth as the most satisfactory formulation. Much was also made by several other authors of the possible transfer of bubonic plague from India to other parts of the world in shipments of jute, e.g. Long & Mostajo (1934) and Machiavello (1947) but Pollitzer (ibid.) concluded that in view of the temperature fleas would be subject to in transit and the length of the sea voyage, such transfer would be most unlikely; gradually the use of HCN fumigation of grain and other cargo only for flea control came to an end. Recent plague events in certain countries and the transfer of air cargo directly from a plague-infected area to destinations half way around the world in a few hours might force a reconsideration of the possibility of plague being transmitted by its vectors and such a matter was seriously considered when shipments were made in the early 70's from Vietnam to the USA.

As has been noted above, the efficacy of fumigants, particularly that of HCN and of burning sulfur for the control of both fleas and their rodent host, was quickly recognized. Unfortunately, aside from the danger in their use to man and domestic animals, fumigants have no persistent action and once a structure is opened and aired, its action ceases. Worse still, many of the buildings, including godowns, buildings and huts in tropical areas were constructed, at least in the past, in such an open manner that they could not be suitably fumigated. Since it was finally recognised that fleas would leave poisoned or clubbed rats, wherever it was possible to fumigate plague-infected premises, this was done so as to kill both the rat and the flea and a search continued to be made for materials that would effectively control 'free fleas'.

Among the measures which were proposed are those described by Wu (1936); if a room was badly infested with fleas, it was suggested that the floor be sprinkled with baked naphthalene and left closed for 24 hours by which time the dead fleas could be swept up. 'Eggs and larvae may be destroyed by washing floors with hot soapsuds, five per cent formalin solution or 1 in 1,000 solution of corrosive sublimate. Where there are carpets, alum may be swept into them. Scattering salt which is later damped down is also useful'. Wu also advised the use of a number of other possible materials including powdered sodium fluoride and an emulsion made with paraffin and soft soap. He suggested that temporary repellents could be used such as oil of pennyroyal, eucalyptus and idioform, some of which he noted were unfortunately as objectionable to human beings as they were to fleas. Many experiments were carried out in different parts of the world with various materials that it was thought might effectively control the flea vectors of plague when outbreaks occurred. Among the materials whose use was reported at one time or another were phenol, lysol, benzine, crushed camphor, a soap-phenol mixture, a mixture of fuel oil and gas oil (this was particularly recommended for use on the sandy floors of African huts), tobacco leaves laid out on the floor, burning hay or straw and sulfuretted hydrogen.

Protection of people trying to control an epidemic of plague was, understandably, always a matter of concern, and Bishopp (1921) urged that where plague was common, workers should be provided with closely fitting clothes to avoid flea bites; since it had long been noted that fleas were attached to a moving object, he suggested that men with their legs wrapped in paper treated with tanglefoot be allowed to walk around in an infested area and in this manner great numbers of fleas could be collected. He further suggested that if plague was suspected to be present at a particular site, animal hosts such as guinea pigs, white rats or rabbits could be used to attract and collect the fleas that were loose in the area and the animals then placed in a jar from where their fleas could be collected or the fur of the animal could be treated with kerosene emulsion or another insecticide. Bishopp cautioned that the 'Indian rat flea and the European rat flea do not go freely to guinea pigs but are caught in great numbers on tanglefoot on man'.

Where premises were infested with adult fleas, Bishopp recommended fumigation with cyanide gas or with burning sulfur, or, when the houses were too open to be effectively fumigated, the use of salt spread onto the floors and frequently wetted down. Flaked naphthalene could be applied at the rate of four or five pounds for each two or three hundred square feet, closing the rooms for a few hours and then sweeping the material to the next room along with the 'stunned fleas'. He also pointed out that pyrethrum powders could be used in the same way, while flea-infested barns, basements and sheds could be treated with a spray of kerosene emulsion or creosote oil.

In 1916, the British Museum first published a brochure entitled 'Fleas as a menace to man and domestic animals' by J.Waterston and by 1936 a third edition, revised by Buxton was issued. The brochure contained a brief section on the control of both nest and vector fleas which summarized the recommendations for control as they were known at the time; these included the use of iodoform, naphthalene, oil of pennyroyal and oil of eucalyptus as repellents and emphasized the importance of cleanliness and preventing free access of flea carriers, i.e. cats, dogs, rats and mice to houses. Control methods proposed included washing dogs and cats with a soap containing some insecticide such as carbolic soap or a warm three per cent creolin solution. Since cats object to being washed, it was suggested that they may be dusted with naphthalene powder (which may make them sick), pyrethrum powder or derris root powder. The use of naphthalene on floors was recommended as the most useful agent available. Outbuildings were to be treated with an emulsion of petroleum or paraffin in soap or painted with whitewash containing alum.

Shortly after, Roubaud in France (1940) screened a number of different chemicals that he thought might be promising against fleas; because of its low price and relatively low mammalian toxicity, he especially examined sodium fluoride dusts which at the time were beginning to find extensive use in agriculture. He tested these powders in tubes in the laboratory against adults and larvae of *Ctenocephalides felis* and *canis* and *X.cheopis* and found that control was achieved for 'several days'. However, he admitted that the speed of action against adult fleas was so slow, taking as long as a day to take effect that the compound was best spread on floors at a dosage of 2 g/m^2 against flea larvae. He also referred to the promise of rotonone against ectoparasites.

From the above review, it can be seen that most of the literature discussing flea control up until the mid-1940's recommended the same chemicals or fumigants; the number of effective compounds was, in actuality, very small, they were often not very easy to apply and frequently of use only under certain specific conditions such as when rooms could be closed for several hours to give naphthalene flakes time to take effect. Pyrethrum powders and Derris powders worked well but were expensive and of only limited persistence. Tinker et al. (1959) described some of the early methods used for rat flea control in the USSR; these included, as elsewhere, pyrethrum at 5 to 12 g per square metre, but this was finally discarded because it gave no lasting results. In the 1940's, repeated applications of fluid clay to the earthen floors of mud huts in the Volga-Ural region was found effective. Vashkov (1956) recommended, for flea control in houses, applications of three to five per cent emulsions of 'K' soap (probably containing bisethylxanthogen) at a rate of 100 to 200 ml per m^2 of floor space. Others found that this substance gave no lasting effect and its unpleasant odour made its use unwelcome.

3 MODERN RAT FLEA CONTROL

The term 'modern' refers to the change in rat flea control and thus plague and murine typhus control that came about when the use of the synthetic organochlorine insecticides DDT and BHC were first introduced. Before the introduction and widespread use of these compounds, flea control was slow and laborious depending either upon the application of those contact poisons described above ranging from pyrethrum to various soaps or kerosene emulsions, or the use of fumigants, particularly HCN released by calcium cyanide dusts. Due to the danger of cyanide gas and the open construction of most tropical houses, fumigant treatment was restricted almost entirely to vessels, granaries or burrows. Thus, while the necessity of undertaking flea control prior to rodent control was recognized by Simond before the beginning of the century (1898) there was little that could be done about it on any scale larger than the treatment of a few 'plague houses' or at best individual graneries (or godowns). The ability to treat large geographical areas for flea control, such as an entire neighbourhood of a city with the rapidity which is necessary when one is faced by an epidemic of plague only came about when DDT, then HCH and later other synthetic, organic insecticides became available.

While DDT was first described by a German chemist, O.Zeidler, in 1874, its insecticidal properties were only realized by Paul Muller in 1939 when he carried out laboratory tests of the compound in Switzerland against the clothes moth, the Colorado potato beetle and a number of other insects.

Very soon after DDT became available in other than small experimental quantities, it was tested against fleas as well; Lindquist et al. (1944) found that dogs were completely freed of several species of fleas by applying four or five per cent DDT dusts to them. Sweetman (1946) also found that dog and cat fleas could be eliminated from both house pets and premises by applying five or ten per cent dusts of DDT and found this less irritating to animals than pyrethrins and rotenone dusts.

Probably one of the first tests of DDT against rat fleas was that carried out by Davis (1945). The first experiment was a very simple one; he placed 10 fleas in a pail 'with a small amount of pure DDT' and all fleas died in four hours. He then carried out a series of other tests including putting naturally infested roof and Norway rats into sacks which contained one gram of different percentages of DDT from 0.02 to 5 per cent, leaving the rats in the sack from 10 to 15 minutes; with the 5 per cent powder, all fleas still on the rats were dead when the rat was chloroformed and combed two days later. Finally rat infested stores were dusted with a 20 per cent DDT powder; the *X.cheopis* indices in the treated stores dropped from an average of 16.7 per rat to 0.6 per rat while at the same time they were rising in the rest of the city. Davis also dusted a shed which was heavily infested with *Pulex irritans* and another shed and a garage that were infested with *Ct. felis;* the insecticide dust promptly cleared all these premises of fleas. Davis concluded his paper by suggesting that while the purpose of the study was to determine if DDT could control the flea vectors of murine typhus, it was also likely that the compound could be used in plague control as well and that it might even be useful in checking plague in wild, burrowing rodent populations. He thought that DDT should probably be dusted before trapping or poisoning rats in a (rat) eradication campaign lest the fleas leave the killed rats and spread the disease.

At about the same time that Davis was carrying out his field trials in San Antonio, Texas, Gouck (1946) carried out studies in the laboratory and field in Savannah, Georgia. Ten per cent DDT dust was applied to rat holes and runways in rat infested poultry stores. Control of *X.cheopis* was achieved for 66 days, *Leptopsylla segnis* was almost completely controlled for 21 days but was only 75 per cent controlled at the end of 66 days. There was complete control of *Echidnophaga gallinacea* for 21 days but no control at the end of 66 days, and this was attributed to the fact that all the chickens had been removed from this treated store and, in the absence of their preferred hosts, they were forced to seek a new host. Gouck did trials with paper cylinders with DDT dust inside of them which were placed near burrows so that rats would crawl through them. Control was good though not as effective as with the general dustings, but less likely to contaminate the grain in the stores.

Ludwig and Nicholson (1947) carried out a large field trial in Savannah, Georgia in 1945 to test the effect of ten per cent DDT dust in pyrophyllite on *X.cheopis* populations on *Rattus norvegicus* as a method of interrupting transmission of murine typhus. The dust was applied in patches to rat runways and insufflated into burrows. The *X.cheopis* index dropped from 27.5 before treatment to 0.2 immediately after, a 99.3 per cent control; and 80 per cent reduction in flea densities was obtained for up to four months after the dustings were made.

The first actual use of DDT in the field to control rat fleas during an epidemic of plague was made in 1944 in Dakar (Lewis, et al. 1945). The epidemic apparently started in April of that year and continued to November with a total of 567 cases and a 91 per cent mortality. By the time the assistance of the US Army Medical Department was requested, the epidemic may well have been on the wane. Nevertheless, the African section of the city was cordoned off by gendarmes every morning and all workers leaving had their clothes dusted by ten per cent DDT dust applied by hand dusters. When this was completed, people inside the cordoned area were treated and eventually 95 per cent of the total population of the quarter were dusted and 'all public houses such as cafes, restaurants, bars, brothels, cinemas, etc., were sprayed with five per cent DDT solution' in kerosene. Of 316 previously flea-infested houses, all but seven were free of fleas two weeks after dusting and five of these had been missed during the preliminary dusting. The use of DDT dust against fleas was decided upon because a rat poisoning campaign was considered too dangerous to children and domestic animals.

Kartman (1946a) found that *X.cheopis,* while common on rats, was 'quite negligible in native huts and villages about Dakar' and suggested on epidemiological grounds that the predominant flea in the African dwellings, *Synosternus pallidus* may possibly have acted as the vector of plague. Since most of the information accumulated at the time had been on the effect of DDT on *X.cheopis,* Kartman (1946b) studied the efficacy of this compound on *S.pallidus.* His field trial was actually carried out in 1944 and his finding that DDT applied either as a five per cent kerosene solution or as ten per cent talc dust at 100 mg of DDT per m^2 to huts gave immediate and persistent kill for at least nine weeks, served as a basis for the subsequent use of this method during the epidemic described by Lewis et al. (1945).

Plague appeared in Casablanca on 20 July 1945. The US Army carried out a combined campaign against the disease which included immunization, quarantine, rat control as well as DDT dustings and spraying. The attack on the flea was considered the desirable immediate objective (Gordon & Knies, 1947) and the combined measures were apparently successful in that only a total of three human cases appeared.

Davis (1949) combatted an epidemic of plague in Ngamiland (Bechuanaland) by dusting 6,000 huts once every four months with DDT using about 225 g of the dust per hut. After the seventh dusting he considered that localized recrudescences of plague had been eradicated and of 500 huts sampled only five fleas were found, all in a single hut.

Plague appeared in Taranto, Italy on 6 September 1945 (Schultz, 1950); since there was no effective local health organization, the British Army began both intensive rodent control measures distributing barium carbonate and later zinc phosphide and arsenious oxide baits and at the same time began spraying with DDT a cargo of imported rags, suspected as the source of infection from which rats spread. A DDT barrier two to three blocks deep was laid down around the suspected source and all persons living in the zone were dusted with DDT weekly. A total of 29 cases and 15 deaths were recorded in the outbreak which lasted until the end of November, 1945. Several thousand dead rats were found and it can be assumed the fleas that escaped from them were quickly killed by the large quantities of DDT applied. In any event, the epidemic was quickly arrested despite the large rodent population that had existed in the city and port.

After initial studies in the USA described above had shown that DDT applied to rat runs, burrows and harbourages reduces the number of fleas found on rats in the area, Davis (1947) carried out a field trial in early 1945 in San Antonio, Texas, to determine the value of reducing rat flea indices for the control of transmission of murine typhus. The southwestern part of the city was dusted with ten per cent DDT in pyrophyllite, applied to all rat runs, burrows and harbourages. Rats were trapped, combed and bled before and after the DDT treatment. The prevalence of antibodies for murine typhus decreased in young rats in the DDT dusted area; there were considerably fewer human cases in the DDT treated area than in the untreated and it was considered that the DDT treatment was an effective control measure against murine typhus.

Based on these promising results and in view of the fact that the incidence of murine typhus was showing an alarming increase with 5,353 cases being reported in 1944 alone in the USA (Eskey & Hemphill, 1948), the use of DDT dust was rapidly incorporated into murine typhus control programmes; from 1945 to 1947, murine typhus declined in the USA and a study was carried out in the state of Georgia to determine if this decline might have been due to factors other than the use of the DDT alone (Hill & Morlan, 1948). This study showed that in an area of the state where the probability of rural residents acquiring murine typhus was equal to or greater than that for urban residents, it was possible to control the disease on a county-wide basis. Through a county-wide application of ten per cent DDT dust to rat runs and harbourages and in the absence of other rodents, rodent ectoparasites, or murine typhus control measures, the incidence of murine typhus was significantly reduced in two counties as shown by comparison with previous experience in the same counties and by comparison from a nearby untreated county.

The DDT dusting operations were found to disturb the normal ecology of rat and rat ectoparasite populations in a variety of ways and this may have altered the epidemiological picture of murine typhus in the treated counties. A significant reduction in the prevalence of typhus complement fixing antibodies in the rat populations of the dusted counties closely followed, and was attributed to the degree of ectoparasite control that was obtained. Comparing levels of populations in the treated and untreated counties, satisfactory control of *X.cheopis* and *L.segnis* was obtained by the DDT dusting. This dusting had only a slight effect on rat mites and rat lice. A later study (Hill et al., 1951) confirmed similar results for a longer period of time, from 1946 through 1949. When dusting operations ceased, effective murine typhus control was apparently achieved for about three years without further effort but *X.cheopis* and typhus antibodies in rats began to increase about seven months after the last dusting was completed (Morlan & Hines, 1951).

The use of DDT dusts against rat flea vectors of murine typhus in conjunction with a rodenticide was taken still further in the work described by Mohr & Smith (1957), when an attempt was made to eradicate murine typhus in two rural foci in two counties of Georgia, USA. *X.cheopis* indices were reduced to zero by the dustings, and rat densities, as measured by a number of infested premises, were also greatly reduced. Murine typhus disappeared from the surviving rat populations and, presumably from the human population as well. Due to the geographically limited area of the trial, 'eradication' is perhaps too strong a claim but the method of treatment was certainly effective.

Nicholson & Gaines (1948) compared the use of five per cent and ten per cent DDT powders in field use in the southern USA and found that *X.cheopis* populations were almost completely eliminated within a week after application of both five and ten per cent DDT dusts in pyrophyllite and that at the end of three months there was no significant degree of recovery nor any significant degree of difference between the two formulations. Later work by the US Center for Disease Control, however, indicated that a better period of persistence was provided by ten per cent DDT dusts.

The early success of DDT in controlling bubonic plague epidemics in North Africa and Europe and in the control of flea vectors of murine typhus in the USA quickly attracted the attention of public health authorities in countries endemic for plague. DDT either alone or in combination with rodent control measures became more and more widely used in plague and later murine typhus control programmes in various parts of the world.

An epidemic of plague broke out in Tumbes, Peru, in September 1945, and continued until DDT powder as a control measure was applied in early 1946. At least 40 human cases were confirmed (Macchiavello, 1946). DDT dust was applied at an average dosage of 2.5 to 3 g of ten per cent DDT dust per m^2 of floor space in treated buildings with an additional application of five per cent DDT dust being made to spaces between ceilings and roofs, beneath floors and to double wall spaces. Four days after the first application of DDT was completed, the epidemic was stopped. There was an 81.6 per cent reduction of the flea infestation on rats and an 87.9 per cent reduction of flea infestations in rats' nests. Rat plague was reduced by 75.6 per cent after the first application of DDT and by 100 per cent after the second. A very satisfactory rat kill was achieved by sodium monofluoracetate, which was only introduced after the third cycle of DDT dusting.

Very soon after the end of World War II, DDT applications to the interior of dwellings as a residual spray for the control of *Anopheles* vectors of malaria came into widespread use; Viswanathan (1949) noticed the complete absence of human cases of plague in the Kanara and Dharwar Districts of India following the use of DDT in the malaria control campaign in that area during 1948. Plague was still present in the rat populations as shown by rat falls and in 1948 the adjoining unsprayed Mysore State had a severe human epidemic of the disease. Simoens & Chatre (1947) considered that the ideal routine in plague control was to immediately treat the focus of an outbreak with DDT and to do no rat baiting or vaccination in the focus. The authors noted that where DDT was being sprayed, it was popular with the villagers as it also controlled bedbugs, flies and other pests as well as fleas. The authors considered the use of cyanogas and kerosene-soap emulsion as 'a too radical destruction of rats' and as having, in any event, merely an ephemeral effect on fleas so that the application of these compounds was occasionally followed by a sudden rise in plague incidence.

This firm belief in the efficacy of DDT alone to control plague in India was further supported by Patel & Rodde (1952) from their observations on the results of field trials in Bombay State between 1947 and 1949. The authors compared several types of treatment including various types of DDT sprayings and dustings in two towns and 180 villages of which 106 had reported human plague cases. They reached the conclusion that DDT when used as an indoor residual spray alone was effective in controlling human plague infection within seven to ten days after spraying and the flea index was also reduced in about this time, and remained low for two months or more. Their experience of four years of anti-plague measures further showed them that 'rat destruction as an anti-plague measure either during the inter-epidemic or epidemic period is not necessary and can be safely discarded. Insufflation of burrows with ten per cent DDT powder as an anti-plague measure during epidemics is also not considered necessary'. One can hardly take so complacent a view as these authors about taking no rodent control measures during inter-epidemic periods, a viewpoint which ignores the economic and public health importance of rats aside from that of plague alone. Experience elsewhere, as will be seen, has also shown that dusting of burrows has an important function in controlling free fleas and hastening epidemic control. Nevertheless, these two papers demonstrate how rapidly great faith developed in the effectiveness of DDT treatments as the sole measure of plague control.

This belief in the ability of DDT to quickly and rapidly control flea vectors of plague was again confirmed, shortly thereafter, in another part of the world when bubonic plague broke out in Haifa, in what was then Palestine, in June 1947 (Pollock, 1948). A large scale control programme was immediately put into operation with the objective of breaking the rat-flea-man cycle by the application of DDT. All the affected areas were sprayed by five per cent DDT and dusting centres utilizing power-dusters were set up which dusted 30,000 people in the first ten days of the programme with ten per cent DDT dust. Seven days after the initiation of the DDT application campaign, human cases abruptly ceased and only a total of 16 human cases appeared, though positive rodents continued to be found. Within three days, the *X.cheopis* index had fallen from around three per rat to less than one per rat. Lorries transporting grain to Affula 30 km distant were apparently responsible for carrying the disease to a flour mill in that town but focal DDT dusting quickly ended the outbreak there. It was concluded on the basis of the evidence from the Haifa campaign that DDT should be the first line of attack and that rat control should be secondary and planned on a long-term basis.

While the paper by Saenz Vera (1953) followed others of the time by opening with the statement 'Until the advent of DDT and its application as an insecticide, no satisfactory solution to the serious problem of the control of plague, particularly rural and sylvatic plague, was possible . . .', nevertheless for the first time a sign of possible trouble in this approach appeared. Speaking of the DDT vector flea control programme carried out in Ecuador since 1946, Saenz Vera mentioned that in 1950 he and his colleagues noticed that in certain areas a number of fleas had apparently survived the systematic application of 10 per cent DDT dust. After ordering a fresh batch of insecticide dust and testing it in the field they found that the fleas still survived the application and by 1951-2 found that fleas were resisting the action of DDT in extensive areas of the country. Furthermore, when visiting the state of Pernambuco in Brazil in 1952, he was informed by plague control workers there that new DDT applications were having little effect. These observations, if confirmed, he observed, would create a grave new problem in plague control and make it urgently necessary to experiment with other insecticides as a replacement for DDT.

Exactly that was being done at the time, though in an effort to find compounds still more toxic to fleas than DDT, rather than to overcome resistance to DDT, whose first appearance in South America had, as yet, not been widely noticed. Smith & Burnett (1948) screened several thousand compounds against several species of ectoparasites including fleas, ticks and lice in the laboratory. From the results of their screening, Smith

(1951) selected 620 compounds which had been most effective against fleas, for further tests as insecticidal dusts against *C.felis* and *X.cheopis.* These tests were conducted by placing a thin layer of loose earth on the bottom of one pint (0,473 l) bottles and sifting in the candidate insecticidal dust in an inert carrier at a rate of 4 grams per ft^2. Twenty fleas less than one day old were exposed in the jars for 24 hours and mortality then recorded. DDT was used as a comparison. At a concentration of 0.5 per cent, 29 compounds were more toxic than DDT, 24 as effective and 566 less so. The outstanding compounds were heptachlor, dieldrin, aldrin and HCH (95 per cent gamma isomer), all of them chlorinated hydrocarbons. Chlordane and parathion were more effective than DDT at concentrations of 0.5 and 0.05 per cent and toxaphene and synergized pyrethrum at 0.5 per cent only. Most of the compounds were more toxic to cat fleas than to rat fleas.

Wagle & Seal (1953) commented, in their comparison of DDT, BHC (HCH) and Cyanogas as plague control measures in India since 1945, that there had been some failures to control plague after dusting and spraying some villages with DDT. They did not, however, suggest that any resistance had appeared in the flea populations and concluded that despite these failures, DDT was a most valuable pulicide and was more effective than HCH or Cyanogas. The latter, in fact, was not found to greatly reduce the flea index in several field trials.

Mercier (1952) utilized a mixture of DDT five or ten per cent dust and HCH five or ten per cent dust in equal parts for the successful control of rat flea vectors of plague on Madagascar. HCH was thought to give a more rapid kill while, once its effect wore off, DDT would continue to act, being more persistent.

Mercier & Razafindrakoto (1953) also examined the effect of the three year programme of DDT use in Tananarive, Madagascar, against the vectors of malaria and plague. They examined the epidemiological data and concluded that as far as plague was concerned, it had apparently been completely eliminated from the Tananarive district, no case having appeared for 34 months at the time (1953) despite the almost continual presence of the disease previously.

The effect of a city-wide anti-malaria campaign on commensal flea populations was also analyzed by Voelckel & Mouchet (1958) in Douala in Cameroon. From 1951 until the time of the article, house sprayings had been carried out with residual applications of DDT, HCH and dieldrin. While the *X.cheopis* indices had not been high before applications began, as measured by a rather unique 'fleas per 100 rats', they fell to more or less 0 per 100 rats as a result of the spraying campaign. The sprayings had little or no effect on other ectoparasites such as mites.

At about the same time a number of Soviet workers were studying the effect of DDT on rat fleas in the laboratory (Fedorov, 1957; Ivanov, 1959) and carrying out field trials of both DDT and HCH in plague endemic areas of the USSR and Odessa and Batum outside the plague endemic areas. The results of this work, which demonstrated that these two compounds were as efficacious in the USSR as elsewhere, have been summarized in some detail by Pollitzer (1966) and the interested reader should refer to his excellent review of plague and plague control in the Soviet Union through 1964.

Probably the largest murine typhus vector control programmes outside of those in the USA was that carried out in Israel in 1953 and 1954 by Gratz (1973), where ten per cent DDT dust in Kaolin was applied as patches or blown into burrows; 12,000 courtyards were treated in Tel-Aviv in 1953 using an average of 300 g per courtyard and more than 3½ tons of formulated dust was used. In 1954, 18,000 courtyards were treated with a total of four tons of dust and focal dustings associated with cases were carried out for a number of years thereafter in Tel-Aviv and foci in Haifa as well. The incidence of murine typhus has since declined considerably in the country and the dustings themselves were highly effective against *X.cheopis.*

Another flea control campaign in the Eastern Mediterranean was that described by Arafa & Salit (1971), carried out in some settlements near Suez, in the port of Tawfik and in the Port of Adabiya in the Suez area, which is considered endemic for murine typhus. Eight per cent HCH dust was applied twice one month apart to all runways and borrows of *R.norvegicus,* and produced a marked reduction in flea indices. The second cycle seemed to last about 40 days.

In the light of the favourable results that he had obtained with two chlorinated hydrocarbon insecticides, aldrin and dieldrin, against mite vectors of scrub typhus in Malaya, R.Traub (in litt.) recommended that field trials of these compounds be carried out against fleas and this was done by Ryckman et al. (1953) in 1951 and 1952 in California. Flea indices before dusting ranged from 111.2 to 318.05 fleas (mainly *E.gallinacea*) per ground-squirrel. Squirrels examined 3, 15 and 35 days after dusting with heptachlor, aldrin and dieldrin showed more than 98 per cent control obtained, the indices being only 2.03 fleas per squirrel. DDT was slightly less effective. Dieldrin applied as a spray to burrows appeared to be more effective than as a dust.

A further field study was conducted by Ryckman et al. (1954) to study the length of time flea populations on ground squirrels could be controlled under natural conditions by 2.5 per cent aldrin, 2 per cent dieldrin, 2.5 per cent heptachlor and five per cent DDT applied as dusts to the squirrel burrows. It was expected that aldrin, dieldrin or heptachlor dusted into burrows over a large area would keep the flea population below a dangerous level for up to nine months. DDT was not successful in controlling the population for this period of time, unlike the other toxicants.

Kartman & Lonergan (1955) expressed concern about the possible indiscriminate use of DDT or other insecticides for the control of wild rodent fleas on Hawaii and sought a more selective way of applying them. They developed a simple bait box which, when visited by rodents for the food inside (either poison-free or a warfarin bait), dusted them with ten per cent DDT and reduced the densities of the vector species, *X.cheopis* and *X.vexabilis hawaiiensis.*

Further tests with the bait box method in California (Kartman, 1958, 1960) demonstrated that it was effective against the fleas, *Malaraeus telchinum* and *Hystrichopsylla linsdalei,* of small wild rodent reservoirs of plague in that state. Barnes & Kartman (1960) carried out a field trial of the bait boxes for control of *Monopsyllus eumolpi* and *M.ciliatus* and chipmunks and *Diamanus montanus* and *Oropsylla idahoensis* on ground squirrels: control was rapid and it was thought that boxes 200 feet (about 60 metres) apart would provide adequate flea control on a far-ranging host. The use of insecticide-treated bait boxes was eventually shown to reduce plague antibody rates in wild rodent populations (Kartman & Hudson, 1971).

The possibility of using bait boxes in Bombay for rat flea reduction was considered by Deoras (1963), who tested both DDT and HCH powders in the laboratory and DDT and pyrethrum powders in the field. Control was claimed to be satisfactory and the use of the boxes was well accepted by the villagers.

While using insecticide-treated bait boxes is probably an effective means of controlling wild rodent fleas and could even be used as an adjunct for the control of commensal rodent fleas, there is obviously a considerable labour effort involved in their preparation, placement and maintenance. Where flea control can be readily carried out by burrow and runway dusting with effective insecticide dusts, this is certainly the method of choice during an epidemic outbreak of bubonic plague, at least in urban areas or villages (Gratz, 1968).

Recent work at the WHO Rodent Control Demonstration Unit in Rangoon, Burma, has shown that bamboo tubes with an attractive bait and treated with an insecticide dust are an effective method of flea control in houses (Brooks, J., report to the WHO, 1977).

4 THE APPEARANCE OF INSECTICIDE RESISTANCE IN FLEA POPULATIONS

The first reports of flea resistance to insecticide were probably associated with the area dustings of DDT for rat flea control; during the summer of 1949 it was reported that DDT dusts were failing to control cat flea infestations (Kilpatrick & Fay, 1952). In several instances high populations of *Ct.felis* persisted even after three applications of five per cent DDT dust but were effectively controlled by five per cent chlordane dust applications. Concerned by the possibility that this apparent resistance could also develop in the rat flea, *X.cheopis,* the authors raised this species in the laboratory and attempted to select it for DDT resistance. Under selection through four minutes of exposure to five per cent DDT dust at a rate of 50 mg/ft^2 increased resistance was obtained through the F_3 generation, but even though further exposure was carried out until the F_7 generation there was no further increase in resistance with the strain tested.

Shawarby (1954) also attempted to select a strain of *X.cheopis* from Egypt for resistance to DDT; by the fourth generation he was able to induce a slight resistance to this compound of about twice that for normal fleas. He also carried out field trials of both DDT and HCH in villages of the Nile Delta where most of the fleas examined were *P.irritans.* HCH at 143 mg of gamma HCH per m^2 applied as a residual spray gave faster and more prolonged control than DDT in the same area. Although the actual persistence of effect of DDT was greater, he thought that the HCH gave a much higher level of initial control, the effects of which lasted longer than those of DDT.

In 1960, Brown reviewed the extent to which insecticide resistance had been reported in flea populations up to that time, and stated that accurate estimation of the presence of DDT resistance was hampered by the lack of a method to carry out tests. In any event, apart from the report by Saenz Vera (ibid.), Brown thought that there was no valid evidence of resistance having developed in *Xenopsylla* species. However, in the same number of the journal in which his review appeared, a case of DDT resistance was reported, confirmed by laboratory studies by Patel et al. (1960). Patel and his co-workers formulated their own test method for determining flea susceptibility (or resistance) to insecticides, partly using insecticide-impregnated papers supplied for testing mosquito susceptibility to insecticides as part of the WHO test kits and partly papers which they themselves impregnated with insecticides. Fleas were exposed to papers impregnated with DDT, HCH and dieldrin. Fleas for testing were taken from one group of villages in the Satara district which had only received a few DDT residual sprays as part of the then malaria eradication project, and another group of villages near Poona where DDT and dieldrin had been sprayed continuously over a 14-year period. The results of their tests showed that *X.*

cheopis populations from the Poona district were 19 to 5,000 times more resistant to DDT than the Satara strain; 8 to 16 times more resistant to gamma HCH, but still susceptible to dieldrin. The authors rightfully pointed out that these results highlighted the importance of carrying out susceptibility tests on *X.cheopis* populations in all areas where extensive use had been made of DDT in malaria control operations, lest DDT fail to control fleas during an outbreak of plague due to resistance, such as had occurred in Ecuador.

In the same year that Patel et al. confirmed the appearance of DDT resistance in *X. cheopis,* the WHO published 'Provisional instructions for determining the susceptibility or resistance of fleas to insecticides' (WHO, 1960). As with most of the other standard test methods, this was – and is – based on the use of insecticide-impregnated papers, impregnated by the method described by Busvine & Nash (1953). Other workers had previously tested impregnated papers against fleas, including Shawarby (1954) (ibid.), Sen (1958) (who found *X.cheopis* in Calcutta still susceptible to DDT) and Smith (1959) (who used them to confirm resistance of *P.irritans* in Tanzania to dieldrin and susceptibility to DDT). The history of the development of the kit has been reviewed by Burden (1966) and Busvine & Lien (1961).

The definitive version of this test procedure was approved by the WHO Expert Committee on Insecticides which met in 1968 (WHO, 1970) and more recently has been modified so that susceptibility levels are now assessed by mortalities produced by different exposure times to a standard concentration rather than to different concentrations at a standard one-hour exposure (WHO, 1975).

With the development of a standard method of testing fleas for susceptibility or resistance to insecticides, numerous reports were published on flea resistance to insecticides, at first mainly to DDT and then, as time went on, to other chlorinated hydrocarbons as well. Most of these reports came from countries in which widespread and intensive indoor residual spray applications were being carried out as part of a malaria eradication programme and illustrated the selective pressure that this type of spraying was having on a number of other insects commonly found in homes, among them fleas, bedbugs and cockroaches. In India, after the report by Patel et al., mixed populations of *X.cheopis* and *X.astia* showed lessened susceptibility to DDT in Mysore State in 1960 (Mohan, 1960) and in 1961 resistance of both *X.cheopis* and *X.astia* to DDT was reported from Delhi (Sharma & Joshi, 1961) and in 1962 high resistance to DDT was found in *X.cheopis* populations from towns in three more states, Punjab, Bihar and Uttar Pradesh (Krishnamurthy & Joshi, 1962), again all in areas which had been subject to regular sprayings of DDT as part of the malaria eradication programme. The authors also observed that the flea indices were higher in towns with insecticide-resistant populations than in those where the flea populations still remained susceptible. In 1962, further tests by Mohan (1962) confirmed the presence of DDT-resistant populations of *X.cheopis* and *X.astia* in Madras and Mysore States and showed that they were also more tolerant to HCH and dieldrin than previously Concerned by the persistence of plague in an area of Mysore State (Krishnaswami et al., 1963a) those authors carried out investigations on the rat and rat flea populations of the area (1963b). They found that despite the withdrawal of DDT malaria spraying activities, the previous year, *X.cheopis* and, to a lesser extent, *X.astia* populations, remained highly resistant to DDT and tolerant to HCH, and felt that the situation could result in serious outbreaks of plague, since the flea indices were quite high. Choudhury (1963) in fact blamed the development of *X.astia* resistance to DDT as reason for increased densities of this species and a subsequent outbreak of plague in Madras State in late 1962.

Although earlier tests in 1956 and 1959 had not indicated the presence of resistance in Madagascar, Brygoo (1966) showed that *X.cheopis* from rats in Tananarive were highly resistant to DDT.

Extensive insecticide susceptibility tests were carried out on *X.cheopis* populations from different parts of Thailand between 1965 and 1967 and reported to the WHO by Kulta-Uthai (in litt.). These showed that DDT-resistance was widespread and, in one area, resistance to dieldrin had also appeared. In 1965, laboratory tests reported to the WHO confirmed the earlier reports from the field of DDT resistance in this species in Ecuador and also showed the presence of dieldrin resistance.

Elsewhere on the Indochina peninsula, DDT resistance had also been reported in flea populations from Burma and Viet-Nam where tests showed only a 51 per cent mortality following one hour of exposure to four per cent DDT papers (Pal, 1966). As will be seen, this last development was to have very serious consequences regarding the transmission of plague in Viet-Nam shortly thereafter.

Starting in the early 1960's, the plague (which has been almost continuously present in Viet-Nam since 1898) began to steadily mount and by 1967 had reached a level of 5,574 cases with 246 deaths reported for that year. Although the outbreak was no doubt associated with the disturbances and mass movements of people that were occurring in the country, nevertheless the spread of the disease was almost certainly facilitated by an abundant rodent population with high flea indices (99 per cent of which are *X.cheopis*) (Cavanaugh et al., 1969) which had by this time reached such a high level of DDT resistance that this insecticide at times caused no appreciable mortality in the laboratory tests (Afifi, 1968; Stasiak et al., 1970) and also failed to achieve control in the field (Cavanaugh, ibid.). Even 24 hours of constant exposure to four per cent DDT papers, i.e. 24 times the normal length of exposure time, caused but 13 per cent mortality in *X.cheopis* populations from Dalat and Nha Trang, while laboratory test also showed that resistance had developed to dieldrin in Viet-Nam which, because of cross-resistance, had apparently caused the failure of an attempt to control fleas in the field with one per cent lindane dusts. Two per cent diazinon dusts provided very good control in the widespread areas where resistance had appeared to the chlorinated hydrocarbons (Chow, 1970).

Still further tests from India indicated that wherever DDT had been used in the malaria programmes, DDT and to a lesser extent HCH, resistance was developing in the flea populations, mainly *X.cheopis* but *X.astia* as well (Bhatia et al., 1971; Sant & Renapurkar, 1971). Kalra & Joshi (1974) carried out a study on the selection of *X.cheopis* for resistance in the laboratory and found that selection for either DDT or dieldrin resistance usually conferred a degree of resistance to the other insecticide, but that there was probably a special defence mechanism in the flea, specific for HCH. R.Turner, in a report to the WHO in 1975, found that *X.cheopis* in the Indonesian plague focus of Boyalali were highly resistant to DDT, although still susceptible to dieldrin and malathion. Resistance in this area may have originated from the extensive use of DDT dusts and sprays specifically for flea control since the 1950's.

The main implication of this widespread development of resistance to DDT in *X.cheopis* populations and increased tolerance to other chlorinated hydrocarbons was that this group of insecticides could no longer be relied upon to provide rapid and effective control of flea-borne diseases. It was thus clearly imperative that laboratory and field trials with alternative insecticide groups be carried out to determine which compounds would be most effective in controlling fleas and serve as a solution to the problem of flea resistance to insecticides.

5 THE DEVELOPMENT OF COUNTER-MEASURES TO INSECTICIDE RESISTANCE

While a number of alternative insecticides had already been tested for flea control beside DDT, for the most part these were either other chlorinated hydrocarbons and a few organophosphorous OP insecticides. In the main, the emphasis had been on the possible development of insecticides which were either more efficient than DDT or perhaps possessed more favourable environmental characteristics. With the recognition that widespread resistance to DDT, and later to other related compounds, was occurring, the search for insecticides that could serve as useful alternatives to DDT and other chlorinated hydrocarbons became much more imperative.

In setting out objectives for laboratory and field trials to be carried out on possible alternative compounds for flea control, one must look for very much the same characteristics that resulted in the successful use of DDT before insecticide resistance began to increasingly limit its use. These characteristics include, of course, a high degree of effectiveness against fleas, a long period of persistent action so as to obviate the necessity for frequent retreatments and a low degree of mammalian toxicity. It would also be very desirable if the compound is bio-degradable and it is essential that the candidate compound have the least possible cross-resistance with the chlorinated hydrocarbons or to any other group to which insecticide resistance has already developed.

6 TRIALS WITH NEWER INSECTICIDES

Concerned by the numerous reports that cat and dog fleas were not being successfully controlled by DDT dusts, Wilson et al. (1957) carried out a series of field trials with different insecticides applied as emulsion sprays to yards with natural infestations of fleas in Orlando, Florida. All the insecticides, except DDT and chlordane – both of which had already been used against fleas – were effective and provided 100 per cent control of seven to nine weeks. These included Diazinon, lindane, malathion, chlorthion, ronnel, coumaphos and dipterex.

Because of the widespread resistance of *X.cheopis* to the chlorinated hydrocarbon insecticides recognized in India, many of the first tests on likely alternative groups were carried out in that country. Krishnamurthy et al. (1963) carried out field trials with a two per cent diazinon spray and dust. They found the spray formulation to be quite ineffective, and the reductions achieved by the dust formulation were not considered significant. Further field trials were carried out with a number of OP and carbamate insecticides as dust formulations (Krishnamurthy et al., 1965) including three per cent fenthion, five per cent malathion, two per cent carbaryl ('Sevin') as well as a 0.3 per cent pyrethrum dust. They found the organophosphorus compounds (OP) were quite ineffective and pyrethrum dusts failed to achieve any reduction at all in the flea indices in the village treated. Carbaryl (Sevin or OMS-29) when applied by blowing it into the rat burrows and by dusting onto rat runs, provided effective control for a period of 12 weeks. The effectiveness of carbaryl against rat fleas was confirmed by laboratory tests on insecticide impregnated papers carried out by Fox et al. (1966) in Puerto Rico; the test insects were strains of *X.cheopis* from San Francisco and from Puerto Rico maintained in the laboratory. Both strains were reported as being tolerant to malathion and fenthion; papers impregnated with temephos and propoxur in the authors' laboratory had no significant effect on the test fleas. Naled and carbaryl were very effective, however. Shortly thereafter, Fox et al. (1968) carried out further laboratory tests in Puerto Rico, this time using *Ct.felis* strains obtained locally from cats in the city of San Juan. Several insecticides were compared, including DDT and dieldrin. Tests were carried out by exposing both larvae and pupae. Neither DDT nor dieldrin produced more than a five per cent mortality against adult fleas nor 12 per cent against larvae. Malathion was also ineffective and propoxur gave only a 35 per cent mortality at the highest dosage tests – 1.6 per cent. Fenthion was slightly more active and the most effective compound was Bayer 41831 (fenitrothion). Further field trials were carried out in India by Krishnaswami in 1966 and reported on to the WHO. He found that dusts of one per cent propoxur, two per cent bromophos and one per cent 3-methyl-5-isopropyphenyl *N*-methylcarbamate all achieved a very high degree of initial control on *X.cheopis* populations in the treated villages, but none had a residual effect longer than four weeks which, he considered, limited the usefulness of these compounds as pulicides.

Burden and Smittle (1968) screened 236 compounds in the laboratory so as to select those which had the greatest promise for field trials against bedbugs and *X.cheopis.* Their first level of screening using insecticide impregnated papers showed that 119 were suitable for further testing, which was carried out in the laboratory by exposing fleas to the test compounds formulated as dusts. Carbaryl at 0.07 per cent had the lowest LC_{50} of the compounds tested.

The efficacy of carbaryl in the laboratory was further studied by Chaturvedi et al. (1969) against *X.cheopis* and *X.astia* collected from several areas in and around Bombay. Carbaryl was compared to malathion, HCH and DDT dusts and in most cases found to be the most effective compound, with DDT the least effective against the field collected fleas.

A very interesting study was made by Olsen (1969) in Viet-Nam during 1967 on the effect of the use of two per cent diazinon dusts on outbreaks of plague. In the instances studied, and from reports he had received, diazinon two per cent dust gave fast and effective control of plague outbreaks within three to six days of application but its persistence of action was no longer than three weeks. In those control operations which were observed by Olsen, the cost of the insecticide represented 24 per cent of the total cost and thus a formulation which would provide greater persistence than the 2 per cent diazinon dust that had been used, would give a considerable saving, not only in the cost of the insecticide but by reducing the cost of the labour required to carry out repeated treatment at shorter intervals in the area under control. It was suggested that the use of a granular formulation of diazinon or of a granular/dust mixture might reduce volitization of the compound and extend its residual life, but it appears that this was never tested.

Miller et al. (1970) carried out a comparative trial of a number of insecticide dust formulations to evaluate their effect on wild rabbit and rodent fleas, mainly because of concern over the undesirable side effects of the chlorinated hydrocarbons rather than to overcome insecticide resistance. These trials, which were done in New Mexico, compared dusts of malathion at 0.9 and 1.34 kg/ha, dursban (chlorpyrifos) at 0.48 and 1.97 kg/ha, carbaryl (Sevin) at 2.36 and 2.41 kg/ha, fenthion at several dosages ranging from 2.41 kg/ha to 4.70 kg/ha and diazinon at 2.20 and 2.40 kg/ha. Although the chemicals were applied at several different dosages and by different methods, a statistical analysis of the results showed that only carbaryl and fenthion produced significant drops in flea densities for a period of up to three weeks. The conditions of the test were probably considerably more rigorous than those which would be obtained in the control of fleas on commensal rodents in most areas. On the other hand, Ryckman (1971), applying dieldrin dusts and sprays in California in an ecologically similar area to Miller's work, obtained a very long period of control of fleas on ground squirrels which probably extended to between 12 and 15 months.

The likely effectiveness of carbaryl also attracted attention in the USSR and Lobacev & Smirnova (1972) tested this compound among several others for immediate action against four flea species: *X.cheopis, Nosopsyllus fasciatus, N.laeviceps* and *Callopsylla caspius* (the last three cited as *Ceratophyllus);* ten per cent locally manufactured carbaryl dusts and an 80 per cent carbaryl formulation from the USA and one per cent neopinamin (tetramethrin) were the most effective against all species.

As effective as dieldrin or any other chlorinated hydrocarbon might be against wild rodent fleas, Barnes et al. (1972) correctly observed that both resistance and environmental measures may well preclude further broadcast use of this group. They tested a two per cent carbaryl dust as an alternative for the control of *Opisocrostis hirsutus* by dusting it into burrows and by applying it in bait stations (Barnes et al., 1974). Both techniques gave excellent control of the vector fleas, the first in prairie dog colonies and the second against *Monopsyllus wagneri* on *Peromyscus maniculatus.*

Because of the increased use of containers in shipping of goods from one country to another, including those which were used for returning cargo by air from Viet-Nam, Maddock et al. (1973) tested the effect of dichlorvos strips against *X.cheopis* in simulated cargo container tests; the dichlorvos impregnated resin strips provided effective control provided that the period of exposure was adequate. This formulation has now been made up as a collar for use on dogs and cats for flea control and has proved very effective for this purpose (Fox et al., 1969a; Fox et al., 1969b) although skin irritation has occurred in some cats. Dichlorvos, which is an effective fumigant insecticide against a number of insect species, was also put under field trial in the form of small dichlorvos impregnated plastic pellets which were put into prairie dog burrows. At a dosage rate of 28 g per burrow good immediate control and persistent action for at least 14 days were obtained against *O.hirsutus* (Bennett et al., 1975). The method was considered particularly applicable for small areas such as public parks where there are many visitors.

Further field trials on a similar formulation were carried out against wild rodent fleas by Cole et al. (1976); the dichlorvos polyvinyl chloride pellets containing 20 per cent dichlorvos were dipped in glue, then rolled in laboratory meal. The baited pellets were put into bait stations from where it was hoped they would be removed and carried into the burrows; here, the dichlorvos emanating from the pellets could then control the burrow flea population, both free and on the rodents. In field tests, this approach was only effective against fleas on *Dipodomys spectabilis* because of its habit of extensively storing food in its burrows. While this approach is very interesting for flea control, it is obviously limited to those species of rodents that customarily hoard food in their burrows and, in addition, will not provide immediate control, it taking some time for indices to decline. It could be considered, however, as a preventive measure against fleas on appropriate reservoir species.

In an effort to reduce the possibility of environmental contamination by insecticides used for flea control, several workers have tested compounds mixed with rat baits for their effect as systemic insecticides against rodent fleas. Clark & Cole (1971) tested 21 compounds against *X.cheopis* as baits fed to flea infested guinea pigs. Seven of these killed all the fleas without noticeably harming the guinea pigs; others were either ineffective or, if effective against the fleas, killed their hosts as well. The most promising of the safer compounds were tested by allowing cotton rats infested with *X.cheopis* to feed on the baits containing the systemic insecticides. Baits containing chlorphoxim gave four to 14 days of 96 to 99 per cent control after a single feeding; Phoxim 95 to 99 per cent control; while trichlorfon did not exceed 79 per cent control (Clark & Cole, 1973). Miller et al. (1975) carried out an open field test with phoxim baits made available to cotton rats infested with *Polygenis gwyni* and found that effective control began 36 hours after feeding on baits and continued as long as the insecticide-containing bait was available to the rodents.

Temephos (Abate) has found wide use in vector control as a mosquito larvicide of very low mammalian toxicity. While generally thought to be a poor adulticide, it has proven to be effective against human body lice, and recent work by Miller & Baker (1975) shows that it will control *Ct.felis* on dogs and cats for up to two weeks when applied as a two per cent powder and will give partial control for as long as four weeks.

Tests on pulicides are now under way at two field research units of the World Health Organization. The Vector and Rodent Control Research Unit in Indonesia will be mainly concerned with carrying out field trials in a focus of plague in central Java, while the Rodent Control Demonstration Unit in Rangoon, Burma is studying the rat flea susceptibility in plague endemic areas of that country and carrying out comparative trials of possible alternative compounds in areas where the flea vectors of plague have developed a high degree of resistance to DDT. Recent results, as yet unpublished, show that pirimiphos-methyl used in bait tubes gave good initial control of fleas on *Bandicota bengalensis, Suncus murinus* and *Rattus rattus.* Several other compounds are now under study at this Unit.

7 FUTURE REQUIREMENTS FOR FLEA CONTROL

It is difficult to predict just how great the need for flea control is likely to be in the future. Certainly the interest in having safe and effective compounds for controlling fleas on domestic animals will continue on a comparatively limited scale. The main concern, however, is the requirements for control of flea vectors of human disease, particularly for the control of plague and murine typhus; clearly, these will depend on the extent to which these diseases are a public health problem and the extent to which vector flea control campaigns are likely to have to be undertaken.

After having reached a peak of over 6,000 cases in 1967, plague has been declining since and only some 737 cases were reported to the World Health Organization in 1976. It must be emphasized, however, that the number of officially reported cases does not necessarily reflect the actual numbers that occur and it is thought that the data from several countries are incomplete. Nevertheless, it may be of interest to see the number of cases that have been reported over the past decade, viz.:

Table 1. Numbers of cases of plague reported to WHO, 1967-1976

	1967	1968	1969	1970	1971	1972	1973	1974	1975	1976
Africa	18	172	147	27	20	128	50	183	144	77
Americas	223	392	424	326	216	297	185	321	521	142
Asia	5,763	4,395	3,882	4,111	4,186	1,408	443	2,230	811	518
Europe				1*						
Total:	6,004	4,959	4,453	4,464	4,422	1,833	678	2,734	1,476	737

* imported.

It will be seen that the year to year variation is considerable and, as indicated above, this is in part due to the vagaries of reporting and in part due to very real cyclic variations in the occurrence of plague. It should be kept in mind that plague still remains endemic in every enzootic focus in which it was previously found and sudden flare-ups of the diseases in the form of epizootics and epidemics can and have occurred and are virtually impossible to predict. It seems prudent, therefore, to have as much information available as possible as to the distribution of flea resistance to the chlorinated hydrocarbon insecticides and as to those compounds which could best be used to replace them if the need arrives. Where vector flea populations remain susceptible to DDT it is of interest to note that this compound is probably the most effective that can be used. Concerned with an increasing trend of plague cases in the USA, the government of that country lifted a ban on the use of DDT in 1976, permitting it to be used to control wild rodent fleas, in those states where wild rodent and human plague were most serious.

Murine typhus is undoubtedly far more widespread than was previously thought to be the case, and recent studies in several countries have shown that the disease is present in areas from which it was previously unreported, and, indeed, may be the cause of significant morbidity in the human population. Due to difficulties in the clinical and laboratory diagnosis, it is certainly under-reported. When the true public health significance of the disease is understood, it is possible that vector-flea control programmes may be carried out to reduce the incidence of the disease.

It is thus likely that effective, economic and safe insecticides for the control of flea vector populations will be required for some time to come. In the long run, the best method of reducing the threat of flea-borne disease is to suppress the rodent populations in those urban areas in which plague and murine typhus are endemic and to maintain a careful surveillance of those rural foci of endemic plague. In all threatened areas, routine insecticide susceptibility tests should be carried out on *X.cheopis* and other vector fleas and adequate stockpiles of the most effective insecticide maintained ready for emergency use.

No less important than having effective materials to utilize in control programmes, is the training of staff to carry out surveillance of all aspects of flea vector, rodent reservoir and human host populations and who are capable of deciding, on the basis of the information obtained, the measures to undertake to combat or prevent outbreaks of flea-borne disease. Such trained staff is unfortunately most lacking in just those countries where the need is greatest, and training should thus have a high priority.

There is also a necessity for continuing research on more effective and selective compounds, many of which have yet to be tested in the field, or methods for the control of fleas, whether as vectors of disease or as pests. While some interest has been expressed in the possibility of genetic or biological control of fleas, there seems little likelihood of either of these approaches showing any promise in the foreseeable future, and most research should, by need, be centered on the development of conventional compounds and improving the manner of their use and formulation.

REFERENCES

Afifi, S.E. 1968, Assessment of insecticide susceptibility of the flea *(Xenopsylla cheopis)* the plague vector in the Republic of Viet Nam. WHO unpublished document WPR/ Plague/3-Rev. 1.

Arafa, M.S. & Salit, A.M. 1971, Trapping,

poisoning and dusting as anti-rat and flea measures in Egypt. J. Egy. Vet. Med. Assoc. 31(3-4): 263-276.

Barnes, A.M. & Kartman, L. 1960, Control of plague vectors on diurnal rodents in the Sierra Nevada of California by use of insecticide baitboxes. J. Hyg. Camb. 58: 347-355.

Barnes, A.M., Ogden, L.J., Archibald, W.S. & Campos, E.G. 1974, Control of plague vectors on *Peromyscus maniculatus* by use of 2% carbaryl dust in bait stations. J. med. Ent. 11(1): 83-87.

Barnes, A.M., Ogden, L.J. & Campos, E.G. 1972, Control of the plague vector *Opisocrostis hirsutus* by treatment of prairie dog *(Cynomys ludovicianus)* burrows with 2% carbaryl dust. J. med. Ent. 9(4): 330-333.

Bennett, W.C., Graves, G.N., Wheeler, J.R. & Miller, B.E. 1975, Field evaluations of dichlorvos as a vapour toxicant for control of prairie dog fleas. J. med. Ent. 12(3): 354-358.

Bhatia, S.C., Renapurkar, D.M. & Deobhanker, R.B. 1971, Note on rapid susceptibility tests with DDT against *Xenopsylla cheopis* and *Xenopsylla astia* in parts of Maharashtra State, India. Patna J. Med., Feb. 1971: 73-79.

Bishopp, F.C. 1921, The life history and control of fleas. In: Richard Badger, Sanitary Entomology, W.D.Pierce, ed., Boston.

Brown, A.W.A. 1960, Present extent of insecticide resistance in fleas. Bull. Wld. Hlth. Org. 23: 410.

Brygoo, E.R. 1966, Epidémiologie de la peste à Madagascar. Arch. Inst. Pasteur, Madagascar, 35(1): 9-147.

Burden, G.S. 1966, Methods of detection of rat flea resistance to insecticides. Unpublished WHO document WHO/VC/66.217: 199-204.

Burden, G.S. & Smittle, B.J. 1968, Laboratory methods for evaluation of toxicants for the bedbug and the oriental rat flea. J. econ. Ent. 61(6): 1565-1567.

Busvine, J.R. & Lien, J. 1961, Methods for measuring insecticide susceptibility levels in bedbugs, cone-nosed bugs, fleas and lice. Bull. Wld. Hlth. Org. 24: 502-517.

Busvine, J.R. & Nash, R. 1953, The potency and persistence of some new synthetic insecticides. Bull. ent. Res. 44: 371-376.

Cavanaugh, D.C., Dangerfield, H.G., Hunter, D.H., Joy, R.J.T., Marshall, J.D., Quy, D.V., Vivona, S. & Winter, P.E. 1968, Some observations on the current plague outbreak in the Republic of Vietnam. Amer. J. Publ. Hlth. 58(4): 742-752.

Chaturvedi, G.C., Deoras, P.J. & Rao, N.R. 1969, Evaluation of DDT, BHC, malathion and sevin as pulicides. Pesticides, June 1969.

Choudhury, D.S. 1963, Investigation on the reappearance of suspected human plague in Gobichetti Palaryamtaluk of Coimbatore district, Madras State, India. Ind. J. Malariol. 17(4): 249-255.

Chow, C.Y. 1970, Present status of insecticide resistance in the Oriental rat flea *(Xenopsylla cheopis)* In Viet Nam. WHO unpublished document WHO/VBC/70.214.

Clark, P.H. & Cole, M.M. 1971, Systemic insecticides for control of oriental rat fleas: preliminary test in Guinea pigs, 1967-1969. J. econ. Ent. 64: 1477-1479.

Clark, P.H. & Cole, M.M. 1973, Oriental rat fleas: Evaluation of three systemic insecticides in baits for control on cotton rats in outdoor pens. J. econ. Ent. 67(2): 235-236.

Cole, M.M., Bennett, W.C., Graves, G.N., Wheeler, J.R., Miller, B.E. & Clark, P.H. 1976, Dichlorvos bait for control of fleas on wild rodents. J. med. Ent. 12(6): 625-630.

Davis, D.E. 1945, The control of rat fleas *(Xenopsylla cheopis)* by DDT. Publ. Hlth. Rpts. 60(18): 485-489.

Davis, D.E. 1947, The use of DDT to control murine typhus fever in San Antonio, Texas. Publ. Hlth. Rpts.62(13):449-463.

Davis, D.H.S. 1949, Current methods of controlling rodents and fleas in the campaign against bubonic plague and murine typhus. J. Roy. Sanit. Inst. 69(3): 170-175.

Deoras, P.J. 1963, Studies on Bombay rats. Curr. Sci. 32: 457-459.

Ebbell, B. 1937, Papyrus Ebers. The greatest Egyptian Medical Document, Copenhagen, quoted in Hoeppli, R. (1969), Parasitic diseases in Africa and the Western Hemisphere. Acta Tropica. Supplement 10.

Eskey, C.R. & Hemphill, F.M. 1948, Relation of reported cases of typhus fever to location, temperature and precipitation. Publ. Hlth. Rpts. 63: 941-948.

Fedorov, V.N. 1957, Laboratory tests of the simultaneous destruction of rodents and their ectoparasites. (In Russian) Med. Parazitol. 26(1): 40-42.

Fox, I., Bayona, I.G. & Armstrong, J.L. 1969b, Cat flea control through use of dichlorvos-impregnated collars. J. Amer. Vet. Med. Assoc. 155(10): 1621-1623.

Fox, I., Rivera, G.A. & Bayona, I.G. 1968, Toxicity of six insecticides to the cat flea.

J. econ. Ent. 61(3): 869-870.

Fox, I., Rivera, G.A. & Bayona, I.G. 1969a, Controlling cat fleas with dichlorvos-impregnated collars. J. econ. Ent. 62(5): 1246-1249.

Fox, I., Rivera, G.A. & Umpierre, C.C. 1966, Toxicity of various insecticides to *Xenopsylla cheopis*. Am. J. trop. Med. Hyg. 15 (4): 611-613.

Gauthier, J.C. & Raybaud, A. 1903, Recherches experimentales sur le rôle des parasites du rat dans la transmission de la peste. Rev. Hyg., Paris 25: 426.

Gordon, J.E. & Knies, P.T. 1947, Fleas versus rat control in human plague. Amer. J. med. Sci. 213: 362-376.

Gouck, H.K. 1946, DDT to control rat fleas. J. econ. Ent. 39(3): 410-411.

Gratz, N.G. 1968, Emergency vector control measures in outbreaks of bubonic plague. WHO unpublished document, WHO/VBC/ 68.83.

Gratz, N.G. 1973, Urban rodent-borne disease and rodent distribution in Israel and neighbouring countries. Isr. J. med. Sci. 9(8): 969-979.

Heiser, V.G. 1922, The practice of medicine in the tropics. London, W.Byam & R.G.Archibald.

Hill, E.L. & Morlan, H.B. 1948, Evaluation of county-wide DDT dusting operations in murine typhus control. Publ. Hlth. Rpts. 63(51): 1635-1653.

Hill, E.L., Morlan, H.B., Utterback, B.C. & Schubert, J.H. 1951, Evaluation of county-wide DDT dusting operations in murine typhus control (1946 through 1949). Publ. Hlth. Rpts. 66(33): 396-401.

Hirst, L.F. 1931, The protection of the interior of Ceylon from plague. Ceylon, Municipal Printing Office, 52 pp.

Hirst, L.F. 1953, The conquest of plague. Oxford, Clarendon Press, 478 pp.

Ivanov, K.G. 1959, Efficacy of the residual insecticidal activity of DDT. (in Russian) Voenn. Med. Zh. 12: 77-79.

Kalra, R.L. & Joshi, G.C. 1974, Studies on the insecticide resistance in rat fleas *Xenopsylla cheopis*. Botyu-Kagaku 39: 110-115.

Kartman, L. 1946a, A note on the problem of plague in Dakar, Senegal, French West Africa. J. Parasitol. 32(1): 30-35.

Kartman, L. 1946b, On the DDT control of *Synosternus pallidus* Taschenberg (Siphonaptera, Pulicidae) in Dakar, Senegal, French West Africa. Amer. J. trop. Med. 26(4): 841-848.

Kartman, L. 1958, An insecticide-bait-box method for the control of sylvatic plague vectors. J. Hyg. Camb. 56: 455-465.

Kartman, L. 1960, Further observations on an insecticide-bait-box method for the control of sylvatic plague vectors: effect of prolonged field exposure to DDT powder. J. Hyg. Camb. 58: 119-124.

Kartman, L. & Hudson, B.W. 1971, The effect of flea control on *Yersinia (Pasteurella) pestis* antibody rates in the California vole *Microtus californicus* and its epizootiological implications. Bull. Wld. Hlth. Org. 45: 295-301.

Kartman, L. & Lonergan, R.P. 1955, Wild-rodent-flea control in rural areas of an enzootic plague region in Hawaii. Bull. Wld. Hlth. Org. 13: 49-68.

Kilpatrick, J.W. & Fay, R.W. 1952, DDT-resistance studies with the Oriental rat flea. J. econ. Ent. 45(2): 284-288.

Krishnamurthy, B.S., Achuthan, C., Rama Rao, T.Z., Chandrahas, R.K. & Krishnaswami, A.K. 1963, Investigation of plague in Koler district part III. Ind. J. Malariol. 17(2-3): 205-214.

Krishnamurthy, B.S. & Joshi, G.C. 1962, Susceptibility status of rat fleas *(X.cheopis)* to DDT, BHC and dieldrin in Punjab, Bihar and Uttar Pradesh. Ind. J. Malariol. 16(2): 137-142.

Krishnamurthy, B.S., Putatunda, J.N., Joshi, G.C., Chandrahas, R.K. & Krishnaswami, A.K. 1965, Studies on the susceptibility of the oriental rat fleas *Xenopsylla* spp. to some organophosphorous and carbamate insecticides. Bull. Ind. Soc. mal. Com. Dis. 2(2): 131-138.

Krishnaswami, A.K., Krishnamurthy, B.S. & Ray, S.N. 1963a, Investigations on plague in Koler district (Mysore State), I. Ind. J. Malariol. 17(2-3): 179-192.

Krishnaswami, A.K., Krishnamurthy, B.S., Ray, S.N., Singh, N.N. & Chandrahas, R.K. 1963b, Investigations on plague in Koler district (Mysore State), II. Ind. J. Malariol. 17(2-3): 193-203.

Lewis, P.M., Buehler, M.H. & Young, T.R., jr. 1945, Plague in Dakar. Bull. US Army Med. Dept. 87: 13-16.

Lindquist, A.W., Madden, A.H. & Knipling, E.F. 1944, DDT as a treatment for fleas on dogs. J. econ. Ent. 37(1): 39.

Lobacev, V.S. & Smirnova, A.S. 1972, Comparative action of some new insecticides and their combined preparations on various flea species. Problems concerning particularly dangerous infections. 2(24): 103-106.

Long, J.D. & Mostajo, B. 1934, Experiencias con pulgas como portadoras de peste bubonica. Bol. Ofic. Sanit. Pan-Amer. 13: 1016.

Ludwig, R.G. & Nicholson, H.P. 1947, The control of rat ectoparasites with DDT. Publ. Hlth. Rpts. 62(3): 77-84.

Macchiavello, A. 1946, Plague control with DDT and '1080', results achieved in a plague epidemic in Tunbes, Peru, 1945. Am. J. publ. Hlth. 16(8): 842-854.

Macchiavello, A. 1947, Reinfeccion pestosa de puertos peruanos por importacion de sacos de yute proveniates de la India. Bol. Ofic. Sanit. Pan-Amer. 26(3): 225-228.

Maddock, D.R., Guerrant, G.O., Schoof, H.F. & Miles, J.W. 1973, Effectiveness of dichlorvos vapor against *Xenopsylla cheopis* in containerized cargo. J. econ. Ent. 66: 689-691.

McCoy, G.W. 1908, Plague in ground squirrels. Publ. Hlth. Rpts. 28: 1289-1293.

Mercier, M.S. 1952, La prophylaxie de la peste au moyen des insecticides organiques de synthèse à Tananarive: premiers resultats. Bull. Soc. Path. Exot. 45: 409-425.

Mercier, S. & Razafindrakoto, J.B. 1953, Bilan de trois années de campagnes de désinsectisation domestique à Tananarive. Bull. Soc. Path. Exot. 46: 463-473.

Miller, J.E. & Baker, N.F. 1975, Insecticidal activity of temephos against *Ctenocephalides felis* on dogs and cats. Amer. J. vet. Res. 36 (9): 1281-1283.

Miller, B.E., Bennett, W.C., Graves, G.N. & Wheeler, J.R. 1975, Field studies of systemic insecticides I. Evaluation of phoxim for control of fleas on cotton rats. J. med. Ent. 12(4): 425-430.

Miller, B.E., Forcum, D.L., Weeks, K.W., Wheeler, J.R. & Rail, C.D. 1970, An evaluation of insecticides for flea control on wild mammals. J. med. Ent. 7(6): 697-702.

Mohan, B.N. 1960, A note on high DDT tolerance in rat fleas collected from Gundlupet town of Mysore State. Ind. J. Malariol. 14 (1): 19-21.

Mohan, B.N. 1962, Susceptibility status of rat fleas *X.cheopis* and *X.astia* to chlorinated hydrocarbon insecticides in some areas of Madras and Mysore States. Ind. J. Malariol. 16(3): 272-282.

Mohr, C.O. & Smith, W.W. 1957, Eradication of murine typhus fever in a rural area. Bull. Wld. Hlth. Org. 16: 255-266.

Morlan, H.R. & Hines, V.D. 1951, Evaluation of county-wide DDT dusting operations in murine typhus control, 1950. Publ. Hlth. Rpts. 66(33): 1052-1057.

Nicholson, H.P. & Gaines, T.B. 1948, A comparison of the effectiveness of 5 and 10% DDT dusts for the control of rat fleas. Publ. Hlth. Rpts. 63(5): 129-136.

Ogata, M. 1897, Ueber die Pestepidemie in Formosa. Centralblatt f. Bakteriol. Parasitenk. 1 Abt. 21: 769-777.

Olson, W.P. 1969, Control of oriental rat flea in Vietnam with diazinon dust. J. econ. Ent. 62(3): 656-660.

Pal, R. 1966, Present status of insecticide resistance in fleas. Unpublished WHO document WHO/VC/66.217: 205-206.

Patel, T.B., Bhatia, S.C. & Deobhenkar, R.B. 1960, A confirmed case of DDT-resistance in *Xenopsylla cheopis* in India. Bull. Wld. Hlth. Org. 23: 301-312.

Patel, T.B. & Rodde, S.T. 1952, Use of DDT as a plague control measure in the Bombay State. Ind. med. Gaz. May 1952: 217-223.

Pollitzer, R. 1954, Plague. WHO monograph No. 22, Geneva. 698 pp.

Pollitzer, R. 1966, Plague and plague control in the Soviet Union. The Inst. Contem. Russian Stud. Fordham Univ. New York. 478 pp.

Pollock, J.S. McK. 1948, Plague controlled in Haifa by the use of DDT alone. Trans. R. Soc. trop. Med. Hyg. 41(5): 657-658.

Roubaud, E. 1940, Emploi du fluorure de sodium dans la lutte contre les puces d'habitations. Bull. Soc. Path. Exot. 33: 96-99.

Ryckman, R.E. 1971, Plague vector studies IV. A comparison of insecticide application techniques for flea control on ground squirrels. J. med. Ent. 8(6): 671-674.

Ryckman, R.E., Ames, C.T. & Lindt, C.C. 1953, A comparison of aldrin, dieldrin, heptachlor and DDT for control of plague vectors on the California ground squirrel. J. econ. Ent. 46(4): 598-601.

Ryckman, R.E., Ames, C.T., Lindt, C.C. & Lee, R.D. 1954, Control of plague vectors on the California ground squirrel by burrow dusting with insecticides and the seasonal incidence of fleas present. J. econ. Ent. 47(4): 604-607.

Saenz-Vera, C. 1953, DDT in the prevention of plague in Ecuador. Bull. Wld. Hlth. Org. 9: 615-618.

Sant, M.V. & Renapurkar, D.M. 1971, Evaluation of susceptibility status of *X.cheopis* to DDT, malathion and dieldrin from Thena district. Pesticides, March 1971: 13-16.

Schulz, K.H. 1950, Control of plague in Taranto, Italy 1945/1946. An account of a successful programme of rodent extermination. Bull Wld. Hlth. Org. 2(4): 675-685.

Sen, P. 1958, Insecticide susceptibility tests on fleas. Bull. Calcut. Sch. trop. Med. 6(1): 14.

Sharma, M.I.D. & Joshi, G.C. 1961, A preliminary note on DDT resistance in rat fleas, *Xenopsylla cheopis* and *Xenopsylla astia* in a sprayed area of Delhi. Bull. Wld. Hlth. Org. 25(2): 270.

Shawarby, A.A. 1954, Laboratory and field trials in the control of fleas and lice. Bull. ent. Res. 44: 377-385.

Simeons, A.T.W. & Chhatre, K.D. 1947, Further observations on plague. Ind. med. Gaz. 82 (8): 447-451.

Simond, P.L. 1898, La propagation de la peste. Ann. Inst. Pasteur 12: 625-687.

Smith, A. 1959, The susceptibility to dieldrin of *Pulex irritans* and *Pediculus humanus corpus* in the Pare area of North East Tanganyika. Bull. Wld. Hlth. Org. 21: 240-241.

Smith, C.N. 1951, Compounds more toxic to fleas than DDT. Am. J. trop. Med. 31: 252-256.

Smith, C.N. & Burnett, D. 1948, Laboratory evaluation of repellents and toxicants as clothing treatments for personal protection from fleas and ticks. Am. J. trop. Med. 28: 599-607.

Stasiak, R.S., Grothaus, R.H. & Miner, W.F. 1970, Resistance of the Oriental rat flea, *Xenopsylla cheopis* to DDT in the Republic of Viet Nam. Bull. Wld. Hlth. Org. 42: 997-998.

Swellengrebel, N.G. 1913, Mededeeling omtrent Onderzoekingen over de Biologie van Ratten en Vlooien en over andere Onderwerpen, die betrekking hebben op de Epidemiologie der Pest op Oost-Java. Geneesk. Tijdschr. v. Ned. Ind. 53: 53.

Sweetman, H.L. 1946, DDT to control cat and dog fleas and dog lice. J. econ. Ent. 39(3): 417.

Thompson, J.A. 1903, Report to the Board of Health on the second outbreak of plague at Sydney, 1902. (with F.Tidswell). Government Printer, Sydney.

Tinker, I.S., Ivanov, I.Kh., Shiranovich, P.I. & Shishkin, A.K. 1959, Results and coming tasks of solving the problems of the fight against fleas as a radical method of plague prophylaxis. In Russian. Priodn. ochagovost. 1959: 305-317.

Verjbitzki, D.T. 1908, The part played by insects in the epidemiology of plague. J. Hyg. Camb. 8: 162.

Vashkov, V.I. 1956, Ruvovodstro po desinfektsii, disinsektsii i deratizatsii. (Handbook on Disinfection, Disinsectization and Deratization) Moscow, Medgiz second edit.

Viswanathan, D.K. 1949, A study of the effects of malaria and malaria control measures on population and vital statistics in Kanara and Dharwar Districts as compared with the rest of the province of Bombay. Ind. J. Malariol. 3(1): 69-107.

Voelckel, J. & Mouchet, J. 1958, Réduction des ecto-parasites du rat par les insecticides à effet remanent à Douala. Med. Trop. 18(6): 904-907.

Wagle, P.M. & Seal, S.C. 1953, Application of DDT, BHC and Cyanogas in the control of plague in India. Bull. Wld. Hlth. Org. 9: 597-614.

Waterston, J. 1936, Fleas as a menace to man and domestic animals. 3rd ed. revised by P.A.Buxton, London, British Museum, 20 pp.

WHO 1960, Insecticide resistance and vector control. Tech. Rpt. Ser. No. 191.

WHO 1970, Insecticide resistance and vector control. Tech. Rpt. Ser. No. 443.

WHO 1975, Instructions for determining the susceptibility or resistance of fleas to insecticides. WHO unpublished document, WHO/VBC/75.588 rev. 1.

WU, C.Y. 1936, Plague – a manual for medical and public health workers. Wu Lien-teh, Chun, J.W.H., Pollitzer, R. & Wu, C.Y., National Quarantine Service, Shanghai Station, China.

HARRY HOOGSTRAAL
United States Naval Medical Research Unit No. 3, Cairo, Egypt

THE ROLES OF FLEAS AND TICKS IN THE EPIDEMIOLOGY OF HUMAN DISEASES*

ABSTRACT

Fleas and ticks, both temporary obligate ectoparasites of vertebrates and often found together on a host, are poles apart epidemiologically. Flea larvae feed on dry blood or debris; pupae are inactive; adults take blood from the host rapidly and repeatedly. Ticks feed as larvae, nymphs and adults on the host for ca. half an hour to several days (depending on stage and species) and imbibe much more blood than fleas. No agent of a human disease is perfectly (symbiotically) adapted to life-long co-existence with fleas. Many agents acquired in a bloodmeal probably perish when large quantities of quickly drying blood are excreted by adult fleas; these agents have little or no chance to invade and to multiply in the hemolymph, salivary glands, reproductive organs, and other tissues of the adult flea. The physical, physiological and biochemical environments in the bodies of the immature and adult stages of numerous tick species are favourable for the survival and multiplication of a large number of variety of viruses, rickettsiae, bacteria and protozoa imbibed with the large quantities of blood consumed and concentrated within the body, and the environment in the tick egg is often favourable for the survival of these agents. The common phenomena of long transstadial and interseasonal survival of agents in the tick body, of transovarial transmission to subsequent tick generations, and of transmission to a variety of new hosts with saliva injected into the feeding site are significant factors enhancing both the reservoir role and the vector potential of many tick species. These phenomena do not occur in fleas.

Fleas and ticks are both temporary obligate ectoparasites of vertebrates. There are some 1,900 taxons of fleas in six or more families but only 800 taxons of ticks in two major families and one monotypic family. The laterally compressed bodies of fleas possess distinctive combs functionally associated with the host feathers or fur pattern. The dorsoventrally compressed bodies of ticks possess, on their appendages, spurs, projections and/or emarginations to assist in their passage through dense feathers or fur toward a favourable site for obtaining a bloodmeal. However, the bloodmeals are subject to different physical

* From Research Project MR041.09.01-0152, Naval Medical Research and Development Command, National Naval Medical Center, Bethesda, Maryland, USA. The opinions and assertions contained herein are the private ones of the author and are not to be construed as official or as reflecting the views of the Department of the Navy or of the naval service at large.

and physiological processing; infectious agents entering these two groups of arthropods with the blood experience vastly contrasting environments. Fleas and ticks may feed together on the same host but are poles apart epidemiologically.

The factors responsible for this epidemiological contrariety begin with the fundamentally dissimilar larval lifestyles and food. Flea larvae feed on debris or on blood excreted by adults; pupae are inactive; only adult fleas are ectoparasites. Ticks are obligate ectoparasites in three postembryonic stages. The tick larva, nymph and adult each feeds on warm circulating blood; the argasid larva and each of the three ixodid stages feed once; the argasid nymph and adult each feeds several times.

Let us postulate that a flea larva *might* become infected when feeding on infected, freshly defecated blood of the adult flea. Can human disease-causing pathogens within the flea larva survive the larval-pupal molt and the pupal-adult molt to live in the adult flea? The answer is generally no. In holometabolous insects such as fleas, the extensive internal changes during molting appear to be deleterious to the survival of most micro-organisms causing human disease. In molting ticks, only salivary gland alveoli are completely replaced and ectodermal derivatives and certain muscle groups undergo histolysis. Most internal cells and organs change gradually throughout the tick's lifetime; thus transstadial survival of an infectious agent from larva to nymph to adult is a common phenomenon in tick-associated viruses, rickettsiae, bacteria and protozoa. The agent originally acquired by the tick larva or nymph may be transmitted to either similar or dissimilar kinds of postlarval or postnymphal hosts. On the other hand, only the adult flea is initially infected with an agent causing disease in humans. During subsequent bloodmeals the flea can only disseminate the agent among the same host group or among a very limited variety of hosts. An exception to this rule is the rodent tapeworm, *Hymenolepis diminuta,* whose eggs, passed in rodent (or rarely, in human) faeces and ingested by flea larvae, develop to the cysticercoid stage and may persist to the adult stage of the flea . . . and be infective when the flea is accidentally ingested.

Parasitism is temporary in both arthropod groups . . . but it is more temporary in most adult fleas than in larval, nymphal, and adult ticks. Adult fleas (except *Tunga*) feed repeatedly but rapidly, in a matter of a few minutes. Engorgement by immature and adult ticks requires at least 30 to 60 minutes (argasid nymphs and adults and some argasid larvae) or several (three to seven) days (immature and adult ixodids and some larval argasids). Ticks are larger than fleas and imbibe enormous quantities of blood from dermal and subdermal pools formed after cytolytic and anticoagulant substances are injected in the tick's watery saliva. Consequently the few minutes of flea feeding versus the hour to days of tick feedings, and the comparative quantities of the blood meals, directly and indirectly affect the intake of infectious agents from the host. The fate of the agent in the body of the flea and in the body of the tick is another story.

The adult flea and each stage of the tick are programmed to imbibe more blood than required for their own life and reproductive processes. The flea rapidly defecates much excess blood, which quickly dries. Many infectious agents (an exception: *Rickettsia mooseri*) perish during the defecation and drying processes. The tick produces relatively little faecal material but eliminates much excess water from the blood through coxal glands (argasids) or through saliva (ixodids). This concentration of the tick bloodmeal does not appear to alter the favourable environment in which an agent may survive, penetrate and multiply in haemolymph and tissues, and migrate to salivary glands and reproductive organs. A large number and variety of viruses, rickettsiae, bacteria and protozoans are transmitted via ixodid and argasid saliva to the vertebrate host. In the flea, on the other hand, infectious agents exist and multiply only in the intestinal tract. These agents never reach the flea salivary glands; they also never reach the reproductive organs. Transovarial transmission therefore has no role in the epidemiology of flea-associated diseases. Conversely, tick larvae often receive viruses, rickettsiae, babesias, theilerias and spirochetes through transovarial transmission from the infected mother; these 'inherited' agents are passed both to subsequent developmental stages of the tick and to vertebrate hosts.

Several flea species parasitize commensal rodents and domestic dogs, cats and fowls, and also feed on humans, but few others find the opportunity to do so. Fleas are primarily adapted to and most successful as rodent parasites; practically no flea species have become adapted to herds of domestic ungulates (cattle, buffalo, sheep, goats) and very few are associated with wild, wandering ungulates. Ticks parasitize the same vertebrate group as fleas but numerous species also thrive on wild and domestic ungulates and are transported far and wide by these hosts. Ticks quest for hosts from vegetation or from the ground surface in forests, steppes, savannas, semideserts, pastures and stables and other buildings, where they may encounter and feed on humans. Therefore, chances for human-tick contact are more varied, and often more numerous, than those for human-flea contact.

As Professor Krampitz reports in this Conference, fleas are hosts of certain trypanosomes (subgenus *Herpetosoma*), of *Grahamella* (Bartonellaceae), and of intraleucocytic hemogregarines (*Hepatozoon,* etc.) infecting small mammals. They also harbour cysticercoids of tapeworms of the domestic dog and wild insectivores and rodents. We have already mentioned the relatively minor role of fleas in the epidemiology of human infections by *Hymenolepis diminuta.* The interrelationships between fleas and pathogens of human and animal diseases were recently reviewed (Bibikova, 1977).

Among the rickettsiae causing human disease, fleas maintain *Rickettsia mooseri (= R. typhi*), the agent of murine or epidemic typhus. However, the relationship between *R. mooseri* and fleas is not biologically perfect, it is one of happenstance. After imbibing the rickettsiae from an infected rodent (and subsequent multiplication of the agent in the gut), the flea voids rickettsiae in its feces, which contaminate abraded or broken host skin or may be inhaled or ingested to cause infection in vertebrates. Conversely, the biological relationships between many tick species and their associated rickettsiae are close to perfect; rickettsemic hosts may not be necessary for stage-to-stage and generation-to-generation survival of the agent in the tick; rickettsial transfer with salivary fluids while feeding is the chief, though not the only, route of transmission to the vertebrate. The chief diseases of humans caused by tickborne *Rickettsia* are Rocky Mountain spotted fever *(R. rickettsi)* of North and South America, boutonneuse fever or tick typhus *(R.conori)* of Africa and parts of Asia, and Siberian tick typhus *(R.siberica)* of temperate Asia and possibly of eastern Europe. *Coxiella burneti,* the agent of Q fever, may represent a tick-associated rickettsia that is more often spread among humans by contaminated animal tissues and products than by tickbite.

I know of no fleaborne arboviruses, though fleas are occasionally reported to be casual hosts of tickborne viruses in the Soviet Union. More than 100 kinds of arboviruses have already been recorded from 115 species and subspecies of ticks. Many of these are true tick viruses; some may be vertebrate viruses surviving in the tick bloodmeal when tested in the virological laboratory. Tick viruses have no other known arthropod vectors and do not multiply in insect tissue cell lines. An interesting example of a truly tickborne agent is Crimean-Congo haemorrhagic fever (CCHF) virus, which has been isolated from 25 species and subspecies of ticks in Asia, Europe and Africa. CCHF virus circulation is enzootic in innumerable foci inhabited by various tick species but may become epizootic in years of high population densities of two or more species of the genus *Hyalomma.* Transstadial and interseasonal survival of the virus in ticks appears to be the rule and transovarial transmission has been demonstrated in *Dermacentor, Rhipicephalus* and two forms of the genus *Hyalomma.* Young rodents, insectivores, leporids and domestic mammal hosts of ticks serve as virus amplifiers. Birds are important hosts and disseminators of some of the chief vector species but do not become viremic with CCHF virus.

Among the other tickborne viruses known or believed to cause illness in humans are the Russian spring-summer encephalitis complex, Powassan and Kyasanur Forest diseases (Togaviridae, Flavivirus), Colorado tick fever (Reoviridae, Orbivirus), Ganjam, Dugbe and Thogoto (Bunyaviridae), and Quaranfil, Soldado, Zirqa, Punta Salinas, Dhori and Tettnang (ungrouped).

The natural history of the plague agent, *Yersinia pestis,* a bacterium which reproduces in the flea intestine and blocks the proventriculus, is well known to all. This pathogen-vector interrelationship pattern is unusual in the epidemiology of arthropodborne disease. Regurgitation of agents in tick saliva has been suggested as possibly a secondary means of transmitting certain tickborne agents, but not those of primary medical interest. Several tick species have been incriminated as reservoirs and vectors of *Y.pestis* from either circumstantial or preliminary experimental evidence in China and the Soviet Union; more precise data are required to make the claims for these phenomena scientifically acceptable.

Francisella tularensis, the bacterium causing tularemia, survives for some days, without reproducing, in the flea digestive tract and may be transferred mechanically when the flea bites or passes feces. However, the flea role in tularemia epidemiology is slight. In North America, *F.tularensis* reproduces in at least six tick species; it survives from one developmental stage to the next and is transmitted in saliva when the tick bites. Transovarial transmission occurs in *Haemaphysalis leporispalustris* and possibly also in other tick species.

All functional and biological factors in the roles of fleas as reservoirs and vectors of agents of human diseases differ distinctly from those of ticks.

REFERENCE

Bibikova, V.A. 1977, Contemporary views on the inter-relationships between fleas and the pathogens of human and animal diseases. Ann. Rev. Ent. 22: 23-32.

DAN C. CAVANAUGH & JAMES E. WILLIAMS
Walter Reed Army Institute of Research, Washington, DC, USA

PLAGUE: SOME ECOLOGICAL INTERRELATIONSHIPS

ABSTRACT

Plague persists in wild rodent populations in extensive, essentially permanent natural foci. While massive rodent surveys have not been continuous, occasional cases in human beings serve as sentinels, attesting to the suitability of various regions for the perpetuation of the disease.

Despite intensive studies, many problems remain. Treatment is not always effective; serious complications may occur or some patients succumb. Vaccination may not prevent chronic disease. Variant *Yersinia pestis* or atypical situations may present difficulties in diagnosis and/or surveillance. Resistance of flea vectors and rodent hosts to pesticides complicates control.

Plague, the black death of history, results from infection with the plague bacillus, *Yersinia pestis.* The disease is deadly. Classical *Y.pestis* organisms are extremely infectious and may be transmitted through contact with the broken skin, mucous membranes, ingestion, or inhalation (58, 72, 83). Bacilli have been isolated from fleas, lice, ticks, many genera of small mammals, and burrow detritus (67, 88, 95). 'Blocked' fleas are undeniably the most significant mode of transmission of bubonic plague (16, 25, 39, 72).

Natural plague foci – Plague is essentially a disease of rodents, and reservoirs of the infection include many genera of wild rodents infested with their individual species of fleas (3, 55, 72, 74, 75, 77, 78, 88). Biological and physical environments in certain locales are suitable for the maintenance of chronic, enzootic plague. These essentially permanent foci of wild rodent plague are classified as 'natural plague foci' (50, 95). The global distribution of areas known or suspected to be natural plague foci is impressive (95).

Persistence of natural plague foci – The persistence of natural plague foci and the likelihood of increased human contact with the plague bacillus with development schemes or other activities has been recognized internationally (95) and is supported by reports from widely distributed natural foci in the western United States (5, 6, 53, 88). Table 1 presents the data on cases of plague resulting from probable contact with wild rodents or their fleas that have occurred in the United States since 1920; the decade in which sylvatic plague was first recognized as a distinct entity (46). By definition, sylvatic or wild plague is a rural disease. Although the rural populations of the nine states listed in table 1 (7) are

Table 1. Plague cases of probable sylvatic origin reported in the United States: 1920-1976

State	Decade 20-29	30-39	40-49	50-59	60-69	70-76	Cases	% total
Arizona				1	3	9	13	10
California	13	7	5	3	1	6	35	26
Colorado				1	3	2	6	4
Idaho			1		1		2	1
Nevada		1					1	1
New Mexico			3	4	21	40	68	51
Oregon		1				2	3	2
Texas					1		1	1
Utah		2			1	2	5	4
Cases	13	11	9	9	31	61	134*	
% total	10	8	7	7	23	45		
Cases/year	1.3	1.1	0.9	0.9	3.1	8.7		
Rural population (thousands)	6 247	7 150	7 939	7 432	6 981	6 553		

* Excludes laboratory infections and urban plague.

Table 2. Age and sex distribution of sylvatic plague cases reported in the United States: 1920-1976

Age	Male	Female	Total	% total
0 - 4	4	9	13	11
5 - 9	11	12	23	19
10 - 14	17	13	30	25
15 - 19	13	2	15	12
20 - 24	5	3	8	7
25 - 29	5	2	7	6
30 - 34	2	3	5	4
35 - 39	6	1	7	6
40 - 44	–	–	–	–
45 - 50	2	2	4	3
> 50	6	3	9	7
Total	71	50	121	
% of total	59	41		100

rather constant, the average number of cases per year shows a rise in incidence. These data may be somewhat conservative. Several plague cases are known to have been misdiagnosed as tularemia (84). Young children appear to be principal victims (table 2), with exposure in the vicinity of the home. The geographic distribution of the cases (fig. 1) shows two pronounced natural foci: one along the Pacific Coast and associated inland mountain ranges; and, one in the general area of the Rocky Mountains. While all regions do not demonstrate the same degree of activity, sporadic cases occurring over the years attest to the persistence of the disease in many of the areas shown. While the total number of cases in the United States remains small, the data may reflect conditions elsewhere in the world.

Potential for human disease – Persistent natural plague foci present a definite, but unpredictable potential for outbreaks of human disease. Obviously, residents in natural foci are

Figure 3. Synecology of a Vietnamese plague focus: abundant, infected rats and fleas; favourable climate; close association with human beings.

Figure 5. An adverse effect of precipitation. Plague infected field rodents are driven into closer association with man.

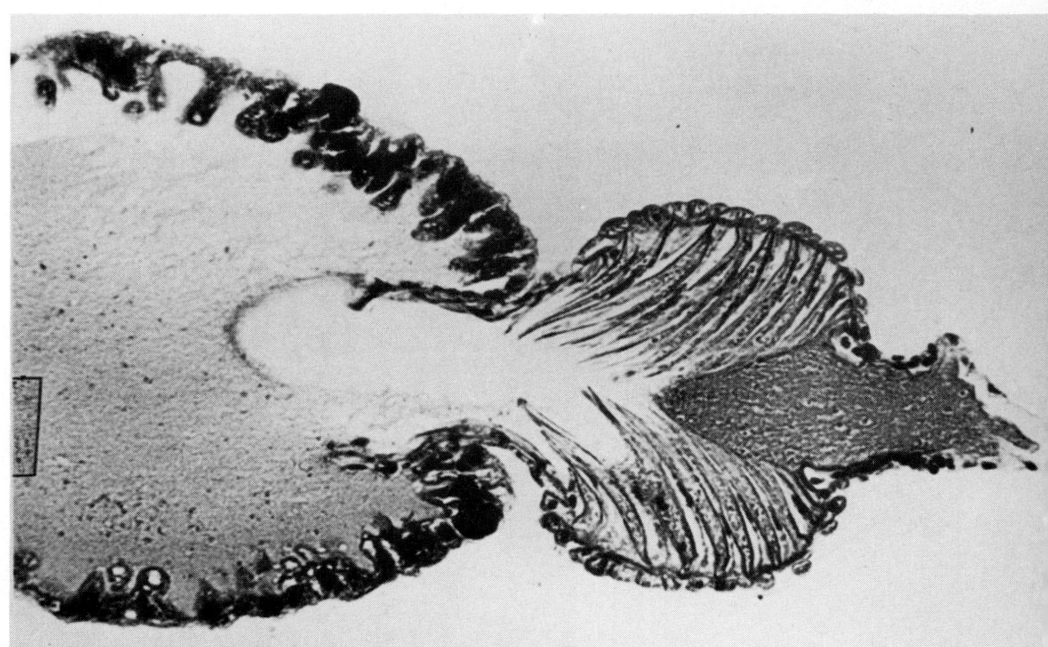

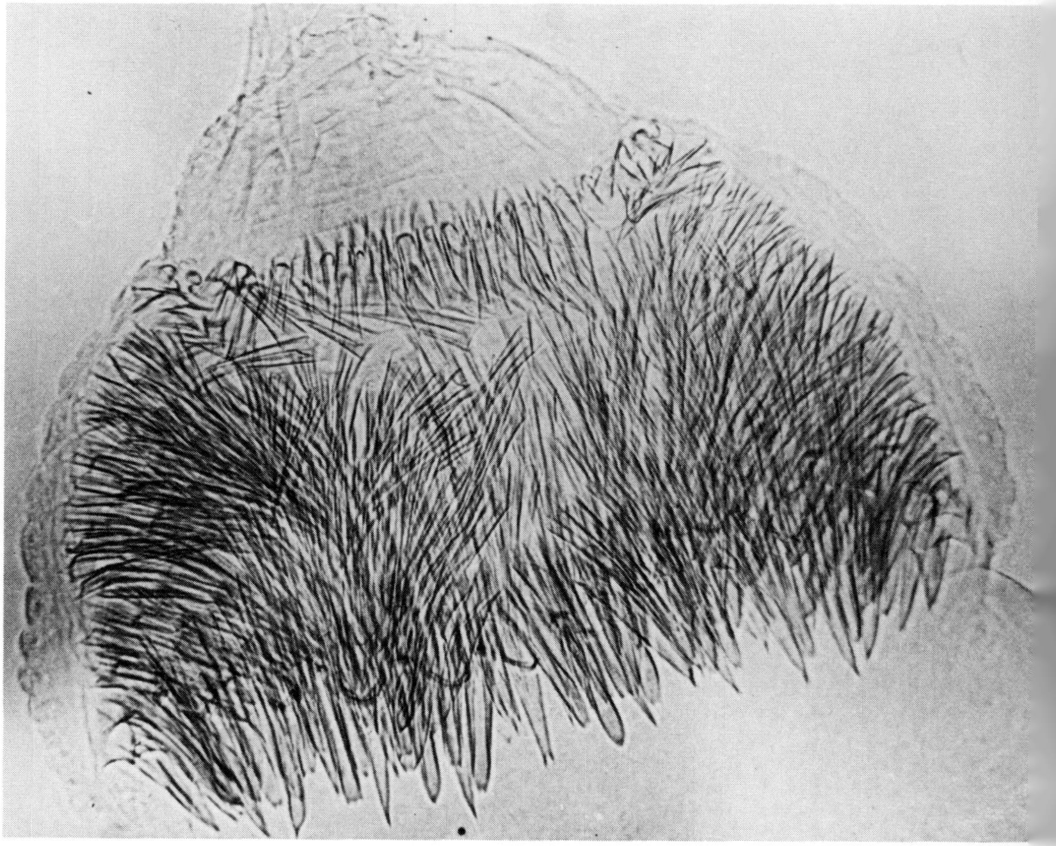

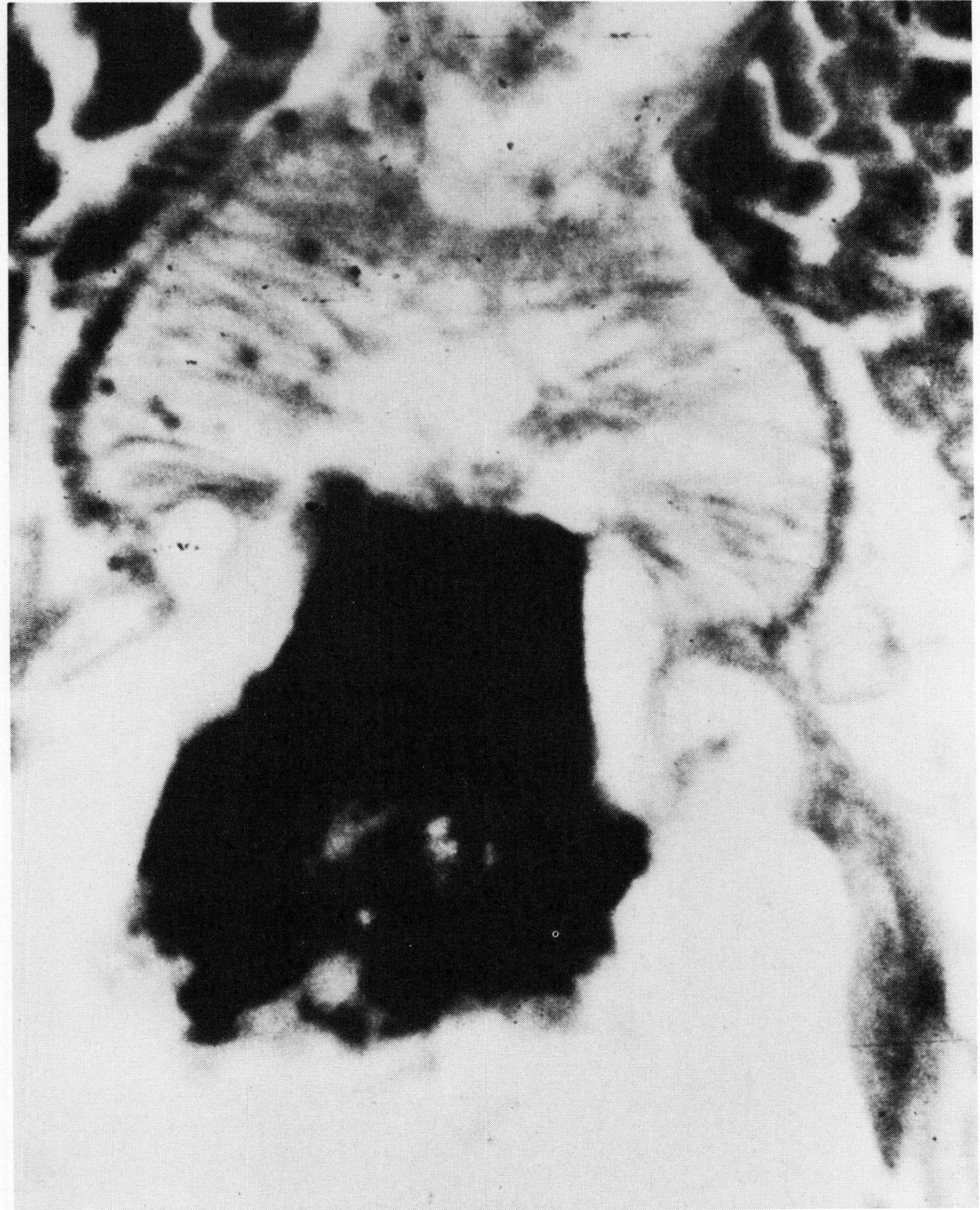

Figure 8. An infected thrombus attached to proventricular spines of *X.cheopis:* longitudinal section. Source: reference 25.

Figure 6. A longitudinal section through the stomach and proventriculus of *X.cheopis.* Source: reference 25.

Figure 7. An isolated proventriculus of *X.cheopis.* Source: reference 25.

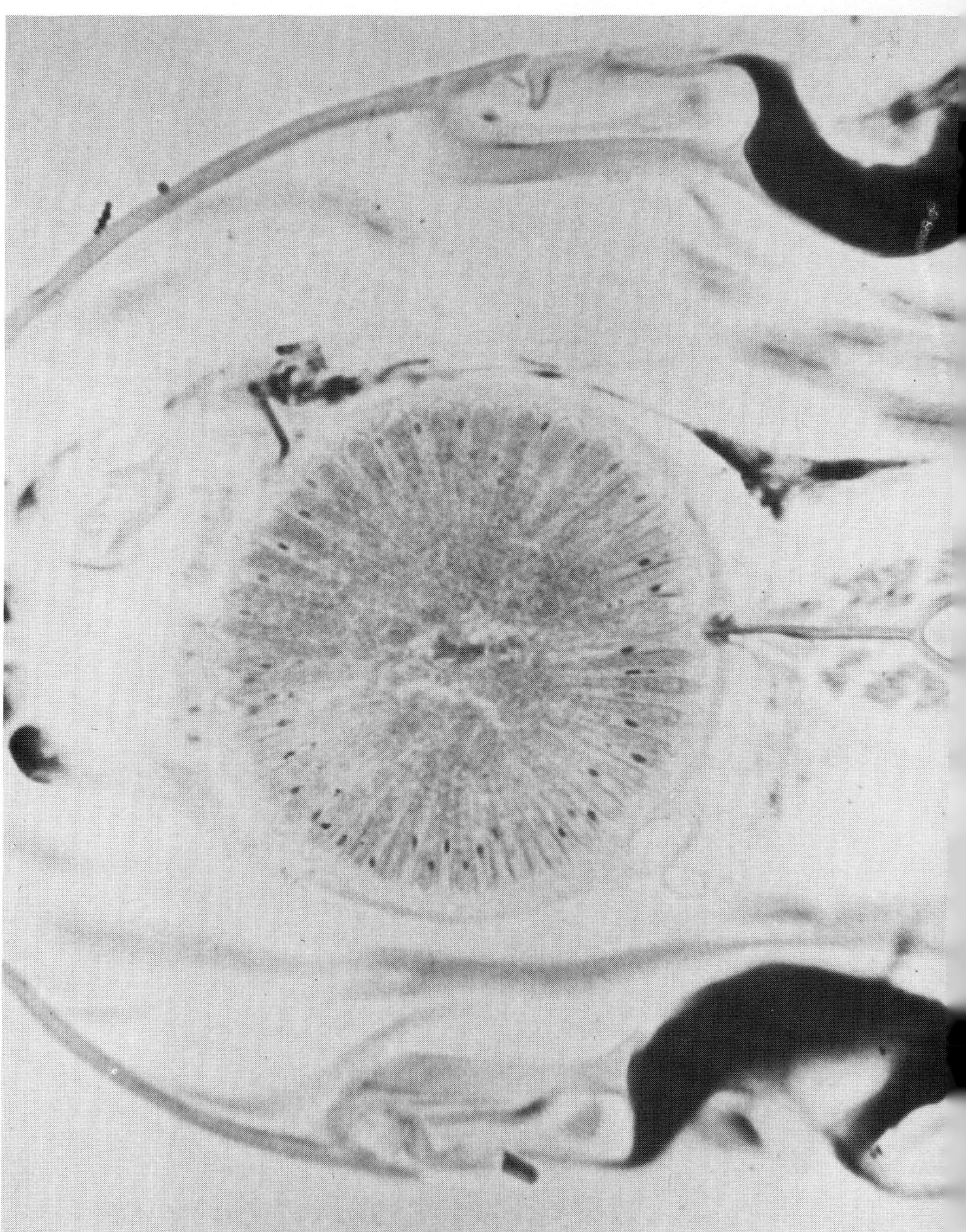

Figure 9. A cross-section of *X.cheopis* proventriculus 'blocked' with *Y.pestis.* Source: reference 25.

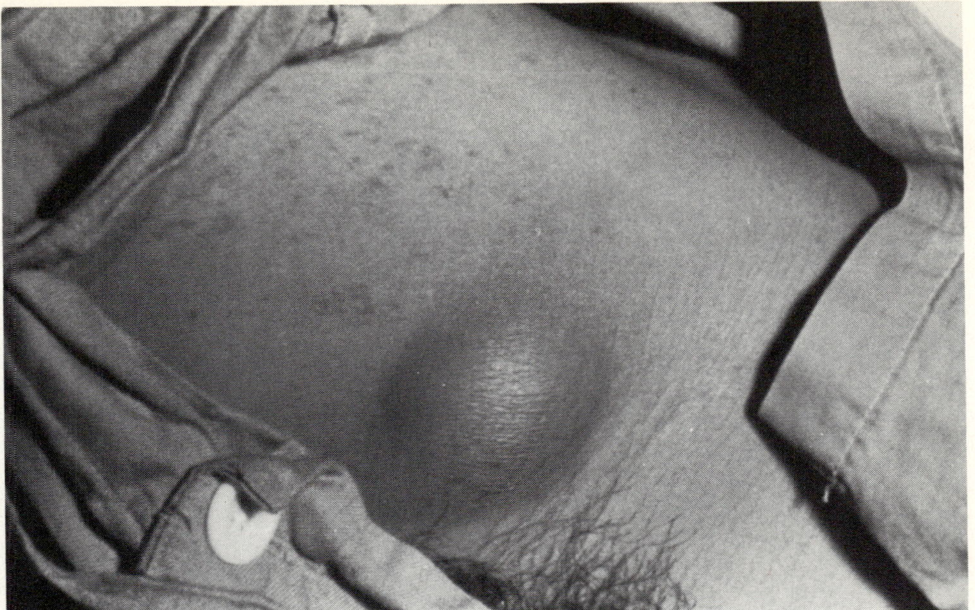

Figure 11. The enlarged, exquisitely tender inguinal bubo of bubonic plague.

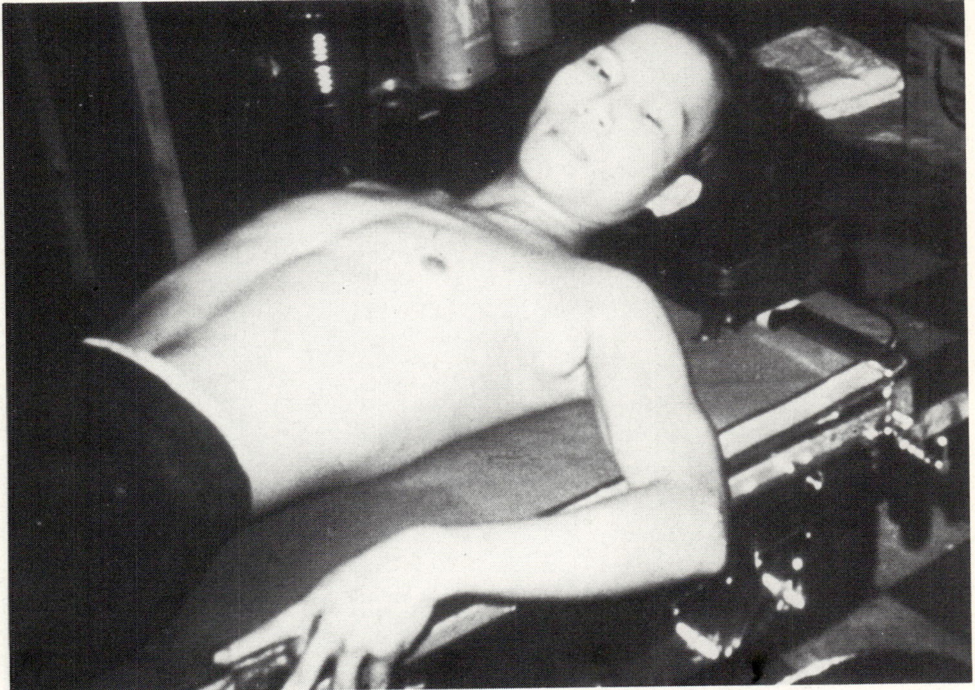

Figure 12. An acutely ill, apprehensive bubonic plague patient with an axillary bubo.

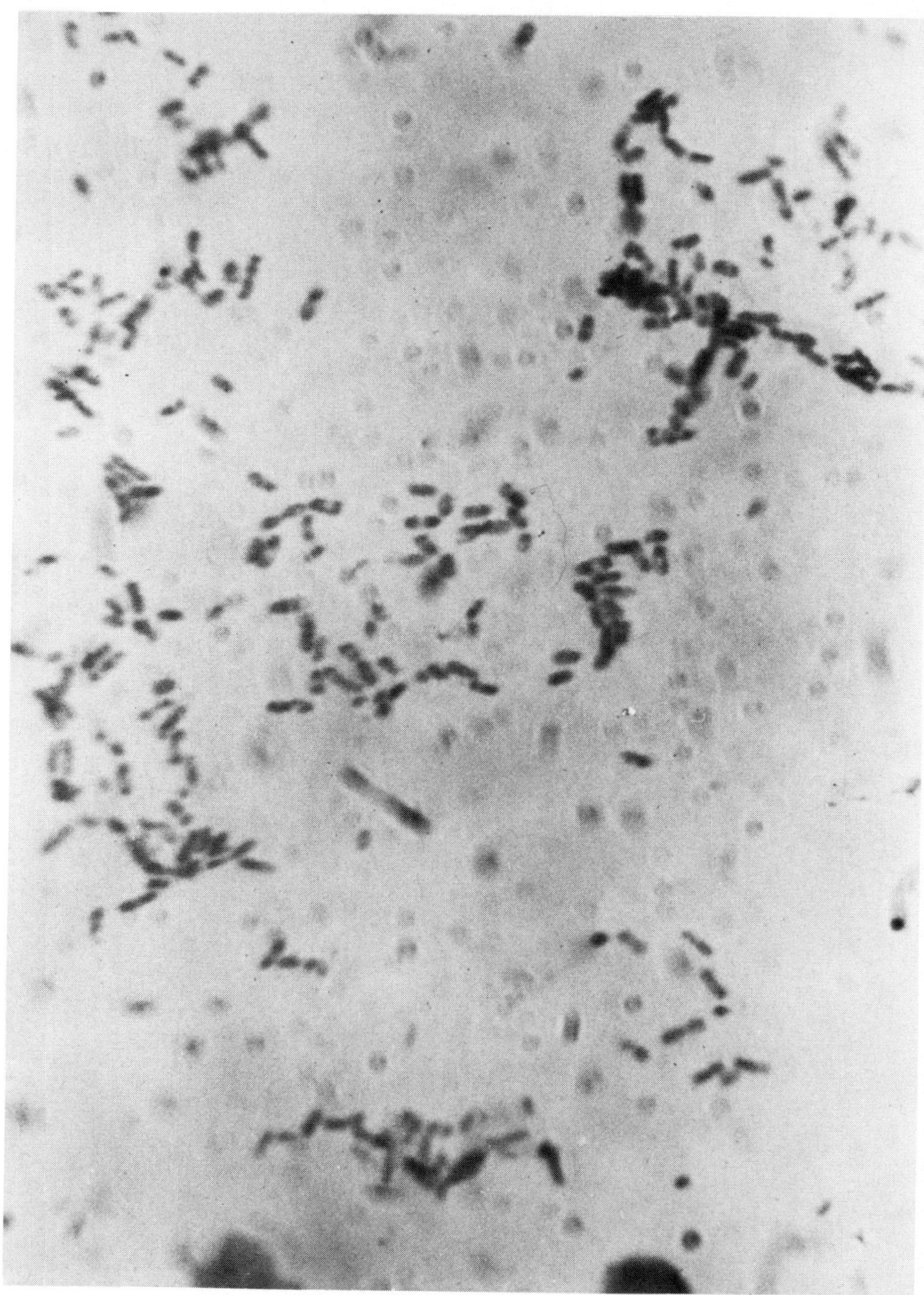

Figure 13. Smear from the peripheral blood of a moribund plague patient. Abundant bi-polar plague bacilli, stained with Wayson's stain are present.

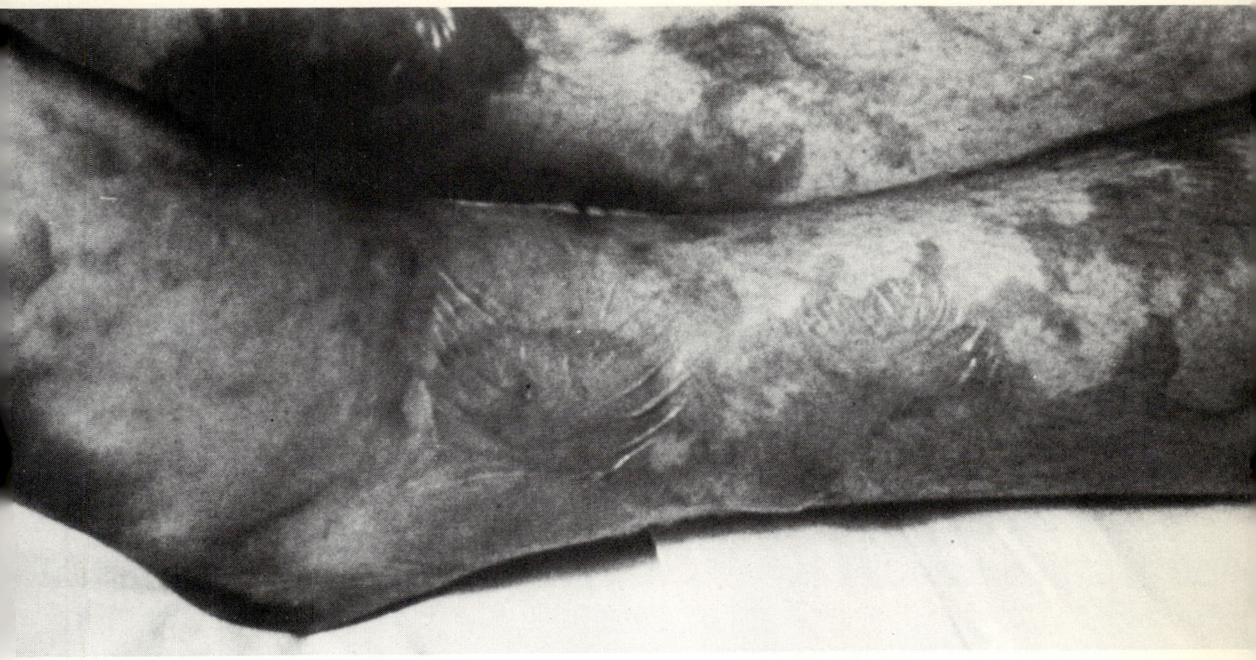

Figure 14. Plague purpura developed in patient surviving the infectious phase of the disease.

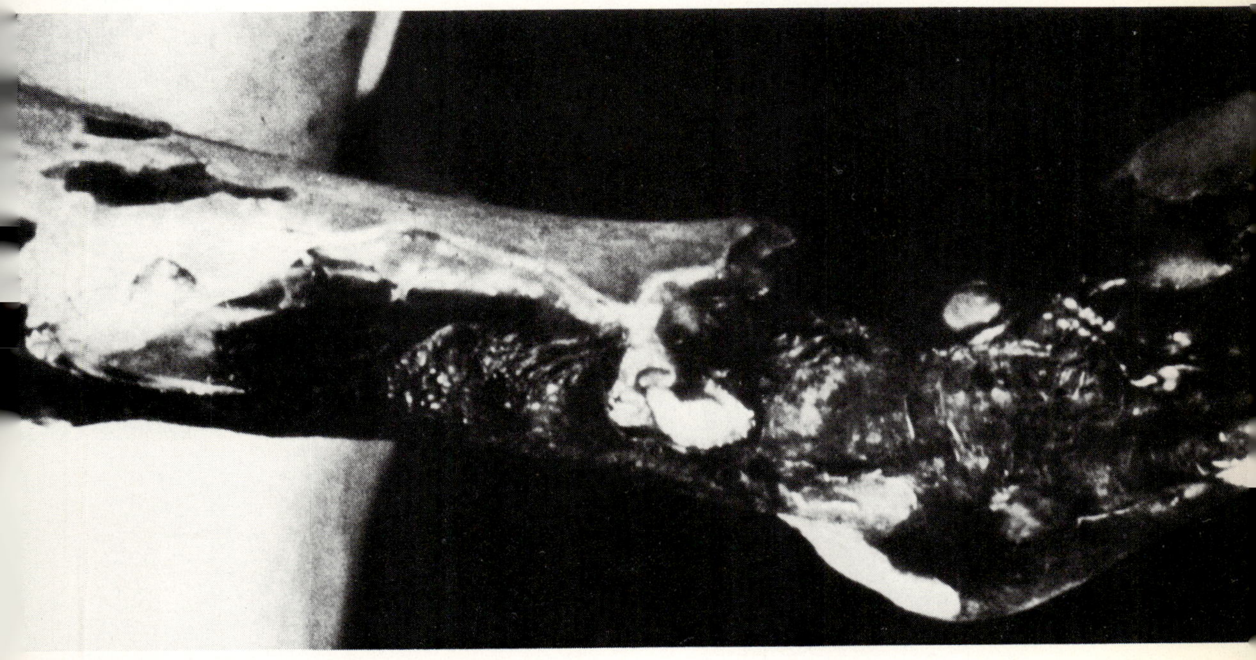

Figure 15. Frank gangrene following plague purpura.

Figure 16. Bloody sputum of a pneumonic plague patient. The specimen teems with *Y.pestis* in their most virulent form.

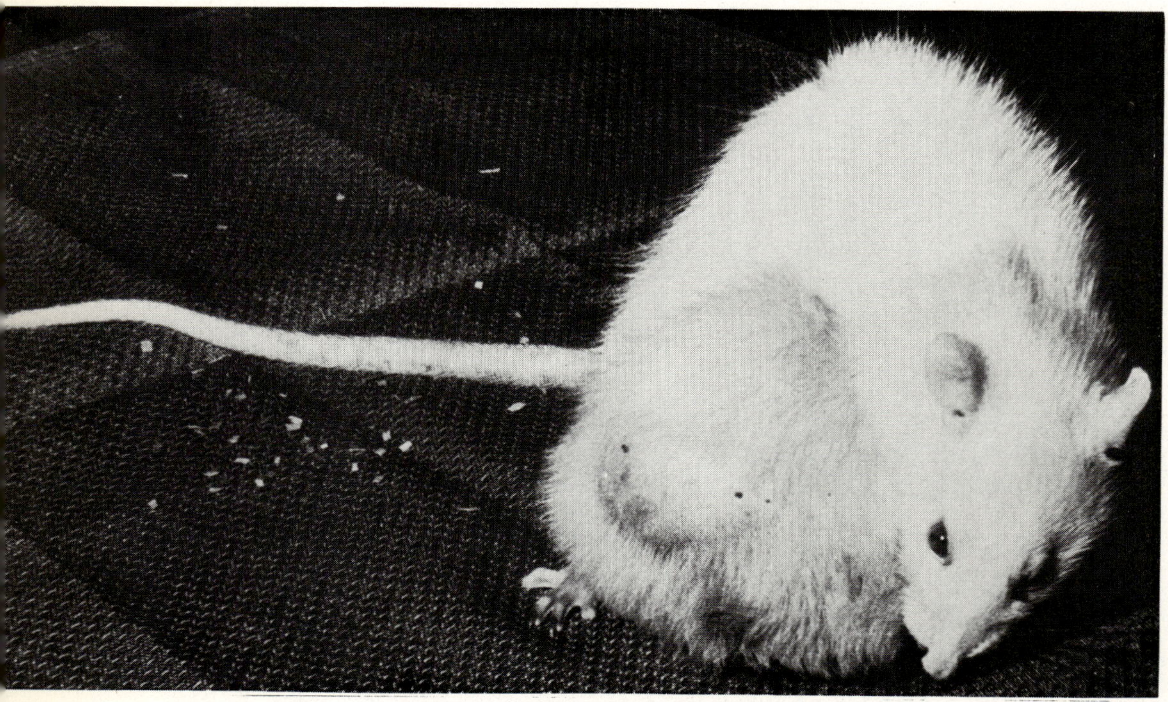

Figure 17. *R.norvegicus* with chronic plague infection. Animal had survived for 24 months post infection. At autopsy, lesions contained virulent *Y.pestis.*

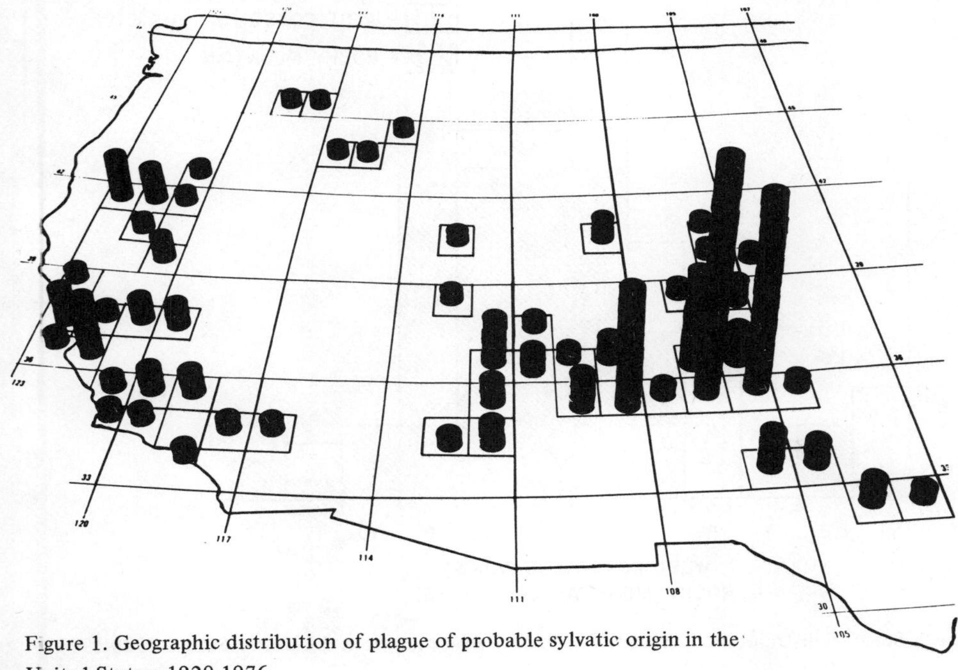

Figure 1. Geographic distribution of plague of probable sylvatic origin in the United States: 1920-1976.

at some risk. Seasonal distribution (fig. 2) of the US cases shows that the threat is present throughout the year. The distinct peaks are characteristic of flea-borne plague (29). 'Off-season' cases may illustrate the potential dangers of handling dead or infected animals (71, 86). Actual contact with natural foci, however, may be unnecessary. A primary danger to man occurs when sanitary conditions in urban areas near natural foci support large concentrations of commensal rats infested with many efficient flea vectors (fig. 3). Introduction of the plague bacillus into such an environment may result in serious epidemics of bubonic plague (66). Further, exposed individuals, developing pulmonary complications, may utilize modern transport, especially air, and initiate explosive outbreaks of inter-human pneumonic plague at their destinations.

Favourable conditions for epidemics of bubonic plague – Certain conditions enhance the possibility of epidemics of bubonic plague in a given locale. Among these are: adequate rodent and flea populations and a suitable climate. Obviously, as greater numbers of rodents and fleas become involved in the epizootic plague cycle, the potential for human exposure becomes greater.

Climate has two influences upon the prevalence of bubonic plague: 1) climate controls the development and survival of fleas; and 2) climate regulates the efficiency with which fleas transmit the plague bacillus (9, 59, 60). For example, the seasonal aspects of bubonic plague in the coastal lowlands of Vietnam are shown in fig. 4. In this area, as in many others, *Xenopsylla cheopis* (29) is the vector, and the population curve for *X.cheopis* closely follows the curve of the epidemic. Humidity, shown as vapour pressure deficit, reflects the requirements of *X.cheopis* larvae for a relative humidity in excess of 65 per

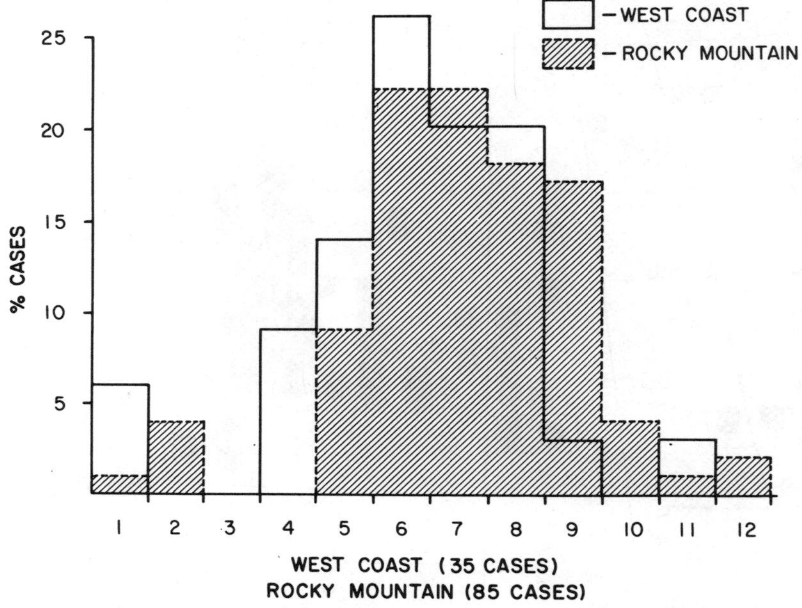

Figure 2. Seasonal distribution of plague of probable sylvatic origin in the United States: 1920-1976.

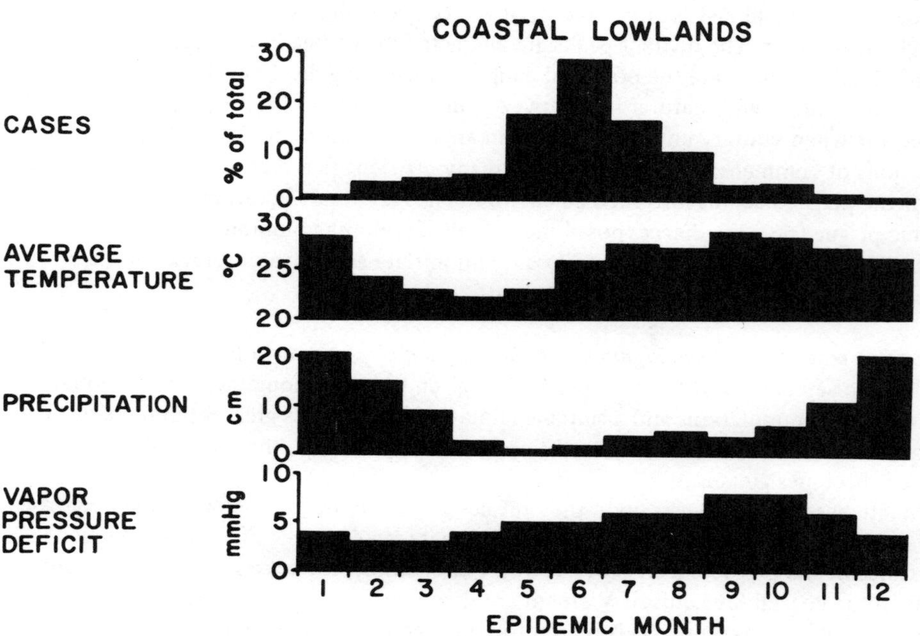

Figure 4. Seasonal aspects of bubonic plague prevalence in the Coastal Lowlands of Vietnam. Data are tabulated to show the sixth month as the month of peak plague prevalence. Source: reference 29.

cent for satisfactory maturation (49). Further, vapour pressure deficit in excess of 7.62 m Hg shortens the life span of adult fleas (9). Excessive precipitation is also harmful to fleas (69). Although the thermal deathpoint for *X.cheopis* approximates 40°C (62), and fleas feed readily at temperatures ranging from 8° to 30°C (47), the ability of fleas to become blocked with *Y.pestis* (47) and to transmit the agent (47, 59, 60) declines as temperatures exceed 27.5°C, corresponding to the relationship between plague prevalence and temperature (29) shown in fig. 4.

It is even possible to classify the severity of bubonic plague epidemics on the basis of mean annual temperature: over 94.8 per cent of cases occur in areas with mean annual temperatures in excess of 13°C (79). Data from over 30 large epidemics that have occurred in various places throughout the world, indicate that bubonic plague occurs with greatest intensity in areas with mean annual temperatures around 24-27°C, with pronounced peak months, some months having as many as 20 per cent of the year's total cases. Where mean temperatures exceed 27°C, far fewer cases occur.

The correlations between humidity, precipitation, temperature and plague prevalence seen in the lowlands of Vietnam are similar to those observed in India by members of the Plague Commission early in this century (17, 37,80). However, infected rodents bearing infected fleas may be driven from fields into houses in some areas (fig. 5). For example, in Java, plague transmission has occurred throughout the year because conditions remain favourable in houses for fleas even when the external climate is adverse.

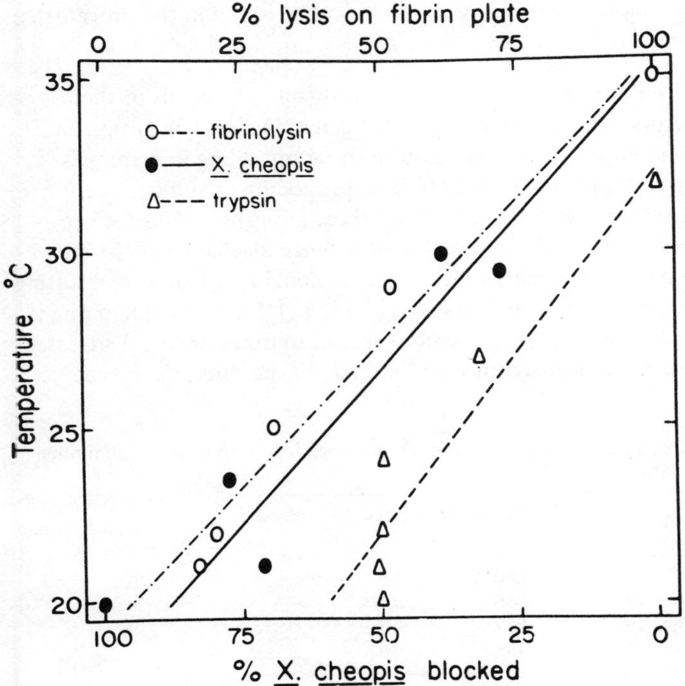

Figure 10. Temperature dependent relationships of fibrinolysis produced by trypsin and *Y.pestis* with respect to the blocking of *X.cheopis*. Source: reference 25.

The bionomics of many potentially important flea vectors are unknown, as is the avidity of these vectors for human blood meals. The complex nature of many flea-host relationships is illustrated by the requirement of certain fleas for the hormones found in the blood of a specific host for the perpetuation of the species (81).

Flea transmission – Blocked fleas are biological vectors of *Y.pestis* (8) transmitting the agent from rodent to rodent, from rodents to other animals, and from rodents to man. The blocking process in *X.cheopis,* identified as a major vector in the pioneer investigations of N.C.Rothschild (59), begins when fleas feed upon moribund, bacteremic rodents. Ingested plague bacilli accumulate and reproduce on the spines of the proventriculus located at the esophageal entrance to the stomach (fig. 6). The spines (fig. 7), referred to as a clot strainer by Wasserburger (quoted by Bibikova (14) are well adapted to snare and retain infectious thrombi until multiplication of *Y.pestis* completely or partially occludes the proventriculus (fig. 8), resulting in an infective flea (fig. 9). When a blocked flea attempts to feed, the powerful voluntary musculature utilized in sucking blood from either capillaries (31) or a wound produced by the flea's mouth parts, impacts blood against the mass of bacilli in the proventriculus and dislodges *Y.pestis,* which enter the tissues of the host, resulting in a new infection.

Several virulence factors or antigens of *Y.pestis* are associated with the blocking process. The virulence factor P^+ (44) produces growths of *Y.pestis* that are more dense and cohesive than those of P^- bacilli, and are thus better adapted for retention in the proventriculus (15, 16). A fibrinolytic factor (96) of the plague bacillus rapidly destroys fibrin at elevated temperatures when the capacity of fleas to block and transmit *Y.pestis* is also declining (fig. 10); thus reducing the probability for retention of bacilli in the proventriculus (25).

Pathogenesis – The plague bacilli introduced into a new host lack, as a result of their incubation in the poikilothermic flea, the capsular F1 antigen (11) and V/W antigens (19, 20) which require incubation temperatures more akin to those prevailing in mammals. While devoid of these antigens, *Y.pestis* are susceptible to phagocytosis and destruction (26). However, some bacilli escape (26, 45) and develop these antigens within a short time, becoming resistant to phagocytosis, a primary host defence mechanism (26). As infection progresses, plague bacilli are concentrated in the regional lymph nodes, resulting in the formation of large, exquisitely tender buboes (figs. 11, 12). Further overcoming the host defences, plague bacilli progress by hematogenous spread to other organs. Untreated mortality rates of 75 per cent are not uncommon in bubonic plague. Bubonic plague

Table 3. Comparative mouse LD_{50} data for three strains of *Yersinia pestis* freshly isolated from pneumonic patients (Arizona or New Mexico, 1975)

Strain	Date	Culture incubated 18 hours at: 25°C*	37°C**
751207	June 1975	390***	9
752675	September 1975	166	19
BAC001884	September 1975	157	3
195/P (Control)		383	18

* flea temperature simulation
** pneumonic plague temperature simulation
*** number of plague bacilli per LD_{50} dose

occurs in all degrees of severity and in some circumstances patients first seen late in the course of the disease may show frank bacteremia (fig. 13), a grave prognostic sign. In some patients, the disseminated intravascular coagulation observed in experimental animals (35, 48) may present a problem (22, 23, 34). In extreme cases, it may lead to purpura and frank gangrene (figs. 14, 15) resulting in loss of extremities (68). Disseminated intravascular coagulation probably results from the actions of either the *Y.pestis* coagulase (33) or endotoxin (4,87). A guarded prognosis is always advisable, as some patients may succumb even when apparently recovering (70, 72).

In some instances, the infection involves the respiratory system (57, 70, 72) and the patient may transmit plague bacilli via aerosol or fomites to his contacts, initiating a transmission chain of man-to-man pneumonic plague (fig. 16). In the case of plague pneumonia, the *Y.pestis* involved have already achieved their most virulent antigenic status and fewer bacilli are required to initiate a fatal infection (table 3). Untreated plague pneumonia is almost invariably fatal.

The epidemiology of primary pneumonic plague is not well understood. Most severe epidemics have occurred in areas where the climate is relatively cool. Primary plague pneumonia is rather rare in the hot tropics, even in the presence of fulminating epidemics of bubonic plague, since high temperatures and humidities are unsuitable for the survival of *Y.pestis* in aerosol clouds. On the other hand, low relative humidity is also associated with the rapid death of plague bacilli in aerosols (18). Aerosolization per se is damaging to *Y.pestis*, increasing both the numbers of bacilli required to infect a host and the time required to multiply on artificial media (13). These deleterious effects may be somewhat overcome by utilizing a suitable proteinacious diluent (94). The fragile nature of *Y.pestis* in small particle aerosols might indicate that, in reality, many of the cases reported as primary pulmonary plague might be initiated by larger droplets, or fomites, containing as they do, abundant material which could exert a protective effect. Heavier particles should 'fall-out' more rapidly. The frequently expressed conclusions of Pollitzer (76) may be appropriate, viz. 'that one of the best methods of reducing the infections of contacts to a pneumonic case is to turn the patient to face the wall'. This observation probably simultaneously reflects both the fragile nature of *Y.pestis* in small particle droplets and the extremely hazardous nature of large particles. In general, cool weather, moderate humidity and close contact between susceptible individuals appear to be most favourable for epidemics of primary plague pneumonia (18, 93).

Treatment with streptomycin, tetracyclines or chloramphenicol has greatly reduced plague mortality (70). However, these drugs do not always result in the complete elimination of *Y.pestis* from walled-off lesions: virulent *Y.pestis* have been isolated from some patients who have been under treatment for prolonged periods of time after treatment (51).

Antibodies to the specific F1 capsular antigen of *Y.pestis* are detectable in the peripheral blood of human beings, rodents and certain other mammals shortly after infection (24, 28, 32). Maternal F1 antibody is also present in the sera of young *Rattus norvegicus* and other rodents born to infected or vaccinated dams (89). F1 antibodies can be demonstrated in the sera of animals collected in plague foci and serum surveys have been widely used for plague detection and surveillance (2, 27, 28, 32, 38, 40, 41, 52).

Both F1 and V/W antibodies are opsonins in that they permit phagocytosis of encapsulated phagocytosis-resistant bacilli and thus represent in the host a developing immune response (26). Also, presence of F1 antibody as measured by the indirect hemagglutina-

Table 4. Persistence of plague infection in some *Rattus norvegicus* surviving challenge with classical plague organisms

Immunization before challenge	Isolation of *Yersinia pestis* Phenotype	Months post-challenge
Living avirulent plague vaccine (EV)	$F1^-:VW^+:P^+:PCF^+$	18
Killed plague vaccine (USP)	$F1^+:VW^+:P^+:PCF^+$	17
Anti-plague serum (polyvalent)	$F1^+:VW^+:P^+:PCF^+$	11
	$F1^-:VW^+:P^+:PCF^+$	24

tion test reflects an immune state which results from: 1) vaccination with either living attenuated (36) or killed (30) plague vaccines, or 2) injections of potent plague antisera (89). *R.norvegicus* with titers in excess of 1:64 are usually highly resistant to acute disease (89). Unfortunately, *Y.pestis* are not always eliminated from the tissues of all seemingly immune rats, and the disease occasionally persists in chronic form (90). Walled-off buboes, containing a spectrum of varying *Y.pestis* phenotypes, continue to enlarge (fig. 17) and may result in the eventual death of the animal. Death in laboratory rats usually occurs when buboes rupture internally and release their highly toxic and infectious contents into the system of the animal. This condition is noteworthy (table 4) because *Y.pestis* of fully virulent phenotype ($F1^+:V/W^+:P^+:PCF^+$) may persist in the tissues of vaccinated animals up to 18 months following challenge infections, and virulent variants of *Y.pestis* ($F1^-:V/W^+:P^+:PCF^+$), selected in vivo in the presence of F1 antibody, have been found 24 months post challenge. The mechanisms for the survival of *Y.pestis* and for the selection of variant bacilli in vivo in immunized rats is not clear, but the data suggest processes which serve to maintain the presence of the disease in natural foci.

Several observations suggest that rodents suffer chronic, latent plague infection, or infections with variant *Y.pestis* selected in chronic lesions in nature. 'Resolving' lesions (54), infection without lesions (54, 64) and latent, inapparent plague infections (65) described in rats or other rodents involved in plague epizootics indicate developing interrelationships between the immune or physiological mechanisms of the host and *Y.pestis.* In some natural foci, atypical *Y.pestis* are encountered. In Utah, for instance, strains are found of low virulence for mice and guinea pigs but of high virulence for local *Peromyscus* (56, 85). In a Brazilian focus, some strains exhibit a requirement for asparagine that

Table 5. Atypical *Yersinia pestis* isolated in nature

Atypical characteristic of *Y.pestis* strain	Source	Locale	Reference
Lacks F1 antigen	Patient	Colorado-Texas?	92
Lacks PCF complex	Patient	Arizona	12
Lacks pigment	Patient	South Africa	43
Lacks pigment	Lagomorpha	New Mexico	82
Lacks pigment	Rodent fleas	New Mexico	82
Lacks pigment	*Spermophilus*	New Mexico	91
Lacks pigment	4 patients	Vietnam	91
V/W antigen variable	4 rat flea pools	Java	91
Lacks Pesticin I	*Peromyscus* fleas. *Eutamias*	Utah	*

* Unpublished data

renders such strains avirulent for guinea pigs, animals which lack plasma asparagine (21). Wide variations in the virulence of strains isolated in the Transcaucasian mountains are also reported (1). Humans, as well as rodents, may occasionally be involved: chronic, relapsing, latent meningeal plague (63) and recurrent plague infections (24) have been reported. Since *Y.pestis* may persist in the buboes of patients seven to fourteen days following treatment (51) the possibility of a chronic infection should receive some consideration; the patient then having a convalescent antibody titer (24, 28) which may exert some selective pressure on the residual plague bacilli (19). Newer laboratory techniques (10) increase the probability of identifying and explaining atypical situations on a basis of a sophisticated laboratory analysis of virulence factors (table 5), protein content (42) or biochemical reactions (61).

CONCLUSIONS

Plague persists in wild rodent populations in extensive, essentially permanent natural foci. While massive rodent surveys have not been continuous, occasional cases in human beings serve as sentinels, attesting to the suitability of various regions for the perpetuation of the disease.

Despite intensive studies, many problems remain. Treatment is not always effective; serious complications may occur or some patients succumb. Vaccination may not prevent chronic disease. Variant *Y.pestis* or atypical situations may present difficulties in diagnosis and/or surveillance. Resistance of flea vectors and rodent hosts to pesticides complicates control.

An overview of the current situation justifies the opinion of Pollitzer (73) that plague is 'An enemy in ambush'.

REFERENCES

1. Akiev, A.K. 1965, Untitled. Weekly Epidemiological Record, World Health Organization, p. 65.
2. Akiev, A.K. 1968, The main principles and techniques used in epizootiological surveys of natural foci of plague in the Soviet Union. WHO Document BS/PL/68.22, 6 p.
3. Akiev, A.K. 1968, The epidemiology and epizootiology of plague. WHO Document BD/PL/68.21, 12 p.
4. Albizo, J.M. & M.J.Surgalla 1970, Isolation and biological characterization of *P.pestis* endotoxin. Infect. Immun. 2:229-236.
5. Anonymous 1920-50, Public Health Reports, Vols. 35-65, in part.
6. Anonymous 1951-77, Morbidity and Mortality Weekly Reports, Vols. 1-26, in part.
7. Anonymous 1975, Historical statistics of the US. Colonial times to 1970. Washington, DC. Government Printing Office.
8. Bacot, A.W. & C.J.Martin 1914, Observations on the mechanism of the transmission of plague by fleas. J. Hyg. (Camb., Plague Suppl.) 3:423-439.
9. Bacot, A.W. & C.J.Martin 1924, The respective influence of temperature and moisture upon the survival of the rat flea. J. Hyg. 23:98-105.
10. Bahmanyar, M. & D.C.Cavanaugh 1976, Plague Manual. Geneva, World Health Organization.
11. Baker, E.E., H.Sommer, L.E.Foster, E.Meyer & K.F.Meyer 1952, Studies on immunization against plague I. The isolation and characterization of the soluble antigen of *P. pestis*. J. Immunol. 68:131-145.
12. Beesley, E.D. & M.J.Surgalla 1970, Pesticinogeny: A characteristic useful for presumptive identification and isolation of *P.pestis*. Appl. Microbiol. 19:915-918.
13. Berendt, R.F. 1957, Some physiological and immunological properties of *P.pestis* recovered from aerosols. J. Bact. 75:217-223.
14. Bibikova, V.A. 1967, Fleas (Aphinaptera) and *P.pestis*. In Biological relationships of blood-sucking arthropods and pathogens of

human disease. Moscow, Medicine Press.
15. Bibikova, V.A. 1977, Contemporary views on the interrelationships between fleas and the pathogens of human and animal diseases. Ann. Rev. Entomol. 22:23-32.
16. Bibikova, V.A. & L.N.Klassovskii 1974, Flea transmitted plague. Moscow, Meditsina. 187 p.
17. Brooks, R.St.J. 1915-1917, The influence of saturation deficiency and of temperature on the cause of epidemic plague. J. Hyg. (Camb., Plague Suppl. 5) 15:881-899.
18. Burmeister, R.W., W.D.Tigertt & E.L.Overholt 1962, Laboratory-acquired pneumonic plague. Report of a case and review of previous cases. Ann. Inst. Med. 56:789-800.
19. Burrows, T.W. 1963, Virulence of *P.pestis* and immunity to plague. Ergebn. Mikrobiol. 37:59-113.
20. Burrows, T.W. & G.A.Bacon 1954, The basis of virulence in *P.pestis.* Comparative behaviour of virulent and avirulent strains *in vivo.* Brit. J. Exper. Path. 35:134-143.
21. Burrows, T.W. & W.A.Gillett 1971, Host specificity of Brazilian strains of *P.pestis.* Nature 229:51-52.
22. Butler, T. 1972, A clinical study of bubonic plague. Amer. J. Med. 53: 268-276.
23. Butler, T., W.R.Bell, Nguyen-Ngoc-Linh, Nguyen-Dinh-Tiep & K.Arnold 1974, *Yersinia pestis* infection in Vietnam I. Clinical and hematologic aspects. J. Infect. Dis. 129: (Suppl.) S78-S84.
24. Butler, T. & B.W.Hudson 1977, The serological response to *Y.pestis* infection. Bull. Wld. Hlth. Org. 55:(in press).
25. Cavanaugh, D.C. 1971, Specific effect of temperature upon transmission of the plague bacillus by the Oriental rat flea, *Xenopsylla cheopis.* Amer. J. Trop. Med. Hyg. 20:264-273.
26. Cavanaugh, D.C. & R.Randall 1959, The role of multiplication of *P.pestis* in mononuclear phagocytes in the pathogenesis of flea-borne plague. J. Immunol. 83:348-363.
27. Cavanaugh, D.C., B.D.Thorpe, J.B.Bushman, P.S.Nicholes & J.H.Rust, Jr. 1965, Detection of an enzootic plague focus by serological methods. Bull. Wld. Hlth. Org. 32:197-203.
28. Cavanaugh, D.C., P.J.Deoras, D.H.Hunter, J.D.Marshall, Do-Van-Quy, J.H.Rust, Jr., S. Purnaveja & P.E.Winter 1970, Some observations on the necessity for serological testing of rodent sera for *P.pestis* antibody in a plague control program. Bull. Wld. Hlth. Org. 42:451-459.
29. Cavanaugh, D.C. & J.D.Marshall, Jr. 1972, The influence of climate on the seasonal prevalence of plague in the Republic of Vietnam. J. Wildlife Dis. 8:76-85.
30. Cavanaugh, D.C. & J.H.Steele 1974, Trends in research in plague immunization. J. Infect. Dis. 129:(Suppl. May) S1-120.
31. Deoras, P.J. & R.S.Prasad 1967, A note on the feeding mechanism of two fleas. Curr. Sci. 36:518-519.
32. Dodin, A. & E.R.Brygoo 1968, La réaction d'hémagglutination conditionnée dans la peste, chez différents hôtes. Rapport entre les anticoprs hémagglutinants et précipitants. I. Les polyosides. Ann. Inst. Pasteur 114: 452-462.
33. Domaradskii, I.V., G.A.Yaromyuk, L.V. Vasyukhina & A.V.Korotaeva 1963, On blood plasma coagulation by plague and pseudotuberculosis microbes. Byulleten Eksperimenta noi Biologii i Meditsiny 56 (7):79-82.
34. Finegold, M.J. 1968, Pathogenesis of plague. Amer. J. Med. 45:549.
35. Finegold, M.J., J.J.Petery, R.F.Berendt & H.R.Adams 1968, Studies on the pathogenesis of plague. Blood coagulation and tissue responses of *M.mulatta* following exposures to aerosols of *P.pestis.* Amer. J. Path. 53:99-104.
36. Girard, G. 1963, Immunity in plague infection. Biol. Med. 52(6):631-731.
37. Greenwood, M., Jr. 1911, XLV. Statistical investigations of plague in the Punjab. 3rd Report: On some factors which influence the spread of plague. J. Hyg. (Camb., Plague Suppl. I.) 12:62-156.
38. Hallett, A.F., D.McNeil & K.F.Meyer 1970, A serological survey of the small mammals for plague in Southern Africa. S. Afr. Med. J. 44:831-837.
39. Hirst, L.F. 1953, The conquest of plague. London, Oxford Univ. Press.
40. Hudson, B.W., S.F.Quan & M.I.Goldenberg 1964, Serum antibody responses in a population of *Microtus californicus* and associated rodent species during and after *P.pestis* epizootics in the San Francisco Bay area. Zoonoses Res. 3:15-29.
41. Hudson, B.W. & L.Kartman 1967, The use of the passive hemagglutination test in epidemiologic investigations of sylvatic plague in the United States. Bull. Wildlife Dis. Assoc. 3:50-59.
42. Hudson, B.W., T.J.Quan, V.R.Sites & J.D. Marshall, Jr. 1973, An electrophoretic and

bacteriologic study of *Y.pestis* isolates from Central Java, Asia and the Western Hemisphere. Amer. J. trop. Med. Hyg. 22:642-653.

43. Isaäcson, M., D.Levy, J.Te.W.N.Pienaar, H.D.Bubb, J.A.Louw & D.K.Genis 1973, Unusual cases of human plague in Southern Africa. S. Afr. Med. J. 47:2109-2113.
44. Jackson, S. & T.W.Burrows 1956, The pigmentation of *P.pestis* on a defined medium containing haemin. Brit. J. Exp. Path. 37: 570-576.
45. Janssen, W.A. & M.J.Surgalla 1969, Plague bacillus: survival within host phagocytes. Science 163:950-952.
46. Jorge, R. 1928, Rongeurs et puces. Paris, Masson et Cie.
47. Kartman, L. 1969, Effect of differences in ambient temperature upon the fate of *Pasteurella pestis* in *Xenopsylla cheopis.* Trans. Roy. Soc. Trop. Med. Hyg. 63:71-75.
48. Kiseleva, I.Ye. 1959, Observations on the mechanism of the hemorrhages in plague. Report I. State of the factors of blood coagulation in experimental plague. Trans. Roster on the Don State Scientific-Research Antiplague Institute 15:97-105.
49. Knülle, W. 1967, Physiological properties and biological implications of the water vapour sorption mechanism in larvae of the Oriental rat flea (*X.cheopis* (Roths)). J. Insect. Physiol. 13:333-357.
50. Kucheruk, V.V. 1965, On the pathogenesis of natural foci of plague. In: B.Rosicky & K.Heyberger (eds.), Proceedings of a symposium. Theoretical questions of natural foci of disease, p. 379-394. Czechoslovak Academy of Sciences.
51. Legters, L.J., A.J.Cottingham, Jr. & D.H. Hunter 1970, Clinical and epidemiologic notes on a defined outbreak of plague in Vietnam. Amer. J. Trop. Med. Hyg. 19: 639-652.
52. Levi, M.I., Y.G.Suchdov, G.M.Orlova, L.G. Gerasyuk, A.M.Shkoda, L.A.Peysakhis, A.N. Stogova, N.F.Lopatina, N.A.Sukharnikova, G.Y.Pak, K.M.Muminov, T.N.Donskaya, L.C.Nassonov, V.I.Weinglat, E.S.Murtazanova, A.I.Sthelman, A.F.Laverentev, N.N. Basova, G.I.Kulov, G.M.Golkovsky, N.I. Salamov & N.I.Zalygina 1964, Significance of serological methods in the epizootological study of plague in wild rodents. J. Hyg. Epidemiol. Microbiol. Immunol. 8:422-427.
53. Link, V.B. 1954, A history of plague in the United States of America. Public Health Monograph No. 26, Washington, DC. Government Printing Office, 120 p.
54. McCoy, G.W. 1911, Studies upon plague in ground squirrels. Public Health Bulletin No. 43, Washington, DC. Government Printing Office, 51 p.
55. Macchiavello, A. 1954, Reservoirs and vectors of plague. J. Trop. Med. Hyg. 57:1-68.
56. Marchette, N.J. 1963, *Pasteurella pestis* of low virulence for mice and guinea pigs. Amer. J. Trop. Med. Hyg. 12:215-218.
57. Marshall, J.D., Jr., Do-Van-Quy & F.L.Gibson 1967, Asymptomatic pharyngeal plague infection in Vietnam. Amer. J. Trop. Med. Hyg. 16:175-177.
58. Marshall, J.D., Jr., D.N.Harrison, J.H.Rust, Jr. & D.C.Cavanaugh 1971, Serological response of rhesus monkeys (*Macaca mulatta*) to immunization and infection with *P. pestis.* Proc. Soc. Exp. Biol. Med. 138:738-741.
59. Martin, C.J. 1911, The spread of plague. Brit. Med. J. 1911:1249-1263, Nov.
60. Martin, C.J. 1913, The Horace Dobell Lectures on insect porters of bacterial infections. Lecture ii. The transmission of plague by fleas. Brit. Med. J. Jan 11:59-68, and Lancet Jan 11:81-88.
61. Martinevskij, I.L. 1968, Propriétés biologiques et génétiques des souches de peste et de pseudotuberculose isolées en divers points du globe. International symposium on pseudotuberculosis, Paris, 1967, Symp. series immunobiol. standard. 9:265-274.
62. Mellanby, B.A. 1932. The thermal death point of a number of insects. J. Exper. Biol. 9:222-231.
63. Meyer, K.F., C.L.Connor, F.S.Smyth & B. Eddie 1937, Chronic relapsing latent miningeal plague. Arch. Int. Med. 59:967-980.
64. Meyer, K.F. & B.Eddie 1938, Persistence of sylvatic plague. Proc. Soc. Exper. Biol. Med. 38:333-334.
65. Meyer, K.F., R.Holdenreid, A.L.Burroughs & E.Jawetz 1943, Sylvatic plague studies IV. Inapparent latent sylvatic plague in ground squirrels in Central California. J. Infect. Dis. 73:144-157.
66. Meyer, K.F. & R.Holdenreid 1949, Rodents and fleas in a plague epizootic in a rural area of California. Puerto Rico J. Publ. Hlth. Trop. Med. 24:201-209.
67. Mollaret, H.H. 1963, Conservation expérimentale de la peste dans le sol. Bull. Soc. Path. Exot. 56:1168-1182.
68. Mullholand, B. 1968, Purpura and dry gan-

grene in plague. 8th Int. Congr. Trop. Med. and Malar. p. 545.

69. Olson, W.P. 1969, Rat-flea indices, rainfall, and plague outbreaks in Vietnam, with emphasis on the Pleiku area. Amer. J. Trop. Med. Hyg. 18:621-628.
70. Poland, J.D. 1972, Plague. In P.D. Hoeprich (ed.), Infectious Diseases. p. 1141-1148. New York, Harper & Row, Inc.
71. Poland, J.D. 1973, Human plague from exposure to a naturally infected carnivore. International Northwestern Conference on Diseases in Nature Communicable to Man. Proc. 27th Ann. Mtg. 28-30 Aug., Banff, Alberta, Canada.
72. Pollitzer, R. 1954, Plague. WHO Monograph 22, Geneva, Switzerland, 503 p.
73. Pollitzer, R. 1956, Plague, an enemy in ambush. Address, World Health Day, April 1956, WHO 56/8.
74. Pollitzer, R. 1960, A review of recent literature on plague. Bull. Wld. Hlth. Org. 23:313-400.
75. Pollitzer, R. 1964, Plague and plague control in the Soviet Union: History and bibliography through 1964. Fordham University, NY. Institute of Contemporary Russian Studies. 478 p.
76. Pollitzer, R. 1964, Personal communication.
77. Pollitzer, R. & K.F.Meyer 1961, The ecology of plague. In: J.M.May (ed.), Studies in disease ecology. New York, Hafner.
78. Pollitzer, R. & K.F.Meyer 1965, Plague in the Americas. p. 433-501. Washington DC.: Pan American Health Organization. 156 p.
79. Robertson, H.McG. 1923, A possible explanation of the absence of bubonic plague in cold countries. Publ. Hlth. Repts. 38:1519-1531.
80. Rogers, L. 1928, The yearly variations in plague in India in relation to climate: Forecasting epidemics. Proc. Roy. Soc. Ser. B 103:42-72.
81. Rothschild, M. & B.Ford 1966, Hormones of the vertebrate host controlling ovarian regression and copulation of the rabbit flea. Nature 211:261-266.
82. Rust, J.H., Jr. & D.C.Cavanaugh 1964, Technique for the isolation of aberrant forms of *Pasteurella pestis* in enzootic plague foci. Bact. Proc. p. 84.
83. Rust, J.H., Jr., D.N.Harrison, J.D.Marshall, Jr. & D.C.Cavanaugh 1972, Susceptibility of rodents to oral plague infection: a mechanism for the persistence of plague in inter-epidemic periods. J. Wildlife Dis. 8:127-133.
84. Sites, V.R., J.D.Poland & B.W.Hudson 1972, Bubonic plague misdiagnosed as tularemia. J. Amer. Med. Assoc. 222:1642-1643.
85. Thorpe, B.D., N.J.Marchette & J.B.Bushman 1963, Virulence studies of *P.pestis* isolates from the Great Salt Lake Desert. Amer. J. Trop. Med. Hyg. 12:219-221.
86. Von Reyn, C.F., A.M.Barnes, N.S.Weber & U.G.Hodgin 1976, Bubonic plague from exposure to a rabbit: a documented case, and a review of rabbit-associated plague cases in the United States. Amer. J. Epidem. 104:81-87.
87. Walker, R.I. 1972, Detection of an endotoxin-like substance during human plague bacteremia. S.E.Asian J. Trop. Med. Publ. Hlth. 3:221-224.
88. Wayson, N.E. 1947, Plague-field surveys in western U.S. during 10 years (1936-1945). Publ. Hlth. Repts. 62:780-791.
89. Williams, J.E., J.D.Marshall, Jr., D.M.Schaberg, R.F.Huntley, D.N.Harrison & D.C. Cavanaugh 1974, Antibody and resistance to infection with *Y.pestis* in the progeny of immunized rats. J. Infect. Dis. 129:(Suppl.) S72-77.
90. Williams, J.E., D.N.Harrison & D.C.Cavanaugh 1975, Cryptic infection of rats with non-encapsulated variants of *Y.pestis.* Trans. Roy. Soc. Trop. Med. Hyg. 69:171-172.
91. Williams, J.E., D.N.Harrison, T.J.Quan, J.L.Mullins, A.M.Barnes & D.C.Cavanaugh 1978, Atypical plague bacilli isolated from rodents, fleas, and man. Amer. J. Publ. Hlth. 68:262-264.
92. Winter, C.C., W.B.Cherry & M.D.Moody 1960, An unusual strain of *P.pestis* isolated from a fatal human case of plague. Bull. Wld. Hlth. Org. 23:408-409.
93. Winter, P.E., J.D.Marshall, Jr., J.H.Rust, Jr. & D.C.Cavanaugh 1971, Plague. In: B.G. Maegraith & H.M.Gilles (eds.), Management and treatment of tropical diseases. p. 377-387. Oxford, Blackwell.
94. Won, W.D. & H.Ross 1966, Effect of diluent and relative humidity on apparent viability of airborne *P.pestis.* Appl. Microbiol. 14:742-745.
95. WHO Expert Committee on Plague. 1970, 4th Report. Wld. Hlth. Org. Techn. Rep. Ser. No. 447.
96. Yaromyuk, G.A. 1962, Observations on the fibrinolytic activity of plague and pseudotuberculosis bacilli. Bull. Irkutsk State Scientific-Research Antiplague Institute for Siberia and the Far East 24:196-201.

V. A. BIBIKOVA* & I. F. ZHOVTYI**

**Institute of Medical Parasitology and Tropical Medicine, Moscow, USSR*

***Irkutsk State Research Antiplague Institute of Siberia and the Far East, Irkutsk, USSR*

REVIEW OF CERTAIN STUDIES OF FLEAS IN THE USSR, 1967-1976***

ABSTRACT

The success of measures on the prevention and control of plague to a great extent depends upon knowledge of the habits, physiology, ecology and systematics of the vectors, fleas, and on their interrelationships with the etiological agent and the reservoirs of the infection. This article summarizes the main aspects of work in the past two years in the USSR on the behaviour, bionomics and host-interactions of fleas of medical importance.

The history of the study of fleas in the Soviet Union can be divided into four periods: 1) Work done before the role of these ectoparasites as vectors of plague had been established. The research was essentially academic and dealt primarily with morphology, systematics and the distribution of the local Siphonapteran fauna. 2) Investigations undertaken after the basic facts about the transmission of plague became known and it was appreciated that a vast fauna of Siphonaptera and rodents were implicated. Studies on classification were intensified, and experiments on transmission and vector capacity were stressed. Simultaneously, fundamental data on the ecology of fleas were sought and obtained. 3) Faunal studies on the fleas (and their hosts) of various areas in our vast country were undertaken, and keys prepared for the identification of fleas occurring in the major regions of concern (e.g. Argyropulo, 1935; Ioff & Tiflov, 1938, 1954; Ioff, et al., 1965; Ioff & Bondar, 1956). 4) Subsequently, investigations on systematics on the faunal aspects became of secondary importance, and more attention was paid to studies on ecology and the development of control measures against fleas, as well as on the interrelations between various species of fleas and the causative organism of plague. It is this last period, a decade to date, which is discussed in the present paper.

1 FAUNAL STUDIES AND SYSTEMATICS

Research on the classification for the faunal aspects of fleas may have been given a lower priority than in the past, with only a few specialists involved, but nevertheless there have been some excellent achievements. The systematics of the larval stage of Siphonaptera

*** This article, originally prepared in Russian, was submitted to us in the form of a literal translation which had not been seen by the authors, and which was then completely rewritten by the Senior Editor and sent to Dr Zhovtyi for comment.

have long been in an unsatisfactory state because scientists everywhere have been frustrated by the dearth of known morphological characters that could be employed in diagnosis and taxonomy. Despite these inherent difficulties, considerable progress has been made in this field (Kiryakova, 1961, 1964, 1965, 1968; Kiryakova & Cotton, 1970; Karandina, 1964). The basic anatomy has been detailed and keys prepared for the identification of the larvae of major fleas in certain areas.

Adult fleas have also been receiving attention, and a series of taxonomic papers and descriptions of new species have been prepared (e.g. Vysotskaya, 1968; Scalon, 1968; Darskaya & Shiranovich, 1971; Murzakhmetova, 1969; Goncharov, 1971; Goncharov & Shiranovich, 1970). The books on the identification, host-relationships and distribution of fleas of various parts of the USSR (Ioff & Tiflov, 1954; Ioff & Scalon, 1954; Ioff & Bondar, 1956; Ioff, et al., 1965; Scalon, 1970) were very successful, and the most recent addition to the faunal series, by Tiflov et al. (1977) will no doubt be equally useful. That opus concludes the planned regional, monographic studies of the Siphonapteran fauna.

As huge tracts of new agricultural lands have been opened, there have been parallel faunal studies on the potential vectors and reservoirs of disease in the areas, particularly as concerns the fleas of Siberia and the Far East. As examples may be cited: 1) Pre-Baikal and Trans-Baikal regions – Emelyanova & Goncharov, 1966; Pauller et al., 1966; Stupina, 1971; Emelyanova et al., 1969. 2) Altai region – Koklyagina, 1969; Vasiliev & Lazereva, 1971; Elistratova et al., 1974. 3) Tuva region – Lyetov et al., 1966; Ustuzhina, 1969. 4) Pre-Amur region – Koshkin, 1966; Kozlovskaya & Tshipanin, 1969. 5) Kamchatka region – Paramonov et al., 1966; Vasiliev et al., 1974. The results of studying the fleas of Ukraine were also summarized (Yurkina, 1961).

2 STUDIES ON THE ECOLOGY OF FLEAS

Research on the ecology of fleas was mainly undertaken in connection with investigations on plague and was primarily concerned with mass rearing and laboratory-studies on the vectors of this disease, which are ectoparasites of various small mammals. The observations were made in the vast territories of the southeastern part of European USSR; in the Caucasus, Central Asia; and the southern region of Siberia.

As is known, the main vectors of plague of *Citellus* are *Citellophilus tesquorum* and *Neopsylla setosa;* on *Marmota – Oropsylla silantiewi* and *Rhadinopsylla ventricosa;* in colonies of *Rhombomys opimus:* fleas of the genus *Xenopsylla* (*X.skrjabini, X.nuttali, X. gerbilli minax, X.gerbilli caspica* and *X.hirtipes*); on the small gerbillines, *X.conformis* and *Nosopsyllus laeviceps* and on *Microtus arvalis, N.consimilis, Callopsylla caspia* and others.

The attention of the investigators at first was directed towards detailed analysis of the annual cycles of the fleas (Darskaya, 1965). Such material was published concerning the following fleas and/or hosts: 1. *Marmota: M.baibacina* (Berendyayeva & Kulkova, 1964); *M.caudata* (Kafarskaya & Lysenko, 1963) and *M.sibirica* (Zhovtyi, 1970). 2. Gerbillinae: (Stepanova et al., 1971; Kunitsky et al., 1971), particularly of the genus *Xenopsylla* (Darskaya, 1970), e.g. *X.skrjabini* (Gerasimova, 1969a, 1971). 3. *Citellus: C.pygmaeus – Neopsylla setosa* and *Citellophilus tesquorum* (Brukhanova & Surkov, 1970a, 1970b) and *C. undulatus – Amphipsylla asiatica* (Vasiliev & Zhovtyi, 1971). 4. *Microtus arvalis: N.consimilis, Amphipsylla rossica, Frontopsylla elata caucasica, Ctenophthalmus wladimiri* (Kosminsky et al., 1966a, 1966b, 1966c, 1970). 5. *Ochotona daurica:* (Vershinina, 1970); *F. luculenta luculenta, Amphalius runatus* (Akinshina & Zhovtyi, 1969a, 1969b). 6. Fleas of *Rattus* (Zhovtyi, 1966c). 7. Fleas of birds (Shchevchenko et al., 1976).

The study of the annual cycles of fleas in our huge country, with its variety of geographical and topographical conditions has only just begun. The results for one of the zones are shown as an example, namely data on the fleas of Siberia. Here the critical phase of the annual cycle for the fleas, as well as for their hosts, is the winter time, and the conditions under which the hosts pass this season exert a tremendous effect upon the fleas. Zhovtyi (1974a) distinguished between four such categories for rodent fleas.

In the first group are the *Citellus* fleas. Both rodent and fleas (all stages) are dormant, and the adults are largely unfed. The *Citellus* awaken intermittently during the winter and then the adult fleas may also become active and perhaps even feed on the host. The ovaries of the female fleas are undeveloped, and reproduction begins only after the host fully awakens from hibernation. Imagos, resulting from the overwintering preimaginal stages, appear in numbers in the spring. The second type consists of *Oropsylla* infesting *Marmota* and *Citellus.* These fleas spend the winter in an active state in all phases of development. The rate of reproduction in *Oropsylla* decreases in the winter, particularly on *Citellus* but does not cease altogether. For this reason, the spring increase in the flea population is considerable and, as a rule, greater than in the autumn. The above two categories are typical of the fleas of hibernating rodents, while the next two types are characteristic of those infesting non-hibernating rodents and Lagomorpha. In the third group are the fleas of synanthropic rodents, wherein both the hosts and fleas spend the winter (usually indoors) in an active state. All the phases of development in the Siphonaptera occur throughout the year. The fourth category consists of wild rodents which do not hibernate, and the lagomorphs as well.

The flea population spends the winter in an active state in the imago and pupal stages, and perhaps also in the egg stage. However, the larvae as a rule, are absent in the winter. In the period of the most intense cold; the pre-imaginal stages cease development and new imagos do not emerge. Females with mature eggs do not occur at this season. Although emergence of fleas from the overwintering pupae commences as early as February-March, there is no noticeable increase in the number of imagos in the spring. However, in some instances in the Far East, the maximum emergence of rodent fleas occurs in the winter. Females with the ripe eggs in rodent fleas appear in the Transbaikalia region in late March-early April. The maximum period of oviposition in fleas of *Microtus maximovezii* and *Apodemus agrarius* occurs in the period February-April. Darskaya (1970) also presented a classification of the seasonal cycles of fleas in dealing with species from rodents, insectivores and birds in the European part of USSR and Transcaucasia, and preliminarily indicated five types of annual cycles, each subdivided into two secondary groups.

Research on seasonal changes was also performed by breeding rodent fleas in the laboratory (Bibikova, 1965; Bgytova, 1963; Afanasieva & Bgytova, 1963a, 1963b; Bgytova & Leonova, 1965; Brukhanova, 1966; Vasiliev & Zhovtyi, 1967; Tepiykh, 1968; Kunitskaya et al., 1969; Chernova, 1971). Concomitantly, quantitative studies were made on population changes (Vishnyakov, 1963; Poliakov & Shchevchenko, 1965; Zagniborodova, 1968; Kosminsky et al., 1970 and others). It was shown, for example, that the characteristic feature of the seasonal rise in the number of the fleas of Siberian rodents is determined by the fact that certain species of rodent-Siphonaptera breed at specific times of the year, which differ from those of yet other taxa (Zhovtyi, 1966a, 1970). In the seasonal curves of the *Citellus* fleas there are two regularly reported rises: 1) autumnal, owing to emerging of the first generation imagos, and 2) a vernal peak, owing to the emergence of adults from the overwintering autumnal pupae. These are followed by drops in population: 1) in the winter, and 2) in the summer. For fleas of non-hibernating wild rodents, almost all the preimaginal phases complete their metamorphosis in the autumnal-winter period, resulting in the maximum annual incidence at that time. In *Oropsylla* of *Marmota* and *Citellus,* there is an annual decline, in the middle of the summer, and one peak, which begins in the autumn and reaches its highest level in the spring. The opposite is the case in fleas of synanthropic rodents, where the increase takes place in the summer and autumn, whereas the decline occurs in winter. The level of these annual fluctuations varies from year to year. The maxima occur when climatological conditions and other factors tend to force the hosts to remain in their lairs for unusually long periods, resulting in unusually favourable circumstances for flea-breeding. Conditions which result in the hosts spending a disproportionate amount of time out of their nests and burrows have the opposite effect.

The sex ratio, ages and other phases of the population cycles have also been studied (Darskaya & Brukhanova, 1972; Zhovtyi & Dubovik, 1967; Zhovtyi, 1969a; Kunitskaya et al., 1971). Generally, for all species that have been studied, there are more females than male fleas. In those years when the population of a particular species is at its lowest ebb, the numbers of males approximates that of the females. However, at certain times of the year, the ratio of males increases greatly, and males may outnumber females, as at the beginning of the period of the emergence from the pupal stage, e.g. middle and late summer. Male fleas have a shorter longevity than females, and hence the proportion of females soon increases and dominates for the rest of the year.

As in the case of many other insects, the age groups within a Siphonapteran population, were studied by determining the physiological age of females, as indicated by the maturity of the ovaries and presence of the yellow bodies (Prokop'ev, 1958; Kunitskyaya, 1968, 1970). The phases and ages of the rodent-flea population varies with the species of flea and host, with their respective breeding cycles, and with the time of year. Annually, there is a period when there is a marked increase in the number of preimaginal stages (generally in summer) and another when the number of imagos rises rapidly (usually in the autumn, winter and spring). In Siberia, this increase in the number of adult fleas of wild rodents is associated with the onset of cold weather, when the activity of their hosts is reduced and they spend more time in the nests. This appears to be an adaptive response which results in the height of the flea population, appearing at a time when there is maximum chance for nourishment. The peak levels of the numbers of preimaginal stages is coincidental with the time of increased rodent-activity, but such stages are less dependent upon the host than are the adults. In fleas of the synanthropic rodents, periodicity is poorly developed. The percentage of young adult fleas increases rapidly at certain times of the year, indicating emergence of the new generation of adults.

Taking into account the more active feeding-requirements of young imagos, and the great augmentation in the population, such periods are evidently potentially more hazardous from the epidemiological point of view.

Conditions of temperature and humidity of fleas inhabiting rodent burrows or nests have also been studied (Bibikova et al., 1963; Gerasimova, 1969b; Ilinskaya et al., 1971c; Ilinskaya et al., 1971a, 1971b). Such factors were also investigated on the animals themselves (Zhovtyi & Vasiliev, 1962; Vansulin & Volkova, 1962; Soldatkin & Levoshina, 1968; Zhovtyi, 1969b) along with the role of the hosts regarding the ecology of their fleas

(Zhovtyi, 1968a, 1968b). Throughout the year, all stages of fleas are subject to a combination of two different sets of environmental conditions: one created by the natural geographic and ecological setting, and another determined by the hosts. As a result, the fleas, like ectoparasites, are subject to what has been denoted as the law of 'double biotope' (Theodorides, 1954) viz, a primary medium established by the host itself and by its actions, and secondarily, the environment to which the host is subject (Zhovtyi, 1968a, 1968b, 1972). The second factor also affects the first through its influence upon the activities of the host.

This concept is reinforced by data upon the overwintering of fleas (Berendyayeva & Kulkova, 1964, 1965; Kizilov et al., 1967; Berendyayeva et al., 1967; Vershinina, 1970; Mialkovskaia & Brukhanova, 1972). Despite the unfavourable external conditions and those in nests and burrows of the hosts, the behaviour of the rodents may create an environment that is suitable for survival of adult fleas (or other stages) in the winter, and sometimes the imagos even remain active. Even in Siberia, fleas like *O.silantiewi* on *Marmota sibirica* (Zhovtyi, 1970) and *Amphipsylla asiatica* on *Citellus dauricus* and *C.undulatus* (Vasiliev & Zhovtyi, 1971) may breed in winter.

For the study of the ecology of fleas and for determining epizootic and epidemic circumstances, as well as for selecting appropriate plague-control measures, it is essential that accurate methods be established for assessing the flea population so that comparable results may be obtained. These points have been studied intensively during the last decade, and various techniques have been suggested for determining the numbers of fleas in domiciles (Gershkovich, 1963) and for fleas infesting *Rhombomys opimus* (Kunitsky & Gauzshtein, 1963a, 1963b; Popova & Sokolova, 1965) and *Citellus pygmaeus* (Mironov et al., 1963).

In their research, the anti-plague organizations have generally employed the so-called 'flea index', also termed the 'index of abundance', as the basic unit when discussing flea populations. This index is defined as the average number of fleas per individual host for the object under study. Another useful unit has been the 'index of occurrence', i.e. percentage of affected mammals, birds or objects that were infested. At first, such calculations were restricted to determining the numbers of fleas on the hosts themselves, but later the technique was broadly applied to collections of fleas at the entrance of a rodent burrow. Calculations of the flea populations in the nests of the hosts could not be done routinely even though it was recognized that the greatest number of fleas (and species) occurred in such habitats. Intensive investigations indicated that in the Kalmyk area, 80-90 per cent of the rodent fleas collected were from the nests of the hosts, and in southeast Transbaikalia, 90-99.9 per cent were from nests (Zhovtyi, 1959, 1963, 1966a; Mironov et al., 1963). As shown in table I, in ten of the 11 species of fleas studied, there was a full and direct correlation between the total number of insects in the micropopulation and the indices of their abundance in the nests, but there was no such correlation with the numbers observed on the host itself or with those collected at the burrow-entrance. The single exception was *N.setosa* on *Citellus pygmaeus,* in which the high degree of correlation was with the numbers of fleas on the animals themselves. It is clear that relying upon the flea index at the burrow-entrance is inadequate for estimating the true population, which can be assessed only by studying the nests and other places of flea-concentration in a microbiotope, e.g. the forage chambers of *Rhombomys opimus* in the Central Asian desert focus of plague.

Because of the shortcomings of the traditional flea-index, Zhovtyi (1960, 1966a) suggested the use of estimates based upon the fleas at the burrow-entrance and in the nests,

Table 1. Correlation coefficient (Σ) of indices of the abundance of fleas on mammals, in burrow-entrances and in nests, with their abundance in a microbiotope

Localities of observation	Hosts	Species of fleas	Nest		Animal		Burrow-entrances		Reference
			Σ	M_2	Σ	M_2	Σ	M_2	
Southeast Transbaikalia	Marmota	*Oropsylla silantiewi*	+0.97	±0.24	+0.37	±0.49	–	–	Zhovtyi 1959, 1960 and 1966
Southeast Transbaikalia	Citellus dauricus	*Citellophilus tesquorum*	+1.00	±0.00	+0.11	±0.37	+0.47	±0.29	
Southeast Transbaikalia	Citellus dauricus	*Frontopsylla luculenta*	+1.00	±0.00	+0.19	±0.36	+0.20	±0.30	
Southeast Transbaikalia	Citellus dauricus	*Neopsylla abagaitui*	+1.00	±0.00	+0.23	±0.36	+0.05	±0.38	
Southeast Transbaikalia	Citellus dauricus	*Neopsylla bidentatiformis*	+1.00	±0.00	+0.18	±0.37	+0.13	±0.37	
Southeast Transbaikalia	Microtus brandti	*Neopsylla pleskei*	+0.98	±0.01	+0.12	±0.29	–	–	
Southeast Transbaikalia	Microtus brandti	*F.luculenta*	+0.99	±0.01	+0.30	±0.27	–	–	
Southeast Transbaikalia	Microtus brandti	*Amphipsylla primaris mitis*	+1.00	±0.00	−0.11	±0.29	–	–	
Southeast Transbaikalia	Meriones unguiculatus	*N.pleskei*	+0.94	±0.05	+0.18	±0.39	–	–	
Lower Volga (Kalmykia)	Citellus pygmaeus	*Neopsylla setosa*	+1.00	±0.00	+0.95	±0.04	+0.65	±0.26	Mironov et al. 1963
Lower Volga (Kalmykia)	Citellus pygmaeus	*C.tesquorum*	+0.94	±0.05	+0.19	±0.43	+0.44	±0.36	

as well. The new index was termed the 'general level of the flea population' (for a designated species), and is computed from the indices of abundance of the fleas in the nests, on the animals, and at the entrances to the burrows. In the method of calculating the index, there is some degree of distortion because several rodents may occupy a single burrow, and also because only one entrance per burrow-system is checked, whereas there are often multiple openings. However, only an insignificant proportion of the micropopulation is actually found in such sites, and the method is regarded as being highly reliable in general. Another index is that of 'domination', i.e. the percentages for the correlation of the various species of fleas in different parts of the rodent-habitation.

Under actual working conditions encountered by the anti-epidemiological units, excavating rodent-nests for the purpose of obtaining such data is usually impractical. Therefore, calculation of the general level of abundance can be performed only under stable or special conditions. As the results with the other indices are comparable when taken under the same conditions and regarding the same host and area, the alternative indices are routinely employed by the epidemiological control-companies. Once monthly indices of probability are established for a fixed area, the calculations for the remainder of the territory can be performed in one of the most accessible habitats of fleas, e.g. on the bodies of the host.

Indices of abundance of fleas in the nests, at the entrances to the burrows, and the index of general level may be calculated on the basis of indices on the hosts, by simple mathematical computation based on the indices of probability obtained in stationary conditions (Zhovtyi, 1966a).

For a more precise method of calculating the general number of fleas in nature, some authors (Novokreshchenova, 1960; Mironov et al., 1963) recommended computation of the stock of fleas per unit of area, summarizing indices of abundance in the nests, on the animals and at the entrances to burrows, and multiplying them according to the number of burrow-systems per hectare. Because of the labour involved, this method can be used only for special needs in research. Despite all the discussion that has taken place regarding quantitative estimates of flea populations, a wholly acceptable method has not yet been devised.

In research concerning the ecology of fleas, radioactive marking has been widely used (Novokreshchenova et al., 1961; Novokreshchenova et al., 1967; Berendyayeva et al., 1969; Sviridov, 1963; Sviridov et al., 1964; Morozova et al., 1969; Soldatkin et al., 1969; Rappoport et al., 1969). As markers Ce^{114}, C^{14} and P^{32} have been used. Depending upon the aims of the investigations, the marking was performed by immersing the insects into the appropriate solution for a few minutes, or else by feeding the fleas upon the host which had been given the isotope with its food or by subcutaneous inoculation. The tagged insects were traced by autoradiography or by counters.

By this method, the period and degree of feeding, e.g. the numbers of fleas sucking blood per unit of time could be determined. In the case of *X.skrjabini,* which is the predominant species infesting *Rhombomys opimus,* it was established in this way that this species feeds more frequently in the spring, during the period of intense reproduction. In the summer and autumn, the frequency of feeding decreases, as per changes in the age groups of the Siphonapteran population. Newly emerged fleas tend to feed less frequently, but imbibe more each time, than do older imagos. The mature adults, at the height of reproductive activity, feed more frequently but imbibe less blood per meal (Novokreshchenova, 1966).

The longevity of tagged fleas was studied under natural conditions (Luk'yanova & Lapina, 1965). It was noted that some solitary specimens of the fleas on Gerbillinae could still be collected after 6,5 months in the autumn and winter, thereby confirming the observations on longevity obtained by analyzing the comparative ages of the various stages of the flea populations (Zhovtyi, 1966a, 1970, 1974b).

Radioactive isotope tags were also used for studying the distance fleas disperse with their hosts, and the territorial movements of plague epizootics. For *Rhombomys opimus,* it was observed that the epizootic may move at the rate of a maximum of one kilometre per month (Rudenchik et al., 1967).

One of the aims of research on the ecology of fleas is to determine the individual characteristics of the various species so as to enhance the possibilities for development of methods for the control of fleas and plague (Zhovtyi, 1966b; Zhovtyi & Kirillov, 1970). The ecological feature which contributes most toward the control of fleas is the close association of these ectoparasites with the lair or burrows of the host. The importance of applying insecticides to the burrow system is therefore obvious. Adulticides of long-term activity are especially effective when all stages of the flea are found in the burrows of a single species of host. The ecological variations noted in different geographical areas should be borne in mind. Thus, in central Asia and the southeastern portion of European

USSR, fleas disperse somewhat from the mouth of the burrows, but this is not the case in Siberia. In the latter area, warm temperatures occur only for a brief period during the year, and only one or two generations of fleas can develop during that short interval. Attention was paid to the annual cycles and fluctuations of the local flea populations during the operations on control of fleas and plague.

The insecticide was applied to the burrows at the time believed to be most appropriate for prevention of the emergence of a new brood of adults or for action against the anticipated season peaks of vectors.

The insecticides employed against fleas were DDT and heptochlor, as aerosols or dusts, and were applied to the burrows by means of special apparatus. Deep dusting of the burrow system was especially effective, resulting in 95-100 per cent control. In Siberia, such high rates of disinsectization persisted for three to five years after application. In the newly cleared agricultural lands, plague foci were eliminated, as indicated in tests of both fleas and rodents for infection with the etiological agent.

3 OTHER ASPECTS OF THE RESEARCH ON SIPHONAPTERA

Many studies were undertaken on the biology and physiology of the fleas and the interrelations with the causal organism of plague. Investigations on life history were made for *X.g.minax* (by Zolotova & Afanasyeva, 1969b), *X.skrjabini* and *X.nuttalli* (Gerasimova, 1969a, 1971; Gerasimova & Gavrikova, 1970), *Pulex irritans* (Zolotova & Pavlov, 1971), *Frontopsylla elata caucasica* (Kosminsky et al., 1966a), *Ctenophthalmus wladimiri* (Kosminsky et al., 1970), *Ceratophyllus tesquorum* (Brukhanova & Surkov, 1970b), *C.trispinus* (Bibikova, 1965), *Xenopsylla* (Darskaya, 1970; Kadatskaya, 1965), *C.farreni* (Gordeeva, 1969).

Observations on the fertility of the fleas *X.gerbilli* and *X.skrjabini,* fed on *Rhombomys opimus* at 22-24°C and 85-90 per cent relative humidity were made by Bibikova et al. (1971). Characteristic differences were noted according to species, season and the age of the fleas. In *X.gerbilli,* an average of 2.7 eggs were laid per day, with the maximum (6.2) observed five to six days after oviposition commenced, i.e. in young fleas. The second species was not as fertile, viz. 0.4-0.5 per day, with the maximum of 1.3-2.0 eggs on the 9-10 day. In both species, egg-production was clearly rhythmical, with increases alternating with decreases. Maximum fertility was observed in females one to two weeks of age. Similar results were obtained in both laboratory-reared and natural populations.

Ovarian changes during the maturation of the eggs of fleas were studied by Prokop'ev (1958) and Kunitskaya (1970). The process has been divided into six phases according to the degree of development of the follicular epithelium, the quantity of accumulated yolk, the appearance of the chorion, and the development of the egg-cells. On the appearance and size of the yellow bodies, which represent empty follicles, five physiological stages have been established. Some authors (Vaschenok, 1966) recognize even more additional stages – as many as a total of ten. Nevertheless, it must be recognized that the actual ovarian process is an uninterrupted one for which research workers have recognized and arbitrarily and subjectively designated various phases. For analysis of the age constitution of a population, for example, the three physiological stages may be denoted: 1) a 'young' one, in which the eggs have not ripened; 2) an 'adult' phase, in which mature eggs are present, and 3) an 'old' one in which development of eggs has terminated and resorption of ovaries has taken place.

Two types of ovaries occur in fleas; panoistic and meroistic (with a polytropic subtype).

The first kind has been found in the majority of species, while *Hystrichopsylla* and *Stenoponia* are in the second category.

After these works were published, there appeared detailed reports about histological features of the digestion and oogenesis of fleas (Vashchenok, 1966; Vashchenok & Solina, 1969). For example, in *Echidnophaga oschanini,* which is a fixed parasite, in the process of blood digestion, three different stages can be distinguished: 1) before disintegration of the regular components of the blood, 2) digestion and absorbtion of nutrients, and 3) formation of haematin and the disappearance of nutrient inclusions from the nutrient cells. Oogenesis in *E.oschanini,* in general follows the same lines as in other insects with panoistic ovaries. The main difference is that multiplying oogonies are absent, the follicular epithelium is single-layered, and deposition of yolk may occur only in one egg per ovariole. There is no cyclical association between the changes in the ovaries and the intake of food. The development of oocytes in *E.oschanini* commences in the pupal stage.

On the basis of the large experimental material, the metabolic processes of rat-fleas were studied intensively on the basis of indices of the degree of oxygen consumption, and this was also done for suslik (*Citellus*) and gerbilline fleas. The index was calculated according to the method of Nielsen (1938) as modified by Shikhbazov (1952). This index and the metabolic processes in fleas in general are determined or affected by a variety of factors, particularly temperature. The intensity of respiration is highest in the period of sexual activity of the fleas. In males, gas metabolism exceeds that in females, but in both sexes the metabolism rate increases after imbibing blood. Females with a large fat-body and which ceased reproduction, use less oxygen than do young, actively reproducing females (Kondrashkina & Gerasimova, 1971).

It is noteworthy that *N.fasciatus* has a high index of metabolism over a rather broad range of temperature, from 0 to 40°C, whereas *X.cheopis,* which is tropical in origin, is active from 20 to 40°C and at the latter temperature does not depress its level of consumption of oxygen. The degree of metabolism is depressed by infection of the fleas with the microbes of plague. For example, *Neopsylla setosa, X.cheopis* and *C.tesquorum,* five days after infection, consumed only one-third to one-half as much oxygen as compared to uninfected fleas of those species. The infected fleas lost one-quarter of their total weight within that period (Kondrashkina et al., 1968).

Within the last decade there have appeared various works which estimate the epizootological importance of the fleas from a general biological point of view, e.g. the interrelations of the parasite (*Yersinia pestis,* the plague organism) with its host flea, pioneered by a series of studies by V.A.Bibikova (Bibikova & Klassovskyi, 1968, 1972), and later by other investigators (Kondrashkina, 1969; Akiyev, 1969, 1970).

Detailed investigations along these lines led to the recognition of mutual adaptations by both groups of organisms. Such modifications are expressed by the presence of the phenotypical variability of the microbes when passing from a warm-blooded host into the flea and vice versa, and by natural selection in the *Y.pestis* population. A dynamic symbiotic association exists between these bacteria and the vectors, serving to link them to the third factor in the chain, the vertebrate host.

The fleas' characteristics as blood-suckers meet the demands of the plague organism as a parasitic species and the modifications in the organism that follow ingestion by the flea may include changes in virulence. Phenotypical variability in the population of the plague *Yersinia* exist and is connected with alternating parasitism in the body of a rodent and its fleas. In consequence, the infection is maintained in nature despite the variety of hosts and vectors that are involved.

REFERENCES

Afanasyeva, O.V. & S.I.Bgytova 1963a, Data on ecology of fleas. Part I. Development of the *Xenopsylla gerbilli minax*. In: O.V.Afanasyeva & V.P.Khrustselevskiy (eds.), Mater. IV Nauch. Konfer. Po Prirodn. Ochagov. I Profilakt. Chumy, p. 11-13. Alma-Ata. (Proc. IV Sci. Conf. Endem. Prophylax. Plague.)

Afanasyeva, O.V. & S.I.Bgytova 1963b, Data on ecology of fleas. Part III. Development of the *Ctenophthalmus dolichus*. In: O.V.Afanasyeva & V.P.Khrustselevskiy (eds.), Mater. IV Nauch. Konfer. Po Prirodn. Ochagov. I Profilakt. Chumy, p. 13-15. Alma-Ata. (Proc. IV Sci. Conf. Endem. Prophylax. Plague.)

Akinshina, T.V. & I.F.Zhovtyi 1969a, Annual cycle of the flea, *Frontopsylla luculenta luculenta* J. & R., 1923, on *Ochotona daurica*, in the southeast Transbaikalia. In: A.P. Markevich (ed.), Probl. Parazit., Tez. Dok. Nauch. Konfer. Ukrain. Respub. Nauch. Obshch. Parazit., p. 111-114. Kiev. (Summaries of papers delivered at the Scientific Conference of the Ukraine Scientific Republic of general parasitology)

Akinshina, T.V. & I.F.Zhovtyi 1969b, Annual cycle of the flea, *Amphalius runatus* J. & R., (Siphonaptera) on *Ochotona daurica*, in the southeast Transbaikalia. Dokl. Irkutsk. Protivochum. Inst. 8:296-299. (Reports of the Irkutsk Anti-plague Institute)

Akiyev, A.K. 1969, Physiological variability of the plague microbe. Probl. Osobo. Opasnykh. Infekt. 6(10):23-25. Saratov. (Problems of particularly dangerous infections)

Akiyev, A.K. 1970, The possible ways of the formation of the causal organism of plague. In: V.E.Tiflov et al. (eds.), Perenoschiki. Osobo opasnykh Infektii i bor'ba s nimi, p. 44-64. Stavropol. (Carriers of particularly dangerous infections and their control)

Argyropulo, A.I. 1935, Fleas of Transcaucasia. Tr. Azerb. Inst. Mikrobiol. Epidem. 5(1): 119-216. (Works of the Azerb. Inst. Microbiol. Epidem.)

Berendyayeva, E.L., D.I.Bibikov, L.P.Rappoport, V.K.Popov & T.A.Varivodina 1969, Experiments on the study of the vitrea-population contacts of the *Marmota baibacina* by the radioactive marking method. Zool. Zhur. 45(3):430-436.

Berendyayeva, E.L. & N.A.Kulkova 1964, Materials to the ecology of the fleas of *Marmota* in the winter. Prirodn. Ochagov. Bolez. I Vopr. Parazit. 4:245-246. Frunze. (Natural foci of infections and questions of parasitology)

Berendyayeva, E.L. & N.A.Kulkova 1965, Ecology of the fleas of *Marmota baibacina* in the winter. Sborn. Ent. Rabot. Akad. Nauk. Kirgiz. S.S.R. Krzg. Otdel. Vsesoyuz. Ent. Obshch., p. 92-98. Frunze. (Entomological collection of the works of the Scientific Academy of Kirghizia, USSR)

Berendyayeva, E.L., N.A.Kulkova & A.A.Mikhailyuta 1967, Survival of a winter by the fleas of *Marmota baibacina* in the various types of burrows in the Tyan-Shan. Mater. V Nauch. Konfer. Protivochum. Uchrezhd. Sr. Azii. I Kazakhst. p. 155-157. Alma-Ata. (Materials of the V Sci. Conf. of the Anti-plague Establishments of Middle Asia and Kazakhstan)

Bgytova, S.I. 1963, Data on ecology of fleas. Part IV. Copulation and egg maturation in the individuals of *Ctenophthalmus dolichus* in the different conditions of the environment. In: O.V.Afanasyeva & V.P.Khrustselevskiy (eds.), Mater. IV Nauch. Konfer. Po Prirodn. Ochagov. I Profilakt. Chumy, p. 15-16. Alma-Ata. (Proc. IV Sci. Conf. Endem. Prophylax. Plague)

Bgytova, S.I. & T.N.Leanova 1965, Data on ecology of fleas. Part V. Critical temperatures for the development of the preimaginal phases of *Xenopsylla gerbilli minax*. In: O.V. Afanasyeva & V.P.Khrustselevskiy (eds.), Mater. IV Nauch. Konfer. Po Prirodn. Ochagov. I Profilakt. Chumy, p. 28-29. Alma-Ata. (Proc. IV Sci. Conf. Endem. Prophylax. Plague)

Bibikova, V.A. 1965, Data on ecology of fleas. Spans of the metamorphosis of *Ceratophyllus trispinus balkhaschensis* Mik., 1958, in the experimental conditions. In: O.V.Afanasyeva & V.P.Khrustselevskiy (eds.), Mater. IV Nauch. Konfer. Po Prirodn. Ochagov. I Profilakt. Chumy, p. 31-32. Alma-Ata. (Proc. IV Sci. Conf. Endem. Prophylax. Plague)

Bibikova, V.A., E.P.Bondar & A.S.Burdelov 1963, Data on ecology of fleas. Burrows of *Rhombomys opimus* in the southern part of Pre-Balkhash. In: O.V.Afanasyeva & V.P. Khrustselevskiy (eds.), Mater. IV Nauch. Konfer. Po Prirodn. Ochagov. I Profilakt. Chumy, p. 23-25. Alma-Ata. (Proc. IV Sci. Conf. Endem. Prophylax. Plague)

Bibikova, V.A. & L.N.Klassovskii 1968, On phenotypic variability of the plague microbe in connection with changing the host. Parazitol. 2(3):209-214.

Bibikova, V.A. & L.N.Klassovskii 1972, Development of the population of the plague microbe in the organism of the flea. Parazitol. 6(3): 229-236.

Bibikova, V.A., S.I.Zolotova, A.Murzakhmetova & T.N.Leanova 1971, Fertility and its dynamics in two species of flea of *Rhombomys opimus* in the experiments. Mater. VII Nauch. Konf. Protivochum. Uchrezhd. Sr. Azii. I Kazakhst. p. 367-369. Alma-Ata. (Materials of the VII Sci. Conf. of the Antiplague Establishment of Middle Asia and Kazakhstan.)

Brukhanova, L.V. 1966, Reproduction and nourishment of fleas of *Citellus.* In: Osobo Opasnykh. Infekt. Kavkaza. p. 37-40. Stavropol. (Particularly dangerous infections in the Caucasus)

Brukhanova, L.V. & L.A.Surkov 1970a, Annual cycle of *Neopsylla setosa* Wagn., 1898, in the Pre-Caucasus. In: V.E.Tiflov et al. (eds.), Perenoschiki Osobo Opasnykh. Infektii i bor'ba s nimi. p. 228-246. Stavropol. (Carriers of particularly dangerous infections and their control)

Brukhanova, L.V. & L.A.Surkov 1970b, Study of the annual cycle of *Ceratophyllus tesquorum* in the Pre-Caucasus. In: V.E.Tiflov et al. (eds.), Perenoschiki Osobo Opasnykh. Infektii i bor'ba s nimi. p. 51-55. Stavropol. (Carriers of particularly dangerous infections and their control)

Chernova, N.N. 1971, Multiplication of *Xenopsylla skrjabini* and attachment of them to the different elements of the burrow of *Rhombomys opimus* in the Mangyshlak. Mater. VII Nauch. Konfer. Protivochum. Uchrezhd. Sr. Azii I Kazakhst. p. 443-444. (Materials of the VII Sci. Conf. of the Antiplague Establishment of Middle Asia and Kazakhstan)

Darskaya, N.F. 1965, The method of the study of the annual cycles of fleas. In: B.N.Mazurmovich (ed.), Parazity i parazitozy cheloveka i zhivotnykh (Parasites and parasitism in man and animals) p. 363-385. Kiev.

Darskaya, N.F. 1970, Study of the annual cycles of the fleas of the genus, *Xenopsylla* Roths. In: V.E.Tiflov et al. (eds.), Perenoschiki Osobo Opasnykh. Infektii i bor'ba s nimi. p. 44-64. Stavropol. (Carriers of particularly dangerous infections and their control)

Darskaya, N.F. & L.V.Brukhanova 1972, Comparison of the methods of determining the physiological age of fleas. In: A.P.Markevich (ed.), Probl. Parazit., Tez. Dok. Nauch. Konfer. Ukrain. Respub. Nauch. Obshch. Parazit., p. 250-252. Kiev. (Summaries of papers delivered at the Scientific Conference of the Ukraine Scientific Republic of general parasitology)

Darskaya, N.F. & P.I.Shiranovich 1971, A new species of the flea of the genus *Ceratophyllus* (Siphonaptera, Ceratophyllidae) from the Pre-Caspian lowland. Zool. Zhur. 50(12): 1827-1834.

Elistratova, N.D., L.A.Lazareva, I.I.Eshelkin, B.V.Lazarev, G.I.Vasiliev & N.K.Burtsev 1974, The new species of flea from the mountainous Altai. Dokl. Irkutsk Protivochum. Inst. 10:228-230. Chita. (Reports of the Irkutsk Anti-plague Institute)

Emelyanova, N.D., V.I.Dubovik, Y.A.Sadkov & T.V.Akinshin 1969, Some new evidence concerning the fleas of the southeast Transbaikalia and neighbouring parts of Mongolia. Dokl. Irkutsk. Protivochum. Inst. 8:273-275. Kyzyl. (Reports of the Irkutsk Anti-plague Institute)

Emelyanova, N.D. & A.I.Goncharov 1966, A new species of flea, *Ceratophyllus zhovtyi* n.sp. from the East Siberia. Izv. Irkutsk. Protivochum. Inst. 26:309-313. Vostoka. (Reports of the Irkutsk Anti-plague Institute)

Gerasimova, N.G. 1969a, Annual cycle of the development of the flea, *Xenopsylla skrjabini.* Zool. Zhur. 48(9):1410-1412.

Gerasimova, N.G. 1969b, Hydrothermic conditions of the development of the phases of two species of the fleas of *Rhombomys opimus.* Parazitol. 3(1):24-33.

Gerasimova, N.G. 1971, Terms of the rates of maturity of adult females of *Xenopsylla skrjabini* Ioff and *Xenopsylla nuttali* Ioff. Zool. Zhur. 50(2):137-139.

Gerasimova, N.G. & E.A.Gavrikova 1970, Span of the life of the fleas *Xenopsylla skrjabini* Ioff and *Xenopsylla nuttali* Ioff. Zool. Zhur. 49(4):385-388.

Gershkovich, N.L. 1963, The methods of discovering and counting fleas in the human dwellings. Dokl. Irkutsk. Protivochum. Inst. 5:156-161. (Reports of the Irkutsk Antiplague Institute)

Goncharov, A.I. 1971, A contribution to the taxonomy of *Nosopsyllus* (Siphonaptera, Ceratophyllidae). Zool. Zhur. 50(2):222-227.

Goncharov, A.I. & P.I.Shiranovich 1970, Finding of the *Neopsylla setosa spinea* Roths., and *Neopsylla teratura rhagesa* Klein in the USSR and keys to the subspecies of some

Neopsylla. In: V.E.Tiflov et al. (eds.), Perenoschiki. Osobo. Opasnykh. Infektii i bor'ba s nimi., p. 386-409. Stavropol. (Carriers of particularly dangerous infections and their control)

Gordeeva, V.P. 1969, Observations on the development cycle and nourishment of the birdflea, *Ceratophyllus farreni* Roths., 1905. In: A.I.Cherepanev (ed.), Perelet. Ptitz. I Ikh. Rol. V Rasporstran. Arbovirus. Izdatel. Nauka. Sibirsk. Otdelen, p. 230-235. Novosibirsk. (Migrating birds and their role in the distribution of arboviruses)

Ilinskaya, V.L., I.V.Morozova, N.J.Savelov, I.P. Kusin & G.I.Makushin 1971a, Experiment of an observation of the temperature in the deep parts of the burrow of *Rhombomys opimus.* Mater. VII. Nauch. Konfer. Protivochum. Uchrezhd. Sr. Azii. I Kazakhst. p. 381-383. Alma-Ata. (Materials of the VII Sci. Conf. of the Anti-plague Establishments of Middle Asia and Kazakhstan)

Ilinskaya, V.L., I.V.Morozova, N.J.Savelov, I.P. Kusin & G.I.Makushin 1971b, Dependence of the temperature regime in the burrows of *Rombomys opimus,* from their position in relief and exposition. Mater. VII Nauch. Konfer. Protivochum. Uchrezhd. Sr. Azii I Kazakhst., p. 384-386. Alma-Ata. (Materials of the VII Sci. Conf. of the Anti-plague Establishments of Middle Asia and Kazakhstan)

Ilinskaya, V.L., V.V.Zubov, A.S.Burdelov & S.K.Baranovsky 1971c, Problem of the micro-climate of the burrows of *Rhombomys opimus* in the central Kara-Kum. Mater. VII Nauch. Konfer. Protivochum. Uchrezhd. Sr. Azii. I Kazakhst., p. 378-381. Alma-Ata. (Materials of the VII Sci. Conf. of the Anti-plague Establishments of Middle Asia and Kazakhstan)

Ioff, I.G. & E.P.Bondar 1956, The Fleas of Turkmenia. Tr. Nauch. Issled. Protivochum. Inst. Kavkaza. I Zakakaz, p. 29-118. Stavropol. (Works of the Scientific Anti-plague Institute of Caucasus and Zakakaz)

Ioff, I.G., M.A.Mikulin & O.I.Scalon 1965, Keys to the Fleas of Central Asia and Kazakhstan. Meditsina, 369 pp. Moscow.

Ioff, I.G. & O.I.Scalon 1954, Keys to the Fleas of Eastern Siberia, the Far East and Adjacent Districts. Acad. Med. Sci., Medgiz. 275 pp.

Ioff, I.G. & V.E.Tiflov 1938, Appliance for Determination of the Fleas of the Southeastern Part of the European USSR. 116 pp. Saratov.

Ioff, I.G. & V.E.Tiflov 1954, Key to Aphaniptera (Suctoria) of the Southeast of the USSR. 201 pp. Stavropol.

Kadatskaya, K.P. 1965, Terms of the metamorphosis of the basic species of the fleas of the Gerbillinae in Azerbaidzhan. Mater. II Nauch. Sessii Ent. Azerbaidzhana. p. 103-107. Baku.

Kafarskaya, D.G. & L.S.Lysenko 1963, Ecology of the fleas of *Marmota caudata* in Tadzhikistan. In: O.V.Afanasyeva & V.P.Khrustselevskiy (eds.), Mater. IV. Nauch. Konfer. Po Prirodn. Ochagov. I Profilakt. Chumy, p. 99. Alma-Ata. (Proc. IV Sci. Conf. Endem. Prophylax. Plague)

Karandina, R.S. 1964, Larvae of the fleas of *Meriones erythrurus* and some other rodents of Azerbaidzhan. Tr. Armyansk. Protivochum. Stants. 3:473-504. (Works of the Armenian Anti-plague Station)

Kiryakova, A.N. 1961, Larvae of the fleas of the family Pulicidae. Akad. Nauk. USSR. Zool. Inst. Otd. Parazit. Sborn. 20:306-323. Moscow-Leningrad. (Scientific Academy of the USSR, Zoological Institute, Parasitological Symposium)

Kiryakova, A.N. 1964, Larvae of the fleas of the family Ctenophthalmidae. Zool. Zhur. 43(4):572-580.

Kiryakova, A.N. 1965, Larvae of the fleas (Aphaniptera) of the family Ceratophyllidae. Ent. Obozr. 44(2):383-395.

Kiryakova, A.N. 1968, External morphology of the flea larvae of the family Pulicidae (Aphaniptera). Parazitol. 2(6):548-558.

Kiryakova, A.N. & M.J.Cotton 1970, External morphology of the larvae of the fleas of the genus *Archaeopsylla* (Siphonaptera, Pulicidae). Ent. Obozr. 49(3):534-540.

Kizilov, V.A., N.M.Semenov & V.S.Korotkova 1967, Ecology of *Marmota caudata* and its fleas in the winter period. Ekologiia Mlekopitajuschikh I Ptic. p. 53-54. Moscow.

Koklyagina, A.T. 1969, Fauna of the fleas of the Steppe Zone of the Altai region. Dokl. Irkutsk. Protivochum. Inst. 8:275-277. (Reports of the Irkutsk Anti-plague Institute)

Kondrashkina, K.I. 1969, Do fleas ail from the plague? Probl. Osobo Opasnykh. Infekt. 5: 212-222. Saratov. (Problems of particularly dangerous infections)

Kondrashkina, K.I. & A.F.Dudnikova 1962, Consumption of the oxygen by fleas of *Citellus* as a physiological test of their viability. Osobo Opasnykh. I Prirodn. Infekt. p. 63-69. Medgiz. (Particularly dangerous natural foci infections)

Kondrashkina, K.I. & A.F.Dudnikova 1968, Consumption of the oxygen by the fleas of the grey rats. In: B.K.Fenyuk (ed.), Gryzuny I Ih Ektoparazity. p. 87-91. Saratov. (Rodents and their ectoparasites)

Kondrashkina, K.I. & N.G.Gerasimova 1971, Dependability of the level of metabolism from the physiological state in two mass species of the fleas of *Rhombomys opimus.* Probl. Osobo. Opasnykh Infekt. p. 32-37. (Problems of particularly dangerous infections)

Kondrashkina, K.I., I.I.Kuraev & G.A.Zakharova 1968, Some problems of mutual adaptation of the plague microbe with the organism of fleas. Parazitol. 2(6):543-548.

Koshkin, S.M. 1966, Contribution towards the fauna of fleas in the district Sovetskaya Gavan. Izv. Irkutsk. Protivochum. Inst. 26: 342-348. Vostoka. (Reports of the Irkutsk Anti-plague Institute)

Kosminsky, R.B., G.A.Avetisyan & A.N.Talybov 1966a, *Ceratophyllus consimilis* Wagn., 1898, in the mountains of Transcaucasia. Osobo Opasnykh. Infekt. Kavcaza. p. 87-90. Stavropol. (Particularly dangerous infections in the Caucasus)

Kosminsky, R.B., G.A.Avetisyan & A.N.Talybov 1966b, Materials to the ecology of the flea *Amphipsylla rossica* Wagn., 1912. Osobo. Opasnykh. Infekt. Kavkaza. p. 91-93. Stavropol. (Particularly dangerous infections in the Caucasus)

Kosminsky, R.B., G.A.Avetisyan, A.N.Talybov & S.G.Tsikhistavi 1966c, Towards the study of the annual cycle of fleas, *Frontopsylla elata caucasica* I. & Arg., 1934, in the Transcaucasus plateau. Osobo. Opasnykh. Infekt. Kavkaza p. 94-97. Stavropol. (Particularly dangerous infections in the Caucasus)

Kosminsky, R.B., G.A.Avetisyan & A.N.Talybov 1970, Annual cycle of the fleas *Ctenophthalmus wladimiri* Is.-G., 1948, on the *Microtus arvalis* in the southeast of Transcaucasus plateau. In: V.E.Tiflov et al. (eds.), Perenoschiki Osobo Opasnykh. Infektii i bor'ba s nimi. p. 79-107. Stavropol. (Carriers of particularly dangerous infections and their control)

Kozlovskaya, O.L. & V.J.Tshipanin 1969, Fleas of the *Apodemus sylvaticus* in Khabarovsk area near Amur. Dokl. Irkutsk. Protivochum. Inst. 8:277-279. (Reports of the Irkutsk Anti-plague Institute)

Kunitskaya, N.T. 1968, Towards the study of the structure of the fleas' ovaries. Summary of the Reports of the XIII Internatl. Congr. Ent. p. 138.

Kunitskaya, N.T. 1970, Structure of the fleas' ovaries. Parazitol. 4(5):444-450.

Kunitskaya, N.T., V.N.Kunitsky & D.M.Gauzshtein 1969, Multiplication and age contents of the populations of fleas of the genera *Coptopsylla* and *Paradoxopsyllus* in the southern area near Balkhash. Mater. VI Nauch. Konfer. Protivochum. Uchrezhd. Sr. Azii. I Kazakhst. 2:74-76. Alma-Ata. (Materials of the VI Sci. Conf. of the Anti-plague Establishments of Middle Asia and Kazakhstan)

Kunitskaya, N.T., V.N.Kunitsky & D.M.Gauzshtein 1971, Age content of the imago in the population of *Xenopsylla gerbilli* and *Xenopsylla hirtipes* in the southern area near Balkhash. Mater. VII Nauch. Konfer. Protivochum. Uchrezhd. Sr. Azii. I Kazakhst. 2:74-76. (Materials of the VII Sci. Conf. of the Anti-plague Establishments of Middle Asia and Kazakhstan)

Kunitsky, V.N. & D.M.Gauzshtein 1963a, Method of the study of numbers of fleas of the genera *Echidnophaga, Xenopsylla* and *Coptopsylla* from the *Rhombomys opimus.* In: O.V.Afanasyeva & V.P.Khrustselevskiy (eds.), Mater. IV Nauch. Konfer. Po Prirodn. Ochagov. I Profilakt. Chumy. p. 122-124. Alma-Ata. (Proc. IV Sci. Conf. Endem. Prophylax. Plague)

Kunitsky, V.N. & D.M.Gauzshtein 1963b, Method of the calculating of the numbers of fleas at the burrow entrances of *Rhombomys opimus.* Zool. Zhur. 42(11):1743-1745.

Kunitsky, V.N., V.M.Volkov & Z.R.Lelikov 1971, Annual fluctuation of the number of fleas of *Rhombomys opimus* in the northeast part of the Caspian lowland. Mater. VII Nauch. Konfer. Protivochum. Uchrezhd. Sr. Azii. I Kazakhst. p. 389-392. Alma-Ata. (Materials of the VII Sci. Conf. of the Anti-plague Establishments of Middle Asia and Kazakhstan)

Lyetov, G.S., N.D.Emelyanova, G.I.Lyetova & A.D.Sulimov 1966, Rodents and their parasites in the populated places of the Tuva region. Izv. Irkutsk. Protivochum. Inst. 26:270-276. Irkutsk. (Reports of the Irkutsk Anti-plague Institute)

Luk'yanova, A.D. & N.F.Lapina 1965, Span of life of the fleas *Neopsylla setosa* Wagn, and *Ceratophyllus tesquorum* Wagn. in natural conditions. Zool. Zhur. 44(6):883-887.

Mialkovskaia, S.A. & L.V.Brukhanova 1972, Fleas of *Citellus pygmaeus* at the time of hibernation of their hosts. Zool. Zhur. 51(2): 308-310.

Mironov, N.D., E.N.Nelzina, I.Z.Klimchenko, D.S.Rezinko, N.J.Chernov, G.M.Danilov, G.P.Samarina & A.V.Rodianova 1963, Distribution of the fleas of the burrows of *Citellus pygmaeus* and rationalization of the methods of the calculation of their numbers. Zool. Zhur. 42(3):384-394.

Morozova, I.V., N.M.Savelov, et al., 1969, Method of the marking of the fleas by radioactive isotopes. Mater. Nach. Konfer. Protivochum. Uchrezhd. Sr. Azii. I Kazakhst. 2: 81-83. Alma-Ata. (Materials of the Scientific Conference of the Anti-plague Establishments of Middle Asia and Kazakhstan)

Murzakhmetova, L. 1969, Two new subspecies of the fleas (Siphonaptera) from Middle Asia and Kazakhstan. Zool. Zhur. 48(4):607-610.

Nielsen, E.T. 1938, Ein einfachr Microrespirationsapparat. Biochem. Zt. 260:1-3.

Novokreshchenova, N.S. 1960, Contribution towards the ecology of the fleas of *Citellus pygmaeus* in connection with their epizootological significance. Trudy Inst. 'Mikrob.' 4: 444-456. Saratov.

Novokreschchenova, N.S. 1966, Question of the methods of the study of the activity of the fleas' nutrition. Dokl. Irkutsk. Protivochum. Inst. 7:237-239. (Reports of the Irkutsk Antiplague Institute)

Novokreshchenova, N.S., I.S.Soldatkin, V.K. Denisenko & L.A.Martens 1961, Use of the radioactive carbon for the marking of the fleas. Med. Parazit. I Parazitar. Bol. 30(1): 72-76.

Novokreshchenova, N.S., I.S.Soldatkin & A.J. Lavoshin 1967, Comparative frequency of the nourishment of the various species of the fleas, determined in the laboratory conditions with the use of radioactive isotopes. In: B.K. Fenyuk (ed.), Gryzuny I Ih Ectoparazity, p. 49-54. Saratov. (Rodents and their ectoparasites)

Paramonov, B.B., N.D.Emelyanova, V.N.Zarubina & V.L.Kontzimavitchus 1966, Contribution to the study of the ectoparasites of rodents and shrews of Kamchatka. Izv. Irkutsk. Protivochum. Inst. 26:331-341. Vostoka. (Reports of the Irkutsk Anti-plague Institute)

Pauller, O.F., N.I.Elshanskaya & I.V.Shretsov 1966, Ecologo-faunistic essay of mammals and birds in the tularemia nidus in the delta of the river Selenga. Izv. Irkutsk. Protivochum. Inst. 26:322-333. Irkutsk. (Reports of the Irkutsk Anti-plague Institute)

Poliakov, V.K. & V.L.Shchevchenko 1965, Distribution and abundance of fleas *Xenopsylla conformis* in the northern margin of the Volga-Ural sands. In: O.V.Afanasyeva & V.P.Khrustselevskiy (eds.), Mater. IV. Nauch. Konfer. Po Prirodn. Ochagov. I Profilakt. Chumy, p. 199-200. Alma-Ata. (Proc. IV Sci. Conf. Endem. Prophylax. Plague)

Popova, A.C. & A.A.Sokolova 1965, Results of the approbation of the methods of calculating the numbers of fleas in the burrows of *Rhombomys opimus* in the Majun-kum near the river Chu. In: O.V.Afanasyeva & V.P. Khrustselevskiy (eds.), Mater. IV Nauch. Konfer. Po Prirodn. Ochagov. I Profilakt. Chumy, p. 201. Alma-Ata. (Proc. IV Sci. Conf. Endem. Prophylax. Plague)

Prokop'ev, V.N. 1958, Method of determination of the physiological age of the *Oropsylla silantiewi* W. and seasonal changes in age groups in the flea population. Izv. Irkutsk. Protivochum. Inst. 17:323-324. (Reports of the Irkutsk Anti-plague Institute)

Rappoport, L.P., E.L.Bezediaev, H.A.Kulkova & A.A.Michailuta 1969, Marking of the fleas by the radioactive isotopes and distribution of them by *Marmota*. Probl. Osobo Opasnykh. Infekt. 3:156-160. Saratov. (Problems of particularly dangerous infections).

Rudenchik, Yu.V., I.S.Soldatkin, E.A.Severova, N.A.Mokiyevich & Z.I.Klimova 1967, Quantitative evaluation of the possibility of the territorial advance of epizooty of plague in the population of the *Rhombomys opimus* (northern Kara-kum). Zool. Zhur. 46(1):117-123.

Scalon, O.I. 1968, A new subspecies of the flea, *Ophthalmopsylla volgensis* from eastern Dzungaria. Zool. Zhur. 47(12):1871-1872.

Scalon, O.I. 1970, Order Siphonaptera (Aphaniptera-Suctoria) – Fleas. In: The Key to the Insects of the European Part of the USSR. 5(2):799-844.

Shchevchenko, V.L., A.K.Grazhdanov, L.K.Zharinova & T.A.Andreyeva 1976, On the ability of the bird flea *Frontopsylla frontalis alatau* Fed., 1946, to transmit rodent plague. Med. Parazit. I Parazitar. Bol. 45(1):49-52.

Shikhbazov, V.G. 1952, Method of study of respiratory trunk of silkworms and other insects. Trudy Inst. Zool. A.N. Ukrain. SSR 7: Kiev.

Soldatkin, I.S. & A.I.Levoshina 1968, To the problems of the duration of the uninterrupted presence of fleas in the fur of *Rhombomys opimus.* In: B.K.Fenyuk (ed.), Gryzuny I Ih Ektoparazity, p. 55-58. Saratov. (Rodents and their ectoparasites)

Soldatkin, I.S., Yu.V.Rudenchik, & E.A.Severov 1969, Improvement of methods of the radioactive marking of the ectoparasites through the blood of their hosts. Mater. Nauch. Konfer. Protivochum. Uchrezhd. Sr. Azii. I

Kazakhst. 2:92-94. Alma-Ata. (Materials of the Scientific Conference of the Anti-plague Establishments of Middle Asia and Kazakhstan)

Stepanova, N.A., N.A.Rachinina & R.A.Urmanov 1971, Study of the annual cycle of the fleas of the genus *Xenopsylla* in central Kyzylkum. Mater. Nauch. Konfer. Protivochum. Uchrezhd. Sr. Azii. I Kazakhst. p. 421-423. Alma-Ata. (Materials of the Scientific Conference of the Anti-plague Establishments of Middle Asia and Kazakhstan)

Stupina, A.G. 1971, Materials to the ectoparasites of the small mammals in tularemia foci in the north Baikal district of Buryat ASSR. Dokl. Irkutsk. Protivochum. Inst. 9:240-241. (Reports of the Irkutsk Anti-plague Institute)

Sviridov, G.G. 1963, Application of the radioactive isotopes in the study of the fleas' ecology. The contact of animals and intensity of the exchange of ectoparasites in the population of *Rhombomys opimus.* Zool. Zhur. 42(6):947-949.

Sviridov, G.G., I.V.Morozova, K.P.Kuluzhenova & V.L.Ilinskaya 1964, Application of radioactive isotopes in the study of several problems of fleas' ecology. Mater. Nauch. Konfer. Protivochum. Uchrezhd. Sr. Azii. I Kazakhst. 4:232-234. (Materials of the Scientific Conference of the Anti-plague Establishment of Middle Asia and Kazakhstan)

Tepiykh, V.S. 1968, Influence of age and conditions of upkeeping on the fertility of some species of fleas. Vestnik Moskau Universitetu Biologiya i Pochvovedenie 5:18-23. (News of the Moscow University, Biology and Soils Science)

Theodorides, J. 1954, Parasitism et ecologie. Biologie Medicale 43(4):440-464.

Tiflov, V.E., O.I.Scalon & B.A.Rostigaev 1977, Identification of Fleas from Caucasus. Stavropol'skoe Knizhnoe Izdatel'stvo. 278 pp.

Ustyuzhina, I.M. 1969, Discovery of *Neopsylla galea* Ioff, 1946, in the Tuva region. Dokl. Irkutsk. Protivochum. Inst. 8:279-280. (Reports of the Irkutsk Anti-plague Institute)

Vansulin, S.A. & L.A.Volkova 1962, The coat of *Rhombomys opimus* Licht. and its influence on the number of fleas in different seasons of the year. Zool. Zhur. 41(1):147-150.

Vaschenok, V.S. 1966, Histological characteristics of oogenesis in fleas, *Echidnophaga oschanini* Wagn. (Pulicidae, Aphaniptera). Zool. Zhur. 45(12):1815-1825.

Vaschenok, V.S. 1968, Morphological and physiological peculiarities of the fleas and different characteristics of their parasitological relationships with their hosts. Resumé address at the VIII Internatl. Congr. Ent. p. 454-455. Leningrad.

Vaschenok, V.S. & L.T.Solina 1969, On the digestion in fleas *Xenopsylla cheopis* Roths., (Aphaniptera, Pulicidae). Parazitol. 3(5):451-460.

Vasiliev, G.I., N.D.Emelyanova, N.J.Elshanskaya, V.N.Zarubina, B.B.Paramonov & V.N.Iakuba 1974, Ectoparasites of mammals and vermin of the west Kamchatka plain. In: F.N.Kirilov (ed.), Simposium Biol. Problemy. Severa 2: 190-192. Yakutsk. (VI Symposium on biological problems of the North)

Vasiliev, G.I. & L.A.Lazareva 1971, Additional data concerning the fauna of the fleas of Sailugemsk range. Bokl. Irkutsk. Protivochum. Inst. 9:220-221. Irkutsk. (Reports of the Irkutsk Anti-plague Institute)

Vasiliev, G.I. & I.F.Zhovtyi 1967, On the biology of the reproduction of *Ceratophyllus tesquorum sungaris* Jord., 1929. In: A.P. Markevich (ed.), Probl. Parazitol., Tez. Dok. Nauch. Konfer. Ukrain. Respub. Nauch. Obshch. Parazit., p. 381-383. Kiev. (Summaries of papers delivered at the Scientific Conference of the Ukraine Scientific Republic of general parasitology)

Vasiliev, G.I. & I.F.Zhovtyi 1971, Annual cycle of *Oropsylla asiatica* Wagn., 1929, (Siphonaptera) on *Citellus undulatus* in the Pre-Baikal region. Dokl. Irkutsk. Protivochum. Inst. 9: 227-229. (Reports of the Irkutsk Anti-plague Institute)

Vershina, O.N. 1970, Ecology of the winter fleas of the *Microtus gregalis* from southeast Transbaikalia. In: V.E.Tiflov et al. (eds.), Perenosshiki Osobo Opasnykh. Infektii i bor'ba s nimi. p. 342-347. Stavropol. (Carriers of particularly dangerous infections and their control)

Vishnyakov, S.V. 1963, Comparative characteristics of the number of fleas and ticks in the nests of *Marmota baibacina* in relation to the stationary locations and inhabitants of the burrows. Zool. Zhur. 42(4):627-628.

Vysotskaya, S.O. 1968, A new species of flea of the genus *Ctenophthalmus* from east Carpathians. Parazitol. 2(4):344-346.

Yurkina, V.I. 1961, Siphonaptera Fauna of the Ukraine. Akad. Nauk. Ukrain. SSR, Inst. Zool. Fauna Ukrainy, 17(4), 152 pp. Kiev. (Scientific Academy of the Ukraine in the USSR, Institute of the Zoological Fauna of the Ukraine)

Zagniborodova, A.E.N. 1968, Many years of

study on the ecology of fleas of *Rhombomys opimus* in the southern part of the central Kara-Kum. In: B.K.Fenyuk (ed.), Gryzuny I Ih Ektoparazity, p. 76-86. Saratov. (Rodents and their ectoparasites)

Zhovtyi, I.F. 1959, Notes on the ecology of fleas in connection with their epizootological significance. Prirodn. Ochag. i Epidem. Osobo Opasnykh. Inf. (Natural foci and epidemiology of particularly dangerous infections) Trudy Inst. 'Mikrob.' p. 170-180. Saratov.

Zhovtyi, I.F. 1960, On principles and methods of calculating the number of fleas in a population. Izv. Irkutsk. Protivochum. Inst. 23: 338-345. (Reports of the Irkutsk Anti-plague Institute)

Zhovtyi, I.F. 1963, Some controversial questions of the ecology of rodent fleas in connection with their epidemiological importance. Dokl. Irkutsk Protivochum Inst. 6:95-106. Kyzyl. (Reports of the Irkutsk Anti-plague Institute)

Zhovtyi, I.F. 1966a, The ecological premises of the method and tactics of ground disinfection at the Siberian natural plague foci. Dokl. Irkutsk. Protivochum. Inst. 7:231-234. Kyzyl. (Reports of the Irkutsk Anti-plague Institute)

Zhovtyi, I.F. 1966b, On possible ways of further perfection of methods of the quantitative study of the population of rodent fleas. Dokl. Irkutsk. Protivochum. Inst. 7:234-237. Kyzyl. (Reports of the Irkutsk Anti-plague Institute)

Zhovtyi, I.F. 1966c, Outlines of the ecology of rodent fleas in Siberia and the Far East. Izv. Irkutsk Protivochum. Inst. 26:282-308. Irkutsk. (Reports of the Irkutsk Anti-plague Institute)

Zhovtyi, I.F. 1968a, Influence of rodents' mode of life on the ecological conditions of their fleas. Izv. Irkutsk. Protivochum. Inst. 27: 195-211. Kyzyl. (Reports of the Irkutsk Anti-plague Institute)

Zhovtyi, I.F. 1968b, A study of conditions of fleas in the burrows of rodents in Siberia and the Far East. Izv. Irkutsk. Protivochum. Inst. 27:212-230. Kyzyl. (Reports of the Irkutsk Anti-plague Institute)

Zhovtyi, I.F. 1969a, Sexual composition of the flea population and regularity of its changing. In: A.P.Markevich (ed.), Probl. Parazit., Tez. Dok. Nauch. Konfer. Ukrain. Respub. Nauch. Obshch. Parazit., 109-111. Kiev. (Summaries of papers delivered at the Scientific Conference of the Ukraine Scientific Republic of general parasitology)

Zhovtyi, I.F. 1969b, Condition of the fleas' nutrition and structure of the rodents' skin. Dokl. Irkutsk. Protivochum. Inst. 8:305-308. (Reports of the Irkutsk Anti-plague Institute)

Zhovtyi, I.F. 1970, Essays on the flea ecology of the rodents in Siberia and the Far East. Fleas of *Marmota.* In: V.E.Tiflov et al. (eds.), Perenoschiki Osobo Opasnykh. Infektii i bor'ba s nimi. p. 253-283. Stavropol. (Carriers of particularly dangerous infections and their control)

Zhovtyi, I.F. 1972, Adaptations of daily and seasonal cycles of the rodent fleas, to the biological cycles of their hosts. In: A.J.Cherepanov (ed.), Fauna I Ekologiia Artropodou V Sibiri, p. 163-170. Novosibirsk. (Fauna and ecology of the arthropods of Siberia)

Zhovtyi, I.F. 1974a, Population ecology of the fleas (Siphonaptera) in Siberia and the Far East. Results and problems. Mater. VIII S.'ezda Vsesoyuznogo Entomol. Obschestva I:224-225. Leningrad. (Materials of the VII Scientific Conference of the Entomological Society of the USSR)

Zhovtyi, I.F. 1974b, The study of fleas (Siphonaptera) in the Far East. Voprosy Ent. Sibiri p. 142-148.

Zhovtyi, I.F. & V.I.Dubovik 1967, Seasonal variations of stage and sexual composition of populations of fleas of *Microtus brandti* in southwest Transbaikalia. In: A.P.Markevich (ed.), Probl. Parazit., Tez. Dok. Nauch. Konfer. Ukrain. Respub. Nauch. Obshch. Parazit. 5:399-400. Kiev. (Summaries of papers delivered at the Scientific Conference of the Ukraine Scientific Republic of general parasitology)

Zhovtyi, I.F. & V.V.Kirillov 1970, Ecological features of fleas and place of disinfection in the prophylactic of plague and improvement of sanitary conditions in the Siberian plague foci. Probl. Osobo. Opasnykh. Infekt. 1:43-47. (Problems of particularly dangerous infections)

Zhovtyi, I.F. & G.I.Vasiliev 1962, Temperature conditions of the flea habitat in the fur of rodents. Dokl. Irkutsk. Protivochum. Inst. 4:152-156. (Reports of the Irkutsk Anti-plague Institute)

Zolotova, S.I. & O.V.Afanasyeva 1969, Biology of *Xenopsylla gerbilli minax* Jord., 1936. Parazitol. 3(4):301-308.

Zolotova, S.I. & A.E.Pavlov 1971, Contribution towards the multiplication of *Pulex irritans.* Mater. Nauch. Konfer. Protivochum. Uchrezhd. Sr. Azii. I Kazakhst. p. 375-377. Alma-Ata. (Materials of the Scientific Conference of the Anti-plague Establishments of Middle Asia and Kazakhstan)

BERNARD C. NELSON* & CHARLES R. SMITH**
* *California Department of Health, Berkeley, Ca., USA*
** *California Department of Health, Redding, Ca., USA*

ECOLOGY OF SYLVATIC PLAGUE IN LAVA CAVES AT LAVA BEDS NATIONAL MONUMENT, CALIFORNIA

ABSTRACT

Lava Beds National Monument, located in northeastern California at the western edge of the Great Basin Biotic Province, encompasses an 18 600 ha area, with elevations between 1 200 and 1 525 m. The vegetation, characterized as sagebrush steppe, is typical of cooler arid regions and is comprised predominantly of sagebrush (*Artemisia tridentata*), juniper (*Juniperus occidentalis*), bitter brush (*Purshia tridentata*), mountain mahogany (*Cercocarpus ledifolius*), rabbit brush (*Chrysothamnus nauseosus*), and bunch grass (*Agropyron spicatum*). Relatively recent volcanic activity has produced extensive lava flows, lava rims and lava tubes. The lava tubes form the nearly 300 known lava caves and the interconnecting canyons that have resulted from the collapsed ceilings of the tubes. These extensive, often complex, cave systems may have at least five levels, with the deeper levels having permanent pools and formations of ice. Lava caves, including those levels with ice, are inhabited by the bushy-tailed wood rat (*Neotoma cinerea*) and the deer mouse (*Peromyscus maniculatus*). Other species of rodents may occupy the talus and twilight zones of these caves, but only these two species use the dark zones (areas of total darkness).

Northeastern California has had a history of major epizootic activity caused by *Yersinia pestis.* Plague epizootics at the Monument have been documented in 1962 (Stark & Kinney, 1969), in 1972-73 (Nelson & Smith, 1976), and in 1976-77 (Smith & Nelson, in prep.). All three epizootics occurred during the colder months of the year, involving *N.cinerea, P. maniculatus,* and their fleas in the lava caves. Until the 1976-77 outbreak, little information was available concerning plague activity outside the cave environment. Laboratory-confirmed evidence of plague activity is now available in surface-dwelling rodents and carnivores. The epizootic in cave-dwelling rodents apparently had preceded that in surface-dwelling rodents.

Following the 1972-73 epizootic, a long term study was begun concerning the ecology of plague in lava caves. Initial observations indicated that *N.cinerea* was an ideal sentinel animal for detecting the presence or absence of plague at the Monument and that its den sites were well suited for surveillance for sign of occupancy. *N.cinerea* suffered extensive mortality due to plague in 1972-73 and in 1976-77, with the few survivors often exhibiting antibody titers to *Y.pestis.* In the 20 caves that are most frequently visited by tourists, approximately 100 den sites are known. Of these, approximately ten dens were occupied at the end of 1972-73 epizootic with three wood rats having antibody titers from 1:1024-1:2048. After the wood rat population increased in 1974 to 1976, it was again reduced in

1976-77 to two wood rats, one of which had an antibody titer of 1:1024. In contrast, little mortality was apparent among the population of *P.maniculatus* inhabiting the same lava caves. After a low population level for *P.maniculatus* in early 1973, this population increased and remained relatively abundant up to and throughout the 1976-77 epizootic, in spite of evidence of plague activity in fleas (*Opisodasys keeni* and *Catallagia sculleni*) taken from deer mice and a titer in one mouse of 1:1024.

A study of the features associated with wood rat dens revealed that changes in them could be used to establish if a den was active, recently abandoned, or old abandoned. The characteristic features of an active den include a platform that has a clean appearance, with fresh sticks added frequently; a nest with a cup that has deep vertical walls and a smooth lining; a food cache with fresh green cuttings; urine stations that have moist, yellow to brown urine with a musky odour; faecal droppings that are soft and olive green; and an assemblage of invertebrates in the nest predominated by fleas and parasitic and predatory mites. In contrast, progressive and predictable changes occur in these features following abandonment: the den takes on a dull appearance followed by growth of fungus on the sticks, nest, and faecal droppings and by the presence of spider webs in the platform; the nest cup lining becomes frayed and filled with debris and later flattens out progressively with time; the clippings in the food cache become dry and yellow; the urine loses its musky odour and becomes dry and varnish-like in texture and brownish-black in colour; the faecal droppings become hard and brown and later grey after the period of fungal growth; and the assemblage of invertebrates goes through a successional change first to predators and scavengers and then to those associated with stored food products and organic detritus. Use of these features in determining den occupancy was instrumental in the early detection of the 1976-77 epizootic and in the determination of the scope of the epizootic in caves. Material collected through trapping of rodents gave direct evidence of plague, verifying the observations.

The focal point of plague transmission in caves appears to be the den, in particular the nest where the fleas live. *N.cinerea,* except during the breeding season, is a solitary, territorial rodent that occupies and rebuilds abandoned dens. Few new den sites are established. Among the dens in the caves are those containing vast accumulations of material. These preferred dens have been used by countless generations of wood rats, perhaps for centuries, and are favoured over the smaller dens. *P.maniculatus* apparently live in close association with wood rats occupying the preferred dens. Furthermore, deer mice may use abandoned dens as home sites; the normal home sites of deer mice are impossible to find within the breakdown and rubble of lava caves. The preference of wood rats for established dens and the predilection of deer mice for abandoned dens bring both species into contact with infective fleas during an epizootic. These behavioural preferences perpetuate the epizootic throughout the lava cave systems and undoubtedly affect most occupant rodents in the caves as well as invading individuals from the surface. The stable environment afforded by lava caves allows at least a known three-month period of survival for infected fleas (*Orchopeas sexdentatus, Phalacropsylla allos,* and *Megarthroglossus* sp.) in abandoned nests.

The distinctive features of the epizootic pattern in lava caves appear to be the differences in susceptibility of the rodent species, the behavioural patterns of these rodent species that perpetuate the epizootic, and the environmental conditions enabling plague to circulate and survive. Although direct evidence is lacking, *P.maniculatus* appears to be a resistant host to the plague organism. In spite of the heavy mortality suffered by *N.cinerea, P.maniculatus* remained abundant in the caves during the epizootic. If an enzootic focus exists at the Monument, it is hypothesized that *P.maniculatus* may be the reservoir host. The behavioural pattern described above ensures circulation of the plague organism through contact with infective fleas by both host species. The stable environment allows relatively long term survival of infective fleas in abandoned nests. It is noteworthy that plague transmission in lava caves occurs at ambient temperatures that are within a few degrees above and below 0°C. In ice caves, transmission occurs at ambient temperatures that are constantly below 0°C. Finally, the preferred den appears to be an ideal medium to test the hypothesis of telluric plague, the survival and transmission of the plague organism in soil.

REFERENCES

Nelson, B.C. & C.R.Smith 1976, Ecological effects of a plague epizootic on the activities of rodents inhabiting caves at Lava Beds National Monument, California. J. Med. Ent. 13:51-61.

Stark, H.E. & A.R.Kinney 1969, Abundance of rodents and fleas as related to plague in Lava Beds National Monument, California. J. Med. Ent. 6:287-94.

S. ITO* & J. W. VINSON**
*Harvard Medical School, Boston, Mass., USA
**Harvard School of Public Health, Boston, Mass., USA

ELECTRON MICROSCOPY OF RICKETTSIAE IN FLEA TISSUES

ABSTRACT

Xenopsylla cheopis infected with *Rickettsia mooseri,* as well as uninfected control fleas, were examined by thin section electron microscopy. Murine typhus rickettsiae multiplied by binary fission in midgut epithelial cells. Some cells contained a few scattered rickettsiae; others were packed with micro-organisms. Rickettsiae, with or without a surrounding thin halo of low density, lay directly in the cytoplasm. Infected cells showed no obvious pathologic changes but in some cells rickettsiae undergoing condensation and disruption had been sequestered in lysosomal vacuoles. Rickettsiae were released on rupture and exfoliation of infected midgut cells.

Symbiotic rickettsia-like micro-organisms, morphologically indistinguishable from *R.mooseri,* were found in the tissues of normal control fleas, as well as in typhus-infected fleas. While a thorough survey of all flea tissues was not made, these rickettsia-like micro-organisms were found in muscle cells surrounding midgut epithelian cells, epithelial cells of the proventriculus and tracheoles, occasionally in both fast and slow contracting muscle cells, and Malpighian tubules. Limited numbers were found in flea larval tissues. In the testes, micro-organisms were limited to supporting cells enclosing sperm bundles. Their presence in all developmental stages of the ovum suggests their transovarial transmission. The rickettsial symbiotes did not seem to cause extensive cellular disorganization, although they produced some focal areas of disruption in striated muscle cells. There was a consistent tendency for rickettsiae to be sequestered within vacuoles, without undergoing degenerative changes.

These observations suggest the usefulness of electron microscopy in revealing the presence of symbiotic micro-organisms which might be unrecognized by other methods, and, in the present case, establishing their relationship to rickettsiae by their morphology and intracellular preference.

1 INTRODUCTION

Symbiotic micro-organisms are known to occur normally in plant and animal tissues. The intracellular microbiota of insects has been the subject of numerous investigations (see Buchner, 1965 and Steinhaus, 1963 for extensive reviews). However, little has been reported regarding intracellular micro-organisms in fleas. An apparent reason for this may be due to the absence of naturally occurring symbiotes that are either large and distinctive or else concentrated in special structures such as mycetomes. On the other hand, it has been amply demonstrated that the oriental rat flea, *Xenopsylla cheopis* (Roth.) transmits *Rickettsia mooseri,* the etiological agent of murine typhus, from rat to rat or to man.[2 6 7 17] The infection of a new host takes place at the site of the flea bite when the highly infectious flea feces is rubbed into the wound.[8] The flea becomes infected when it imbibes an infectious blood meal and abundant multiplication of the rickettsiae occurs in the midgut epithelial cells.

Since we recently detailed some of the ultrastructural features of *R.mooseri*-infected fleas[14] this aspect will be only briefly herein illustrated to show the ultrastructural features of a known pathogenic rickettsia. During the course of these studies, it was found that some tissues of uninfected control fleas contained intracellular micro-organisms which resemble *R.mooseri.* In this presentation, we will illustrate rickettsiae or rickettsia-like micro-organisms in various flea tissues and note their relationships to their host cell and any notable pathological effects induced.

Although the flea has been studied extensively by light microscopy, the presence of these small symbiotic micro-organisms are not readily distinguished from mitochondria and other cellular structures and therefore may be overlooked. On the other hand, transmission electron microscopy of adequate thin sectioned flea tissues readily permits recognition of the micro-organism. Furthermore, the ultrastructure of rickettsiae is distinct from chlamydiae or mycoplasmas. In this report, we will refer to micro-organisms as rickettsiae whether they are known pathogenic rickettsiae or unidentified micro-organisms intracellularly located and having characteristic ultrastructural features of rickettsiae.

2 MATERIALS AND METHODS

Our colony of *X.cheopis,* originally obtained from the US Department of Agriculture, Entomology Research Division, Gainesville, Florida, was maintained at Harvard School of Public Health by Dr A.Spielman. The procedure used to infect the fleas with *R. mooseri* has been described previously. In brief, mice infected by intraperitoneal inoculation of *R.mooseri* in yolk sac suspensions were used to feed fasted fleas. The midgut of these fleas was fixed at various intervals and prepared for electron microscopy as described below.

Uninfected normal adult fleas in various feeding states were immobilized by chilling; flea tissues were then dissected out on a wax plate in a drop of fixative. A variety of fixatives were surveyed to obtain tissues that were generally favourably preserved. A mixture of 3 per cent formaldehyde (made up from paraformaldehyde), 4 per cent glutaraldehyde, and 0.05 per cent trinitrocresol in 0.1 N cacodylate or phosphate buffer pH 7.3[14] was used for most of these observations. To make up the fixative, 1.0 ml of a 25 per cent purified glutaraldehyde solution was mixed with 0.05 ml of a 50 per cent solution of hydrogen peroxide according to Peracchia et al. This mixture was immediately added to the formaldehyde and trinitrocresol mixture. The solutions were kept in an ice bath at 4°C. After 1 to 2 hours, the tissues were washed in 0.2 M cacodylate buffer pH 7.3 and osmicated with 1 per cent OsO_4 in the same buffer for 1 hour. Tissues were then treated with 1 per cent uranyl acetate in maleate buffer pH 5.2 for 1 hour at

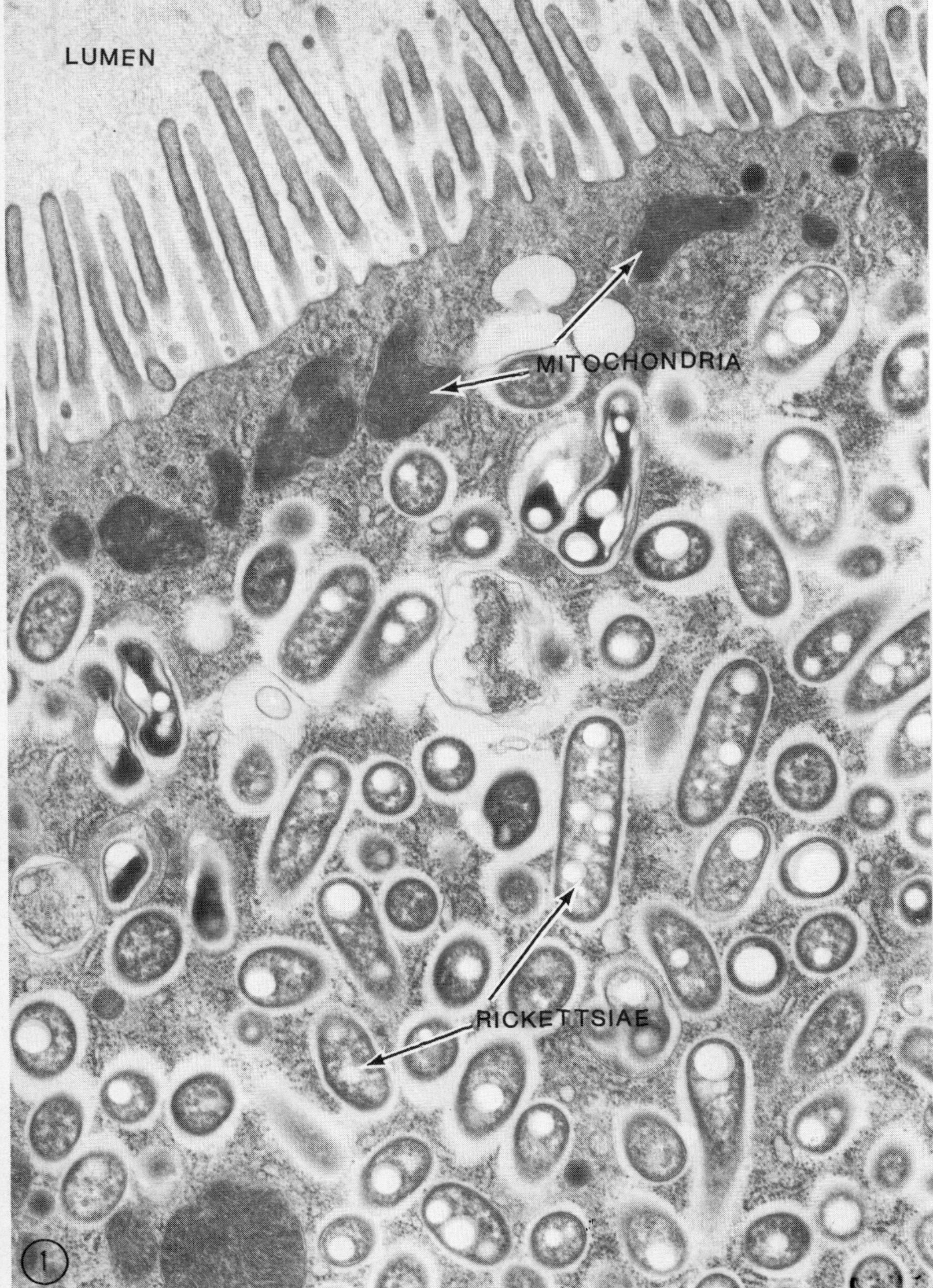

Figure 1. An electron micrograph of the apical part of a flea midgut cell infected with a large number of *Rickettsia mooseri.* The rod-shaped rickettsiae lie directly in the host cell cytoplasm and in some preparations are surrounded by a halo of low-density. Numerous clear vacuoles of unknown significance or content are present in the rickettsiae. Some micro-organisms enclosed within a membrane compartment are undergoing condensation and apparent destruction. X35,000.

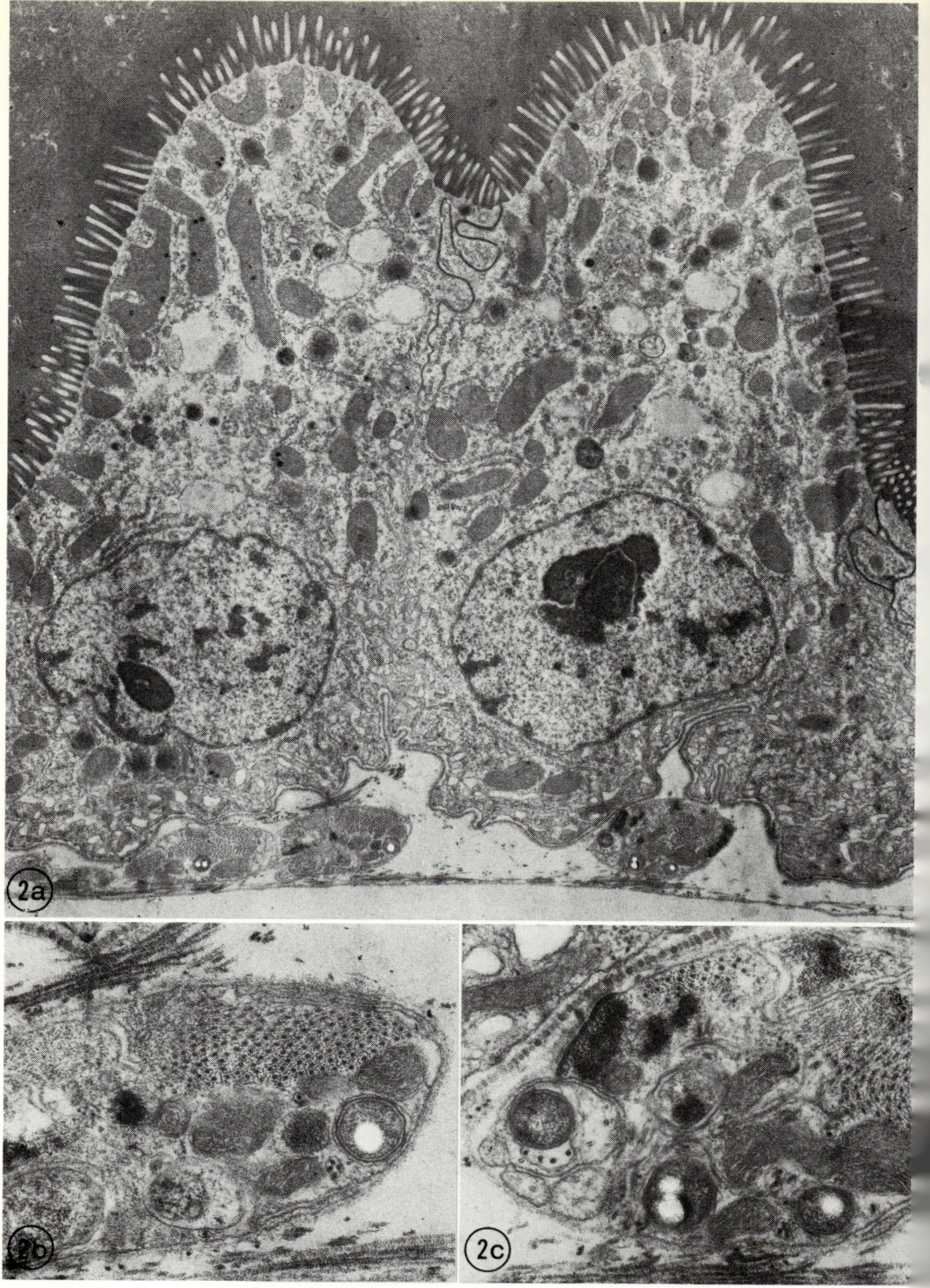

Figure 2a, b, c. A low power micrograph (2a) illustrates the entire width of a flea midgut cell from a normal adult *Xenopsylla cheopis.* The lumen of the gut is filled with the dense remains of a blood meal and the epithelial cells contain the usual cytoplasmic organelles. Beneath the basal lamina of the midgut cells and the basal lamina limiting the hemocoel, there are a number of striated muscle cells containing symbiotic rickettsiae. X7,200. In the two insets (2b, 2c) enlarged portions of two muscle cells from the same preparation illustrate several rickettsiae, some with the characteristic clear vacuoles among mitochondria and muscle filaments. The characteristic punctate basal lamina of this flea epithelium and the fibrous basal lamina at the hemocoel surface are also included. X36,000.

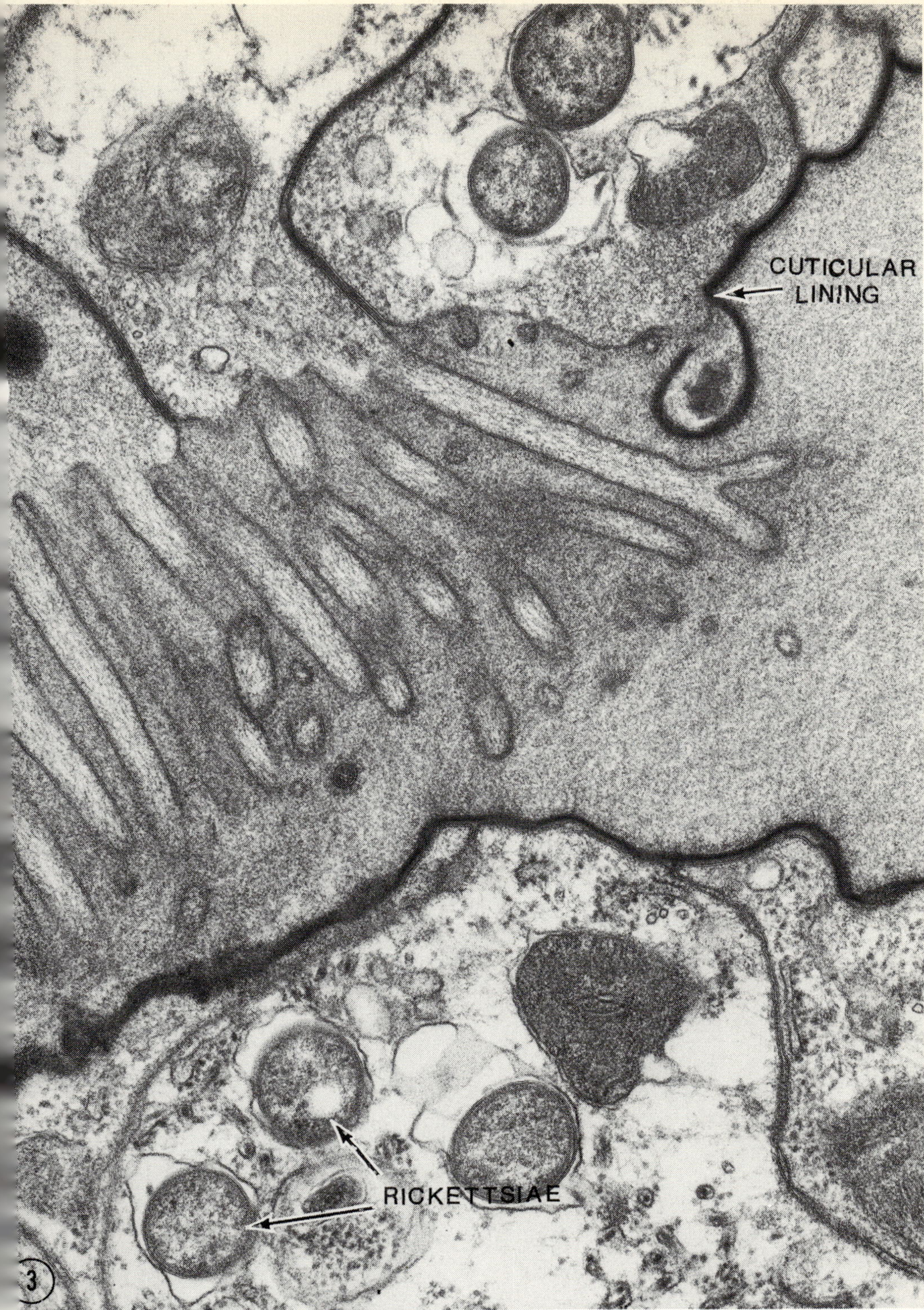

Figure 3. A higher power view of the junction between the midgut and proventriculus of the flea. The midgut epithelium has numerous long microvilli and is joined by continuous junctions rather than typical tight junctions. In contrast the proventricular epithelium is lined with a dense cuticle just lumenal to the apical plasma membrane and the cells are joined by what appear to be tight junctions, gap junctions and septate junctions. The proventricular epithelium contains several typical rickettsiae but the midgut cells were apparently free of these symbiotes. Note at this magnification the rickettsial cell wall and the inner plasma or cell membrane of the rickettsia are visible. X65,000.

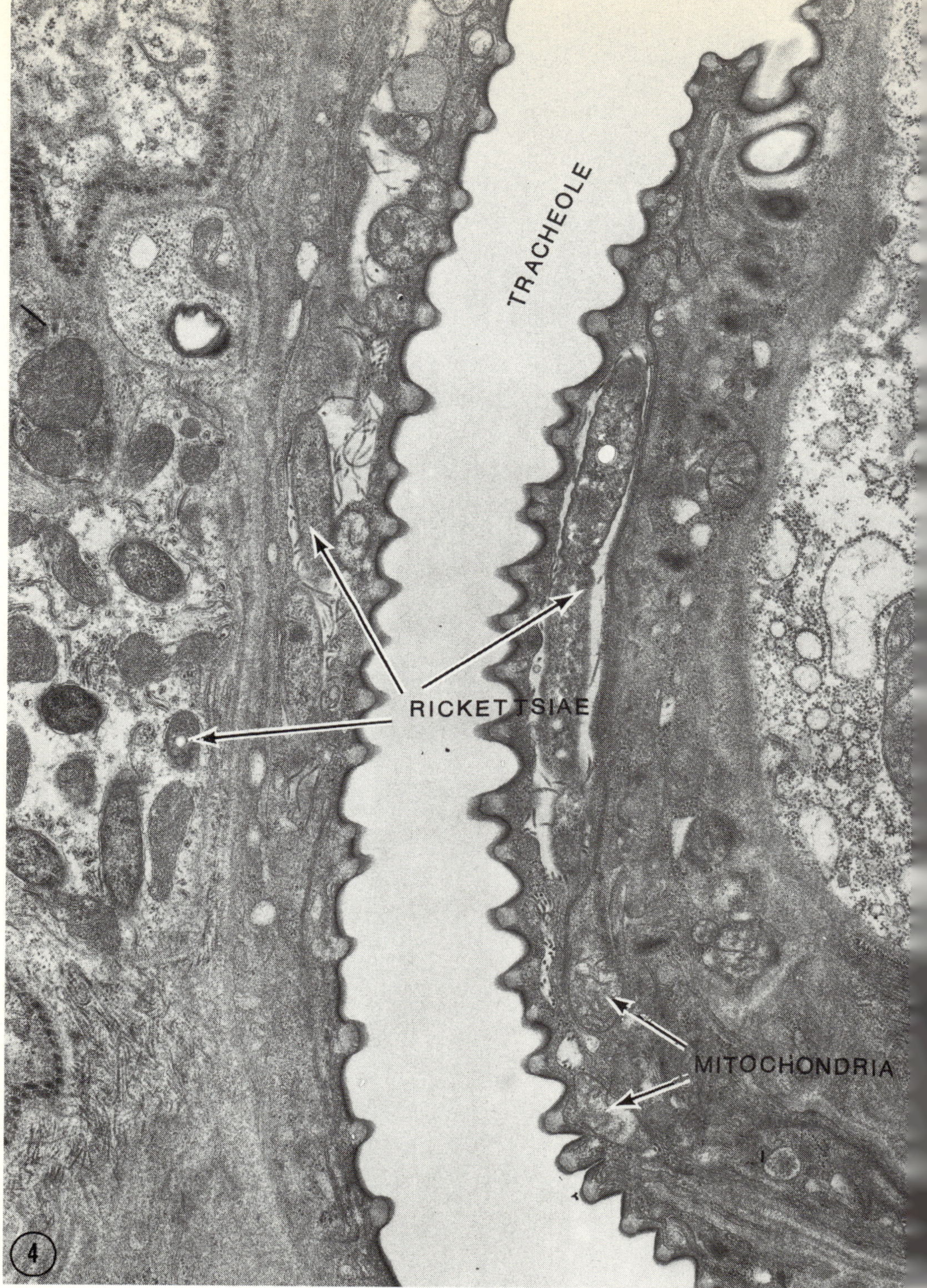

Figure 4. A longitudinal section of a moderate sized tracheole beneath the midgut cell of the flea. Within the epithelial cells of the tracheole and in the surrounding muscle cell (left) there are a number of typical rickettsiae. The long profile in the right wall appears to be in the initial stages of fission. The size and shape of rickettsiae closely resemble mitochondria. X20,000.

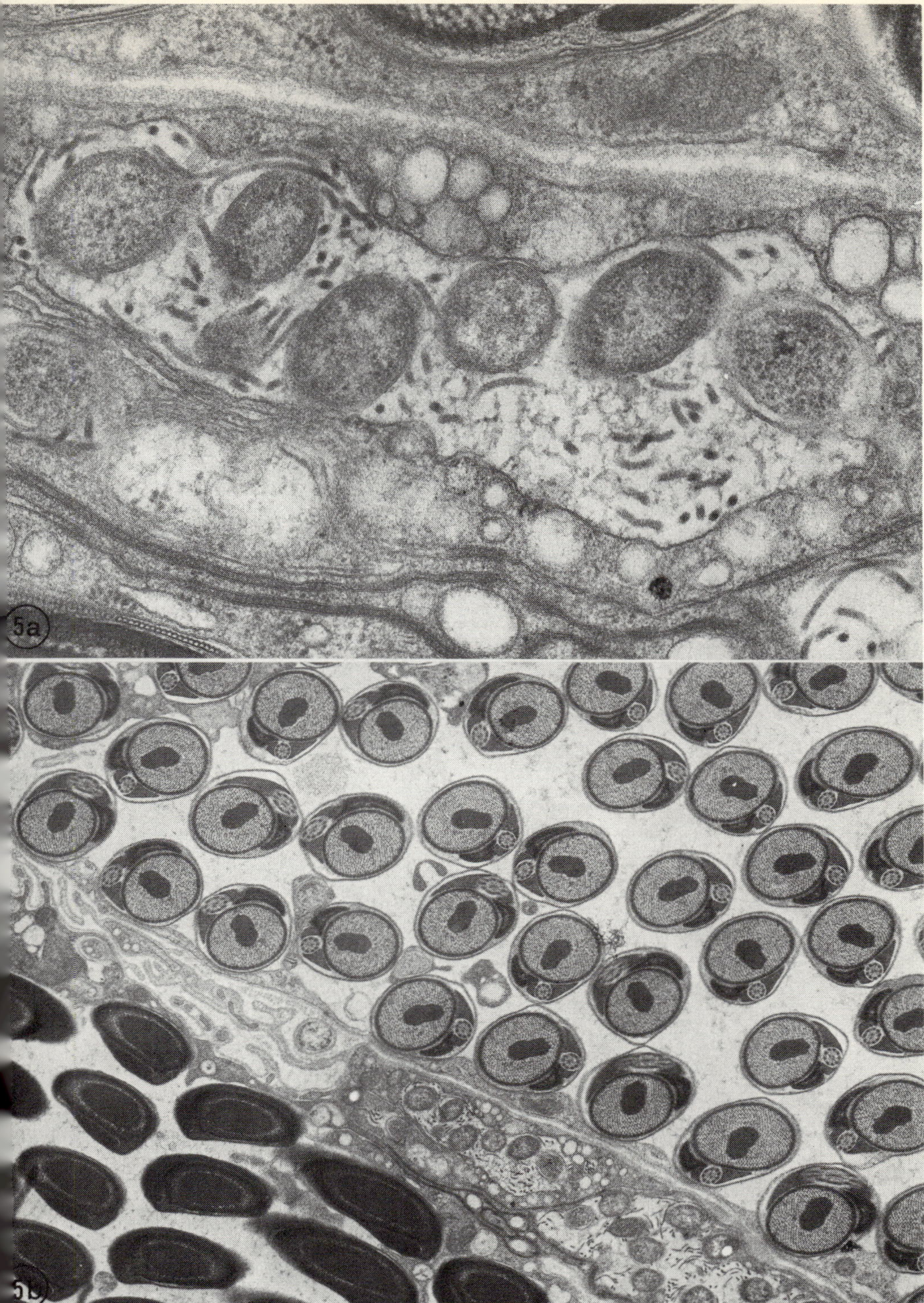

Figure 5a, b. The lower micrograph is a view of part of the testis of a flea including cross-sections of the mid-piece at the upper right and sections through the dense head at the lower left. In the septal cells between the sperm bundles a number of rickettsiae may be seen. The upper illustration is a high magnification of a portion of the same section. Several rickettsiae with scattered profiles of coarse fibres which may be remnants of myosin filaments are shown. 5a: X60,000. 5b: X9,000.

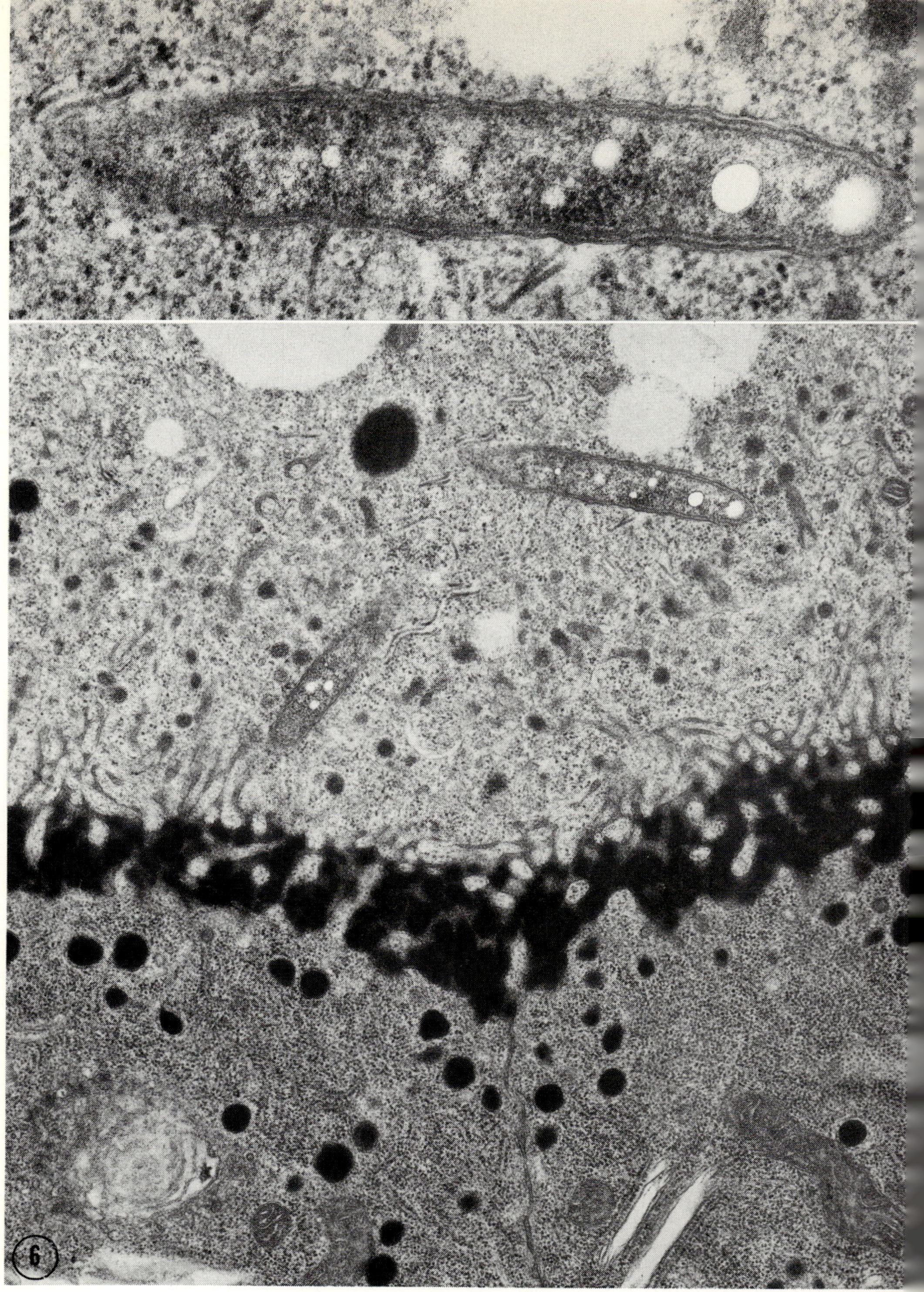

Figure 6. Part of an intermediate sized developing egg of *Xenopsylla cheopis.* The nurse cells at the bottom of the illustration contain many ribosomes in addition to dense granules and mitochondria. The oocyte above the forming boundary layer of the egg contains two rickettsiae in addition to its expected cytoplasmic contents. The upper inset is a high magnification of one of these rickettsiae illustrating its typical features. X26,000. X57,000.

room temperature and washed thoroughly in maleate buffer. After rapid dehydration in graded, cold ethanol solutions, the tissues were embedded in a mixture of the low viscosity, embedding medium made up of diepoxyoctane and dodecenyl succinic anhydride according to Luft (1973). After polymerization, thin sections were stained with equal parts saturated uranyl acetate and acetone for about 15 seconds and then for a similar period in lead citrate before examination by transmission electron microscopy.

Identity of *R.mooseri* infection in the flea was established by means of the indirect immunofluorescent test performed by standard methods. Appropriate macerated flea tissues, used as antigen placques, were tested with normal and murine-typhus convalescent human sera. Tissues from uninfected control fleas were invariably unreactive.

2.1 *Rickettsia mooseri in midgut epithelium*

As described in detail previously [14] infected flea midgut cells either contained a few scattered *R.mooseri* or else were packed with hundreds of micro-organisms. In the host cell, the rickettsiae lie directly in the cytoplasm freely intermingling with mitochondria and other cell components. The rickettsia may or may not have a thin halo of low density surrounding the micro-organism (fig. 1). The variability of this clear space is attributed to the loss of the surface coat substance on the cell wall of the rickettsiae during specimen preparation and possible further exaggeration during tissue processing. Despite the infection, the midgut cell does not show obvious pathological changes; but some indications were seen of sequestration of rickettsiae in lysosomal vacuoles where they were undergoing condensation and disruption (fig. 1).

Although the infection of a flea is not always assured after imbibing an infectious blood meal, once inside the host cell the rickettsiae proliferate by binary fission and are eventually released when the midgut cell ruptures and is exfoliated. It is not yet clear whether the burden of infection results in a foreshortened life span of the epithelial cells or whether the cells maintain a normal life span.

Rickettsia mooseri are not distinctive and resemble all other recognized rickettsiae in their fine structure. This is true whether they are cultivated in chick yolk sacs, cultured cells, or in mammalian or arthropod hosts. Rickettsiae are usually 0.2-0.45 μm in width and about 1-1.5 μm in length. Some dividing forms may be over 2 μm long. The cell wall is about 10 μm thick and is formed by an outer leaflet 2 μm thick, a clear space of 3 μm and an inner more dense component 5 μm thick. Beneath the cell wall, there is a trilaminar plasma membrane about 5 μm in thickness. The two envelopes are separated from each other by a space of about 10 μm. The structure of the cell wall and cell membrane of rickettsiae are similar to other gram negative bacteria.

The cytoplasm of rickettsiae is devoid of membranous structures and contains a meshwork of fine DNA filaments and granules believed to be ribosomes. The only other structures present are clear vacuolar areas up to 150 μm in diameter. These vacuoles are particularly prominent in rickettsiae proliferating in arthropod cells and only infrequently seen in infected yolk sacs, cultured cells or mammalian cells.

2.2 *Symbiotic rickettsiae in flea tissues*

In our specimens of uninfected control flea tissues we found rickettsia-like microorganisms in muscle cells surrounding midgut epithelial cells.[14] An example of such a specimen is shown in figure 2, which includes parts of two typical epithelial cells from a flea whose midgut still contained the dense remains of a blood meal. The epithelial cells with their striated border, basal nuclei, and large nitochondria are normal in appearance and do not contain rickettsiae. However, the sectioned profiles of three striated muscle cells lying beneath the basement lamina contain rickettsiae which appear prominent even at low magnification because of their clear vacuolar inclusions. Two of these muscle cells have been enlarged (figs. 2b, c) to show the rickettsiae to better advantage.

At the junction between the midgut and proventriculus (fig. 3) there is an abrupt transition in the epithelium from the cuticle-lined epithelium and the microvillous-lined midgut cells. The epithelial cells of the proventriculus contain a number of micro-organisms. These appear identical to the rickettsiae in the muscle cells (fig. 1) and most of them are enclosed within a distinct membranous envelope. Of incidental interest is the presence of typical zonula continua between the midgut cells and at the junction with the proventricular epithelium while the epithelial cells of the proventriculus are joined by what appear to be tight junctions.

In the epithelium of tracheoles, it was not uncommon to find rickettsiae. Figure 4 illustrates a moderate sized tracheole with a longitudinal profile of an extraordinary long micro-organism some 3.5 μm long, possibly entering a division process, and two other rickettsiae in the left wall. The presence of irregularly arrayed coarse filaments in the clear space around the rickettsiae is common and may also be noted in figures 2, 3 and 5. Note that the mitochondria of the tracheolar cells have some resemblance to the rickettsiae. In figure 4, the muscle cell at the left centre margin also harbours a number of rickettsiae.

The testis and ovary illustrated in figures 5 and 6 were also found to contain rickettsiae. In the case of the testicular tissue the micro-organism was limited to the supporting cells enclosing the sperm bundles. The coarse homogenous filaments seen among the rickettsiae are unidentified but could be myosin filaments. The egg cytoplasm illustrated in figure 6 contains two profiles of rickettsiae. These organisms contain numerous clear vacuolar areas. A higher magnification of one of the rickettsiae reveals a bounding membrane which closely envelopes the micro-organisms. Most of the ovarian rickettsiae were observed in the egg cell but a few were seen in the follicular cell as well.

Although a thorough survey of all the tissues in the flea was not made occasional rickettsiae were seen in both the fast and slow contracting skeletal muscle cells and in Malpighian tubules. Limited numbers of micro-organisms were also found in some larval flea tissues.

3 DISCUSSION

Although electron microscopical studies of flea tissues are relatively scarce, Reinhardt (1976) has examined the midguts of *Xenopsylla cheopis, Echidnophaga gallinacea* and *Tunga penetrans* in great detail. An earlier study on the flea midgut basal lamina of the flea, *Ctenophthalmus,* was reported by Richards & Richards (1968). In these studies, the presence of symbiotic micro-organisms in flea tissues was not reported and it is not known whether our colony of oriental rat fleas is unusual or exceptional in consistently possessing indigenous rickettsiae. In our specimens, they were never found in large numbers but were encountered in virtually all individuals examined.

The question of how to classify these micro-organisms – as rickettsiae, rickettsia-like or wolbachiae – is an obvious matter of some concern. A number of recent reports on transmission electron microscopy of invertebrate tissues[4 5 8 12 13 23] and plant cells[1 11 16] have referred to intracytoplasmic micro-organisms as rickettsiae or rickettsia-like micro-organisms. Similar intracellular micro-organisms have been designated as mycoplasma-like in *Drosophila paulistorum* by others.[9] The problem of appropriate designation or classification is not the aim of this study. The purpose of this report is to describe the ultrastructural localization of micro-organisms in some flea tissues. As far as we know these indigenous symbiotes are neither responsible for any recognizable disease in their mammalian hosts nor have any pronounced effect on the flea itself.

It is apparent that in the absence of evidence for their pathogenicity and the difficulties of light microscopic methods to reveal their presence, their role in the flea may easily be overlooked. Thus electron microscopy, with its many acknowledged limitations, serves as an important tool in revealing the presence of these micro-organisms. We have elected to consider the micro-organisms in flea tissues as rickettsiae, because they satisfy all of the ultrastructural characteristics found in known rickettsial micro-organisms. The rickettsial symbiotes found in *X.cheopis* do not seem to cause extensive cellular disorganization. However, they may cause focal areas of disruption in the striated muscle cells where they are most often found. The normally regularly ordered actin and myosin filaments are often distorted and destroyed. Furthermore, there is a consistent tendency for rickettsiae to be sequestered within a vacuolar membrane as though the host cell response endeavoured to segregate the rickettsiae from the rest of the cell. Somewhat surprising, however, is that the symbiotic rickettsiae observed in flea tissues were not seen undergoing degenerative changes as in the case of rickettsiae in the midgut cells experimentally infected with murine typhus rickettsiae. This relationship may suggest the evolution of a more compatible symbiotic relationship of these rickettsiae in the normal life of the flea.

The presence of rickettsiae in all stages of development of the flea ovum strongly suggests that these micro-organisms are transferred transovarially as is the case for the tick,[45] *Drosophila,*[9] mite,[18] and mosquito.[23] In the flea, there is no evidence that trans-spermatozoid transfer of rickettsiae occurs as reported for many leafhoppers.[12]

With improved techniques for the detection of micro-organisms and advances in the study of intracellular parasitism,[22] there is little doubt that much more information on this aspect of cellular biology will become available in the future.

REFERENCES

1. Auger, J.G., T.A.Shalla & C.I.Kado 1974, Pierces' disease of grapevines: evidence for a bacterial etiology. Science V 184: 1375-1377.
2. Blanc, G. & M.Baltazar 1936, Longévité du virus du typhus murine chez la puce *(Xenopsylla cheopis).* C.R. Acad. Sci. 202: 1461.
3. Buchner, P. 1965, Endosymbiosis of animals with plant micro-organisms. New York, Interscience Publisher.
4. Burgdorfer, W. 1963, Investigation of 'Transovarial Transmission' of *Rickettsia rickettsia* in the Wood Tick, *Dermacentor andersoni.* Exp. Parasit. V 14: 152-159.
5. Burgdorfer, W. & L.P.Brinton 1975, Mechanisms of transovarial infection of spotted fever rickettsiae in ticks. Ann. N.Y. Acad. Sci. 266: 61-72.
6. Ceder, E.T., R.E.Dyer, A.Rumreich & L.F. Badger 1931, Typhus fever: typhus virus in feces of infected fleas *(Xenopsylla*

cheopis) and duration of infectivity of fleas. Publ. Hlth. Rept. 40: 1-9.
7. Dyer, R.E., A.Rumreich & L.F.Badger 1931, Typhus fever. A virus of the typhus type derived from fleas collected from wild rats. Publ. Hlth. Rept. 46: 334-338.
8. Dyer, R.E., E.T.Ceder, W.G.Workman, A. Rumreich & L.F.Badger 1932, Typhus fever. Transmission of endemic typhus by rubbing either crushed infected fleas or infected flea feces into wounds. Publ. Hlth. Rept. 47: 131-133.
9. Ehrman, L. & R.P.Kernaghan 1972, Infectious heredity in *Drosophila paulistorum.* In: Pathogenic Mycoplasmas. Ciba Foundation Symposium, North Holland Publishing Co., Elsevier Excerpta Medica, p. 227-250.
10. Harshbarger, J.C., S.C.Chany & S.V.Otto 1977, Chlamydiae (with phages), mycoplasmas, and rickettsiae in Chesapeake Bay bivalves. Science 196: 666-668.
11. Hopkins, D.L. & H.H.Mollenhauer 1973, *Rickettsia*-like bacterium associated with Pierces' Disease of grapes. Science 179: 298-300.
12. Maillet, P.L. 1971, Observations ultrastructurales et biologiques sur des microorganismes de type Rickettsien trouvés chez divers Homoptères Auchenorhynques. Bull. Biol. 2: 95-111.
13. Irving-Bell, R.J. 1974, Cytoplastic factor in *Culex pipiens* complex mosquitoes. Life Sciences 14: 1149-1151.
14. Ito, S., J.W.Vinson & T.J.McGuire 1975, Murine typhus rickettsiae in the oriental rat flea. Ann. N.Y. Acad. Sci. 266: 35-60.
15. Luft, J.H. 1973, Embedding media – old and new. In: J.K.Koehler (ed.), Advanced Techniques in Biological Electron Microscopy. New York, Springer Verlag. p. 1-34.
16. Mohr, W., F.Weyer & E.Aschauer 1972, Murines Fleckfieber. In: O.Gsell, St. Gallen & W.Mohr (eds.), Infektionskrankheiten. Rickettsiosen und Protozoenkrankheiten, Vol. IV. Berlin, Springer Verlag.
17. Mollenhauser, H.H. & D.L.Hopkins 1974, Ultrastructural study of Pierces' Disease Bacterium in Grape Xylem Tissue. J. Bact. 119: 612-618.
18. Rapmund, G., R.W.Upham, jr., W.D. Kundin, C. Manikumaran & C.T.Chan 1969, Transovarial development of scrub typhus rickettsiae in a colony of vector mites. Trans. Roy. Soc. Trop. Med. Hyg. 63: 251-258.
19. Reinhardt, C.A. 1976, Ultrastructural comparison of the midgut epithelia of fleas with different feeding behaviour patterns (*Xenopsylla cheopis, Echidnophaga gallinacea, Tunga penetrans,* Siphonaptera, Pulicidae). Acta Tropica 33(2): 105-132.
20. Richards, A.G. & P.A.Richards 1968, Flea *Ctenophthalmus:* Heterogeneous hexagonally organized layer in the midgut. Science 160: 423-425.
21. Steinhaus, E.A. 1963, Insect Pathology. New York, Academic Press.
22. Trager, W. 1974, Some aspects of intracellular parasitism. Science 183: 269-272.
23. Yen, J.H. & A.R.Barr 1971, New hypothesis of the cause of cytoplasmic incomparability in *Culex pipiens* L. Nature 232: 657-658.

ROBERT TRAUB, CHARLES L. WISSEMAN, JR & ABDULRAHMAN FARHANG-AZAD
University of Maryland School of Medicine, Baltimore, Md., USA

THE ECOLOGY OF MURINE TYPHUS*

ABSTRACT

Although murine typhus occurs on all continents and cases have occurred by the thousands annually, there still remains many questions about its ecology, including the precise means of transmission and the definitive reservoir. This lack became strikingly clear during the preparation of a critical review of the subject prepared in connection with studies we have been pursuing in Ethiopia, based in Addis Ababa, in the laboratories of the US Naval Medical Research Unit No. 5. That extensive review, including original ideas derived from analysis and synthesis of the available literature, and upon certain aspects of our current studies, is in press in the Tropical Diseases Bulletin and provides over 500 bibliographic citations, obviating the need to document the present abstract.

Even though the literature is replete with the names of various kinds of small mammals (theraphions) and ectoparasites as being involved in this rickettsiosis, much of the data cannot be evaluated properly because of the particular serological techniques or criteria employed. Nevertheless, despite such shortcomings and the many yet-unresolved questions, certain generalizations may be made, viz.: 1) commensal rats of the subgenus *Rattus* are deeply implicated; 2) human cases, as well as infection in theraphions, are associated with the indoors; 3) ectoparasites of *Rattus* play an important role even though transmission is not by the bite of the arthropod; and 4) serious consideration should be given to the possibility that transmission may occur via the aerosol route. Although the point has not been adequately made previously, it should be borne in mind that the natural cycles of *Rickettsia mooseri* infection in theraphions may involve different vectors, modes of transmission, and sources of infection than those responsible for the disease in man, who enters as an intruder.

Commensal rats, such as *R.(R.)rattus* and *R.(R.)norvegicus* are critical factors in the ecology of this rickettsiosis, and at least insofar as concerns human cases, may be the ultimate source of the *R.mooseri* in question. While such rats have been recognized as

* This study was supported by Grant No. AI-04242 of the National Institutes of Health, Bethesda, Maryland, USA, with the Department of Microbiology, University of Maryland School of Medicine, Baltimore, Maryland, USA. The field-studies in Ethiopia mentioned in this article were sponsored by Contract N00014-76-C-0303 of the Office of Naval Research and the Navy Medical Research and Development Command, Washington, DC. The opinions and assertions herein are not to be construed as necessarily reflecting the views of the Department of the Navy or the National Institutes of Health.

occurring in all foci reporting such cases (although instances have been noted where they have been scarce), they extend far beyond the known range of the etiologic agent. Murine typhus infection in theraphions is intimately associated with commensalism in man's domiciles or with contact with buildings. Thus, the only other mammals besides commensal *Rattus* that have been seriously implicated have either been: 1) fully peri-domestic species like house mice *(Mus musculus)*; 2) facultative species acting like commensals, such as *Bandicota* and *Cricetomys* rats and *Suncus* shrews in certain regions; 3) fully domestic pets living indoors, e.g. cats; or else 4) theraphions that may enter buildings, like opossums. All the species in these categories can come into contact with commensal *Rattus,* and in each instance only relatively few individuals have been found naturally infected with *R.mooseri.* Moreover, where the etiological agent has been demonstrated in such hosts, the infection has been self-limiting, and failed to involve theraphions living outside of the particular focus. The role of the house mouse requires clarification because *Mus musculus* have been regarded as being of importance in a few instances where or when commensal *Rattus* were purportedly scarce. Nevertheless, frequently there has been no sign of *R.mooseri* in house mice, whether in endemic foci or not, or whether collected indoors or outdoors.

Despite the large number of countries where cases have occurred, murine typhus infection is often limited in extent, such as being essentially or wholly restricted to port areas or along routes of commerce, or wherever else commensal *Rattus* are prevalent, bespeaking the close connection between the two. The vast majority of human cases have presumably been acquired indoors, with the only exceptions occurring in proximity of houses and buildings, where *Rattus* or *Mus* were also abundant.

Mammals that are sylvan or campestral do not play a significant role in the ecology of this rickettsiosis. The few reports suggesting the possibility of such involvement do not bear critical scrutiny, e.g. an isolation from a field mouse that had been in the laboratory for weeks, or one from an opossum – a mammal that readily enters buildings.

The available evidence indicates that man and other mammals acquire the rickettsiae by contact with the feces or crushed bodies of infected arthropods such as fleas, and not directly by the bite of such an ectoparasite. It is believed that such infective material may perhaps be transmitted as an aerosol of dust, but firm data are lacking, although it is known that rickettsiae in dried flea feces may remain viable for years and still be capable of causing infection. If such aerosol transmission is indeed an important factor in the ecology of murine typhus, then there are two points whose significance has been overlooked: 1) species which have been dismissed from consideration as potential vectors because they do not bite man may nevertheless serve as a major source of infection to humans (and theraphions) via the aerosol route and 2) the seasonal cycles of abundance of such an ectoparasite need not be obviously correlated with the incidence of disease or infection, since there may be a considerable lag between the time of deposition of rickettsiae in arthropod feces and the actual acquisition of the infection via an aerosol.

Natural infection with *R.mooseri* has been reported for six genera and seven species of fleas, while one species of another genus has been implicated because of experimental data. Of these, the major vector is imputed to be *Xenopsylla cheopis,* a species which can acquire the infection by feeding, remain infected for life, and pass viable rickettsiae in its feces – all without its life-span being overtly affected. Epidemiological considerations support this contention. In general, there has been a good correlation between the seasonal cycles of this flea and the incidence of the disease in man. Moreover, marked decline in the number of cases followed the tremendous reduction in the *X.cheopis* population effected by the sustained DDT-dusting programme in the USA which commenced in 1945. The disease (or infection in theraphions) is not known to occur in the documented absence of *X.cheopis,* although there have been instances of where it was present only in very small numbers. On the other hand, *X.cheopis* can be extremely abundant in areas or habitats where this rickettsiosis has not been recorded, and a few reports have pointed out an inverse correlation between incidence of cases and numbers of fleas.

The laboratory data on *X.astia, X.brasiliensis* and *Pulex irritans* as vectors of murine typhus has been comparable to that for *X.cheopis.* The other species implicated in varying degrees are *Ctenocephalides felis, Leptopsylla segnis, Echidnophaga gallinacea, Nosopsyllus fasciatus* and *Monopsyllus anisus.* All of these have been found on *Rattus,* but *P. irritans* and *C.felis* only occasionally; the rest are common on those hosts. Of these particular species, only the last two and perhaps *L.segnis* (along with *X.cheopis*), may be expected to have occurred wherever murine typhus has been reported. Critical as commensal *Rattus* and *X.cheopis* are to the ecology of this rickettsiosis, the key to endemicity is obviously the concomitant presence of *R.mooseri.* There are areas which are presumably free of *R.mooseri* infection and where these rats co-exist with any of the above fleas, or even with the majority of them.

The potential role of other ectoparasites associated with murines or man is unclear, but all that are mentioned below may perhaps serve as a source of infection to theraphion or man by means of their infected feces and transmission via the aerosol route, if not through direct contact and consequent contamination. *Polyplax* and *Hoplopleura* rat-lice have been found naturally infected and shown to be capable of transmitting experimental infection, and hence may at least be significant intra-murid vectors. It has been suggested that human lice may acquire *R.mooseri* from a patient and transmit the disease to other people, but this has never really been proven. Although natural infection and experimental transmission have been reported for various bloodsucking mites, their substance as vectors has been denigrated, disparaged or disputed on epidemiological grounds, or because of further laboratory studies. Natural infection in rat-chiggers have been noted in Java, but these mites have never been properly studied as potential vectors, even though the species in question is common on rats in domiciles throughout southern Asia and many Pacific archipelagos and we now know it can feed on man. Ticks and bedbugs are not believed to be of any consequence in the ecology of murine typhus.

Minifoci of murine typhus have been noted wherein rats, mice, humans and/or several species of ectoparasites were found to harbour *R.mooseri* infection simultaneously, and cats in such foci have also acquired the rickettsiosis. The factors involved in such outbreaks are unknown, but *Rattus* is regarded as the most probable source of infection in such restricted foci.

The association between *Xenopsylla* and *Rattus* is a phenomenon of historical times, and it is only through the agency of man that commensal rats and their ectoparasites (and this rickettsiosis) have reached so many parts of the world. Man otherwise is not directly involved in the perpetuation of murine typhus and only accidentally enters the cycles of infection. Even though there are many unanswered questions concerning the mode of transmission and ecology of murine typhus, it is clear that strenuous application of known methods for the control of rats and their ectoparasites would effectively reduce the incidence of this infection, which is undoubtedly more common and important than the available data indicate.

CLUFF E. HOPLA
University of Oklahoma, Norman, Okla., USA

FLEAS AS VECTORS OF TULAREMIA IN ALASKA

ABSTRACT

The field aspects of the study were carried out in the taiga and tundra regions of Alaska. Five species of fleas were found naturally infected with tularemia. Four of these species are confined to voles, the fifth infests hares and lynx. One species of vole flea was successfully established in the laboratory. Experimental data confirmed that fleas become infected with tularemia organisms, retain the infection for several weeks, but do not transmit the organisms to susceptible animals when a clean method of feeding is employed. Susceptible animals were sometimes infected when fleas were given free access to voles. Other data obtained suggest mechanical transmission through interrupted feeding is the most important method of transmission. A concept of strain B tularemia organisms existing as a natural cycle in voles and fleas is proposed. This cycle can exist in the absence of lagomorphs and ticks.

1 INTRODUCTION

Although Siphonaptera were the first arthropods reported associated with *Francisella tularensis* (McCoy & Chapin, 1912), their importance as vectors of tularemia organisms is not clearly understood. In general, various authors (Bell, 1965; James & Harwood, 1969; Jellison, 1977) have indicated that fleas are poor vectors and of little or no significance. As biological vectors involved in the direct transmission of tularemia organisms to man this concept is undoubtedly correct (Hopla, 1974).

The significance of fleas in the ecology of tularemia in the Nearctic region is more appropriately addressed with regard to their possible significance in the maintenance of tularemia organisms in nature among the cricetid and sciurid rodents, and possibly the lagomorphs (Jellison, 1959). Experimental data have thus far indicated that fleas do not usually transmit the organisms during feeding. Having this ability would enhance the fleas' capabilities as vectors of tularemia organisms; but it would not be the only way transmission could be accomplished. Data suggest other possible means of infection, e.g. infected fleas ingested by the host while grooming itself, or by contamination with flea fecal material that may remain in the fur of the host. The importance of interrupted feeding (mechanical transmission) is not well understood but should not be overlooked.

In the review of previous literature, if the scientific name of the flea, as cited by the author, is obsolete or incorrect, the original version is presented in parenthesis following the currently accepted name. (It is important to remember that there is no universal agreement concerning the taxonomy of fleas.) Names of Nearctic species used in this study follow Holland (1958) and Hopla (1965).

2 MATERIALS AND METHODS

The data reported are but a small part of an overall study concerned primarily with arthropod-borne zoonoses in Alaska from 1955-1972. The methods used were adapted from the standard procedures used in attempting to isolate tularemia organisms from tissues of animals and other arthropods (Hopla, 1966).

For this study, four approaches were used: 1) the isolation of organisms from fleas obtained from the wild animal fauna in Alaska; 2) experimental infection of laboratory-reared Alaskan fleas; 3) ascertaining the ability of fleas to transmit tularemia organisms during the course of a normal blood meal on susceptible hosts not previously associated with tularemia organisms; 4) mechanical transmission by interrupted feeding on an infected host, and then permitting the fleas to complete their feeding on a normal host.

1. The fleas were removed from the host after capture; both were lightly anesthetized with chloroform. The fleas from each individual host were placed in a separate glass vial which contained 1 ml of sterile skim milk and assigned the same number as the host. The vials were placed in containers of liquid nitrogen which were shipped on regular basis from Alaska to my laboratories at the University of Oklahoma for microbiological assay. The vials containing the fleas were removed from the liquid nitrogen; allowed to stand at room temperature for a short time to partially thaw the contents and then emptied into a sterile petri dish. The plate was placed on a modified chill table on the stage of a dissecting microscope to establish specific identification of the fleas.

Fleas of the same species removed from the same species of host in the same trap-setting were then pooled to match tissue which had been removed and pooled from these hosts while in the field. In other words, the individual pools of Siphonaptera consisted of one species only. The fleas were rinsed twice in sterile skim milk. Because of the potential of destroying tularemia organisms which possibly were on the mouth parts, the fleas were not immersed in 5 per cent formalin as was done with ticks. The fleas were next placed in a grinding tube containing 1.5 ml of sterile skim milk which served as the diluent for the teflon grinders. As a matter of precaution, 1 000 units of penicillin were added to each tube.

Guinea pigs weighing 250 to 500 grams were the animals of choice for routine isolation of tularemia organisms. 0.5 ml of the resultant suspension was injected intraperitoneally into the animal. All guinea pigs were bled prior to inoculation and the serum frozen to be checked in the event of a valid rise in temperature occurred and/or antibodies or organisms were subsequently observed. The remainder of the flea skim milk suspension was sealed in ampules and stored at −60°C in low temperature cabinets for subsequent studies if needed.

Rectal temperatures of the guinea pigs were measured daily at a specific time for three weeks. If a reading of 40°C or above was encountered for three days the animals were bled and transfers made of the blood into embryonated hens' eggs. Such animals were then sacrificed and the spleen removed. These spleens were checked for the characteristic pathology and then homogenized in the teflon grinders as previously described and injected into fresh animals. Additional samples of the suspension were cultivated on cystine-glucose-blood agar and thioglycolate broth. Each of two mice received 0.5 ml of the same spleen suspension intraperitoneally as a more delicate test than culture media for *F. tularensis.*

After three weeks, all animals not showing an elevation in temperature were bled and the sera tested by the standard agglutination test. The animals were then necropsied, the spleen removed and processed as above. Tissues from animals showing an antibody titre only for tularemia were watched with special care after the reinoculation.

All bacteria isolated were identified on the basis of morphological and biochemical characteristics (Hopla, 1955, 1966). Slide preparations of bacteria having the colony morphology of *F. tularensis* were stained by Gram's method and examined for Gram-negative organisms with morphology similar to tularemia organisms. If the organisms fulfilled these criteria, slide agglutinations were carried out with known positive antisera.

Upon the completion of the identification of the organism as *F. tularensis,* its virulence and glycerol fermenting capacity was tested as described by Owen et al. (1961). Virulence capacities of isolates were made by doing standard LD_{50} titrations in white mice and comparing them with such strains as Chu, B259C No. 4, and Jap, which are of high, moderate, and low virulence respectively. The glycerol fermenting capacity was undertaken to establish whether the isolate was of the A or B strain.

Adequate controls were maintained to detect laboratory errors which might arise. For example, in the serological procedures, known negative controls were introduced at random. Animal controls consisted of specimens injected only with the diluent. All experimental animals purchased were quarantined for two weeks prior to starting experimental procedures. The fertile eggs were purchased from a local source which used a diet formulated without antibodies.

2. Sufficient adult fleas were obtained from rodent nests to attempt establishing colonies for experimental transmission of *F. tularensis.* However, reliable rearing results were obtained only for one species, *Malaraeus penicilliger dissimilis* Jordan. The other species attempted were *Catallagia dacenkoi fulleri* Holland, *Hoplopsyllus glacialis lynx* (Baker), *Megabothris calcarifer gregsoni* Holland and *Peromyscopsylla ostsibirica longiloba* (Jordan). F_1 adults not less than 72 hours and not more than 96 hours old, were divided into 20 groups of 20 each and placed upon the shaved abdomen of laboratory-reared *Clethrionomys rutilus* (Pallas), or *Microtus oeconomus* Pallas, both of which are virtually equally susceptible to *F. tularensis.* These voles had been injected with 100-1 000 organisms intraperitoneally 48-60 hours previously.

The inoculate was prepared from a serial dilution of a standardized suspension (Hopla, 1955). The strain used was originally isolated from *M. oeconomus* from Manley Hot Springs and had an LD_{50} of $10^{-4.7}$ in *M. oeconomus* and *C. rutilus* and 5.1 for random-bred white Swiss mice. The voles were usually moribund at 48 hours and cardiac blood, plated at that time, indicated 85 per cent should have been circulating organisms. Adult fleas were maintained at a temperature of 15°C plus or minus 2°C.

Upon completion of the infective feeding by the fleas, two blood glucose cysteine plates were inoculated with the blood of each vole presumed to have had a systemic infection at the time the fleas fed. These plates were read at 24 and 48 hours after inoculation, checking for the characteristic colony morphology of *F. tularensis.* From plates having characteristic morphology, slides were prepared and Gram's stain used. If the organisms

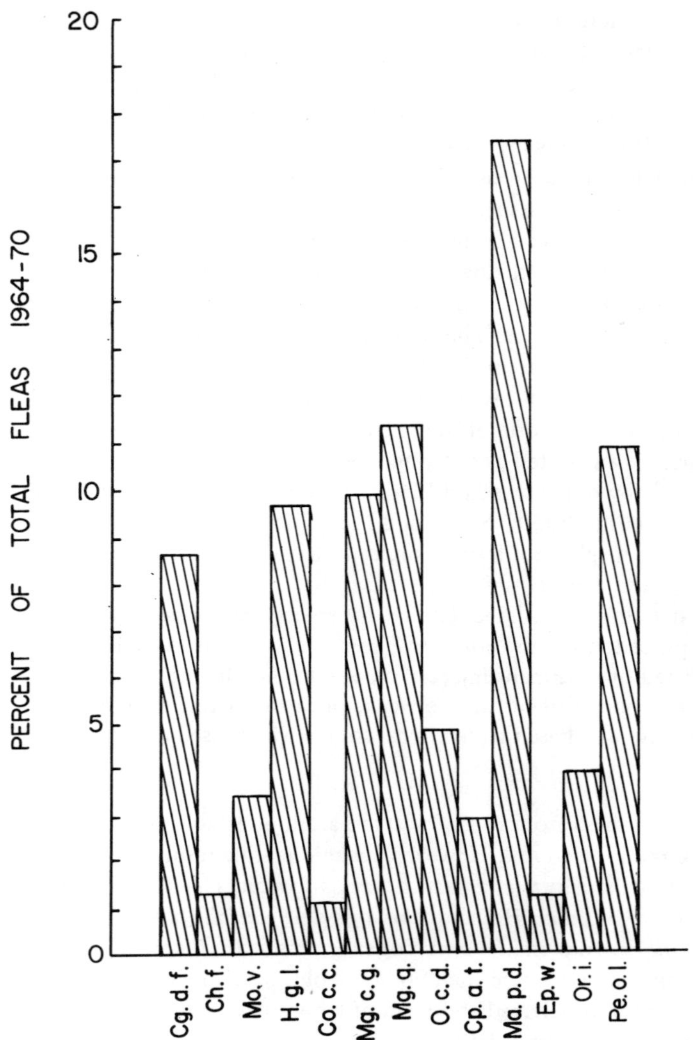

Figure 1. Graphic representation of the 13 commonest species of Alaskan Siphonaptera collected from 1964-1970. Cg.d.f. *Catallagia dacenkoi fulleri,* Ch.f. *Chaetopsylla floridensis,* Co.c.c. *Corrodopsylla curvata curvata,* Cp.a.t. *Ctenophyllus armatus terribilis,* Ep.w. *Epitedia wenmanni,* H.g.l. *Hoplopsyllus glacialis lynx,* Ma.p.d. *Malaraeus penicilliger dissimilis,* Mg.c.g. *Megabothris calcarifer gregsoni,* Mg.q. *Megabothris quirini,* Mo.v. *Monopsyllus vison,* O.c.d. *Orchopeas caedens durus,* Or.i. *Oropsylla idahoensis,* Pe.o.l. *Peromyscopsylla ostsibirica longiloba.*

fulfilled these criteria, slide agglutinations were carried out with known positive antisera. These data on systemic circulation of organisms were usually obtained in retrospect, for the feeding experiments had been completed before the time cultures could be read. It was valuable, however, in that if the vole's blood was negative for organisms, usually none of the fleas became infected.

Of the 400 fleas used in this plase of the study, ten from each of the 20 pools were further divided into two groups of five. These fleas were ground and 0.5 ml of the resultant suspension injected into the white mice to determine the percentage of the pools infected. Plates were streaked from using the spleen tissue impression smear technique; the splenic tissue was removed at necropsy from each mouse as a further confirmation of tularemia organisms.

3. The remaining fleas from the original pools were allowed to feed on the shaved abdomen of voles to ascertain if the fleas could transmit the organisms during a regular feeding period (one hour). These fleas were held 48 hours before placing them on the new uninfected host. This time was thought to virtually remove the possibility of mechanical transmission.

4. Mechanical transmission associated with interrupted feeding was accomplished by infecting a series of 30 pools each consisting of 20 fleas. As described in 2, the same method was used to ascertain the percentage of positive pools. The interrupted feeding experiments were established on the basis of allowing the fleas to feed a short time on the infected host for 5 to 7 minutes, then placing them on normal hosts to feed until replete.

The fleas in this and the previously mentioned feeding experiments were confined to feeding capsules developed to prevent the escape of fleas but at the same time to allow the fleas to move at random within the feeding area defined by the capsule.

3 RESULTS

1. Table 1 summarizes the principal data for part 1. To some it may come as a surprise that the lynx (*Lynx canadensis* Kerr) is listed as associated with *H.g.lynx* instead of the varying hare (*Lepus americanus* Erxleben). However, in the winter months dense populations of this flea were encountered on the lynx. For example, out of a total of 10 782 fleas for all species collected from the native mammalian fauna of Alaska from 1964-1968, 12 per cent were *H.g.lynx* (Hopla, 1969). The high numbers came from a small number of lynx; not the varying hare. The varying hare was virtually absent through these years, being at the bottom of the population curve. The flea was encountered but 36 times during the five-year period.

Figure 1 provides a resumé of the commonly collected fleas 1964-1970. With the exception of *Epitedia* and *Orchopeas,* the other genera are Holarctic in distribution. *Epitedia wenmanni* (Rothschild), *Megabothris quirini* (Rothschild) and *Orchopeas caedens* (Jordan) are examples of what are probably post-Pleistocene migrants northward into Alaska (Holland, 1958; Hopla, 1965).

Four of the lynx from which the fleas were removed had a low titre for tularemia ranging from 1:20-1:80. As to whether the fleas were responsible for the antibody titres such as they were, could or cannot be ascertained, because of the feeding habits of the lynx. Old trappers have stated to me that they get 'lousy with lynx fleas' sometimes when skinning a lynx and that the fleas remain with them for two weeks or so. Over the years, six collections of fleas have occurred from these individuals and each time the fleas were indeed *H.g.lynx.* In two instances, serological evidence indicates that trappers possibly acquired tularemia from the fleas. One trapper had a titre of 1:40 at the time we removed

Table 1. Alaskan fleas found naturally infected with tularemia organisms

Species	Host	Locality	LD_{50}
Hoplopsyllus g.lynx	*Lynx canadensis*	Paxson	$10^{-4.3}$
Hoplopsyllus g.lynx	*Lynx canadensis*	Circle Hot Springs	$10^{-5.2}$
Malaraeus p.dissimilis	*Microtus oeconomus*	Manley Hot Springs	$10^{-4.7}$
Malaraeus p.dissimilis	*Microtus oeconomus*	Central	$10^{-4.8}$
Megabothris c.gregsoni	*Microtus oeconomus*	Central	$10^{-5.2}$
Megabothris c.gregsoni	*Clethrionomys rutilus*	Delta Creek	$10^{-4.7}$
Megabothris c.gregsoni	*Dicrostonyx groenlandicus*	Nome	$10^{-2.7}$
Megabothris quirini	*Microtus pennsylvanicus*	Fairbanks	$10^{-5.6}$
Peromyscopsylla o.longiloba	*Microtus oeconomus*	Manley Hot Springs	$10^{-5.1}$

a small number of fleas from him and certain personal effects. Three weeks later the titre had risen to 1:320. The others had titres of 1:10 and 1:80. However, the trappers were skinning a variety of animals, including beaver, fox, lynx, marten, mink and an occasional varying hare.

The other four fleas listed in table 1 comprise four of the five species most commonly associated with cricetine rodents in the taiga of Alaska. Two of them (*Ma.p.dissimilis* and *Mg.c.gregsoni*) are the most common species encountered on voles in the tundra. At Central, *Ma.p.dissimilis;* at Nome, *Mg.c.gregsoni;* and at Manley Hot Springs, *P.o.longiloba* were collected from hosts that were found not to be infected. The other six pools were collected from hosts experiencing an infection of tularemia organisms at the time.

The situation at Nome was far more complex than these brief data indicate. In 1968 tularemia organisms of such low virulence were encountered that frequently the guinea pigs showed no significant rise in titre and infection was not discovered until necropsy at the end of three weeks. Also, through an inadvertent error in shipping a nitrogen chest, on the part of the air transport company, many of the fleas collected at Nome were lost for microbiological examination during 1968. Nonetheless, it is of interest to report that despite the low numbers sampled, the number of isolates of *F.tularensis* from pools of mammalian tissues obtained from that single area in 1968 amounted to 20 per cent of the total for the previous four years from all of Alaska. Indeed, it was a costly error for I am convinced more pools of fleas would have otherwise been found infected.

2. Of the 20 pools of the F_1 *Ma.p.dissimilis* that fed upon an experimentally infected host, 18 or approximately 90 per cent of the pools were positive for the presence of tularemia organisms. No reason can be given as to why the other two pools – comprising 40 fleas – were negative. It is thought that perhaps sufficient organisms were not circulating in the peripheral blood of the host at the time of feeding, because the organisms were definitely present in heart blood and in the peripheral circulation.

A series of *Ma.p.dissimilis* (ten pools of 20 each) proved that tularemia organisms could be retained over a period of from 15 up to 60 days when stored at 14°C and 6°C respectively.

3. Attempts to transmit tularemia organisms during the experimental feeding of *Ma.p.dissimilis* were unsuccessful. The method of feeding confined the fleas within the capsule for one hour, after which the fleas and capsule were removed and the shaved abdomen of the host washed lightly with detergent and water. In no instance was the organism transferred

to the host. However, ten pools of ten fleas each, placed in an equal number of gallon jars in which a vole resided, resulted in three of the ten voles becoming infected. These fleas had been associated with the voles for 96 hours. It may well be that the fleas should have been left longer, but at the same time, if left longer, the voles would have soiled the bedding excessively with urine and so altered the environment within the container that no knowledgeable control over the ecology and life span of the flea or the vole would have been maintained. Only a small amount of bedding (autoclaved sand and dry grass), two inches deep was placed in the jars to make it easier to recover the fleas, but even so, only 75 per cent of the fleas were recaptured. There was no way the fleas could have escaped for although the metal of the center area of the lid had been removed, it had been replaced with 70 mesh bronze screen which was securely soldered in place. The litter was checked bit by bit with the aid of a stereoscopic microscope to insure no fleas were missed. Voles are fastidious creatures, at least equal to the murid rodents in respect to grooming. Oral ingestion possibly accounted for the missing fleas, and cannot be ignored.

4. Mechanical transmission studies were conducted by using a total of 30 pools consisting of 20 specimens each of *Ma.p.dissimilis* (600 fleas) which fed for a short time (five to seven minutes) upon infected voles. Ten fleas from each pool were used as described in paragraph 2 of Methods to ascertain the percentage of positive pools. These pools were used to ascertain if the infected fleas could transmit tularemia organisms to non-infected voles by using interrupted feedings over varying intervals of time. All pools were reduced to ten fleas after the samples were removed to ascertain the percentage infected.

The first series consisted of 15 pools each transferred to individual capsules already in place upon 15 uninfested animals to allow the fleas to continue feeding. This was done within one to two hours after the feeding was interrupted. Out of this series, eight of the voles developed an infection from which *F. tularensis* was recovered. All 15 pools were ground and injected into guinea pigs after this feeding and three were found uninfected and it was assumed they were uninfected during feeding, despite the earlier testing indicating that all but one of these pools were infected. This points out the shortcomings of working with pools of fleas as compared to individuals.

The second series of 15 pools were used 24 hours after feeding on the infected animal. This was an attempt to gain some impression that if fleas functioned as mechanical vectors, how long this capacity was retained. Five of the voles subsequently developed tularemia, indicating mechanical transmission of tularemia organisms had occurred. On injecting these fleas into experimental animals after the completion of the feeding, five were found uninfected with tularemia organisms although only three were indicated as uninfected after feeding on the infected animals. As in the previous series, these two were considered as uninfected at the time of feeding.

Briefly, these phases of the study utilizing 30 pools of fleas totaling 600 specimens revealed that in five of the pools tularemia organisms could not be demonstrated after feeding on infected animals. Subsequently four additional pools of fleas were discovered to be uninfected after the completion of the interrupted feeding studies. These four pools were presumed not to have been infected at the time of the engorgement. Eight voles and three voles from the first and second series respectively developed infection with tularemia organisms. Although there was a difference in the rate of infected pools, it appeared that the fleas lost some ability to transmit tularemia organisms within a 24-hour period.

In most, but not all of the voles that developed tularemia as a result of the interrupted feeding experiments, minute white-focal lesions developed on the external surface of the epidermis of the vole in the area where the fleas had fed. This was reminiscent of the condition encountered several years past when working with lice, rabbits and *F. tularensis* and which promoted the use of a water soluble detergent on the feeding area shortly after the feeding experiments were completed as an effort to reduce chance contaminations. Culture for tularemia organisms had been attempted from some of these lesions on the rabbits – and somewhat to my surprise, rich cultures of tularemia organisms were obtained. Some hemolytic *Streptococcus* were also encountered, but were not identified as to species at the time.

4 DISCUSSION

Before discussion of the results is undertaken, a resumé of the literature concerning fleas and tularemia organisms is relevant. Sazanova (1965) stated that approximately 124 species of fleas had been found naturally infected with plague organisms up to 1960. The species of fleas found naturally infected with tularemia organisms appear to be approximately a dozen.

The first scientific paper on tularemia (McCoy, 1911) reported recovery from a flea then known as *Ceratophyllus acutus* Baker; now *Diamanus montanus* (Baker). Specimens of this flea were taken from sick or dead ground squirrels and tested in guinea pigs. McCoy (1911) attempted to transmit the disease by placing healthy animals in the containers with flea-infested sick squirrels. In several experiments, transmission was effected, but in one experiment presences of buboes in the cervical region is said to imply transmission by ingestion rather than by flea bite. Knowing something of the feeding habits of this flea, it would seem that the cephalic region of the host would be a likely place for the fleas to feed. Therefore, the ingestion route may not be as evident as he inferred.

In one report of the Minnesota wildlife disease investigation, Green et al. (1938) recorded the recovery of *F. tularensis* from one lot of four fleas removed from a snowshoe rabbit, *L. americanus* and from three lots of one, nine, and three fleas respectively from the cottontail rabbit, *Sylvilagus floridanus.* All had been tested by animal inoculation. In no instance was infection recovered from fleas when it was not demonstrated in the host, either snowshoe hare or cottontail rabbit. The fleas concerned in these tests were referred to as *Spilopsyllus cuniculi* (Dale), which is the European rabbit flea whereas the common flea on the snowshoe hare is *H. g. lynx* and the characteristic fleas on the cottontail in the west north central portion of the United States are *Cediopsylla simplex* (Baker) and *Hoplopsyllus glacialis affinis* (Baker). *Odontopsyllus multispinosus* (Baker) is found occasionally. Waller (1940) recovered tularemia organisms from *C. simplex* removed from a sick cottontail rabbit in Iowa.

Tularemia was very infrequently recovered when wild rodent fleas were tested for plague at the San Francisco Plague Laboratory. The earlier findings were usually announced in Public Health Reports without authorship. One such report deals with the recovery of infection from fleas of prairie dogs, *Cynomys,* collected in Wyoming (1941) and another with fleas from ground squirrels in Alberta, Canada (1942). The species of fleas was not indicated in either instance.

Woodbury & Parker (1954) found fleas removed from the antelope ground squirrel, *Citellus leucurus* (Baker) were naturally infected with tularemia organisms although the identification of the fleas was not reported. Thorpe et al. (1965), in reviewing the results of their efforts, reported 32 149 fleas comprising 1 437 pools were used in attempts to recover tularemia organisms. They were successful with two of the pools consisting of *Thrassis bacchi* (Rothschild) removed from the antelope ground squirrel. In these studies, it was not stated whether the host was infected.

Olsuf'iev (1963) listed the following six species of fleas as having been found naturally infected within the USSR and cited the appropriate authorities. 1) *Ctenophthalmus assimilis* (Taschenberg); 2) *Ctenophthalmus acuminatus* Ioff & Argyropulo; 3) *Ctenophthalmus pollex* Wagner & Ioff; 4) *Megabothris calcarifer* (Wagner) (as *Ceratophyllus*); 5) *Malaraeus penicilliger* (Grube) (as *Ceratophyllus*); 6) *Nosopsyllus consimilis* (Wagner) (as *Ceratophyllus*).

How the flea transmits tularemia organisms to the voles is not clear. Experimental studies by Prince & McMahon (1946) indicated that *Xenopsylla cheopis* (Rothschild) and *Nosopsyllus fasciatus* (Bosc d'Antic) were unable to transmit the organisms when placed on appropriate laboratory animals. The work of Parker (1957) and Parker & Johnson (1957) working with *Cediopsylla inaequalis* (Baker), *Orchopeas leucopus* (Baker), *Pulex irritans* Linnaeus and *T.bacchi,* agree with that of Prince & McMahon.

Olsuf'iev (1963) and Sazanova (1965) review the numerous studies concerned with fleas and tularemia within the USSR. The former author concludes that ticks and rodent lice are extremely active vectors of tularemia whereas transmission through fleas to rodents is difficult. In experimental studies, Olsuf'iev & Tolstukhina (1941) placed infected *C.assimilis* on laboratory animals and produced acute infection. These same authors reported positive results when feeding ten *N.fasciatus* 12 hours after having fed on an infected animal. However, a large series of other experiments produced negative results. The fleas were still infected when removed from the 'healthy' animals several days later. This would tend to indicate that infection following the ingestion of fleas by the host was not as common as with certain gamasid mites (Hopla, 1951; Petrov, 1971).

According to Sazanova (1953), *F.tularensis* is excreted in the faeces of *Ct.assimilis* from the first to the 14th day and up to the 30th day with *Ma.penicilliger* (as *Ceratophyllus*). The last species is the same as the one in my own studies reported above. Tiflov (1959a) considers that flea faeces offer two routes of infection to the rodents: 1) by scratching induced by the reaction of the flea bite and contaminating it with flea faeces; 2) via the mucous membranes when the rodents groomed their coat by licking, thereby encountering the infected faeces, or by actually crushing and ingesting the infected fleas.

The data presented in the present study indicate that Tiflov's concepts are valid, for in part 3 of this study, voles did become infected when infected fleas were given free exposure to them for 96 hours. By contrast, no infection was produced when the fleas had a limited access and were confined to a clean feeding procedure as in part 1. However, the results in part 3 clearly indicate that ingestion of infected fleas and contamination by voles play a part, but probably of a limited capacity. Were it truly important, the disappearance of 25 per cent of the fleas could, under the closed system used, logically be ascribed to the voles destroying them, most probably by oral grooming. Tiflov's (1959a) huge study of 227 trials, in which 34 or 15 per cent were positive, support this concept.

Several of the authors cited previously found that adult fleas, once infected, remained so for varying lengths of time. In part, the variability is probably caused by the response of the various genera and species used, and more importantly, the temperature at which the infected fleas are stored. Tiflov (1959a) reported that *Ctenophthalmus wagneri* Tiflov retained tularemia organisms up to 335 days when stored at temperatures of 0° to 15°C. Other authors report periods of much shorter length, for example, Parker & Johnson (1957) up to five days but temperature was not stated. Studies by Sazanova (1953) indicate *Ma.penicilliger* remained infected 99 days at 14°C. This agrees with present findings with the Alaskan form of this species. A difference of 8°C extended the retention from 15 days to 60 days.

In studies carried out by Tiflov (1959b), larvae of *N.consimilis,* fed a mixture of tularemia organisms in the diet, retained organisms in the alimentary canal up to 15 days when maintained at a temperature of 13° to 14°C. Some larvae were free of organisms within three days and most of them by six days.

Hood and Molyneux (1970) working with *X.cheopis* larvae fed on dried blood (in which tularemia organisms are known to survive poorly), could detect no *F.tularensis* after 24 hours. Using *N.fasciatus* larvae but feeding them artificially on a dilution of tularemia organisms in citrated human blood, Molyneux (1967) noted that a significant number of organisms survived up to three days. Tiflov (1959a) and Hood & Molyneux (1970) reported no tularemia organisms in the adult fleas after the metamorphosis of those larvae.

The experimental data presented in 4 concerning mechanical transmission by interrupted feeding, appears to be the first serious study from this standpoint. The increase in the transmission rate by the pooled fleas to at least 66 per cent as compared to 30 per cent when the fleas had free access to the voles, is significant. Those fleas held 24 hours before continuing the blood meal still retained a 30 per cent transmission rate. If the percentages were corrected for the three pools found not infected after the completion of the interrupted feeding, the rate would be higher. Preliminary data indicate that this type of mechanical transmission is influenced somewhat by temperature. For example, at 6°C, a higher percentage of units was able to transmit over a longer period of time. Unfortunately, no significant data have been obtained that would ascertain if multiple interrupted feedings on uninfected hosts were important.

Ma.p.dissimilis was a reliable experimental model with which to work. Its Holoarctic distribution enhances the significance in such studies. Lack of adequate culture methods for the other Alaskan vole fleas was disappointing, especially regarding *Mg.c.gregsoni,* for field observations indicated it to be a flea of considerable interest and probably at least as important as *Ma.p.dissimilis,* not only because it has as broad a distribution, both geographically and host-wise, but because it undoubtedly was equally involved with tularemia organisms.

A common vole flea that is Holoarctic in distribution, *C.d.fulleri* was not found naturally infected with tularemia organisms even though it was at times collected from infected voles at the same time as the two species just mentioned as having been found infected. Laboratory rearing results were poor, but what data I have indicated it was less apt to become infected in the laboratory than the other two species. While sufficient data are not available, it is tempting to speculate that there is a species or generic difference in susceptibility to tularemia organisms.

A flea confined to the Nearactic region, *Mg.quirini* was the second most common species collected, as indicated in figure 1. It was present on some voles positive for tularemia organisms from the Manly Hot Springs study area. In fact, on at least two occasions, it was found in collections containing *Ma.p.dissimilis* and *P.o.longiloba* that were positive for *F.francisella,* yet the specimens of *Mg.quirini* were not. Out of approximately 1 200 specimens of this flea, only one pool from the Fairbanks area was found infected. In view of the abundance of *Mg.quirini* it would seem that it should have been selected as a possible experimental model. *Mg.c.gregsoni* was chosen as the representative of the genus because it was present in the taiga and tundra whereas *Mg.quirini* was found almost entirely within the taiga.

Kartman, Prince & Quan (1959), reporting on a large-scale outbreak of tularemia in voles in Oregon during 1957-1958, stated that voles were shown to have the highest rate of infection with *F.francisella* during the winter months, but at the same time lice and fleas were found essentially negative to infection with this organism. As spring developed, the epizootic in the voles was reduced; but conversely, the ectoparasites showed an increased rate of infection with tularemia organisms. By early summer the incidence of *F.tularensis* in ectoparasites and in voles was decreased to the point that this organism was seldom encountered. In Alaska, for the years mentioned, voles and fleas were at peak populations during late summer, fall, and usually early winter. Most of the flea and rodent isolates of tularemia organisms were from specimens collected during this period.

Reports of latent strains of tularemia associated with fleas occur in the literature, for example, Sazanova (1965) and Tiflov (1959a). Apparently material was passaged several times before the organism was isolated. As a general rule, during the study, fleas and tissues of voles were not passaged more than twice in the original attempts at isolation. With the procedures used, more did not seem warranted. It is not known if continued passage would have produced a higher rate of recovery, but it is seriously doubted if it was worth such effort. In personal correspondence on the occurrence in Nature of attenuated tularemia strains, Academician H.G.Olsuf'iev (September 1977), states, 'Nobody has isolated such strains'.

From field and laboratory studies, it is clear that fleas can become infected with tularemia organisms. Fleas seldom transmit tularemia organisms in the process of routine feeding upon voles but interrupted feeding must be considered of some importance in the natural spread of tularemia among these rodents. On the other hand, when fleas are in their natural environment the difference between 'normal' and 'interrupted' feedings likely is more apparent than real. Fleas which must fast for several days lose their infection prior to securing the new hosts. The chance of a vole becoming infected by ingestion of flea faeces of the flea itself during grooming is of some relevance. My experience at Nome (Hopla, 1969) indicates that two species of voles, one species of lemming and *F.tularensis* that contaminated water, were all part of a natural cycle in that study area. The low virulence of some of the B strains encountered here did not produce marked changes in the vole and lemming population.

The flea-microtine rodent-water cycle just mentioned with strain B organisms is of fundamental importance. As one compares the concepts by Olsuf'iev (1964) concerning the paleogenesis of tularemia, the flea-microtine rodent-water cycle provides an alternate hypothesis. Olsuf'iev (1964) contends that tularemia has been primarily associated with Leporidae and ticks, especially *Dermacentor* with a secondary involvement in rodents. I think that the fleas and microtine rodents are the primary association in nature and that ticks and Leporidae are secondary. It is interesting that the tick-lagomorph cycle is of greater importance in parts of North America. Tularemia organisms are better adapted to a cool moist habitat than a hot dry one, but once established in ticks and lagomorphs, this secondary adaptation opened new avenues for geographical spread of the disease, and

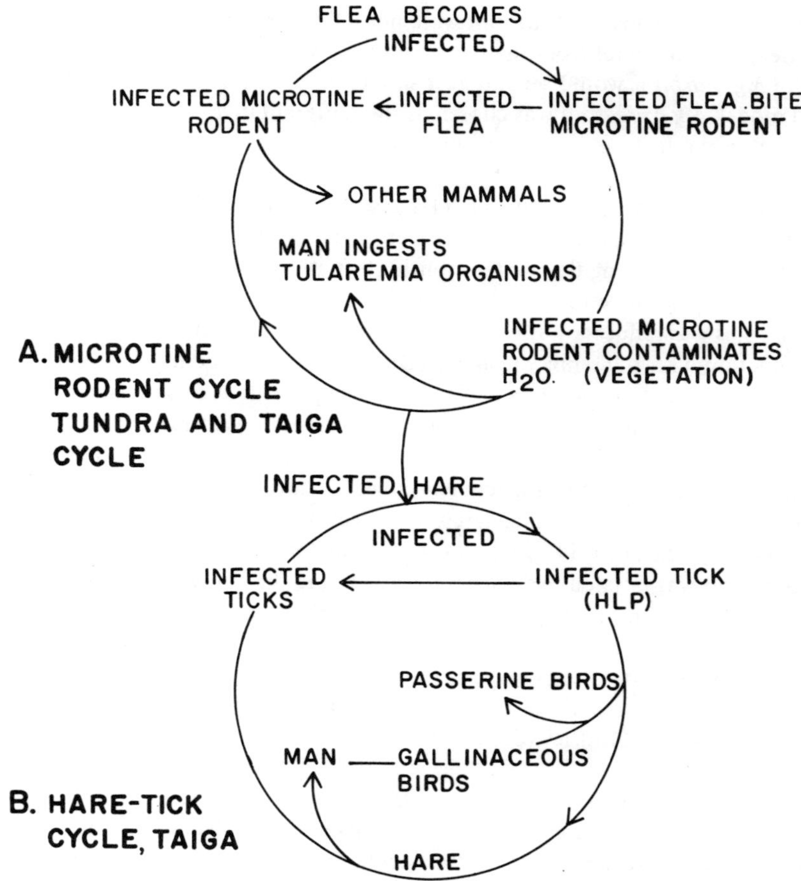

Figure 2. The natural cycle of tularemia in Alaska. The microtine rodent cycle in the tundra is less complicated due to the smaller number of different species involved. It should be kept in mind that the snowshoe hare intrudes a certain distance into the tundra. In the taiga, passerine birds, especially during the low phase of the hare cycle, are important hosts for the immature stages of *Haemaphysalis leporispalustris.* The gallinaceous birds are important in this respect, the grouse and ptarmigan can be a direction source of infection in man. Because the rabbit tick, *H.leporispalustris* (HLP) is parasitic upon birds and the varying hare and is the only tick known in the taiga, tick transmission of tularemia organisms is limited to these animals.

an opportunity to become associated with a whole new series of vertebrate animals, including frequent contact with man.

Two other groups of ectoparasites, mites and lice, occur on the rodents mentioned; unpublished data indicate that mites are involved to some extent, but sufficient information is simply not available to ascertain to what degree they are involved in the three areas mentioned. Lice were only rarely used in attempts to isolate tularemia organisms during the course of this study, in part because they were frequently overlooked by the field biologists and hence there were no meticulous records of the type obtained for fleas. Furthermore, the breadth of the study required that some priorities be established and fleas were selected over the other two groups of ectoparasites.

No observations similar to that of Tiflov (1959a) indicating a multiplication of tularemia organisms in the flea resembling a flea-plague organism relationship was ever encountered although laboratory-reared fleas, experimentally infected, were examined with some care. Field-captured fleas were screened as well for such phenomena, but were negative.

The review of the data presented, and other data collected during the course of this investigation (manuscript in preparation), strongly suggest that with strain B tularemia organisms, fleas, voles and lemmings have a meaningful relationship in maintaining a cycle of enzootic tularemia in the absence of lagomorphs. Figure 2 summarizes this concept in graphic fashion. In some of the study areas (Anatuvuk Pass, Kotzebue, and Nome) within the tundra in Alaska, populations of hares have not been recorded for 20-30 years. For example, in the 20 years my research programmes were in Alaska, arctic hares were not seen or captured in the three areas mentioned. Some remains of one hare were obtained from an Eskimo hunter in the summer of 1957 who shot the animal during the previous winter, several miles east of Kotzebue. It was the only specimen observed by him at that time. To my knowledge, that is the only hare known to have come from any of the three study areas. A total absence of ticks on terrestrial mammals in these areas during this same period of time was observed; the ticks have likely been absent from these tundra areas throughout recent history, yet tularemia organisms continue to be found in voles, lemmings and fleas.

ACKNOWLEDGEMENT

Supported, in part, by funds from the Faculty Research Council, University of Oklahoma. The skilled technical assistance of Mona Reyes and Barbara Rimkus is acknowledged.

REFERENCES

Anonymous 1941, Tularemia infection found in fleas from prairie dogs in Wyoming. U.S. Publ. Hlth. Repts. 56:1521-1522.

Anonymous 1942, Plague and tularemia infections of fleas. U.S. Publ. Hlth. Repts. 57: 1358-1359.

Bell, J.F. 1965, Ecology of tularemia in North America. J. Jinsen Med. 11(2):33-44.

Green, R.G., C.A.Evans, J.F.Bell & C.L.Larson 1938, The role of fleas in the natural transmission of tularemia. Minnesota Wildlife Dis. Invert. Rpt. pp. 25-28.

Holland, G.P. 1958, Distribution patterns of northern fleas. Proc. 10th Congr. Ent. 1: 645-658.

Hood, A.M. & D.H.Molyneux 1970, Survival of *Pasteurella tularensis* in flea larvae. J. Med. Ent. 7:609-611.

Hopla, C.E. 1951, Experimental transmission of tularemia by the tropical rat mite. Amer. J. Trop. Med. 31:768-782.

Hopla, C.E. 1955, The multiplication of tularemia organisms in the lone star tick. Amer. J. Hyg. 58(1):101-118.

Hopla, C.E. 1965, Alaskan hematophagous insects, their feeding habits and potential as vectors of pathogenic organisms. I. The Siphonaptera of Alaska. Arctic Aeromed. Tech. Rpt. 64(12) 267 p. Fairbanks, AK 99701.

Hopla, C.E. 1966, Ecology and epidemiology research studies in Alaska: A report of field collections and diagnostic assay. Univ. Okla. Res. Inst., proj. 1577, pp. 1-50. Norman, OK 73019.

Hopla, C.E. 1969, Ecology and epidemiological research studies in Alaska: A resumé of field collections and laboratory diagnostic assay from 1964-1968. Univ. Okla. Res. Inst., proj. 1577, pp. 137. Norman, OK 73019.

Hopla, C.E. 1974, The ecology of tularemia. Adv. Vet. Sci. Comp. Med. 18:25-53.

James, M.T. & R.F.Harwood 1969, Hermes' Medical Entomology, 6th Ed., pp. 484. New York, McMillan Company

Jellison, W.F. 1959, Fleas and disease. Ann. Rev. Ent. 4:389-414.

Jellison, W.F. 1977, Tularemia in North America 1930-1974. Print. Dept. Univ. Montana, Missoula, pp. 80-82.

Kartman, L., F.M.Prince & S.F.Quan 1959, Epizootiologic aspects. In: The Oregon Meadow Mouse Irruption. (F.E.Price, dir.), pp. 43-53. Federal Cooperative Extension Service. Oregon State College.

McCoy, G.W. 1911, Studies on plague in ground squirrels. II. A plague-like disease of rodents. Publ. Hlth. Bull. 43:53-71.

Molyneux, D.H. 1967, Feeding behavior of the larval rat flea, *Nosopsyllus fasciatus* Bosc. Nature 215:779.

Olsuf'iev, N.G. 1963, Tularemia. In: Human Disease with Natural Foci. E.N.Pavlovsky (ed.), pp. 219-281. Foreign Languages Publishing House, Moscow.

Olsuf'iev, N.G. 1964, On the paleogenesis of natural foci of tularemia. Zool. Zh. 43(3): 355-370.

Olsuf'iev, N.G. & E.M.Tolstukhina 1941, The role of *Ctenophthalmus assimilis* Tasch. in the transmission and preservation of tularemia infection. Arkh. Biol. Nauk. 63(1-2): 81-88. Leningrad, USSR.

Owen, C.R., J.B.Bell, W.L.Jellison, E.V.Buker & G.J.Moore 1961, Lack of demonstrable enhancement of virulence of *Francisella tularense* during animal passage. Zoonoses Res. 1-75.

Parker, D.D. 1957, Attempted transmission of *Pasteurella tularensis* by three species of fleas. J. Econ. Ent. 50:724-726.

Parker, D.D. & D.E.Johnson 1957, Experimental transmission of *Pasteurella tularense* by the flea, *Orchopeas leucopus* (Baker). J. Infect. Dis., 101:69-72.

Petrov, V.G. 1971, The role of the mite *Ornithonyssus bacoti* Hirst as a reservoir and vector of the agent of tularemia. Parazitologiya 5(1):7-14.

Prince, F.M. & M.C.McMahon 1946, Tularemia: Attempted transmission by each of two species of fleas: *Xenopsylla cheopis* (Roths.) and *Diamanus montanus* (Baker). Publ. Hlth. Repts. 61:79-85.

Sazanova, O.N. 1953, On the transmission and preservation of tularemia by fleas of the common field vole. Vop. Kravev. Obshch. Eksp. Parasit. Med. Zool. 3:157-163.

Sazanova, O.N. 1965, Fleas (Insecta, Aphaniptera). In: Vectors of Diseases of Natural Foci. P.A.Petrischcheve (ed.), pp. 166-182. Israel Program for Scientific Translations. Jerusalem, Israel, S.Monson.

Thorpe, B.C., R.W.Sidwell, D.E.Johnson, K.L. Smart & D.D.Parker 1965, Tularemia in the wildlife and livestock of the Great Salt Lake Desert Region, 1951-1964. Amer. J. Trop. Med. Hyg. 14(4):622-637.

Tiflov, V.E. 1959a, The role of fleas in epidemiology of tularemia. Trudy Nauch.-Issled. Protivoch Inst. Kavk. Zakavk. 2:363-392.

Tiflov, V.E. 1959b, *B.tularense* in the phases of flea development. Trudy Nauch.-Issled. Protivoch Inst. Kavk. Zakavk. 2:393-398.

Waller, E.F. 1940, Tularemia in Iowa cottontail rabbits and in a dog. Vet. Student 2:54-55, 73.

Woodbury, A.M. & D.D.Parker 1954, Studies of tularemia, *Pasteurella tularensis.* Ecology of the Great Salt Lake Desert. Ecolog. Res. Univ. of Utah, Spec. Rept. 2, 14 pp. Salt Lake City, UT 81101.

ROSAMOND SHEPHERD
The Keith Turnbull Research Institute, Frankston, Victoria, Australia

THE EUROPEAN RABBIT FLEA *SPILOPSYLLUS CUNICULI* (DALE) IN AUSTRALIA – ITS USE AS A VECTOR OF MYXOMATOSIS

ABSTRACT

Spilopsyllus cuniculi was first introduced into Australia in 1966. A field release was made in 1970 in the semi-arid region of Victoria. Flea and antibody to myxoma virus data show that the study can be divided into 'pre flea' and 'post flea', the rabbit breeding season of 1973 being the transition point.

The initial spread of *S.cuniculi* was slow, but once every rabbit carried *S.cuniculi,* a population explosion occurred and spread increased rapidly.

Prior to the establishment of *S.cuniculi,* myxomatosis epizootics occurred each summer-autumn and, in association with nutritional stress, reduced the rabbit population by 80 to 90 per cent. Late winter-spring myxomatosis first occurred in 1973 when the mean number of *S.cuniculi* per doe and buck were approximately 30 and 20 respectively. Some outbreaks have been observed each year since then, usually beginning in September.

The severity of these outbreaks has varied from very few observed cases to generalized myxomatosis when up to 80 per cent of survivors carried antibody. It appears that rabbits must carry a mean of at least 30 to 40 *S.cuniculi* before a general outbreak occurs.

When rabbits were first introduced into Australia in 1859, they did not carry *Spilopsyllus cuniculi,* the European rabbit flea. *S.cuniculi* was introduced into Australia in 1966 by Dr Sobey, CSIRO Sydney (Sobey & Menzies, 1969) to act as a permanent vector of myxomatosis, and thus help effect a control of these pests. Since then these fleas have been released in a wide number of different geographical and climatic areas of Australia – areas which range from semi-arid central Australia with a rainfall of approximately 180 mm per annum, to coastal areas with an annual rainfall of approximately 800 mm (Johnston, 1973, Shepherd & Edmonds, 1976, Sobey & Conolly, 1971, Williams, 1971). *S.cuniculi* have bred successfully whenever rabbits breed but have shown a variable degree of success as a vector of myxomatosis.

Myxomatosis epizootics in northern Victoria have always been seasonal and dependent on mosquito vectors, usually *Anopheles annulipes* and *Culex pipiens australicus* (Fenner & Ratcliffe, 1965). The main study area was a semi-arid region, Pine Plains, with a mean annual rainfall of 275 mm, where rabbit numbers fluctuated markedly each year.

The rabbit breeding season may be up to six months long, usually winter-spring, with no breeding during the hot, dry summer. The population entering the summer was always high and caused severe pasture losses.

Epizootics occurred for about four to six weeks each summer-autumn, January to April, and were associated with an 80 to 90 per cent reduction in the rabbit population. Stickfast fleas *Echidnophaga myrmecobii* and *E.perilis* were commonly found on rabbits during those months, but although they can act as vectors of myxoma virus, they are not very effective (Fenner & Ratcliffe, 1965, Shepherd & Edmonds, unpublished data).

At Pine Plains 6,500 *S.cuniculi* were released during 1970 (Shepherd & Edmonds, 1976). These were reared by Dr Sobey in Sydney (Sobey, Menzies & Conolly, 1974), but since then *S.cuniculi* have been bred under laboratory conditions at Frankston. Monthly visits have been made to Pine Plains since December 1970 to monitor their establishment and spread, and to check for the occurrence of myxomatosis by sighting and by taking blood samples to test for antibody to myxoma virus.

Establishment depended on the time of release and the density of the rabbit population. Spread was slow during the first few years. The maximum spread was one kilometre in the first year, 3.5 km by the second year, 5 km by the third, 12 km by the fourth and 35 km by the fifth year (Shepherd & Edmonds, 1976). The rate of spread and numbers per rabbit were highly erratic so that it was possible in 1975 to find a rabbit 35 km from a release point carrying several hundred *S.cuniculi* while some rabbits close to a release site carried few. *S.cuniculi* did not spread along a front. Once nearly every rabbit in an area carried *S.cuniculi,* there was a 'population explosion' and a rapid increase in spread was noted. This occurred especially in 1974 when the rainfall was double the mean and there was eight months of rabbit breeding. Since then nearly all rabbits within the study area have carried *S.cuniculi,* except very recently littered does and very young kittens.

Data on fleas and antibody to myxoma show that the study can be divided into 'pre flea' and 'post flea', with the breeding season of 1973 being the transition point.

Myxoma virus has been introduced experimentally since 1970, either by innoculation or on flea mouth-parts (Shepherd & Edmonds, in press). The strain used has been Lausanne, a virulent strain cloned by Dr Sobey in Sydney. This is a stable persistent strain with very florid symptoms and therefore distinguishable from 'field strains' which originated from the virulent but less florid Glenfield strain.

During the 'pre flea' period the introduction of virus into a susceptible rabbit population was not sufficient to cause a myxomatosis outbreak. Individual rabbits may have been killed but not much transmission occurred. There were neither sufficient *S.cuniculi* present per rabbit, nor enough rabbits carrying *S.cuniculi* to cause an epizootic (figs. 1 and 2). During these years, despite introductions of virulent virus during the winter, epizootics of myxomatosis occurred during the normal months, January to April, only. The population crash associated with each epizootic was quite spectacular, with numbers over a 30 km transect falling from 1,322 in December 1970 to 98 in May 1971, a decrease of about 90 per cent; 1,184 in December 1971 to 210 in April 1972, a decrease of about 80 per cent and from 890 in November 1972 to 110 in March 1973, a decrease of about 80 per cent (fig. 3).

Until the beginning of the 1973 breeding season, the mean number of *S.cuniculi* was low with the highest monthly mean of 36 on does and 19 on bucks.

In 1973 the breeding season was longer and the mean number of *S.cuniculi* rose and the normal pattern of myxoma outbreaks changed. The first spring outbreak of myxomatosis was recorded in September-October. The population decreased by 60 per cent and the normal increase in the rabbit population did not occur. The antibody occurrence rose from about 20 per cent to 70 per cent. By October the mean number of *S.cuniculi* per doe had increased to 117. By December there was little active myxoma virus present but the antibody occurrence reached approximately 80 per cent. This outbreak of myxomatosis had continued throughout the breeding season as a continuous presence rather than the spectacular epizootic of the previous summer months (fig. 3).

An early myxoma outbreak occurred in June 1974, with a rise in antibody occurrence. Although it occurred during the breeding season few young rabbits were observed to enter the population. The outbreak was apparently confined within the warrens. *S.cuniculi* numbers rose, but no spectacular increase was found until late in the breeding season – September-November, when the mean number of *S.cuniculi* on does was 121 and 70 on bucks. Maxima were 502 and 300. By then a second outbreak of myxomatosis had occurred, starting in mid-September and continuing into December when the mean *S.cuniculi* numbers dropped to 14. During this outbreak, the count decreased from 700 to 110, a reduction of about 80 per cent (figs. 1 and 2). The antibody occurrence rose from 20 per cent in September to about 90 per cent in February. There was no evidence of the normal, later outbreak of myxomatosis in the summer-autumn. Presumably there were not sufficient susceptible rabbits present (fig. 3).

1975 and 1976 were dry years and breeding did not begin until July and June respectively. The 1975 spring outbreak was observed as a slight rise in antibody occurrence during September, the time *S.cuniculi* numbers began to rise. However, by early December, antibody occurrence was the lowest it had been since September 1971, and the population was high. An ineffective epizootic was first observed in December when the mean number of *S.cuniculi* per doe and buck was 30 and 19 respectively. It is probable that this epizootic was mosquito borne although *S.cuniculi* may have played a minor role as a vector.

During 1976 there was a steady rise in *S.cuniculi* numbers to peak in October on both does and bucks with a mean of 110 and 41. This was the first year of the survey to show a steady rise in *S.cuniculi* numbers throughout the rabbit breeding season and then a sudden decrease as a result of the heavy mortality of *S.cuniculi* after a number of breeding cycles (figs. 1 and 2). This pattern is similar to that found by Mead-Briggs, Vaughan and Rennison (1975) for *S.cuniculi* in Britain.

Some myxomatosis occurred in July-August. It was observed as an increase in antibody occurrence from approximately 40 per cent in July to 80 per cent in August (fig. 3). Very few infected rabbits were seen above ground. The outbreak was short-lived and probably confined to kittens and breeding does in the warrens. Heavy spring rains resulted in rabbits breeding into early summer and numbers peaked at 1,539 over the 30 km transect. Another 'trickle' myxomatosis was observed from September to March 1977. Although *S.cuniculi* numbers were high until October, no epizootic occurred. By March mean numbers had fallen to 8 and 14 per doe and buck (figs. 1 and 2).

It seems that although a stable host-parasite relationship was probably reached in 1976, rabbit breeding did not occur early enough for *S.cuniculi* to reach the numbers required to initiate and maintain an outbreak of myxomatosis. When the rabbit population increased, and antibody occurrence dropped, in October-December, the number of *S.cuniculi* present was only sufficient to maintain a trickle of myxomatosis.

Effective spring outbreaks such as occurred in 1974 will only happen when high *S.cuniculi* numbers and low antibody occurrence coincide. The decrease in rabbit numbers resulting from a spring epizootic means there is little stress by the rabbit on the pasture over the dry summer period. However, if little rain falls, the low numbers of *S.cuniculi* will mean ineffective spring myxomatosis and any summer reduction in the rabbit population could be a combination of trickle myxomatosis, poor nutrition and desiccation. Fortunately, these conditions are not very frequent and so useful spring epizootics should occur in perhaps seven years out of ten.

The effect of *S.cuniculi* on the control of rabbits in Australia is variable and hard to estimate. It has proved to be useful in some areas in some years – but it must be stressed that it is just one more means of controlling rabbits, and one which may be of limited use in some regions of Australia.

REFERENCES

Fenner, F. & Ratcliffe, F.N. 1965, Myxomatosis. Cambridge University Press.

Johnston, G.C. 1973, The European rabbit flea in Tasmania. Tasmanian J. Agric. 44: 157-8.

Mead-Briggs, A.R., Vaughan, J.A. & Rennison, B.D. 1975, Seasonal variations in numbers of the rabbit flea on the wild rabbit. Parasitology 70: 103-118.

Shepherd, R.C.H. & Edmonds, J.W. 1976, The establishment and spread of *Spilopsyllus cuniculi* (Dale) and its location on the host, *Oryctolagus cuniculus* (L.), in the Mallee Region of Victoria. Aust. Wild. Res. 3: 29-44.

Shepherd, R.C.H. & Edmonds, J.W. in press, Myxomatosis: the transmission of a highly virulent strain of myxoma virus by the European rabbit flea *Spilopsyllus cuniculi* (Dale) in the Mallee region of Victoria. J. Hyg. Camb.

Sobey, W.R. & Conolly, D. 1971, Myxomatosis: the introduction of the European rabbit flea, *Spilopsyllus cuniculi* (Dale) into wild rabbit populations in Australia. J. Hyg. Camb. 69: 331-346.

Sobey, W.R. & Menzies, W. 1969, Myxomatosis: the introduction of the European rabbit flea, *Spilopsyllus cuniculi* (Dale) into Australia. Aust. J. Sci. 31: 404-406.

Sobey, W.R., Menzies, W. & Conolly, D. 1974, Myxomatosis: some observations on the breeding of the European rabbit flea, *Spilopsyllus cuniculi* (Dale) in an animal house. J. Hyg. Camb. 72: 453-465.

Williams, R.T. 1971, Observations on the behaviour of the European rabbit flea, *Spilopsyllus cuniculi* (Dale), on a natural population of wild rabbits, *Oryctolagus cuniculus* (L.) in Australia. Aust. J. Zool. 19: 14-51.

Figure 1. The mean number of *S.cuniculi* present per month, December 1970 to February 1977, on female rabbits. Shaded months indicate breeding season, open months indicate non-breeding season. Percentage of rabbits infested with *S.cuniculi* is shown. 'Pre flea', December 1970 to June 1973. 'Post flea', July 1973 to February 1977.
NC no collection made.
* *S.cuniculi* present, but numbers not counted.

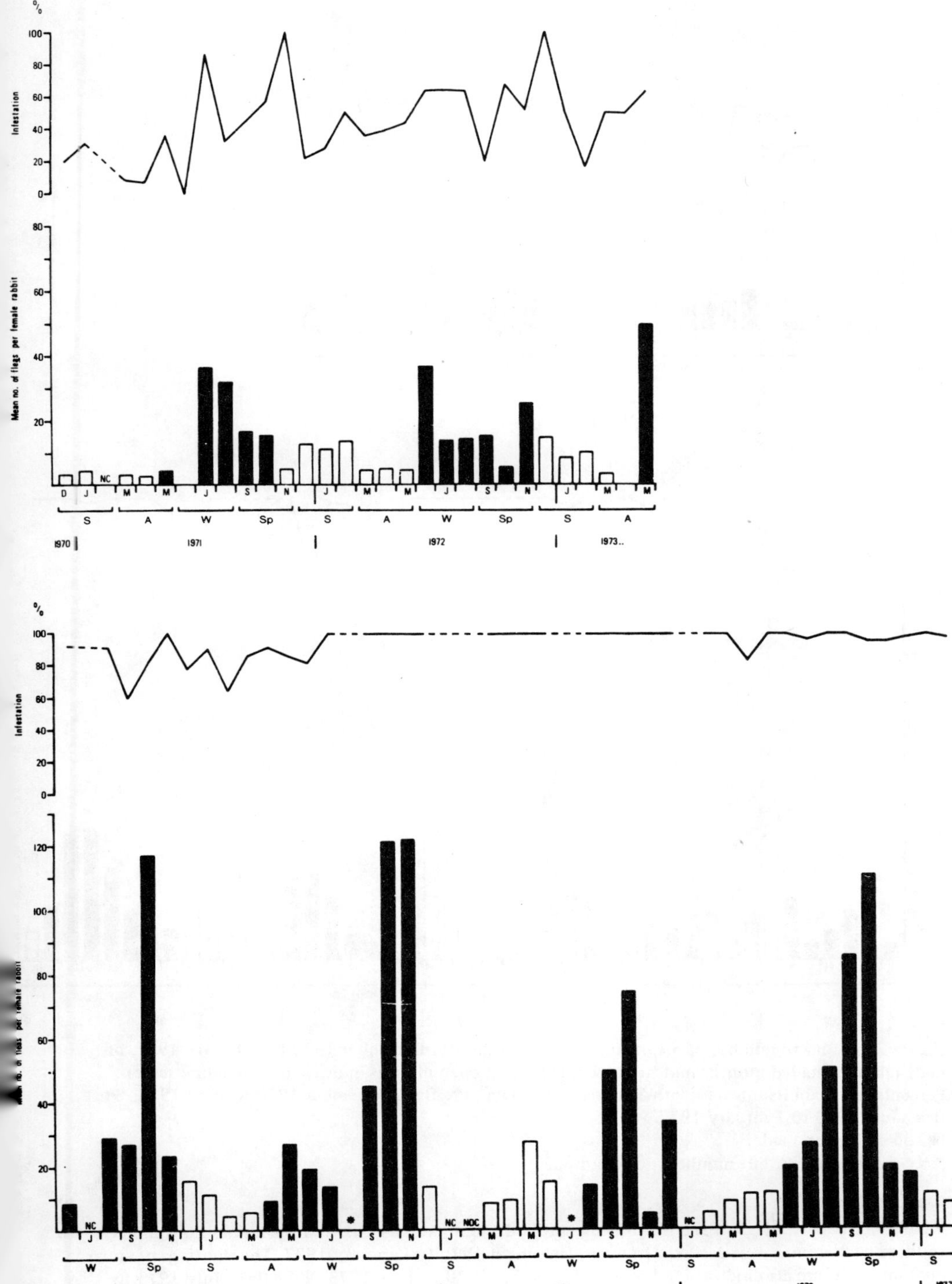
%
Infestation
Mean no. of fleas per female rabbit
NC
D J M M J S N J M M J S N J M M
S A W Sp S A W Sp S A
1970 1971 1972 1973..
NC NOC
W Sp S A W Sp S A W Sp S A W Sp S
..1973 1974 1975 1976 1977

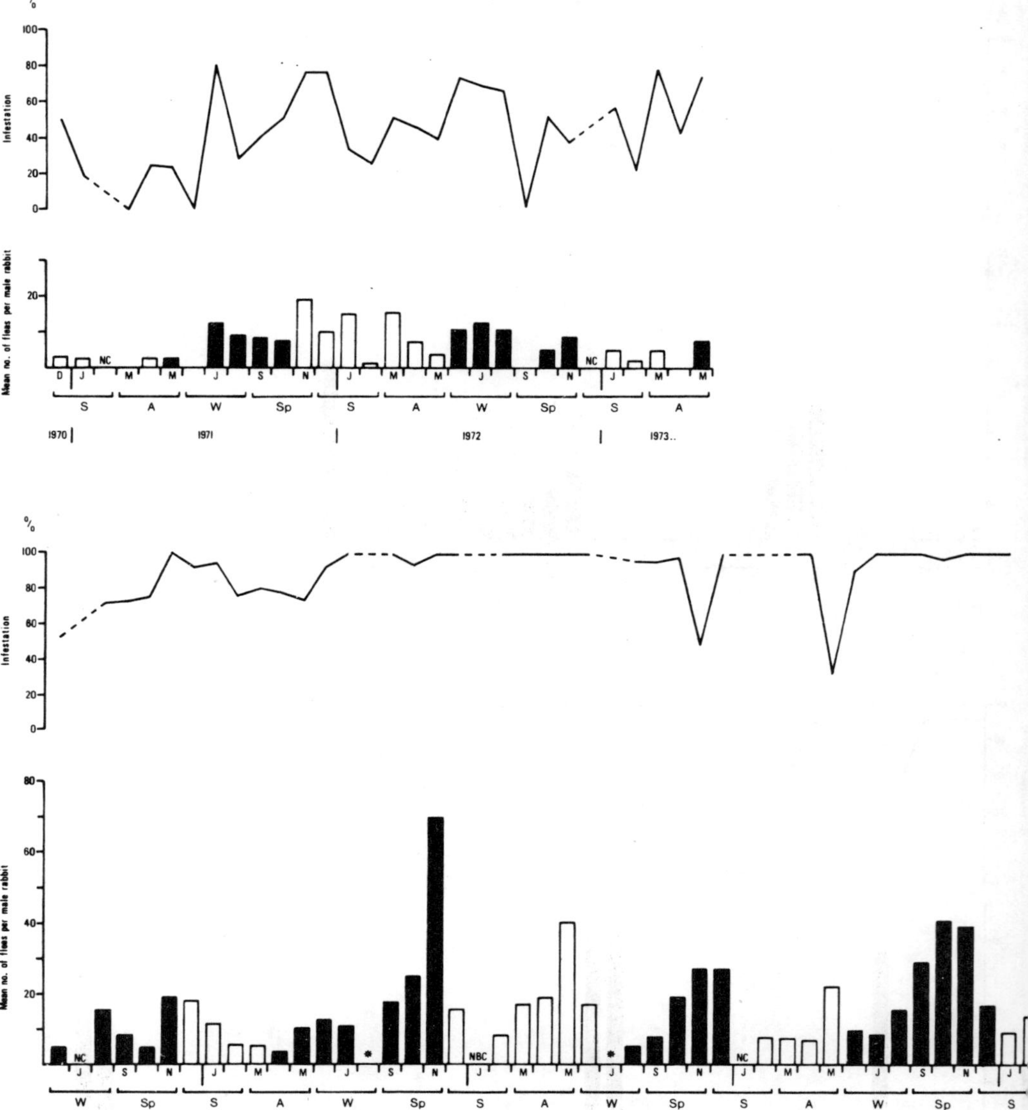

Figure 2. The mean number of *S.cuniculi* present per month, December 1970 to February 1977, on male rabbits. Shaded months indicate breeding season, open months indicate non-breeding season. Percentage of rabbits infested with *S.cuniculi* is shown. 'Pre flea', December 1970 to June 1973. 'Post flea', July 1973 to February 1977.
NC no collection made.
* *S.cuniculi* present, but numbers not counted.

Figure 3. The percentage of rabbits showing antibody to myxoma virus, and the number of rabbits counted each month over a 30 km transect, December 1970 to February 1977. The sightings of active myxoma virus are also indicated. 'Pre flea', December 1970 to June 1973. 'Post flea', July 1973 to December 1976.

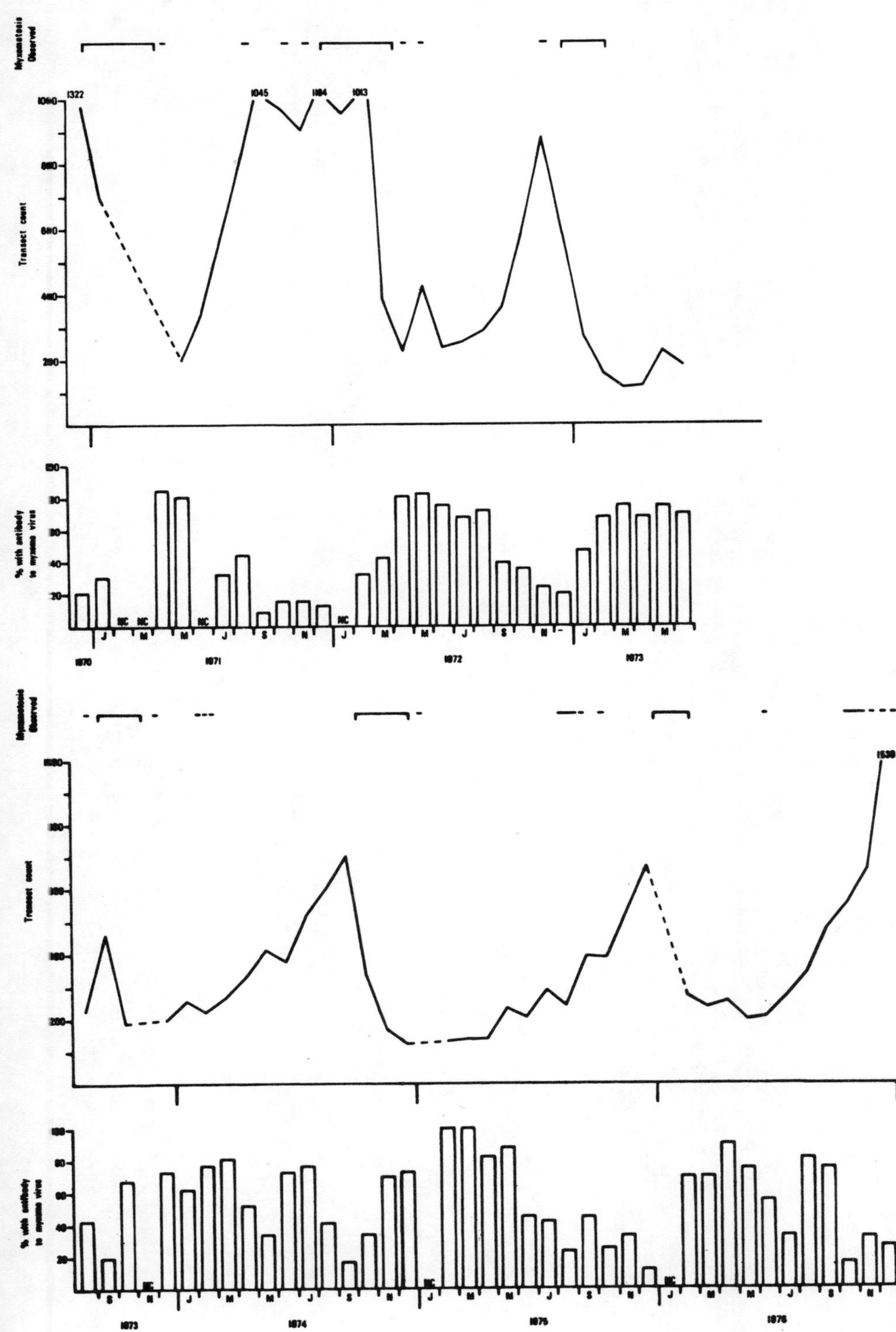

Myxomatosis Observed
Transect count
% with antibody to myxoma virus
1322
1045
1184
1013
1530
NC
1970
1971
1972
1973
1974
1975
1976

A. R. MEAD-BRIGGS & J. A. VAUGHAN
Pest Infestation Control Laboratory, Worplesdon, Guildford, Surrey, UK

THE IMPORTANCE OF THE EUROPEAN RABBIT FLEA *SPILOPSYLLUS CUNICULI* (DALE) IN THE EVOLUTION OF MYXOMATOSIS IN BRITAIN

ABSTRACT

The rabbit flea is the principle vector of myxomatosis in Britain. The establishment of the disease as an enzootic one was accompanied by the rapid appearance of moderately attenuated virus strains. An experimental procedure which has demonstrated relative differences in the transmissibility by the flea of these strains is described. The results are analysed and their importance discussed in terms of the likely evolution of the disease as a major mortality factor for wild rabbits. It is postulated that with the appearance of rabbits which are innately more resistant, a reversal of the trend for selection of less virulent viral strains should occur.

The first suggestion that the rabbit flea *(Spilopsyllus cuniculi)* might be of major importance as a vector of myxomatosis in Britain was made by Rothschild (1953). The initial field studies done in Britain in 1954-55 by Muirhead-Thomson (1956a, b, c) showed that mosquitoes were of little significance and confirmed Dr Rothschild's opinion of the important role of the rabbit flea. Since that time much research has been done on the breeding biology of the flea and on its host relationships, including studies on seasonal trends in its abundance, on its ability to transfer between individual free-living hosts and on its host-finding behaviour (for review, see Mead-Briggs, 1977).

Prior to the first epizootic of myxomatosis the European wild rabbit (*Oryctolagus cuniculus* (L.)) was a major pest of British agriculture, causing an annual loss then estimated at 40 to 50 million pounds. The initial spread during 1954-55 was marked by extremely high morbidity and mortality rates. With both rates probably in excess of 99 per cent and the rabbit the only available susceptible host, it was perhaps surprising that the disease survived to become enzootic. That it did so was mainly attributable to the development of mutant 'strains' of myxoma virus that produced a somewhat more protracted and less virulent disease. (The strains of myxoma virus that have developed in the field (in Australia and Europe) as the result of viral mutations can be classified, by the method of the Australian workers, Fenner & Marshall (1957) into grades of virulence. The basis of the classification is the mean survival time of groups of six domestic rabbits challenged by a standardised procedure.) The survival time is inversely proportional to the virulence of the sample (table 1).

Within a few years the virus strains most frequently isolated in the field caused 70 to 95 per cent mortality amongst fully susceptible (laboratory) rabbits; other strains with

Table 1. Basis of classification of mutant field strains of myxoma virus into grades of virulence by mean surivival time of groups of laboratory rabbits, and the approximate mortality rate they cause

Virulence grade	Mean survival time (days)	Estimated case-mortality rate (%)
I	<13	>99
II	13-16	95-99
IIIA	17-22	90-95
IIIB	23-28	70-90
IV	29-50	50-70
V	over 50	<50

even lower virulence were also present. Since in the wild population, those rabbits that recover from infection are immune for life; since immune females pass a transient immunity to their progeny and since there is the possibility of selection for animals of reduced susceptibility (enhanced resistance), a rapid evolution of the disease to a relatively innocuous one seemed likely. Consequently, it was necessary to assess the long-term influence of myxomatosis as a mortality factor for rabbits.

As the likelihood of any strain of virus being perpetuated depends upon its capacity to be transmitted, any differences between strains in their transmissibility by rabbit fleas could be of marked significance in the evolution of the disease. Mead-Briggs & Vaughan (1975) have shown that there are in fact differences in transmissibility between strains of differing virulence. This is the case although transmission is simply mechanical; transmission involves the inoculation into the susceptible host's skin, during probing by the flea, of infectious virus particles adhering to the flea's laciniae when they were withdrawn from an infective (donor) host. In preliminary studies nine strains of myxoma virus, selected as representative of high, medium and low virulence grades, were used and a total of 45 experiments was done. The basic experimental procedure was to determine how many of the fleas in samples of 24, taken at intervals during the course of the infection of a 'donor' wild rabbit, were infective when assayed by allowing them to probe successively on the flanks of two domestic rabbits. The procedure is illustrated in fig. 1.

Each experiment was started by infecting a susceptible wild rabbit with one of the selected strains of virus and infesting it with fleas. Samples of 24 fleas were removed at stages during the course of the ensuing disease. The fleas in each sample were starved for three days at 4℃, 95 per cent RH and then given the opportunity of probing for up to 30 minutes on the shaved flanks of a sedated New Zealand white rabbit. The posi-

Table 2. Mean percentages of fleas transmitting myxoma infection, during assay feeding on domestic rabbits, after removal from wild rabbits at various times during the progress of the disease. Results grouped by survival times of the wild rabbit, not by the strains of virus used to initiate the disease

	'Donor' wild rabbit survival time groups (days)					
	10-16	17-23	24-30	31-37	38-44	Survivors
Virulence grades of infecting virus	I	III	III	III	III	III & V
Number of rabbits in groups	8	4	5	5	3	20
Number of fleas assayed	504	312	576	600	360	2,208
Mean percentage of infective fleas	12.0	49.6	52.8	29.8	36.1	8.3

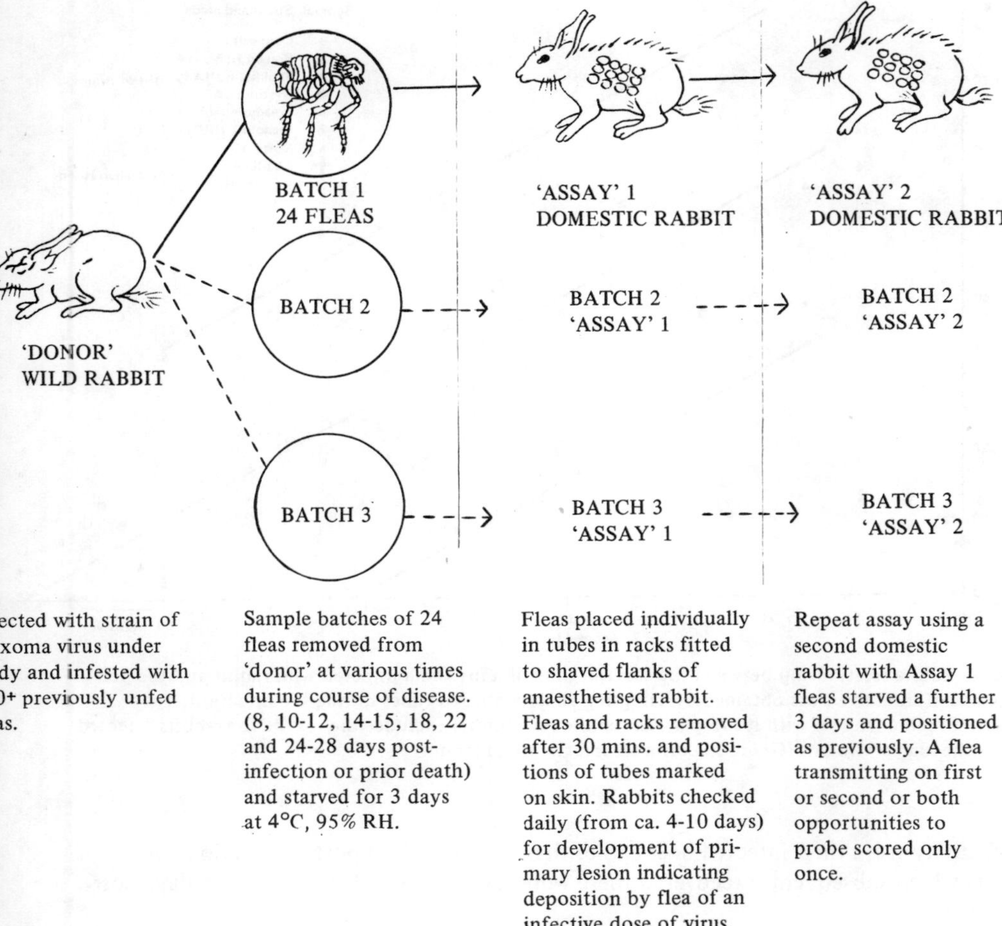

Figure 1. Diagram of the experimental procedure used to assess the efficiency with which the rabbit flea transmits different strains of myxoma virus.

tions on the flanks where the fleas had been individually confined and might have probed were marked with an indelible pen. The procedure was repeated using a second assay rabbit after the fleas had been starved for a further three days. The assay rabbits were examined daily from about four to ten days after the probings to score the number of fleas that had caused the development of primary lesions. Fleas which transmitted at the first, second or both feedings were scored only once. From among 4,324 of the fleas assayed, 701 (16 per cent) transmitted at the first feed, 513 (12 per cent) at the second. Of the latter, 304 (7 per cent) had not transmitted at the first opportunity and 115 (2.7 per cent) of them had not taken a blood meal, nor presumably probed the skin, at that time.

Analysis of the results (table 2, fig. 2) showed that the number of fleas that were infective depended upon three inter-related factors:

Firstly, the stage of the disease when the fleas were sampled. Very few fleas acquired

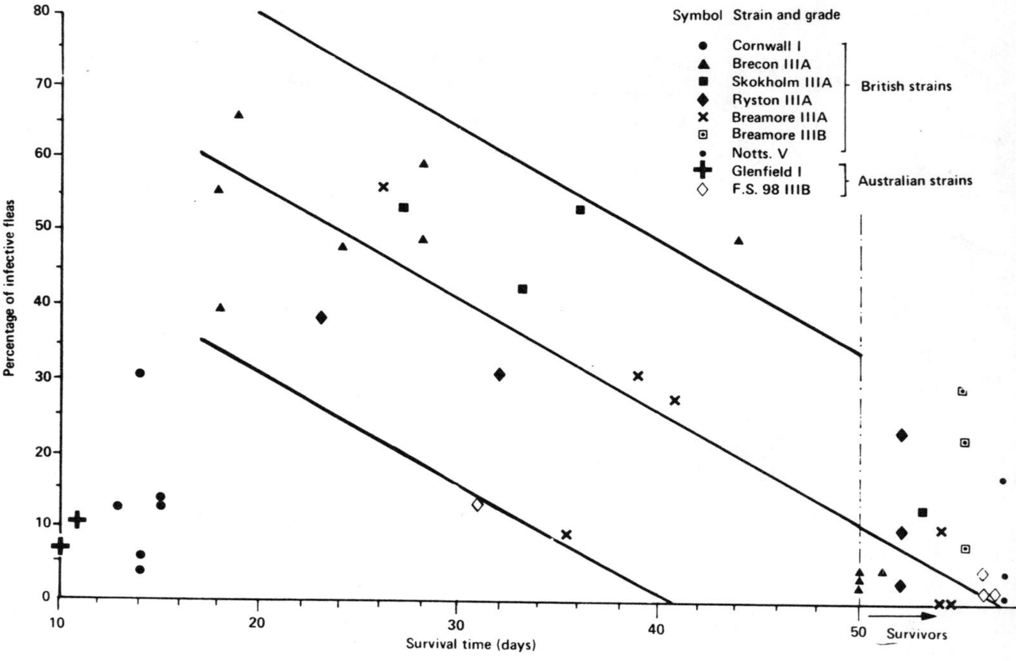

Figure 2. The relationship between the survival time of a myxoma-infected wild rabbit and the percentage of infective fleas obtained by sampling at standardised times during the infection. The regression line (shown with its 95 per cent confidence limits) excludes the results for rabbits infected with the highly virulent (Grade I) Cornwall and Glenfield strains.

infectivity from their infected host earlier than 10 to 12 days post-inoculation. On hosts which subsequently recovered there were few infective fleas beyond 18 days post-inoculation.

Secondly, the virulence of the infecting strain of virus. Moderately virulent (Grade III) strains generally led to the production of the highest proportions of infective fleas. 'Donors' infected with highly virulent (Grade I) strains gave rise to relatively low numbers of infective fleas, as also did animals infected with highly attenuated (Grade V) strains.

Thirdly, the response of the individual ('donor') rabbit to the infection, as measured by its survival time and, more importantly, whether it recovered from the infection. Significantly fewer infective fleas were obtained throughout the sample period from donors which survived infection with any of the attenuated strains, than from those donors which succumbed to the infection with the same strains.

The proportion of infective fleas sampled from each individual donor rabbit is shown in fig. 2 plotted against the survival time of the rabbit. The linear regression line only relates to the seven attenuated strains. The eight points in the lower left-hand corner refer to rabbits infected with one or other of the two highly virulent strains used, and in no case did they generate a very high proportion of infective fleas. For purposes of calculation survivors were given an arbitrary survival time of 50 days (although for clarity they are mostly shown in the diagram scattered beyond 50 days). In nearly all cases there were low numbers of infective fleas in the samples from those donors which survived.

It is apparent that those individual rabbits which were killed by Grade III strains of myxoma virus during the period 20 to 30 days after infection provided the highest proportions of infective fleas. It is likely that the development of the disease in these rabbits provided the optimum as regards availability of virus accessible to probing fleas, i.e. probably a prolonged duration with a sufficiently high titre of virus in the superficial 200 to 300 μm of the skin – the flea's laciniae are of this length.

It might thus appear that a stable situation has been reached in the rabbit host-flea vector-myxoma virus association favouring the transmission, and hence dominance in the field, of moderately attenuated strains – strains causing a mortality of 70 to 90 per cent amongst non-immune, susceptible rabbits. However, there is now evidence that wild rabbits in Britain are developing an increased (genetic) resistance to myxomatosis (Ross & Sanders, 1977). When infected, rabbits with such an increased resistance undergo a more protracted illness with a greater chance of recovery and longer mean survival time for those individuals that finally succumb. The symptoms associated with maximum infectivity of the host for fleas are likely to be produced in a more resistant host only by a more virulent strain of virus, than that required in a fully susceptible rabbit. Thus with the development among wild rabbits in Britain of increasing genetic resistance to myxomatosis we expect to find an increasing proportion of more virulent viral strains. The result of such a selection for more virulent strains should maintain for the time being a fairly constant mortality rate amongst non-immune, but increasingly resistant animals.

REFERENCES

Fenner, F. & Marshall, I.D. 1957, A comparison of the virulence for European rabbits *(Oryctolagus cuniculus)* of strains of myxoma virus recovered in the field in Australia, Europe and America. J. Hyg., Camb. 55: 149-191.

Mead-Briggs, A.R. 1977, The European rabbit, the European rabbit flea and myxomatosis. In: T.H.Coaker (ed.), Applied Biology, Vol. 2, p. 183-261. London and New York, Academic Press.

Mead-Briggs, A.R. & Vaughan, J.A. 1975, The differential transmissibility of myxoma virus strains of differing virulence grades by the rabbit flea *Spilopsyllus cuniculi* (Dale). J. Hyg., Camb. 75: 237-247.

Muirhead-Thomson, R.C. 1956a, The part played by woodland mosquitoes of the genus *Aedes* in the transmission of myxomatosis in England. J. Hyg., Camb. 54: 461-471.

Muirhead-Thomson, R.C. 1956b, Field studies of the role of *Anopheles atroparvus* in the transmission of myxomatosis in England. J. Hyg., Camb. 54: 472-477.

Muirhead-Thomson, R.C. 1956c, Observations on the European rabbit flea *(Spilopsyllus cuniculi)* in relation to myxomatosis in England. Unpublished Report to the Scientific Sub-Committee of the Myxomatosis Advisory Committee, Ministry of Agriculture, Fisheries and Food. (Restricted circulation)

Ross, J. & Sanders, M.F. 1977, Innate resistance to myxomatosis in wild rabbits in England. J. Hyg., Camb. 79: 411-415.

Rothschild, M. 1953, Report to the Myxomatosis Advisory Committee, Ministry of Agriculture, Fisheries and Food (Private circulation).

J. GOOSE
Fisons Ltd., Chesterford Park Research Station, Saffron Walden, Essex, UK

THE USE OF BENDIOCARB TO CONTROL FLEAS

ABSTRACT

Bendiocarb 80 per cent WP (FICAM W) has been available to professional pest control operators as a general public health insecticide for five years. In this time, it has become highly regarded as an extremely effective and safe residual insecticide which is odourless and which leaves no unsightly deposit on treated surfaces. The compound was highly active against fleas *(Ctenocephalides felis)* in vitro and had a longer in vitro residual effect than malathion. Bendiocarb 0.1 per cent dust was effective against fleas on infested cats in an initial laboratory trial and provided a similar degree of protection from reinfestation as 4 per cent bromocyclen dust. However, the use of bendiocarb to control fleas on animals is still under development and no approval for animal treatment has yet been sought. Spray treatment of flea infested premises *(C.felis)* with 0.25 per cent bendiocarb eradicated the infestation in a series of six trials. This efficacy has been well confirmed in the field in many countries of the world.

1 INTRODUCTION

Bendiocarb is the B.S.I. common name for the carbamate insecticide 2,2-dimethyl-1,3-benzodioxol-4-yl N-methyl carbamate.

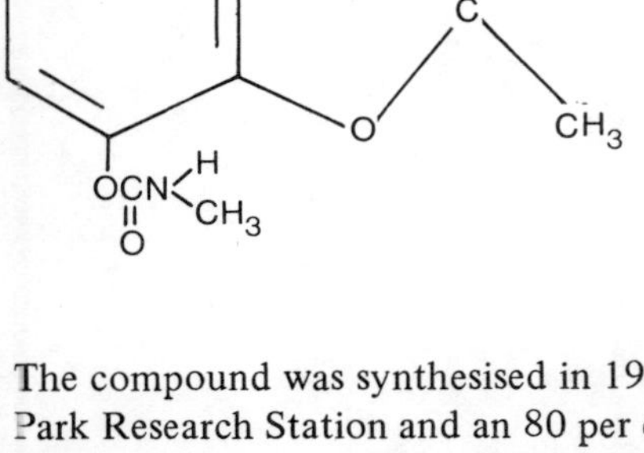

O CH$_3$
C
O CH$_3$
OCN H CH$_3$
O

The compound was synthesised in 1967 at Fisons Agrochemical Division's Chesterford Park Research Station and an 80 per cent WP was introduced commercially under the trade name Ficam W as a public health insecticide in 1972. Bendiocarb has a very broad spectrum of activity and examples of some of the known susceptible insect species are listed in Table 1.

Table 1. Species of insects known to be susceptible to bendiocarb

Order	Species
Thysanura	*Lepisma saccharina*
Orthoptera	*Acheta domesticus*
	Schistocerca gregaria
Dermaptera	*Forficula auricularia*
Dictyoptera	*Blattella germanica*
	Blatta orientalis
	Periplaneta americana
	Periplaneta australasiae
	Periplaneta japonica
	Supella longipalpa
	Leucophaea maderae
Siphunculata	*Pediculus humanus corporis*
Hemiptera	*Cimex lectularius*
	Rhodnius prolixus
	Triatoma phyllosoma
Lepidoptera	*Tineola bisselliella*
Hymenoptera	*Monomorium pharaonis*
	Tetramorium caespitum
	Iridomyrmex humilis
	Lasius niger
	Messor structor
	Acanthomyops flavus
	Dorylus nigricans
	Vespula spp.
	Apis mellifera
Diptera	*Musca domestica*
	Fannia canicularis
	Stomoxys calcitrans
	Calliphora erythrocephala
	Lucilia serricata
	Culex pipiens quinquefasciatus
	Aedes aegypti
	Aedes taeniorhynchus
	Anopheles stephensi
	Anopheles albimanus
	Anopheles quadrimaculatus
Siphonaptera	*Xenopsylla cheopis*
	Ctenocephalides felis
Coleoptera	*Anthrenus flavipes*
	Anthrenus verbasci
	Attagenus megatoma
	Attagenus pellio
	Dermestes maculatus
	Oryzaephilus surinamensis
	Tribolium castaneum
	Tribolium confusum
	Sitophilus granarius
	Ptinus tectus

The compound's spectrum of activity as a residual contact and stomach poison against a wide range of public health and agricultural pests was described by Lemon (1971). Grayson (1970) tested bendiocarb under the code NC 6897 and found it the most effective of ten new materials included in the 1969 cockroach control programme at Virginia Polytechnic Institute. Attributes of bendiocarb which make it specially suitable for use in restaurants, homes, hospitals, ships and aircraft, etc. are its non-staining and non-odour properties. These desirable properties make it possible to treat hospital wards without disturbing patients or staff (Wintringham, 1973). The use of bendiocarb in public health and its potential in this area has been reviewed by Story (1972, 1975). This success of bendiocarb in public health has encouraged extension of use to other pest species, particularly adult houseflies (Lemon & Bromilow, 1977).

Bendiocarb's mode of action as an insecticide is by cholinesterase inhibition, with quite rapid knock-down effect and no recovery once effects on the insect have begun. By contrast, toxic effects of sub-lethal doses of bendiocarb in mammals are rapidly reversible. The safety of the product is evidenced by the complete absence of any report of toxic effects on operators, or other persons coming into contact with the sprayed areas, since its introduction five years ago.

2 EFFICACY TRIALS AGAINST FLEAS

2.1 *In vitro experiments*

2.1.1 *Activity against Ctenocephalides felis* – In these tests bendiocarb was compared with the standard insecticides BHC, propoxur, malathion and bromocyclen. All the compounds were used as technical material dissolved in acetone with the exception of propoxur, which was used as a 50 per cent wettable powder.

Filter papers (9 cm Whatman No. 1) were treated with the test material and allowed to dry for 24 hours. They were then folded and placed in 75 mm x 25 mm glass tubes. Approximately 12 fleas were placed in each tube and mortality was assessed after 24 hours, the lethal concentration for 50 per cent (LC_{50}) was calculated for each compound from these mortality figures.

The results in table 2 show an LC_{50} of 50 ppm for bendiocarb with LC_{50} values between 39 and 400 ppm for the other materials. The carbamate insecticide propoxur had an LC_{50} of 150 ppm in this test.

2.1.2 *Residual efficacy against Xenopsylla cheopis* – Experiments comparing the residual effects of bendiocarb and malathion were also performed. Strips of Whatman No. 1 filter paper 15 mm x 30 mm were impregnated with 0.13 mls of an acetone solution containing 0.031 per cent insecticide. This treatment rate produces an equivalent of 50 mg a.i./m^2. After drying, the papers were placed in 18 mm x 150 mm glass tubes. The papers were bioassayed one, two and four weeks post-treatment and at four weekly intervals thereafter up to 24 weeks. For bioassay 10 *X.cheopis* were placed in each tube and their percentage mortality recorded after 24 hours.

Table 3 shows that bendiocarb gave 100 per cent mortality for 16 weeks and a high level of mortality at 24 weeks. Malathion gave 100 per cent mortality at eight weeks, a somewhat lower degree of control than bendiocarb at 12, 16 and 20 weeks and considerably poorer control at 24 weeks.

Table 2. Activity of bendiocarb and other insecticides against *Ctenocephalides felis* in vitro

	LC_{50} ppm
Bendiocarb	50
BHC	39
Propoxur	150
Malathion	330
Bromocyclen	400

Table 3. Residual activity of bendiocarb and malathion against adult *Xenopsylla cheopis*

		% mortality: weeks residual effect				
Insecticide		8	12	16	20	24
Bendiocarb	50 mg a.i./m^2	100	100	100	95	80
Malathion	50 mg a.i./m^2	100	80	90	85.	45

2.2 *Flea control on animals*

Bendiocarb formulated as 0.1 per cent dust was compared with 4 per cent bromocyclen dust for control of *C.felis* on cats. The animals were infested with approximately 50 adult fleas 20 minutes prior to treatment. After application of 5 g dusting powder, the cats were placed for four hours in small, mesh-floored cages over trays lined with white paper. In this way, dead or dying fleas falling from the animals were easily collected and counted. The cats were examined for live fleas and flea feces at intervals throughout the test. The cats were challenged with approximately 50 fleas at seven day intervals after treatment until fleas became established, indicating that the protection afforded by the treatment was lost.

Table 4 shows the results of this trial. Treatment with 0.1 per cent bendiocarb dust caused dead or dying fleas to fall from the cats within two hours of treatment. Daily examination of the animals for seven days after treatment showed no signs of flea activity on the animals during this time. No dead fleas were recovered after challenge of these cats seven days after treatment and fleas of this challenge infestation became established on the animals. Bendiocarb 0.1 per cent dust was therefore highly effective against fleas present at the time of treatment but less than seven days protection from reinfestation was obtained.

The results with bromocyclen 4 per cent dust were rather similar to those obtained with bendiocarb, less than seven days protection being obtained. It was noted that very few fleas dropped from one of these cats within 24 hours of treatment, perhaps indicating slow knock-down.

2.3 *Control of fleas in infested premises*

The following series of trials is presented as an example of the use of bendiocarb to control fleas in infested homes, etc. This work was carried out in the USA by Prof. Gary Bennett of Purdue University. All the fleas encountered were cat fleas, *C.felis.* Assessment of infestation levels was made by counting the number of adult fleas which jumped onto the legs of two observers as they walked through the infested area. This assessment was facilitated by wearing white, knee-high socks over the shoes and lower

Table 4. Number of dead fleas recovered from treated cats within 24 hours of challenge

Treatment	Cat no.	Time of challenge after treatment 0 days	7 days
Bendiocarb 0.1 %	1	40 (–)	0 (+)
Dust	2	29 (–)	0 (+)
Bromocyclen 4 %	1	6 (–)	2 (+)
Dust	2	66 (–)	0 (+)

(–) no live fleas or flea faeces seen on the cat
(+) live fleas and/or flea faeces seen

trousers; this made the fleas much easier to see. All assessments were made by the same people in a standard manner. After the pre-treatment count was complete, the rooms were sprayed with bendiocarb 80 per cent WP at 0.25 per cent a.i. in aqueous suspension using a fine fan spray nozzle, particular attention was paid to areas more attractive to fleas, e.g. rugs and areas of little disturbed carpeting. Post-treatment assessments were made 1, 7 and 30 days after treatment. Table 5 contains the pre- and post-treatment counts for six sites. It can be seen that good initial flea control was obtained in all six trials and that this single treatment eradicated the infestation from four of the sites. The two other sites deserve special mention.

In trial 1, the increase in count on day seven was found to be due to an area in which cats were sleeping and which had been missed on the initial bendiocarb treatment. Treatment of this area with malathion by the local pest control officer completed eradication. In trial 4, it was found that flea infested cats were moving freely in and out of the treated area, treatment of these animals controlled re-infestation. The sites were free of infestation for at least four months after treatment.

Many other similar trials in the UK, Australia and other parts of the world have shown bendiocarb to be extremely effective in controlling fleas in infested premises when applied as a structural residual spray. Although bendiocarb appears from initial laboratory trials to be similarly effective in controlling fleas on infested animals, this use of the compound is still under development and no official approval for this use has yet been sought. However, it is important, for bendiocarb as with other public health insecticides, that any animals frequenting flea infested premises be treated, with a product approved for flea control, at the same time as the area is sprayed.

Table 5. Cat flea control with 0.25 per cent Ficam W

Trial no.	Pre-treatment count	Post-treatment counts 1 day	7 day	30 day
1	100+	15	55*	0
2	15	2	3	0
3	40	1	1	0
4	305	5	20**	0
5	47	0	0	0
6	50	0	0	0

* Untreated area found – treated malathion on day 7
** Untreated cats present – flea collared on day 7

REFERENCES

Grayson, J.McD. 1970, Virginia Polytechnic Institute research on cockroaches. Pest Control 38(2): 31-35.

Lemon, R.W. 1971, 2,2-dimethyl-1,3-benzodioxol-4-yl N-methyl carbamate, a new broad spectrum experimental insecticide. Proc. 6th British Insecticide and Fungicide Conf. 2: 570-576.

Lemon, R.W. & C.R.Bromilow 1977, Further development of bendiocarb for the control of disease vectors with particular reference to the housefly, *Musca domestica* L. Pesticide Science 8: 177-182.

Story, K.O. 1972, Control of cockroaches and other domestic pests with a new carbamate insecticide. Internatl. Pest Control 17(6): 6-10.

Story, K.O. 1975, The international potential of bendiocarb insecticide in relation to pest control operations. Proc. 4th British Pest Control Conf., St. Helier 8th Session, Paper No. 17.

Wintringham, M.G. 1973, Pharaoh's ants – the modern plague. Environ. Hlth. 81(1): 131-132.

PHYSIOLOGY AND MORPHOLOGY

RACHEL GALUN
Israel Institute for Biological Research, Ness-Ziona, Israel

THE SPECIFICITY OF THE FEEDING RESPONSE OF FLEAS AND ITS BIOLOGICAL SIGNIFICANCE

ABSTRACT

Fleas showed the highest specialization among insects – ATP was the only nucleotide which had phago-stimulatory properties, among a wide range of nucleotides which stimulate other insects. The advantage of such specialization is discussed, and possible mode of action of ATP at the chemoreceptor level is proposed.

In vitro feeding experiments have shown that nucleotide phosphates, particularly those based on adenine, appear to act as phagostimulants in a wide variety of haematophagous arthropods. Thus, soft ticks in addition to GSH and amino acids, are stimulated to gorge by ATP, ADP, ITP and DPNH, in the presence of glucose (Galun & Kindler, 1968). *Rhodnius* responds to a wide range of nucleotides and the presence of adenine moiety is not essential, since in addition to adenine nucleotides phosphates of inosine, guanosine, cytidine and uridine are also stimulatory, but in decreasing order of effectiveness, as their structure increasingly differs from that of ATP. The exception is cAMP which is much more effective than might be predicted from its molecular structure (Friend & Smith, 1971).

Nucleotides not containing adenine were not effective in mosquitoes (Hosoi, 1959; Galun, 1967) nor in tsetse flies (Galun & Margalit, 1969). ATP, ADP, DPN, FAD and AMP are all phagostimulatory in descending order.

The specificity of the feeding response of the rat flea, *Xenopsylla cheopis,* is much greater than of any other arthropod studied. Only ATP will elicit a feeding response above that obtained with plasma or serum alone (Galun, 1966).

As in all the insects studied, most effective stimulation was achieved by ATP; it seems that throughout evolution, selection has favoured the development of increasingly more efficient mechanisms. Thus, parasites have become more and more specialized in their feeding response, assuming that the specialized insects can perform much better.

In natural situations specialized species have a better chance of survival – the gustatory chemors of the flea, being tuned only to ATP, detect a reliable clue to a suitable food – fresh blood. However, there is a built-in danger in specialization – when conditions change, the species may have lost its flexibility to adapt to the new condition. This will happen sooner or later because environment is always changing. In such a situation, the species will become extinct, unless it is preadapted to new situations.

ATP, the only phagostimulant of the flea, is very abundant in blood cells, but if the blood is shed, or the animal dies, it will decompose by dephosphorilation and deamination to non-stimulatory compounds. If the flea relied upon only this chemical cue for recognition of the proper diet, it should die from either dessication or starvation – when living hosts are not available. Perhaps the flea can afford to be so highly specialized in its feeding response, because it possesses additional mechanisms which induce gorging of liquids other than blood. Water is imbibed to a small extent, and if the flea encounters solutions with the proper tonicity, uptake increases. It is postulated that the phagostimulatory effect of osmotic solutions is due to possession of osmoreceptors by the fleas (Galun, 1966). This probably may explain why fleas fed readily on aged thawed blood which was hemolized and presumably does not contain any ATP (Kartman, 1954).

Hematophagy, which is common in many groups of arthropods, has evolved independently in the various groups, probably from plant feeding ancestors. It is therefore quite surprising that all of them develop along the same line in their gustatory preference, i.e. being stimulated to feed mainly by adenine nucleotides. If chemical activation of feeding response is an adaptive characteristic, the selection of adenine nucleotides would presumably be logical, because they are abundant in blood cells of all animals, and have small enough molecules to be easily detected by chemoreceptors. Yet, the nucleotides are confined to the blood cells, and therefore are not freely available to the chemoreceptors of the parasites. Thus, mosquitoes, tsetse flies (Galun & Rice, 1971) or ticks (Galun & Kindler, 1968) are not stimulated to feed by intact red blood cells (RBC), if precautions are taken to remove all blood platelets from this RBC fraction. No experiments to test this aspect in fleas were performed, as in 1966 when I studied the feeding response of fleas, it was not known that the stimulatory effect of RBC was in fact due to blood platelets always accompanying this fraction in the regular procedures used to separate plasma and cells (Galun & Rice, 1971). Blood platelets aggregate in the bitten region and release adenine nucleotides which are detected by the insect chemoreceptors. This detection mechanism probably required a series of evolutionary adaptations. Therefore, it may be logical to assume that the adaptive value of adenine nucleotide stems from their high efficiency at the molecular level of their interaction with the chemoreceptor membrane. This assumption is further supported by the chemotactic role played by adenine nucleotides in various living systems.

An increasing number of phytophagous insects are being found to be stimulated by adenine nucleotides which are abundant in plant tissues as well as in animal tissues (for review, see Galun, 1976). Adenine nucleotides were found to play a chemotactic role in aggregation of slime mold (Konijn et al., 1967), in attraction of nematodes to their food (Ward, 1973) and in flavour enhancement for man (Kuninaka, 1960). These compounds are also chemotactic in suborganistic level – aggregation of platelets (Käser-Glanzmann et al., 1962) and attraction of peritoneal neutrophils (Rivkin & Becker, 1976).

Our knowledge of the chemoreceptor organs which are exclusively concerned with recognition and control of ingestion of blood is not extensive. The epipharyngeal sensilla in the roof of the food canal of *Rhodnius* and *Triatoma* are the most likely ones to respond to ATP (Friend & Smith, 1971). These sensilla were described by Bernard (1974) as 11 tiny domes – about 1 μ in diameter, each having a slit in the centre.

In all Diptera, there are always four chemoreceptor-like basiconic sensilla situated most proximal on the posterior cibarial walls (for ref. see Rice, 1973). These are the final sense organs to make contact with the food prior to its entry to the foregut and thus they are the animal's ultimate faculty to determine the nature of the material they ingest. In the tsetse fly each of these four sensilla possesses a single neuron – which is probably stimulated by ATP – and the impulses produced as a result of this stimulation are fired to the tritocerebrum from which motor impulses pass out to instigate continuous cibarial pumping (Rice et al., 1973).

The four cibarial sensilla basiconica are not unique to Diptera. Similar sensilla were described for the flea *Ctenocephalides canis* by Wenk (1953), who called them 'cibariale Geschmacksinnesorgane'.

Because of the small size and their internal location, no direct electrophysiological recording has so far been done with any of these sensilla, and the evidence for their function and specificity is based on behavioural observations. The only electrophysiological evidence for specific ATP receptors was recently presented by Mitchell (1976), who recorded impulses from labial sensilla which arise from sockets situated between the prestomal teeth of the tsetse labella (Rice et al., 1973a). There are eight such sensilla on the labella, each containing two chemoreceptors and one mechanoreceptor. These sensilla are well situated to taste tissues prior to piercing and therefore Rice et al. (1973) considered them to be involved in monitoring tissue ATP. Mitchell (1976) has shown that one of the chemoreceptor cells in these sensilla responds to ATP, and the range of sensitivity and structural specificity of the stimuli agree fairly well with the behavioural results of Galun & Margalit (1969).

While further electrophysiological data may be expected from additional species in the foreseeable future, the chances of obtaining receptor membrane preparations of ATP chemoreceptors is almost nil – mainly due to the small number of neurons involved in phago-detection of ATP – in all the insects studied. Thus the speculation on the molecular interaction of the ATP and the receptor are based on behavioural observations. In these studies, a variety of nucleotides was offered to insects in the presence of various ions, or inhibitors, and the feeding response was noted (for review, see Galun, 1975).

As a result of such observations on the feeding behaviour of *Rhodnius* Smith & Friend (1972) suggested the possible role of adenylate cyclase as a basis of the gustatory transduction mechanism of the ATP chemoreceptors. Adenylate cyclase was demonstrated and its role in chemoreception in olfactory mucose of cow and porcine tongue (Menevse et al., 1974, 1976), and the oral disc of sea anemone (Gentelmen & Manseur, 1974) was implied. This hypothesis assumes that like in many hormonal systems, also in sensory systems, cAMP is used as an internal cell messenger. In *Rhodnius* chemoreceptors are presumably stimulated by nucleotides temporarily binding with receptor sites on the exposed surface of the cell. Such binding activates adjacent internally-membrane-bound adenyl cyclase, converting intracellular ATP to cAMP. Changes in ionic permeability of the receptor membrane in a separate region of the cell occur, and consequent ionic currents act as generators to initiate spike potentials. Smith & Friend (1972) believe that cAMP, which was exceptionally effective as a stimulant, penetrates the receptor and acts by raising intracellular concentration of cAMP.

Their hypothesis is supported by enhancement of gorging response to nucleotides in the presence of the drugs theopheline and caffeine. These drugs inhibit the action of phospho-diesterase normally responsible for cAMP hydrolysis. The level of cAMP in the cell represents a balance between the activities of two opposing enzymes: adenyl cyclase catalyzes the formation of cAMP from ATP, which the phosphodiesterase catalyzes its hydrolysis to 5'AMP. Potentiation of the effect of the phagostimulatory nucleotide by theophylline can therefore be taken as a good presumptive evidence that the nucleotide acts by stimulating adenyl cyclase.

Since ATP is the most effective phagostimulant for blood-sucking insects, it is very tempting to assume that the adenyl cyclase itself is the enzymatic receptor protein in the chemoreceptor membrane, and the stimulus receptor interaction occurs by ATP being the best substrate for adenyl cyclase. A similar mechanism was proposed for the sucrose receptors in the blowfly where the α-glucosidase of the chemosensory sets appears to split sucrose to glucose and fructose (Hansen, 1974; Koizumi et al., 1974). Smith and Friend (1976) examined this hypothesis by testing the potencies of combined doses of nucleotides as gorging stimulants for *Rhodnius.* They assumed that the nucleotides which are less effective than ATP, could act as competitive inhibitors to ATP – for the enzymatic activity of the adenyl cyclase. Since they did not observe any competitive inhibition, they concluded that a physical fit between the stimulating molecule and the chemoreceptive protein, rather than enzymatic modification of the stimulant is the mechanism of reception of these compounds.

There is a lot of room for misinterpretation – when gross behavioural observations are used to explain stimulus-receptor interaction, and as long as we will not be able to obtain receptor membrane preparations – the explanations for the mode of action of the 'blood chemoreceptors' will have to be very speculatory. Unfortunately, obtaining receptor membrane preparation from the small number of highly specialized sensilla in the flea seems to be unrealistic.

REFERENCES

Bernard, J. 1974, Etude électrophysiologique de récepteurs impliqués dans l'orientation vers l'hôte et dans l'acte hématophage chez un Hémiptère: *Triatoma infestans.* Ph.D. Thesis – Université de Rennes.

Friend, W.G. & J.J.B.Smith 1971, Feeding in *Rhodnius prolixus:* potencies of nucleotide phosphates in initiating gorging. J. Insect Physiol. 17: 1315-1320.

Galun, R. 1966, Feeding stimulants of the rat flea *Xenopsylla cheopis.* Life Sciences 5: 1335-1342.

Galun, R. 1967, Feeding stimuli and artificial feeding. Bull. World Health Org. 36: 590-599.

Galun, R. 1975, The role of host blood in the feeding behaviour of ectoparasites. In: A. Zuckerman (ed.), Dynamic Aspects of Host-Parasite Relationships. 2: 132-162.

Galun, R. 1976, The physiology of hematophagous insect/animal host relationships. Proceedings, XV Congress of Entomology, Washington DC 1976 (in press).

Galun, R. & S.H.Kindler 1968, Chemical basis of feeding in the tick *Ornithodoros tholozani.* J. Insect Physiol. 14: 1409-1425.

Galun, R. & J.Margalit 1969, Adenine nucleotides as feeding stimulants of the tsetse fly *Glossina austeni.* Nature 22: 583-584.

Gentelmen, S. & T.E.Mansour 1974, Adenylate cyclase in a sea anemone. Implication for chemoreception. Biochimia et Biophysica Acta 343: 469-476.

Hansen, K. 1974, α-glucosidases as sugar receptor proteins in flies. In: L.Jaenicke (ed.), Biochemistry of sensory functions. p. 207-233. Springer.

Hosoi, T. 1959, Identification of blood components which induce gorging of the mosquito. J. Insect Physiol. 3: 191-218.

Kartman, L. 1954, Studies on *Pasteurella pestis* in fleas. I. An apparatus for the experimental feeding of fleas. Exp. Parasit. 3: 525-537.

Käser-Glanzmann, R. & E.F.Lüscher 1962, The mechanism of platelet aggregation in relation to hemostasis. Thrombosis et diathesis haemorrhagica 7: 480-490.

Koizumi, O., H.Kijima & H.Morita 1974, Characterization of α-glucosidase at the tips of the chemosensory setae of the fly *Phormia regina*. J. Insect Physiol. 20: 925-934.

Konijn, T.M., J.G.C.van de Meene, J.T.Bonner & D.C.Barkley 1967, The acrasin activity of cAMP. Proc. Nat. Acad. Sci. 58: 1152-1154.

Kuninaka, A. 1960, Studies on taste of ribonucleic derivatives. J. Agric. Chem. Soc., Japan 34: 489.

Menevse, A., L.Cheng, B.Menco & G.Dodd 1974, Adenyl cyclase activities of olfactory and gustatory plasma membranes. Proc. 1st ECRO Congr. Univ. Paris Sud, July 1974.

Menevse, A., G.Dodd & M.Poynder 1976, Adenylate cyclase as the basis of olfactory transduction mechanism. Proc. 2nd ECRO Congr. Univ. of Reading, UK, Sept. 1976.

Mitchell, B.K. 1976, Physiology of ATP receptor in labellar sensilla of the tsetse fly *Glossina morsitans*. J. Exp. Biol. 65: 259-271.

Rice, M.J. 1973, Cibarial sense organs of the blowfly *Calliphora erythrocephala*. Internat. J. Insect Morph. Embryol. 2: 109-116.

Rice, M.J., R.Galun & J.Margalit 1973a, Mouthparts sensilla of the tsetse fly and their function. II. Labial sensilla. Ann. Trop. Med. Parasit. 67: 101-107.

Rice, M.J., R.Galun & J.Margalit 1973b, Mouthparts sensilla of the tsetse fly and their function. III. Labrocibarial sensilla. Ann. Trop. Med. Parasit. 67: 109-116.

Rivkin, J. & E.Becker 1976, Effect of exogenous cAMP and other adenine nucleotides on neutrophil chemotaxis and motility. Internat. Arch. Allergy & Appl. Immunol. 50: 95-102.

Smith, J.J.B. & W.G.Friend 1972, Chemoreception in the blood-feeding bug *Rhodnius prolixus;* a possible role of cyclic AMP. J. Insect Physiol. 18: 2337-2342.

Smith, J.J.B. & W.G.Friend 1976, Potencies of combined doses of nucleotides as gorging stimulants for *Rhodnius prolixus*. J. Insect Physiol. 22: 1049-1052.

Ward, S. 1973, Chemotaxis by the nematode *Caenorabditis elegans:* Identification of attractants and analysis of the response by use of mutants. Proc. Nat. Acad. Sci. USA 70: 817-821.

Wenk, P. 1953, Der Kopf von *Ctenocephalides canis* Aphaniptera. Zool. Jahr. Anat. 73: 103-164.

PETER WENK
University of Tübingen, Germany

HOW BLOODSUCKING INSECTS PERFORATE THE SKIN OF THEIR HOSTS

ABSTRACT

In bloodsucking Diptera, the mechanical forces used for the protraction of the stylets are formed (created) either by bending or by torsion of peculiar skeletal structures. This is accomplished when the respective retractor muscles are contracted. In fleas, on the other hand, resilin between the hypopharynx and the lacinial bar is compressed by a protractor, but action results from a slight shift of the hypopharynx which is provided with strikingly strong muscles.

Bloodsucking arthropoda first have to perforate the skin of their hosts to reach the source of their food. Insects use mechanical forces applied, however, on principles that differ considerably. In Culicidae, e.g. *Anopheles* and *Theobaldia,* the mandible is a long slender stylet with a sharply pointed tip. In *Anopheles* there are three muscles, a protractor (fig. 1,1) and two retractors originating from the tentorium (2) or from the head capsule (3). Surprisingly, the retractors are bigger and therefore must be stronger than the single protractor. Even the protractor is inserted close to the mandible joint, i.e. it acts through a short arm (fig. 2, Ap) moving a long arm (S). This means the short lever arm has to lift the long arm, an arrangement of rather poor efficiency. The long arm (S) is linked to the stylet of the mandible (MdB) at the same point where the retractors (2, 3) insert. Looking on the mandible joint itself, we detect that it is formed by a solid construction (fig. 2). The rod inside of the gena (GeL) and the basic bar of the mandible are fused to form a triangular structure which will be contorted when the retractors are contracted. Thus they store energy for a rapid protraction when the retractors are released and the blade of the mandible is shot like an arrow by the bow and directed forward by the labrum and maxillary stylets working like the effector cable of a photographic release.

In the genus *Theobaldia* the same mechanical principle is used; however, there exists a real joint between the genal rod and the mandibular bar (fig. 3). Storing of energy by contraction of the tentorial retractor (2) is effected by a solid connection (St) between the triangle shaped bar (S) and the blade of the mandible (MdB). The shape of the basal parts of the mandible (S, St) suggests that the bar rolls forward when the mandible is retracted. The position of the skeleto-muscular arrangement of the mandible is shown for *Anopheles* (fig. 4). In *Aedes* and *Culex,* the same muscle equipment and a construction of the (primary) mandibular joint similar to that in the genus *Theobaldia* is observed (Wenk, 1961).

The distal parts of the maxillary stylets are provided with numerous hooks directed

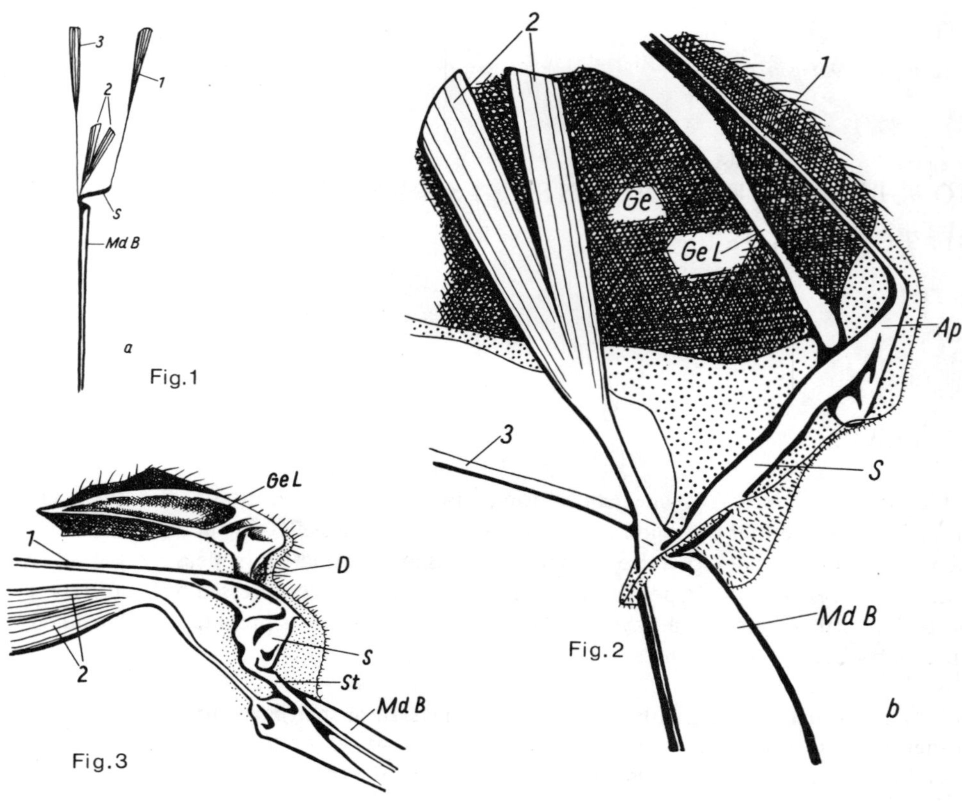

Figure 1. *Anopheles maculipennis* ♀ (Culicidae)
Stylet of mandible (MdB) bearing a small basal bar (S), one protractor (1) and two retractors (2, 3). From Wenk (1961).

Figure 2. *Anopheles maculipennis* ♀ (Culicidae)
Linkage of the mandible (left) with the gena (Ge) of the cranium by solid connection of a genal rod (GeL) and the basal bar (S) of the mandible. Movable joint between the bar S and the basis of the stylet (MdB) beside the insertion of the retractors (2, 3). The protractor (1) acts through a short apodeme (Ap), moving the long arm S.
From Wenk (1961)

Figure 3. *Theobaldia annulata* ♀ (Culicidae)
Basal parts of the mandible (left). The triangular structure S articulates with a socket (D) of the genal rod (GeL). The connection St is bent when the retractor (2) is contracted. Other abbreviations as in figs. 1 and 2.
From Wenk (1961)

Figure 4. *Anopheles maculipennis* ♀ (Culicidae)
Head capsule opened demonstrating the position of muscles and sclerites of the mandible. Ge gena, GeL genal rod, Lb labium, Lr labrum, MdB stylet of mandible, Pge postgena, S basal bar of mandible, Tent tentorium.
From Wenk (1961)

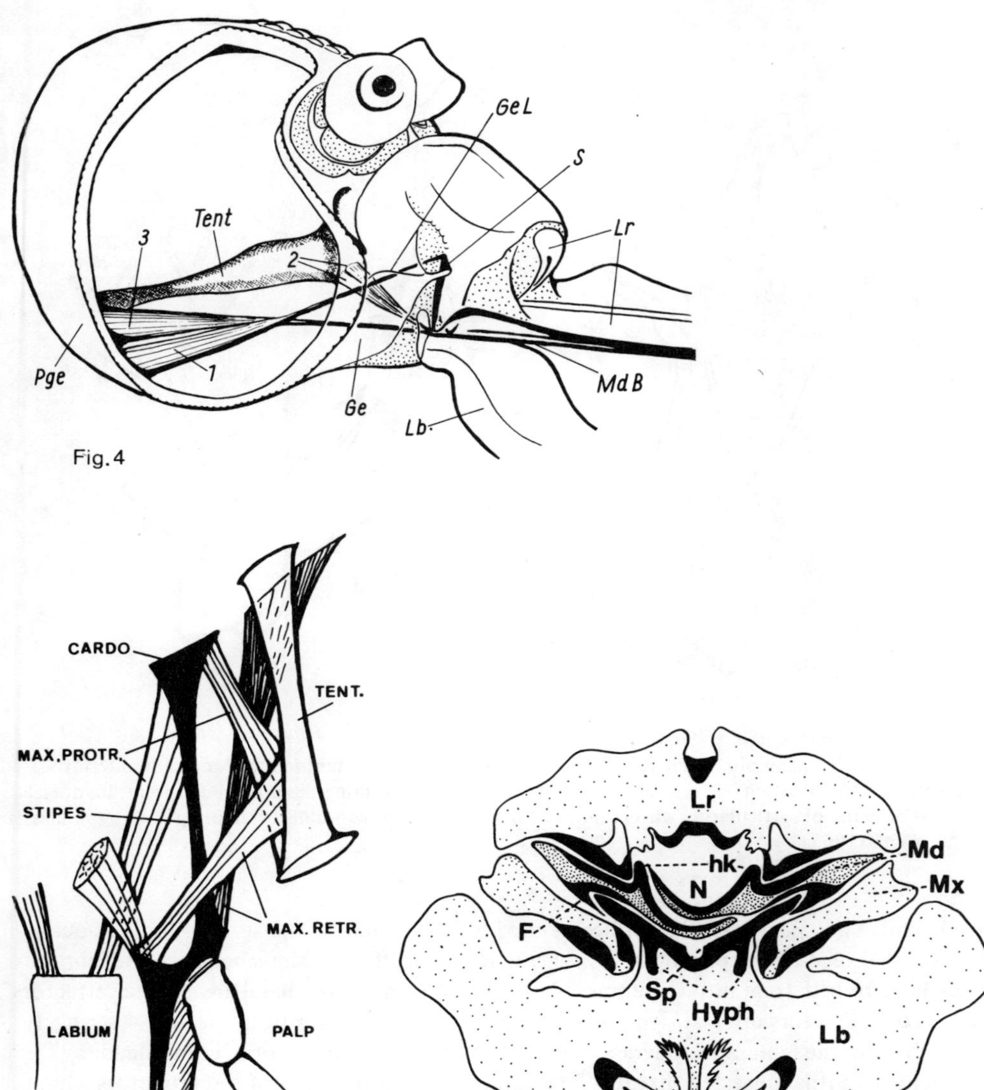

Figure 5. *Aedes aegypti* ♀ (Culicidae)
Muscles of the maxilla.
From Snodgrass (1959)

Figure 6. *Wilhelmia equina* ♀ (Simuliidae)
Proboscis transversal, hk dorsal process of mandible, Hyph hypopharynx, Lb labium, Lig ligula, Lr labrum, Md mandible, Mx maxilla, N food canal, Sp salivary canal, F lateral flange of mandible.
From Wenk (1962)

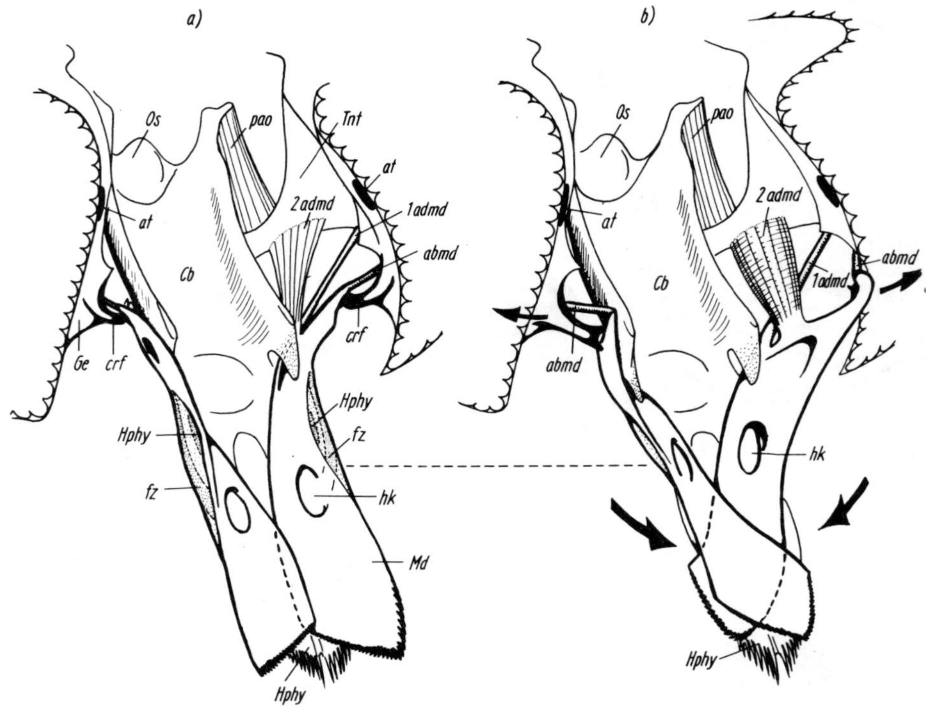

Figure 7. *Wilhelmia equina* ♀ (Simuliidae)
Movement of the mandible. abmd abductor, 1, 2 admd cranial resp. tentorial adductor, at anterior tentorial pit, Cb cibarium, crf cranial process of the gena (Ge) forming the mandibular joint, hk dorsal process of mandible articulating with labrum, fz lateral flange of mandible articulating with hypopharynx. From Wenk (1962)

backward. They penetrate the surface of the host's skin and tie the stylet bundle as quasi the screw-thread of the photographic release. Kept together by the labellae of the labrum, they have to penetrate into the upmost layer of the skin. At the basal area, three retractors and two protractors insert, originating on the tentorium, the head capsule and – strangely enough – on the labium (Snodgrass, 1959). The cardo and stipes form a long slender apodeme and it is suggested that it is bent laterally by contraction of the retractors which store energy for a strong punch forward to penetrate the hard keratinized layer of the epidermis (fig. 5).

The mandibles of female Simuliidae act like scissors. They are kept in position by the labium, which does not penetrate into the skin but bends forward pressing the mandibular blades towards the hypopharynx, which from behind is supported by the maxillary stylets (fig. 6). In order to enable sideways movement, a torsion of the mandibular blades, and at the same time, a gliding back and forth, is necessary. Direct observation of the living fly mounted by its tergum on the table of a dissecting micriscope proves that the sockets bearing the mandibular joint (fig. 7a, crf) are bent inward in the abduction position. The energy stored thereby is released during adduction (fig. 7b) when the host tissue is snipped by the rows of fine, medially directed teeth on the front part of the mandibles.

Exactly the same principle is realized in Ceratopogonidae, where the mandibles are

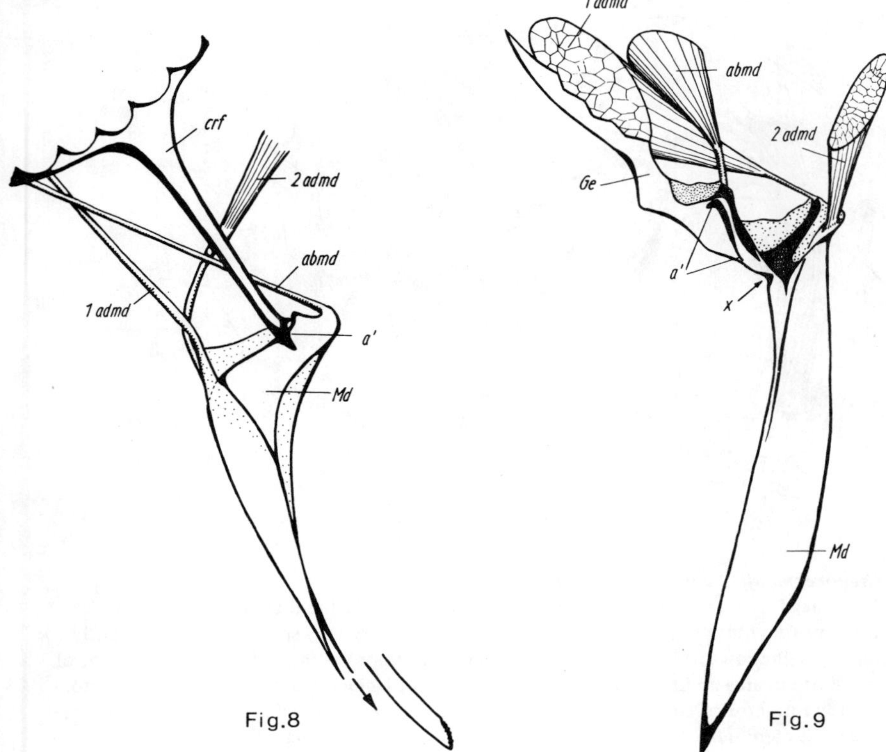

Figure 8. *Culicoides* spec. ♀ (Ceratopogonidae)
Mandible (right) and its articulation (a') with the cranium by means of a flexible elongated cranial process (crf).
From Wenk (1962)

Figure 9. *Haematopota* spec. ♀ (Tabanidae)
Mandible (anterior) and its connection with the cranium. a' primary (morphological) joint of the mandible, x point of flexion.

provided with an extremely prolonged cranial process (fig. 8, crf). In Tabanidae, the mandible is continuous with the strongly sclerotized head capsule (fig. 9). The mandibular blades move solely laterad-mediad when the proboscis is already pierced deeply in the host tissues (Dickerson et al., 1959). In dissected specimens, one observes a particular point of skeletal inflexion (fig. 9, X). The tentorial adductor (2 admd) inserts mediad; the cranial adductor (1 admd) inserts on the posterior and the abductor (abmd) on the anterior end of a bow-like sclerotized area therefore belonging to the basal parts of the mandible. The primary mandible joint has to be located morphologically close to the insertion of the abductor (fig. 9, a'). Although it is difficult to decide which kind of possible movements the cranial adductor (1 admd) and the abductor (abmd) actually provide, it is certain that an inflexion of the sclerotized parts is essentially involved (Wenk, 1962).

The jumping capacity of fleas is related to the presence of resilin, a highly compressible rubber-like substance originally found for the first time in the wing articulation of flying

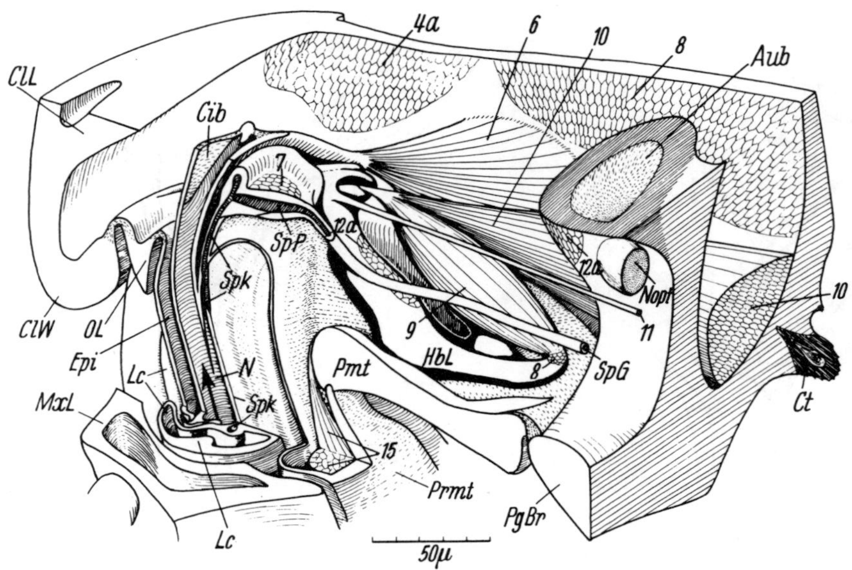

Figure 10. *Ctenocephalides canis* ♀ (Aphaniptera)
Basal part of the head. Articulation of the lacinia (Lc) by means of a basal bar (HbL) which links slightly with the postmentum (Pmt). The extensors of the lacinia (9, 10) spread the stylet slightly to provide fastening of the rows of hooks in situ. 6 retractor of hypopharynx, 7 insertion of dilator of salivary pump, 8 origin area and insertion area of protractor of lacinia, 9 extensor and 10 retractor of lacinia, 11, 12a retractor of lacinia, 15 flexor of labium, AuB eye, Cib cibarium, ClL, ClW clypeal ridge, resp. edge, Ct ctenidium, Epi epipharynx, Hbl lacinial bar, Lc lacinia, MxL maxillary lobus, N food canal, Nopt optic nerve, OL labrum, PgBr postgenal bridge, Pmt postmentum, Prmt prementum, SpG salivary duct, SpK salivary canal, SpP salivary pump.
From Wenk (1953)

insects by Weis-Foch (1960, 1961). It has characteristic staining properties and is therefore easily distinguished from usual membranes, e.g. when hematoxylin Delafield is used (Wenk, 1953, p. 131, 137). In front of the lacinial bar (fig. 10, Hbl) which acts against the posterior surface of the hypopharynx laterad of the salivary pump (SpP) there is a strong retractor of the hypopharynx (6). This keeps the hypopharynx in position against the lacinial bar when it is lifted by the protractor of the lacinia (8), inserting at the final caudal end of the bar (fig. 10) and originating on the surface of the gena below the eye (Wenk, 1953, Abb. 31a, b). Another muscle (adductor laciniae interior) (9) originates on the bar and has a common tendon together with the adductor laciniae exterior (10). The tendon merges into the body cavity of the lacinia and inserts at its lateral sclerotized blade. It enables the flea to spread the tips of the laciniae laterad, as observed by M.Rothschild (pers. comm.).

Since the laciniae are the only stylets in the flea's proboscis, Snodgrass (1946) presumed that they act like a pneumatic drill. Probably two phases occur (fig. 11): firstly the lacinial bar is slightly lifted by contraction of the protractor lacinia (8). While the retractor hypopharyngis (6) is contracted, energy is stored by compression of the resilin membrane layers. In the second phase, the position of the hypopharynx may be slightly corrected by its retractor (6) and the bar glides downward (fig. 11, Click!). These movements probably

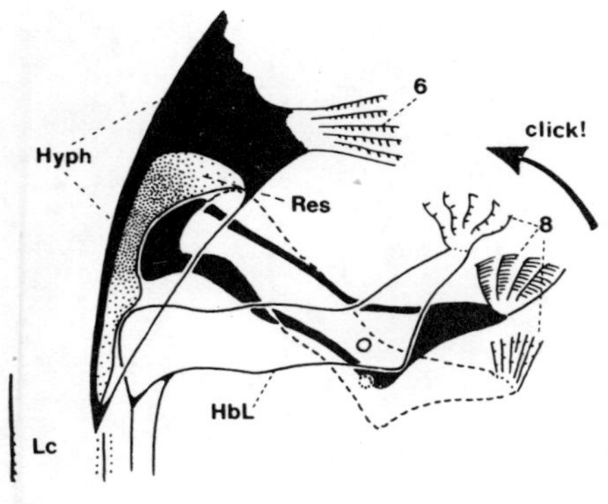

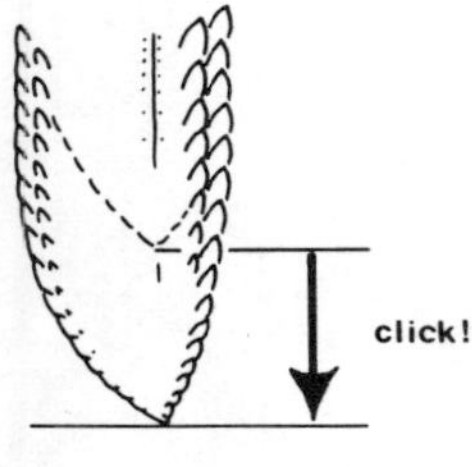

Figure 11. Movement of the flea's stylet (diagrammatic). The basal bar of the lacinia (HbL) is lifted by the protractor laciniae (8) and the resilin membrane (Res) is compressed. The hypopharynx (Hyph) is kept in position by its retractor (6) and the energy is stored by tension. Probably the alternating protrusion of the lacinia (Lc) is produced by a slight correction of the position of the hypopharynx on either side. Circle: point of slight linkage between the postmentum (fig. 11, Pmt) and the lacinial bar (HbL).

alternate on the left and right side and are repeated with high frequency. Thus the fleas use a different principle for perforation of the skin of their host: compression of resilin instead of bending of sclerites. The skeleton of the flea is highly sclerotized to withstand bites and other action of the host since these parasites generally spend most or all of their adult lives as inhabitants of the host, whereas the above mentioned Diptera spend only the few seconds required for feeding on their host.

REFERENCES

Dickerson, G. & M.M.J.Lavoipierre 1959, Studies on the methods of feeding of bloodsucking arthropods. III – The method by which *Haematopota pluvialis* (Diptera, Tabanidae) obtains its blood-meal from the mammalian host. Ann. Trop. Med. Parasit. 53:465-472.

Snodgrass, R.E. 1946, The skeletal anatomy of fleas (Siphonaptera). Smiths. Misc. Coll. 104 (18), 89 p.

Snodgrass, R.E. 1959, The anatomical life of the mosquito. Smiths. Misc. Coll. 139(8), 87 p.

Weis-Foch, T. 1960, A rubber-like protein in insect cuticle. J. exp. Biol. 37:889-907.

Weis-Foch, T. 1961, Molecular interpretation of the elasticity of resilin, a rubber-like protein. J. Molec. Biol. 3:648-67.

Wenk, P. 1953, Der Kopf von *Ctenocephalus canis* (Curt.) (Aphaniptera). Zool. Jahrb. Anat. 73:1-186.

Wenk, P. 1961, Die Muskulatur der Mandibel einiger blutsaugender Culiciden. Zool. Anz. 167:254-259.

Wenk, P. 1962, Anatomie des Kopfes von *Wilhelmia equina* L. ♀ (Simuliidae, syn. Melusinidae, Diptera). Zool. Jahrb. Anat. 80:81-134.

ROBERT C. KING & MELANIE TEASLEY
Northwestern University, Evanston, Ill., USA

INSECT OOGENESIS: SOME GENERALITIES AND THEIR BEARING ON THE OVARIAN DEVELOPMENT OF FLEAS

ABSTRACT

During the development of the insect egg, ribosomes are synthesized and subsequently stored in the ooplasm. The transcription of the required ribosomal RNA takes place in the oocyte nucleus unless this cell is connnected to sister 'nurse' cells. In this case, ribosomes synthesized by the endopolyploid nurse cells are transferred to the oocyte through a system of interconnecting canals. Fleas like *Spilopsyllus cuniculi* have panoistic ovarioles, and since the oocyte nuclei are multinucleolate it is probable that the genes coding for rRNA are amplified extra-chromosomally. Fleas like *Stenoponia tripectinata* produce oocytes which are attached to an anterior cluster of 'nurse' cells. Their small size and nuclear appearance suggests that they are not active in rRNA synthesis. The transcription of rRNA apparently occurs in the multiple nucleoli of the large oocyte nucleus, as in *Spilopsyllus.* It may be incorrect to place *Stenoponia* ovarioles in the meroistic polytrophic class, since the oocyte and 'nurse' cells may not represent a clonally-derived cluster of interconnected cells.

During oogenesis four basic processes take place. (1) Oocyte chromosomes undergo meiosis. (2) The machinery necessary for the synthesis of proteins and the appropriate blueprints for their construction are set aside for use during early embryogenesis. In molecular terms this means that ribosomes, transfer RNAs, and messenger RNAs must be synthesized and stored in the ooplasm. (3) Reserves of raw materials are built up. In most insect eggs these consist of populations of protein-rich yolk spheres, glycogen deposits, and lipid droplets. (4) A vitelline membrane and chorion are laid down about the egg. These coverings are secreted by the ovarian follicle cells.

The protein-rich yolk spheres found in mature oocytes provide sources of amino acids to be utilized during embryonic development. They appear at the periphery of the vitellogenic oocyte at the time its membrane becomes pinocytotically active (Roth & Porter, 1964). The proteins of the yolk spheres are referred to as vitellins, and their precursor serum proteins are called vitellogenins. The fat body synthesizes vitellogenins and secretes them into the hemolymph (Pan, Bell & Telfer, 1969). Vitellogenins are acidic proteins which also contain carbohydrates and lipids. They are large, nearly spherical molecules with diameters of 70-75 Å. Each is usually made up of two different polypeptide subunits, and presumably the subunits are encoded by different genes (Kunkel & Pan, 1976). Vitellogenins are taken up selectively by the oocyte, because they bind specifically to

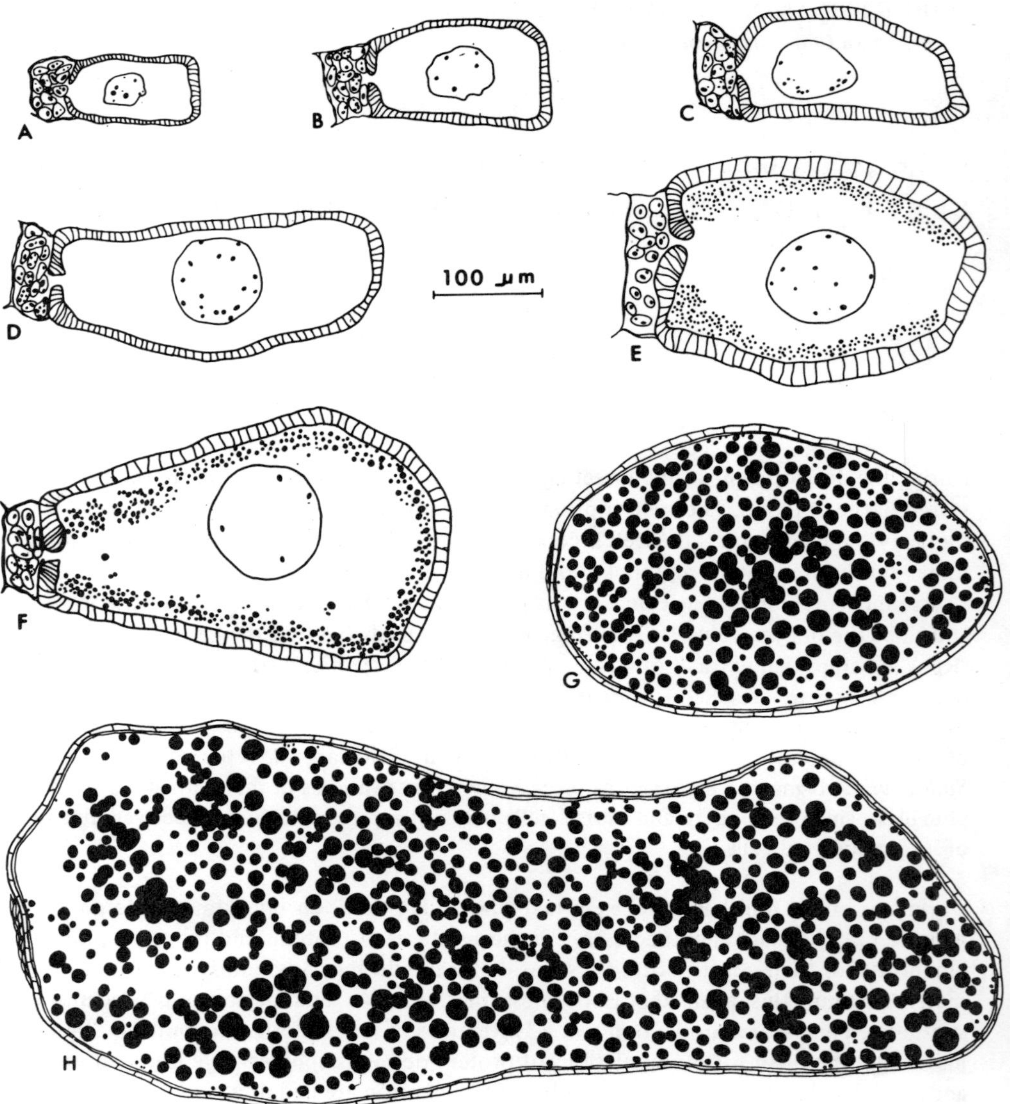

Figure 1. Representative stages of oogenesis of *Stenoponia tripectinata.* Note the multiple nucleoli in all oocyte nuclei, the nurse chambers – each containing 13 nuclei, and the canals penetrating the layer of columnar follicle cells adjacent to the nurse cells. The nurse cell plasma membranes and the follicle cell nuclei have not been drawn. Vitellogenesis begins in stage E. Between stages F and G the oocyte nuclear envelope breaks down. The meiotic tetrads were not included in the sections at our disposal. These stages should be compared with stages 7-10B of *Drosophila melanogaster* (King, 1970, his figs. II-9 through II-13).

receptor protein molecules that are embedded in the oocyte membrane (Roth, Cutting & Atlas, 1976). In fleas the globin portions of ingested hemoglobin molecules presumably are broken down in the gut to their individual amino acids. These are carried in the hemolymph to the fat body, where they are reassembled to form vitellogenin molecules.

One of the major synthetic activities occurring during oogenesis is the synthesis of ribosomes for storage in the mature oocyte. In those insects with panoistic ovarioles (*Acheta domesticus* is an example) ribosomal DNA is often amplified extrachromosomally (Lima de Faria et al., 1973). In such cases, only the genes transcribing the 18 and 28S ribosomal RNAs are replicated during pachynema, and these amplified DNA sequences transcribe the ribosomal RNAs in the nucleoplasm of the oocyte. After processing, these RNA molecules pass into the ooplasm through the nuclear pores. Flea species with panoistic ovarioles may produce rRNAs in a like manner.

In the case of insects with meroistic polytrophic ovarioles (*Drosophila melanogaster* is an example) the nurse cells synthesize billions of ribosomes and transfer them to the oocyte through a system of canals. Hundreds of tandem copies of the genes transcribing the RNA components of the ribosomes are localized at specific chromosomal sites. In the nurse cell nuclei the chromosomes undergo endomitosis and are replicated hundreds or thousands of times. As a result, the genes transcribing rRNA often undergo a million-fold amplification during oogenesis (King, 1970).

A small controversy has developed with respect to certain fleas. In most entomology texts, *all* fleas are said to possess panoistic ovarioles. This statement is certainly true for fleas of the Superfamily Pulicoidea. The European rabbit-flea (*Spilopsyllus cuniculi*), for example, contains six panoistic ovarioles in each ovary (Mead-Briggs, 1964). However, according to Kunitskaya (1960) each ovary of *Stenoponia insperata,* a member of the Superfamily Ceratophylloidea, contains three meroistic polytrophic ovarioles.

Diagrams of representative stages in oogenesis of fleas belonging to a related species (*Stenoponia tripectinata*) are shown in fig. 1. These drawings were prepared from reconstructions made from slides containing 80 serial sections (each 8 μm thick) through two paraffin-embedded females from the slide collection of Miriam Rothschild. The fleas were collected in Lebanon from nests of desert gerbils (*Meriones tristrami*) by Robert Lewis during the winter of 1965.

Since nurse cells and oocytes are normally generated during cystocyte mitoses that are followed by incomplete cytokinesis, the number of nurse cells is constant and generally follows the 2^n-1 rule. For example, in *Drosophila* where four consecutive cystocyte divisions occur, the nurse cell number is 15(2^4-1). In *S. tripectinata* the number of nurse cells is generally (but not invariably) 13 (see fig. 1), so the 2^n-1 rule is not obeyed. Furthermore, it has not been demonstrated that each oocyte and its adjacent nurse cells represent a clone of interconnected cells. This would require reconstructions of thinner serial sections from plastic-embedded ovaries.

The relative growth rates of the nuclei of oocytes and nurse cells of *S. tripectinata* and *D. melanogaster* were compared throughout a synthetic period during which the volume of ooplasm doubled 5.5 to 6 times (table 1). In the fruitfly, each oocyte nucleus doubled its volume only 1.5 times; whereas each nurse cell nucleus underwent between 4 and 5 doublings. In the flea, the situation was reversed. The volume of the oocyte nucleus doubled 5 times, while each nurse nucleus only doubled its volume once during the interval studied. In *Drosophila,* endomitotic doublings of nurse cell DNA are accompanied by doublings in the volumes of these nuclei. In *Stenoponia,* since the nurse nuclei underwent

Table 1. A comparison of the growth rates for follicle cells, nurse cells and oocytes of *Stenoponia* (S) and *Drosophila* (D) during comparative stages of oogenesis

Change in oocyte volume (μm^3)			
	S	2.6×10^5 to 1.1×10^7	(5.5 doublings)
	D	5×10^3 to 3×10^6	(6 doublings)
Change in nuclear diameters (μm)			
follicle	S	4.5 to 7	
cell	D	4.5 to 8	
nurse	S	9 to 12	
cell	D	17 to 50	
oocyte	S	30 to 100	
	D	14 to 21	
Change in nuclear volumes (μm^3)			
follicle	S	91 to 216	(1.5 doublings)
cell	D	48 to 268	(2.5 doublings)
nurse	S	680 to 1 300	(1 doubling)
cell	D	2.5×10^3 to 7×10^4	(4.5 doublings)
oocyte	S	1.7×10^4 to 5.8×10^5	(5 doublings)
	D	1.5×10^3 to 5×10^3	(1.5 doublings)

Data for *Drosophila* were obtained for stages 7 through 10B from King, 1970 (his fig. II-19)

but one doubling of their volumes during the period studied, it is unlikely that they reached high ploidy levels. Therefore, it is likely that they play a minor role in ribosome synthesis, unless their rDNA is replicated differentially. On the other hand, the oocyte nucleus grows at a rate similar to that seen in *S.cuniculi,* a flea with panoistic ovarioles, and much faster than the rate seen for *Drosophila* oocytes. Furthermore, while the *Drosophila* oocyte lacks a nucleolus during the period under study, multiple nucleoli are present in the oocyte nuclei of both flea species. It follows that the oocyte nucleus is the primary site of rRNA transcription during oogenesis in *Stenoponia,* and their 'nurse' cells do not function in a manner comparable to those of *Drosophila.*

REFERENCES

King, R.C. 1970, Ovarian development in *Drosophila melanogaster.* Academic Press, New York, 227 pp.

Kunitskaya, N.T. 1960, On the reproductive organs in female fleas and determination of their physiological age. (in Russian) Med. Parazit., Moskva 29(6):688-701.

Kunkel, J.G. & M.L.Pan 1976, Selectivity of yolk protein uptake: comparison of vitellogenins of two insects. J. Insect Physiol. 22: 809-818.

Lima-de-Faria, A., H.Jaworska, T.Gustafsson & S.Daskaloff 1973, Amplification of ribosomal DNA in *Acheta.* III. Release of DNA copies from chromomeres. Hereditas 73:163-184.

Mead-Briggs, A.R. 1964, The reproductive biology of the rabbit flea *Spilopsyllus cuniculi* (Dale) and the dependence of this species upon the breeding of its host. J. Exp. Biol. 41:371-402.

Pan, M.L., W.J.Bell & W.H.Telfer 1969, Vitellogenic blood protein synthesis by insect fat body. Science 165:393-394.

Roth, T.F. & K.R.Porter 1964, Yolk protein uptake in the oocyte of the mosquito *Aedes aegypti* L. J. Cell Biol. 20:313-332.

Roth, T.F., J.A.Cutting & S.B.Atlas 1976, Protein transport: A selective membrane mechanism. J. Supramolec. Struct. 4:527-548.

F.G.A.M.SMIT
British Museum (Natural History), London, UK

MISSING AND FLOATING GENITALIA IN MALE FLEAS

ABSTRACT

A survey is given of the occurrence of male fleas with genitalia either completely missing – due to an unknown cause affecting the development of primary phallic lobes – or in an unattached floating condition where for some reason the parameres did not become attached to tergum IX. These abnormalities may help to elucidate the relationship between tergum IX and the parameres.

Snodgrass (1957:11) remarked that 'The great structural diversity in the male genitalia of insects is the delight of taxonomists, the despair of morphologists'.

The male clasping organs of fleas, which form such an excellent example of diversity obscuring homologies of various constituent structures, were for a long time believed to be modifications of tergum IX. In descriptions of new taxa they have often been included under 'modified abdominal segments'. However, it seems now generally accepted that in fleas, as in other higher orders of insects, the male genitalia are derived from a pair of ectodermal invaginations which Snodgrass (1957) termed the primary phallic lobes and which appear early in the third larval stage. These lobes are well illustrated in figures 64-69 of Sharif's brilliant 1937 treatise on the internal anatomy of the third-instar larva of the rat flea *Nosopsyllus fasciatus.* By the end of the third larval stage, the prepupa, or early on in the pupal stage, each primary lobe divides into two secondary lobes, the phallomeres, the two inner ones being the mesomeres, the two outer ones the parameres. These phallomeres are situated just behind segment IX. The two inner lobes, the mesomeres, subsequently unite with each other to form the phallosome while the lateral lobes, the parameres, become the clasping organs which secondarily in their development become two-segmented, the basal segment being the basimere, the distal – often movable one – the telomere. Both mesomeres and parameres have apodemes, those of the former joining to form the aedeagal apodeme, the ones of the latter are the manubria. The basimeral part of the parameres, requiring firm attachment to a part of the flea's exoskeleton, joins the inner side of the ventral part of tergum IX. This junction can be quite complete, leaving no suture or such, and one of the difficulties encountered in the study of parameres is to determine the extent to which terga IX and X are part of the clasping organ.

When attempting to understand a normal condition, it may be useful to study abnormal conditions.

Missing genitalia – On rare occasions something drastic happens, presumably at the latest during the very early third larval stage, which prevents the genitalia from developing at all. I used to refer to the resulting anomaly as a case of 'complete castration' (Smit, 1953) but, as genitalia that presumably have never been cannot be castrated, it would be better to refer to this condition as one of 'completely undeveloped genitalia' or, simply, as one of 'missing genitalia'. There is no evidence that this abnormal condition is caused by the action of parasitic nematodes. The abdomen of macerated specimens without genitalia looks quite empty (figs. 1, 2). One or a few of the dorsal setae on tergum VIII may be at least somewhat malformed but the segment itself is quite normal. The apodeme of tergum IX is still fairly well developed but the lower end of the tergum – not being required – is reduced. In Ceratophylloid specimens with missing genitalia, sternum IX has not received the caudal end of the aedeagal apophysis and the point of attachment is now a short rod, usually displaced somewhat upwards. In this Superfamily (Ceratophyllidae, Leptopsyllidae, Ancistropsyllidae, Ischnopsyllidae and Xiphiopsyllidae) we are in no or little doubt about the constituent parts of the clasper; the pattern is the same throughout the Superfamily, which accounts for about 45 per cent of all known fleas. In normal Hystrichopsylloid and Pulicoid specimens, however, tergum IX often amalgamates with the basimere and tergum X much more intimately, which makes delineating the extent of that tergum practically impossible (figs. 8, 9). In those fleas the aedeagal apophysis does not secondarily connect with sternum IX, hence this sternum does not develop a probing rod. (Sometimes only a small part of the genitalia manages to develop and, for instance, a malformed basimere may connect with tergum IX; such specimens are not dealt with in this paper.)

Floating genitalia – On occasions equally rare as those of missing genitalia (see table 1) the mesomeral/parameral organ fails – during the pupal stage – to join up with tergum IX, often because these genitalia are in a reversed position. This abnormal condition, in which the unattached genitalia are lying as one package inside the abdomen, is that of

Table 1. Specimens with genitalic abnormalities as referred to in the text

CERATOPHYLLIDAE				HYSTRICHOPSYLLIDAE			
Ceratophyllus gallinae	A	B		*Dinopsyllus lypusus*	A		
Ceratophyllus farreni farreni	A	B		*Ctenophthalmus assimilis assimilis*	A		
Ceratophyllus hirundinis	A	B		*Ctenophthalmus calceatus cabirus*	A		
Ceratophyllus garei	A			*Ctenophthalmus nobilis vulgaris*			C
Ceratophyllus borealis			C	*Palaeopsylla minor*		B	
Amalaraeus penicilliger mustelae			C	*Rhadinopsylla integella*		B	
Amalaraeus penicilliger pyrenaicus	A	B		PYGIOPSYLLIDAE			
Megabothris turbidus	A			*Notiopsylla enciari enciari*		B	
Megabothris walkeri	A						
Megabothris rectangulatus		B		RHOPALOPSYLLIDAE			
Myoxopsylla laverani		B		*Tetrapsyllus rhombus*	A		
Dasypsyllus gallinulae gallinulae			C	*Tiamastus cavicola*	A		
LEPTOPSYLLIDAE				PULICIDAE			
Peromyscopsylla fallax	A			*Spilopsyllus cuniculi*		B	

A = completely undeveloped genitalia;
B = floating genitalia (inside abdomen);
C = genitalia in about normal position but parameres unattached.

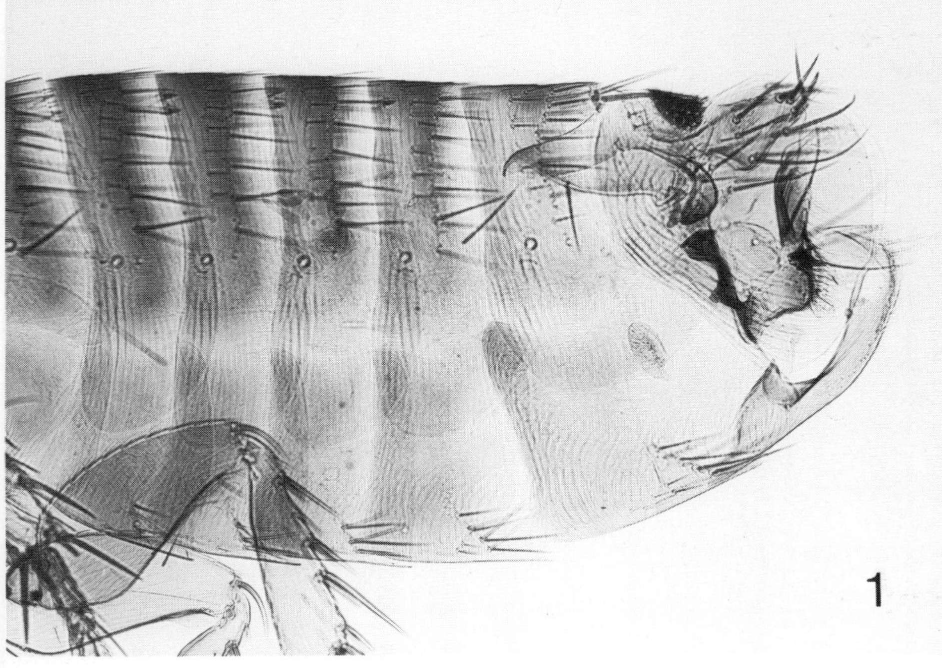

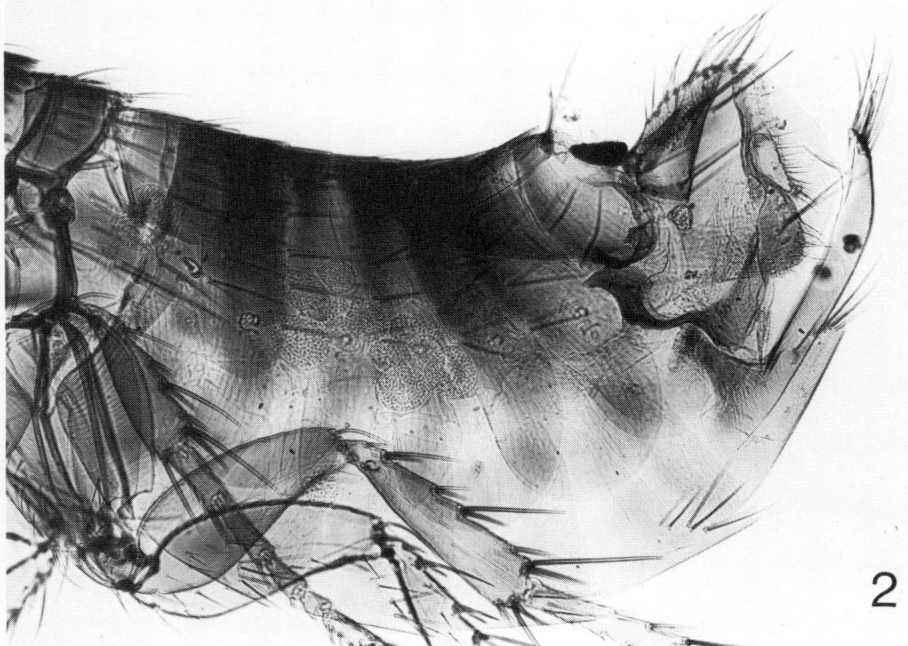

Figures 1, 2. Completely undeveloped genitalia in males of 1) *Megabothris turbidus* (Rothschild) (Col de la Givrine, Switzerland), 2) *Ceratophyllus farreni farreni* Rothschild (Haddo, Scotland).

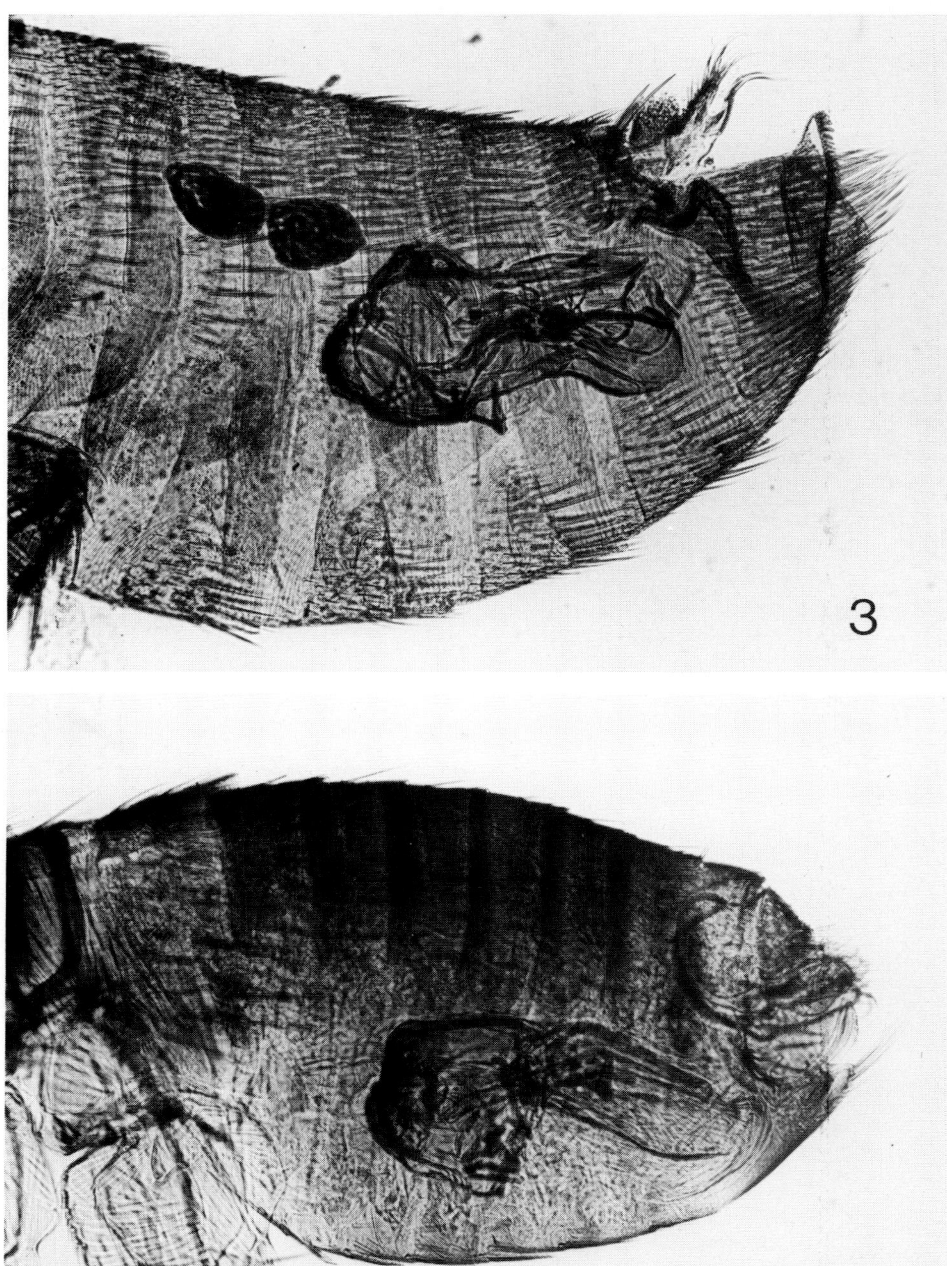

Figures 3, 4. Floating genitalia in males of 3) *Notiopsylla enciari enciari* Smit (Auckland Islands, New Zealand), 4) *Spilopsyllus cuniculi* (Dale) (Ickwell Green, England).

so-called 'floating genitalia' (Smit, 1972). Even more rarely segment IX fails to develop properly, giving the genitalia (lying in a more or less normal position) no chance to link up (fig. 5). The abdomen of specimens with floating genitalia looks rather filled (figs. 3, 4). As is to be expected, segment IX develops to the same degree in specimens with unattached genitalia as in those with completely undeveloped ones. In these abnormal specimens, too, one or a few setae on tergum VIII may be malformed. In some specimens with unattached genitalia these organs are in approximately the normal final position but they are at least slightly deformed.

A survey of specimens which I have seen with the types of genitalic abnormalities referred to is given in table 1.

The one Pulicoid specimen with floating genitalia which I have seen is a male of the rabbit flea *Spilopsyllus cuniculi* (figs. 4, 7) which belongs to one of the Pulicid subfamilies in which the basimere bears a pseudotelomere dorsad of the telomere proper. In the abnormal rabbit flea the apodeme of tergum IX bears a ventral setose lobe (fig. 7). As the genitalia package inside the abdomen (fig. 4) is complete with basimeres, telomeres and pseudotelomeres, this setose lobe must therefore belong to tergum IX. It obviously corresponds with the shorter setose lobe below the basimere in normal specimens (fig. 6), a condition which is found in several related genera.

In the Pulicid subfamily Xenopsyllinae the homology of clasper processes has especially been obscure; when there are three setose processes these have conveniently been referred to as P1, P2 and P3 (fig. 8). While P1 is the basimere and P2 the telomere, the homology of P3 has remained obscure. It now seems clear that, in analogy with what we have seen in the rabbit flea, this process is presumably the lower end of tergum IX. Such a downward extension of tergum IX can also be seen in other Pulicids (fig. 9).

Specimens with abnormal genitalic developments can thus confirm the correctness of the presently held views on the ontogenetic development of the genitalia and they may show the approximate extent of tergum IX for some groups. They do not, of course, elucidate the fusion of tergum IX with tergum X, good examples of which are shown in figs. 8 and 9. But here it might be useful to consider that normally one tiny sensory organ, often accompanied by one or two minute setae, situated just below the sensilium, belongs to tergum X. It is of some interest to give an example of tergum IX-X fusion in the opposite sex (where genitalia never seems to be absent and always floating); fig. 10a-f shows the various degrees of relationship between tergum IX and tergum X in females of members of one genus, *Chaetopsylla.*

Acknowledgement – Dr J.C.Beaucournu has my best thanks for kindly enabling me to study various abnormal specimens from his collection.

REFERENCES

Sharif, M. 1937, On the internal anatomy of the larva of the rat-flea, *Nosopsyllus fasciatus* (Bosc.). Phil. Trans. R. Soc. (B) 227(547): 465-538.

Smit, F.G.A.M. 1953, Monstrosities in Siphonaptera IV. Ent. Ber., Amst. 14(342): 393-400.

Smit, F.G.A.M. 1972, On some adaptive structures in Siphonaptera. Folia parasit., Praha 19(1): 5-17.

Snodgrass, R.E. 1957, A revised interpretation of the external reproductive organs of male insects. Smithsn. misc. Coll. 135(6): 1-60.

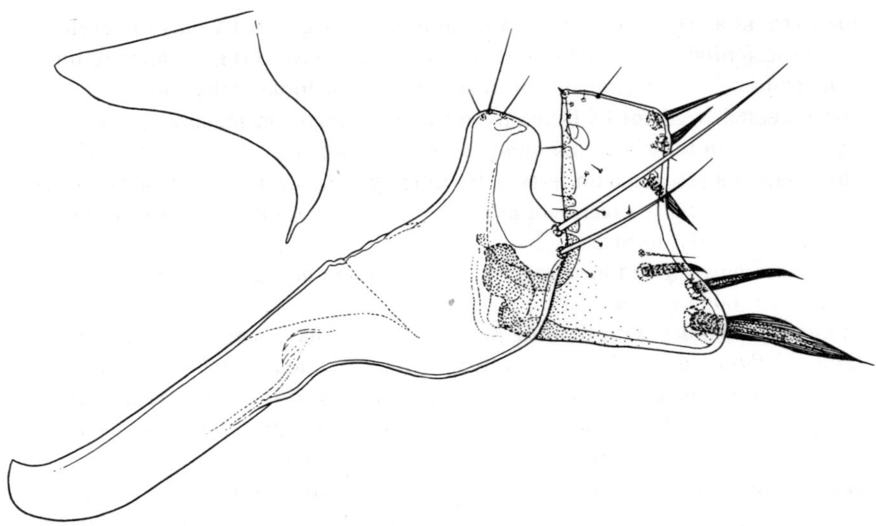

Figure 5. *Dasypsyllus gallinulae gallinulae* (Dale) (Crewkerne, England), showing ill-developed tergum IX and unattached slightly malformed paramere.

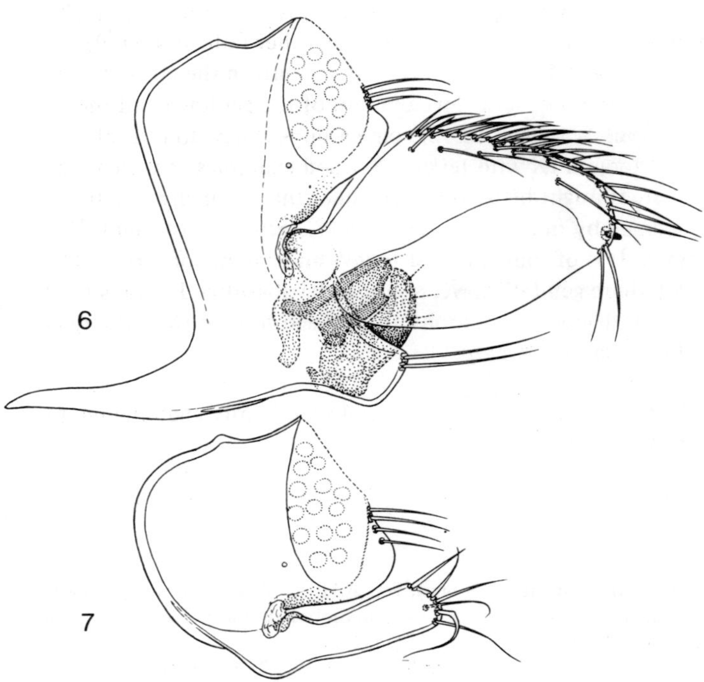

Figures 6, 7. *Spilopsyllus cuniculi* (Dale): 6) paramere, terga IX and X of a normal male (Herm, Channel Islands, UK), 7) terga IX and X of specimen with floating genitalia as shown in figure 4.

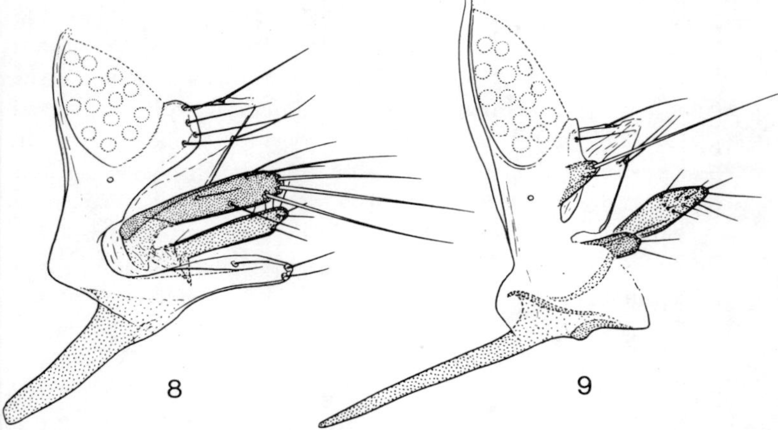

Figure 8. *Xenopsylla gerbilli caspica* Ioff (Shardara steppe, Kazakhstan), paramere, terga IX, X and segment XI of a normal male.

Figure 9. *Xenopsylla hirsuta hirsuta* Ingram (nr Cape of Good Hope, South Africa), paramere, terga IX, X and segment XI of a normal male (note that in males of the *hirsuta*-group of *Xenopsylla* tergum X bears a structure which is homologous with the anal stylet of tergum X in most female fleas; moreover, the anal sternum is connected with the terga IX-X fusion).

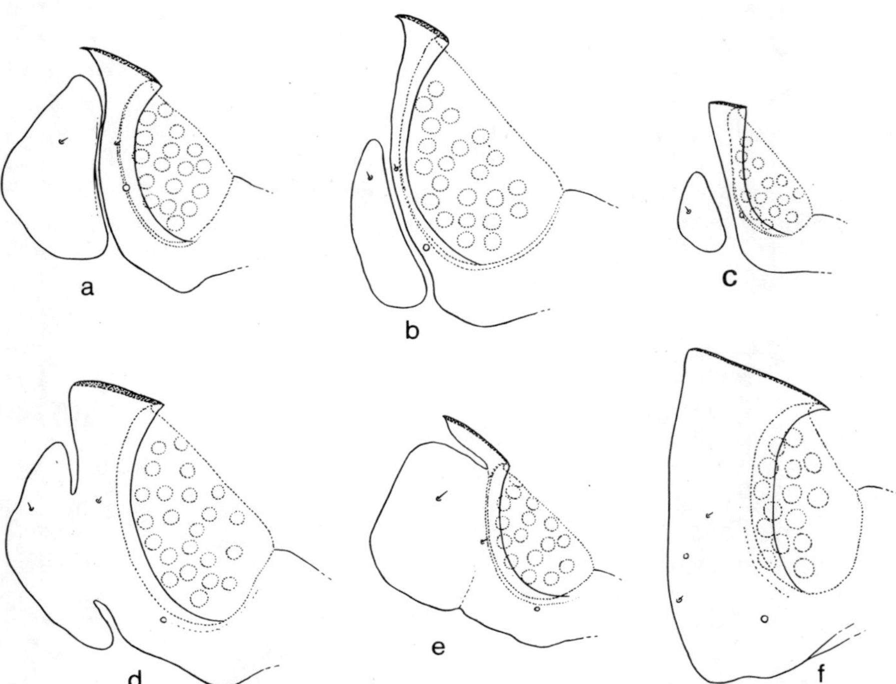

Figure 10. Terga IX and X in females of *Chaetopsylla:* a) *mikado* Rothschild, b) *setosa* Rothschild, c) *alia* Ioff, d) *trichosa* Kohaut, e) *homoea* Rothschild, f) *tuberculaticeps* (Bezzi).

GLENN E. HAAS
State Dept. Health & Social Services, Anchorage, Alaska, USA

UNUSUAL ALASKAN FLEAS

Examples of morphological curiosities include partial castration and inter-sexuality in *Ceratophyllus styx riparius* Jordan & Rothschild, two spermathecae in *Ceratophyllus adustus* Jordan and *Euhoplopsyllus glacialis lynx* (Baker), and a mesonotal ctenidium in *Catallagia dacenkoi* Ioff.

In three of four specimens of *C.adustus* with two spermathecae, the smaller organ is visibly connected to the ductus obturatus. A female of *C.dacenkoi* has a ctenidium on the mesonotum as well as on the pronotum. Neither ctenidium is complete; the mesonotal one appears to be part of the missing lower left portion of the pronotal one.

HEINZ E. KRAMPITZ* & VARUNEE WONGCHARI**
** Universität München, Germany*
*** Chulalongkorn University, Bangkok, Thailand*

THE DEVELOPMENT OF *HEPATOZOON ERHARDOVAE* IN EXPERIMENTAL MAMMALIAN AND ARTHROPOD HOSTS

1. THE EVALUATION OF SUITABLE ARTHROPOD VECTORS***

ABSTRACT

In experiments on the determination and roles of arthropod hosts of *Hepatozoon erhardovae* Krampitz, 1964, a parasite of the European bank vole *Clethrionomys,* observations in test areas near Munich have confirmed that fleas commonly serve as the host stage for sexual reproduction of the protozoan. This *Hepatozoon* can develop to the sporocyst stage in all five species of fleas tested: *Xenopsylla cheopis, Nosopsyllus fasciatus, Megabothris turbidus, Ctenophthalmus agyrtes* and *C.assimilis,* and the sporocyst index (per cent of fleas found infected with that stage) range from 95 per cent in *X.cheopis* and *C. agyrtes* down to 17 per cent in *C.assimilis. X.cheopis* is a suitable experimental host for the observation of sporogeny and its time sequence. Sexual reproduction lasts 14 days and takes place in the coelom of the arthropod. The size of a mature oocyst is up to 330 x 280 μm and contains an average of 10-12 (9-36) mobile sporozoites. Sporogeny can take place in both mature adult fleas (14 days after emergence) and in young specimens. Transovarial transmission, or excretion of infective sporozoites with the faeces of fleas, could not be established. The vitality of infected fleas is not markedly affected: their life span is not shortened nor their fecundity reduced. Evidently vertebrates become infected exclusively by ingestion of the ectoparasites. The use of Siphonaptera as arthropod hosts for *Hepatozoon* species seems to be more an exception than a common rule within the genus. Obviously, mites can also serve as definitive hosts, at least locally, for a species of *Hepatozoon* of bank voles which is morphologically indistinguishable from *H.erhardovae.*

ZUSAMMENFASSUNG

Die geschlechtliche Entwicklung von *Hepatozoon erhardovae* Krampitz, 1964 in wirbellosen Wirten ist experimentell überprüft und dargestellt. Die Sporogonie geht in Flöhen vor sich. Die Vektoreignung von *Xenopsylla cheopis, Nosopsyllus fasciatus, Megabothris turbidus, Ctenophthalmus agyrtes* und *Ct.assimilis* für den Rötelmaus Parasiten wurde bestimmt. Alle geprüften Floharten sind als Wirte geeignet, am besten gedeiht der Parasit jedoch in *X.cheopis, M.turbidus* und *Ct.agyrtes.* Der tropische Rattenfloh diente als bevorzugtes Versuchstier. Die Sporogonie erfolgt in der Bauchhöhle des Flohes. Reife Oozysten können mehr als 300 μm durchmessen und mit unbewaffnetem Auge sichtbar sein. Sporozysten sind in ihnen stets zu Hunderten enthalten, jede mit 25 x 15 ± 3 μm

*** Supported by Grant Kr 72/15 of the Deutsche Forschungsgemainschaft.

und enthält durchschnittlich 10-12 Sporozoiten. Flöhe jeden Alters sind als Endwirte geeignet. Der Wirbeltierwirt infiziert sich per os durch Zerquetschen oder Fressen von Flöhen. Transovarielle Übertragung oder ein natürliches Ausgeschiedenwerden von Sporozoiten durch den Floh konnten nicht nachgewiesen werden. Es gibt in der Natur örtlich auch Sporogonien in blutsaugenden Milben (Gamasoidea). Die Sporozyste in ihnen ist abgebildet. Es wird jedoch offengelassen, welche biologische Position dieser zukommt und ob vielleicht noch eine zweite Hepatozoonart in Rötelmäusen vorkommt.

1 INTRODUCTION

Intracellular parasites of the genus *Hepatozoon* Miller, 1908 are commonly found in blood and tissues of vertebrates. Killick-Kendrick (1974) listed 67 species of rodents as hosts for *Hepatozoon* in the world, but the exact number of species within the genus is still unknown. In Europe they are fairly abundant in small mammals, mostly in voles (Microtidae). By testing sufficient numbers of animals it became clear that almost all vole species can act as hosts of these parasites (Krampitz, 1964; Killick-Kendrick, 1974; Šebek, 1975, 1977). Slight differences in size and structure of the bloodstream and tissue forms are not always statistically significant and therefore such characters are not very useful for species differentiation. Despite the morphological similarity, especially of the gametocytes in the blood, many differences occur in the frequency and intensity of the infections, in the specificity for mammalian hosts, and in the organ which a given species prefers for asexual reproduction. It therefore appears more involved than merely a *Hepatozoon microti* complex which includes several biotypes that have not yet reached the species level. As the experimental data have clearly demonstrated, many different behavioural patterns and separate ways of life are possible, and these factors should also be considered as a basis for species differentiation.

One of the most common indigenous species which could be separated from the 'complex' by transmission experiments was named *H. erhardovae* by one of us in 1964. Many ecological factors may be responsible for its nigh continent-wide distribution in bank voles (*Clethrionomys*), but the distribution and abundance of the arthropod host definitely is of utmost importance. It is well-known that bank voles are always more heavily infested with all groups of ectoparasites than are other small mammals in the same area. Of the 66 species of fleas reported in Germany, no less than 20 were collected from bank voles or their nests (Peus, 1970). It has also been repeatedly described that the sporogeny of *H. erhardovae* takes place in fleas (Siphonaptera) (Krampitz, 1964, 1971, 1972, 1973; Krampitz et al., 1976; Beaucournu & Deunff, 1975; Frank, 1977a). The parasite is transmitted by fleas, but other ectoparasites, which are usually always present, are also capable of acting, at least occasionally, as vectors (Frank, 1977b). The experimentally proven vector-capacities of several ectoparasitic arthropods known to occur naturally on bank voles are dealt with below.

2 MATERIAL AND METHODS

Using suitable host animals, five species of fleas belonging to three families were reared continuously in captivity: *Xenopsylla cheopis, Nosopsyllus fasciatus, Megabothris turbidus, Ctenophthalmus agyrtes* and *C. assimilis.* Except for *X. cheopis,* all strains were derived from specimens collected from naturally infested, local wild animals, and were checked regarding monospecificity (Krampitz, 1974). The cultures always contained, for testing purposes, both newly emerged hungry fleas and specimens that had fed previously.

To infect hungry fleas with *Hepatozoon,* a blood meal was arranged on unanaesthetized parasitemic *Clethrionomys.* In order to reduce the loss of fleas by host defences during the development of the haemogregarina in the experimentally infected arthropods, the infected fleas were subsequently fed entirely on other kinds of hosts, namely the particular species of rodent which experience has shown to offer the specific flea the best chance of survival. Such 'nursery' hosts consisted of laboratory rats, hamsters, *Mastomys* (multimammate rats) and long-tailed field mice (*Apodemus*).

Since the development of *Hepatozoon* in the fleas is not affected by subsequent feedings on normal blood from appropriate hosts, the nursery method proved to be more successful than the storing of the infected fleas in a humid chamber for several weeks while depriving them of food. Furthermore, such changes of host prevent the acquisition of new gametocytes and the consequent and simultaneous development of several generations of oocysts in the same flea as can be seen in fig. 5. Infected fleas were kept together with their hosts at a room temperature of 25°C (= 77°F) and 60-80 per cent relative humidity.

Quick tests for detection of sporocysts and living motile sporozoites were carried out by dissecting fleas in a drop of saline and examining smears of the body either as a fresh specimen under a coverslip or stained with Giemsa. The localization of a particular sporogeny stage in the arthropod's body could be detected in paraffin sections stained with Haematoxylin-Eosin or with a modified Gomori stain.

In order to study arthropod-specificity in *Hepatozoon,* which is a strictly host-specific parasite, large samples of various kinds of ectoparasites were collected from infected voles in the test areas. The pools were macerated, suspended in saline, and administered orally to laboratory-born *Clethrionomys.*

There is no doubt that more than one flea species may serve as vector, and determining the infection rate of various species of fleas does not permit clear-cut conclusions as to which of the species is the most important vector. Therefore the vector-capacity of several species of fleas were tested as follows: five eight-week old bank voles were given equal infective doses of a pool of 73 naturally infected fleas of various species collected in a test area. (Infection was determined by staining and isolation of the agent.) When, between the 20th and 23rd day, post-infection (p.i.), gametocytes appeared in the tail blood of the receptor voles, newly emerged, hungry fleas of each of five species of fleas were fed on individual voles for three consecutive days. Fourteen days after the last feeding, the numbers of surviving fleas were tested for sporocysts.

In order to get a synopsis about the sequence and duration of the development of *H. erhardovae* in fleas, we infected a large number of young *X.cheopis.* At periodical intervals between three hours and 30 days p.i. five fleas at a time were dissected and examined to ascertain the parasites' stage of development. An additional ten fleas from each group were given per os to a bank vole. The blood of the receptor voles was tested between the third and 30th day p.i.

In tests undertaken to determine whether the age of the flea affects sporogeny, sexually mature *X.cheopis* were retained for 14 days in humid chambers at 25°C and then permitted to feed for about 24 hours on *Clethrionomys* carrying gametocytes. Fleas from newly-emerged colonies served as a control. The sporocyst indices were checked after 14 days.

The rodents used in these experiments were bred in captivity and were free of ectoparasites and *Hepatozoon* infection prior to experimentation. The highly reproductive breed of bank voles has been maintained in the laboratory since 1959. Its history of domestication is described by Löbmann (1968) and Runzheimer (1969). Experimental infections with *Hepatozoon* were carried out orally by the aid of a stomach tube with 0,2-0,5 ml of a saline suspension of ground, infected arthropods. Since thick blood films are unsuitable for the diagnosis of *Hepatozoon* gametocytes, we applied successfully the simple slide concentration method of white cells from the peripheral blood as recommended by Sachtleben and Albrecht (1967).

3 RESULTS

3.1 *Evidence of spontaneous* Hepatozoon *infections in ectoparasites of free-living voles and their biological significance*

Despite a high incidence of infection in rodent tissues throughout the year, the development and spread of *Hepatozoon* is correlated with the breeding period of the mammalian host, as are many other parasites of rodents. In the *Clethrionomys* of our surveyed areas, i.e. in the surroundings of Munich, gametocytaemia occurs more frequently in the summertime than in the cold season. The chance of finding infected ectoparasitic arthropods is augmented with the increasing rates of sub-adults in the vole population. A high density of these parasites in the blood or lymph imbibed by ordinarily unsuitable arthropods like ticks or larval chigger mites, can initiate a transitory occurrence of gametocytes in the gut. Evidently the parasites can resist the digestive enzymes longer than the constituents of the blood.

Only when mature sporozoites were found microscopically or were demonstrated indirectly by the positive result of an inoculation experiment, was the vector role of an ectoparasite species considered to be proven. If other, not yet infective, stages of the sexual part of the protozoan life cycle were observed in the preparation, the arthropod was suspected to be a vector.

In tests on host specificity in the various ectoparasites, only those animals which received suspensions of fleas invariably developed an infection. With lice and ticks positive results were never obtained. The same was true for most of the mite pools, and some of the exceptions must be mentioned. One member of the superfamily Gamasoidea collected from a test area outside Bavaria is obviously able to infect bank voles when orally administered as a suspension. Furthermore, oral application of the larvae of the chigger mite *Neo-*

Table 1. Infection rates of five species of fleas simultaneously fed on bank voles infected with *H.erhardovae*

Flea species	Total no. examined	No. infected with sporocysts	% positive = sporocyst index
Xenopsylla cheopis	113	107	95
Nosopsyllus fasciatus	97	19	20
Megabothris turbidus	130	63	48
Ctenophthalmus agyrtes	108	102	95
Ctenophthalmus assimilis	48	18	17

trombicula zachvatkini can result in a very scanty and transitory gametocytaemia in receptor hosts. This, however, can easily be overlooked since this kind of transmission never results in multiplication of the parasite. It was not possible to pass mite-borne strains through fleas or flea-strains through mites.

The low infection rate in fleas collected from highly infected wild hosts is remarkable. The sporocyst index never exceeds 8 to 10 per cent. Even a rate of positive fleas of only 2 to 3 per cent, as is normally seen in the test areas near Munich, may be characteristic for a holo-endemic situation in a vertebrate host population. In order to take into account the possibility of very low infection rates in fleas or other ectoparasites, the number of specimens tested was always over 300.

3.2 *Experimental vector capacity of five different species of fleas*

The results of the tests on comparative vector efficiency are shown in table 1, whence it can be seen that although all five species of fleas can acquire the Bavarian strain of *H. erhardovae* by feeding, the incidence of sporocyst-development varied considerably, viz. from 17 to 95 per cent.

Bank voles can be successfully infected with sporozoites from all of these species of fleas, suggesting there are no differences in the infectivity of the sporozoites derived from different sources. The parasite burden does not influence in any appreciable degree the vitality, life span, or fecundity of the flea.

3.3 *The sporogeny of* H.erhardovae *in* Xenopsylla cheopis

The most difficult questions to deal with in the elucidation of the sexual part of a *Hepatozoon* life cycle are always the demonstration of the terminal maturation of the gametocytes, the place, time and morphology of the syngamy and the zygote formation. Since our observations on the initial phases of sporogeny are still incomplete, they are reserved for a later presentation. The gametocytes taken up with blood meals leave their host monocytes within a few minutes. Subsequently they become elongated and motile (fig. 13). Such 'vermicular' forms are on an average 12-19 x 3 μm in size and they disappear from

Table 2. Time sequence of the sporogony of *H.erhardovae* in *Xenopsylla cheopis*

Serial no. of receptors	Time post infection	No. of fleas with sporocysts	Parasitaemia of the receptor voles (Days post infection) 3	7	10	14	17	20	24	30	35
7413	3 hours	0	o	o	o	o	o	o	o	o	o
7414	6 hours	0	o	o	o	o	o	o	o	o	o
7415	12 hours	0	o	o	o	o	o	o	o	o	o
7416	24 hours	0	o	o	o	o	o	o	o	o	o
7417	3 days	0	o	o	o	o	o	o	o	o	o
7418	7 days	0	o	o	o	o	o	o	o	o	o
7419	10 days	0	o	o	o	o	o	o	o	o	o
7420	14 days	1/5	o	o	o	o	o	+	+	+	+
7421	17 days	3/5	o	o	o	o	o	o	+	+	+
7422	20 days	2/5	o	o	o	o	o	+	+	+	+
7423	24 days	5/5	o	o	o	o	o	+	+	+	+
7424	30 days	5/5	o	o	o	o	o	o	o	+	+

the lumen of the intestinal tract within 24 to 28 hours after the uptake of infected blood. Most of them may be digested and discharged, but a few, however, seem to penetrate in an obscure way the flea's gut wall in order to invade the abdominal cavity. Somewhere on this migration route, the sexual contact takes place, probably in the coelom. Anyhow, the zygote settles and develops here (fig. 5). The result of the studies on sporogeny are summarized in table 2, wherein it is clear that this aspect of the life cycle is completed within two weeks.

Stages younger than the mature sporozoites are non-infective, even in the specific rodent host. There is a three-week incubation period in the receptor vertebrate. The complete life cycle of the protozoan takes a minimum of five weeks, and variations may be observed under the same experimental conditions, e.g. plus or minus four days for the duration of sporogeny, and irregularities in the same order of magnitude in the developmental time can occur in the rodent. The reasons for these variations are unknown.

In all species of fleas, the entire sporogeny takes place in the coelom. The sporocysts never occur near the intestinal tract but are commonly situated close to the flea's body wall. Fat bodies and muscles may be more or less displaced by the growing oocyst (figs. 1, 2). In an individual flea several oocysts can develop side by side (fig. 5). The largest measured mature oocyst when unfixed and unstained had a diameter of 330 x 280 μm and was just visible with the naked eye. In spite of the very thin oocyst membrane (fig. 9), the big agglomeration of sporocysts sticks together tightly. In some cases, it was possible to separate the intact oocyst from the flea's abdominal cavity (fig. 6).

The size of the oocyst seems to vary according to the number of sporocysts but it is independent of the flea species. Usually there are hundreds of sporocysts in each oocyst (figs. 3, 4). Mature sporocysts measure 25 x 15 ± 3 μm (figs. 10, 11). The number of sporozoites observed within a sporocyst varied from 9-36, the average being 10-12 sporozoites. When transferred into a liquid isotonic medium the sporozoites are quite motile, mainly by rotation (fig. 14).

3.4 *The effect of the age of the vector upon sporogeny*

In the tests reported above, newly emerged, hungry fleas were employed exclusively, but it is important to know how sexually mature fleas compare as vectors, and the results of such experiments are shown in table 3.

The indices varied between 4 and 94 per cent, evidently depending on variables that were difficult to standardize. The experiments, however, clearly suggest that mature fleas

Table 3. Comparison of the vector capacity of newly emerged and mature *X.cheopis* for *H.erhardovae*

Serial no.	Origin of *Hepatozoon*	Age of fleas	No. of pools with carriers of sporocysts	% positive (sporocyst index)
730	Wild caught carrier	Mature	46/50	92
732	Wild caught carrier	Young	2/50	4
794	Wild caught carrier	Mature	40/50	80
807	Wild caught carrier	Young	18/50	36
7 422	First passage of strain	Mature	25/50	50
7 421	First passage of strain	Young	2/50	4
7 503	First passage of strain	Mature	35/50	70
7 500	First passage of strain	Young	47/50	94

(two weeks after emergence) are generally far superior to newly emerged specimens as vectors of this protozoan. The ability of fleas to acquire *Hepatozoon* sporocysts may thus increase with the maturity and sexual development of the flea.

3.5 *The possibility of transovarial transmission of* Hepatozoon *in fleas and the excretion of infective stages with the faeces of the flea*

Inasmuch as the sporozoites do not liberate themselves from the sporocyst while in the flea, neither transovarial transmission nor voiding of sporozoites with the faeces of the flea would be expected to occur. The data bear this out. Large numbers of fleas were successfully reared from infected parents, since the procreative ability of the vector is not adversely affected by the protozoan. No sign of infection with any stage of *Hepatozoon* was ever detected in the larval or adult progeny of infected fleas, regardless of species of Siphonaptera tested. Suspensions of faeces from sporocyst-carrying fleas failed to produce infection in susceptible voles, and the same negative result was obtained when eggs of parasitized fleas were administered per os to the rodents. These observations suggest that the ability of *Hepatozoon* to locomote and penetrate tissues is limited to the stages in the vertebrate. As in the case of the rodent, there is no laboratory evidence to indicate that passage of infection occurs from one generation to the next.

4 DISCUSSION

In that it uses fleas as the arthropod host, *H. erhardovae* differs markedly from other *Hepatozoon* species parasitizing small vertebrates and whose life cycles have so far been clarified. There is no doubt that voles normally become infected by ingestion of flea-borne sporozoites as a result of the host defence mechanism, whereby the vole kills fleas with its teeth, thus releasing the infective sporozoites. It seems to be a rule that insects infected with entomophilic protozoa never die rapidly from the infection, and that this kind of parasitism 'can play an important role in systems of integrated control' (Pramer & Rabiai, 1973). However, there is no evidence that *Hepatozoon* markedly handicaps the vector. Nevertheless, even a slightly handicapped flea would be more prone to capture by the host and thus favour considerably the protozoan's transmission and spread would be facilitated. The role of fleas as vectors of *Hepatozoon* of European rodents has been suspected for a long time (Splendore, 1920; Dasgupta & Meedeniya, 1958). Krampitz (1964) confirmed the circumstantial evidence by transmission experiments, at least for *H. erhardovae.* We believe, however, that flea-borne species may be the exception rather than the rule, since most of the *Hepatozoon* species in reptilian and mammalian hosts are known to be transmitted by mites, ticks and even mosquitoes. The exceptional but stable transmission by fleas can be considered as the result of a secondary evolutionary adaptation to a successful ectoparasite.

In some test areas, a morphologically indistinguishable form of *Hepatozoon* occurs which is easily transmitted to bank voles by gamasid mites (fig. 12). This may be a second, and perhaps sibling species, and should be investigated. It is interesting to note that mite-transmission, a phenomenon known to occur in many species of rodent *Hepatozoon* (Miller, 1908; Miyairi, 1934; Brumpt, 1946; Hoogstraal, 1961; Furman, 1966; Redington & Jachowski, 1972; Frank, 1977a) may be found in the same host and area as vectorship by fleas.

Basically the morphology of the sporogenic forms in fleas seems to be closely related to the respective developmental stages of the species in reptiles which use culicids as the arthropod host (Ball et al., 1967; Landau, 1970; Ball & Oda, 1971; Pessoa et al., 1971, 1974). In both cases, the sporogeny, and, probably, the syngamy, takes place in the coelom of the arthropod, and the vertebrate becomes infected by ingestion of the vector. Finally, more than one species of the respective group of bloodsucking arthropods may be involved in the transmission of a single *Hepatozoon* species. As far as *H. erhardovae* is concerned, several members of a series of families of the order Siphonaptera can serve as the arthropod host.

Our experience with fleas as vectors for *H. erhardovae* in *Clethrionomys* encourages a few general and speculative remarks. It is surprising that *X. cheopis,* an introduced and southern species which does not occur in the area, seems to be as efficient a vector, judged by the sporocyst index, as the main natural, local vector, *C. agyrtes* (table 1). It also seems much superior to *M. turbidus,* another characteristic, local flea of the bank vole. In contrast, the other native fleas seem to vary in susceptibility in proportion to their abundance on the bank vole. It therefore appears that the protozoan has adapted to the major fleas of *Clethrionomys.*

Transovarial and congenital transmission of parasites in arthropod hosts have been mainly reported for Acari (Burgdorfer & Varma, 1967; Fine, 1975). Little experimental work has been done on vertical transmission in vectors other than ticks. In view of the lack of evidence indicating these phenomena occur in insects, our negative results with *Hepatozoon* in fleas are perhaps not surprising.

We express our thanks to Robert Traub for revising this manuscript from the linguistic aspect.

REFERENCES

Ball, G.H., J.Chao & K.S.Telford, Jr. 1967, The life history of *Hepatozoon rarefaciens* (Sambon & Seligmann, 1907) from *Drymachon corais* (Colubridae) and its experimental transfer to *Constrictor constrictor* (Boidae). J. Parasit. 53:897-909.

Ball, G.H. & S.N.Oda 1971, Sexual stages in the life history of the hemogregarine *Hepatozoon rarefaciens* (Sambon & Seligmann, 1907). J. Protozool. 18:697-700.

Beaucournu, J.-C. & J.Deunff 1975, Intéret de l'examen extemporane des Siphonaptères pour l'étude de leurs parasites. Ann. Parasit. 50:831-835.

Brumpt, E. 1946, Contribution à l'étude d'*Hepatozoon muris.* Utilisation du xénodiagnostic pour l'identification des espèces d'hémogrégarines. Ann. Parasit. 21:1-24.

Burgdorfer, W. & M.G.R.Varma 1967, Transstadial and transovarial development of disease agents in arthropods. Ann. Rev. Ent. 12:347-376.

Dasgupta, B. & K.Meedeniya 1958, The vector of *Hepatozoon sciuri.* Parasitology 48:419-422.

Fine, P.E.M. 1975, Vectors and vertical transmission. An epidemiologic perspective. Ann. N.Y. Acad. Sci. 266:173-194.

Frank, C. 1977a, Ein Beitrag zur Biologie von *Hepatozoon erhardovae* Krampitz, 1964 in Rötelmäusen aus der Südweststeirmark und dem Neusiedlerseegebiet (Burgenland). Z. Parasitenk. 53:251-254.

Frank, C. 1977b, Über die Bedeutung von *Laelaps agilis* C.L.Koch 1836 (Mesostigmata: Parasitiformae) für die Übertragung von *Hepatozoon sylvatici* Coles 1914 (Sporozoa: Haemogregarinidae). Z. Parasitenk. 53:307-310.

Furman, D.P. 1966, *Hepatozoon balfouri* (Laveran, 1905): Sporogonic cycle, pathogenesis and transmission by mites to jerboa hosts. J. Parasit. 52:373-382.

Hoogstraal, H. 1961, The life cycle and incidence of *Hepatozoon balfouri* (Laveran, 1905) in Egyptian jerboas (*Jaculus* sp.) and mites (*Haemolaelaps aegyptiacus* Keegan, 1956). J. Protozool. 8:231-248.

Killick-Kendrick, R. 1974, Parasitic protozoa of the blood of rodents. II. Haemogregarines, malaria parasites and piroplasms of rodents: Annotated checklist and host index. Acta Trop. 31:28-69.

Krampitz, H.E. 1964, Über Vorkommen und Verhalten von Haemococcidien der Gattung *Hepatozoon* Miller, 1908 (Protozoa, Adeleidea) in mittel- und südeuropäischen Säugern. Acta Trop. 21:114-154.

Krampitz, H.E. 1971, Haemogregarines in European rodents. C.R. I. Multicoll. Europ. Parasit. Rennes 1971, p. 242.

Krampitz, H.E. 1972, Die sporogonie von *Hepatozoon* in Flöhen. Z. Parasitenk. 39:49.

Krampitz, H.E. 1973, The development of bank vole's *Hepatozoon erhardovae* Krampitz, 1964 in normal and unusual hosts. Prog. Protozool. Abstr. Papers 4. Internat. Congr. Protozool. Clermont Ferrand 1973, 2. 229.

Krampitz, H.E. 1974, Reproduction studies on some common European fleas reared in captivity. Proc. 3. Internat. Congr. Parasitol. München, p. 880.

Krampitz, H.E., V.Wongchari & O.Geisel 1976, *Hepatozoon erhardovae* Krampitz, 1964: Zellparasitismus im Vertebraten und Evertebraten. Z. Parasitenk. 43:202.

Landau, J. 1970, Mise en évidence d'un double mode de transmission chez un *Hepatozoon* des reptiles malgaches. C.R. Sci. Nat. D 270, 2308-2310.

Löbmann, P. 1968, Quantitative Untersuchungen an der Rötelmaus, *Clethrionomys glareolus* (Schreber, 1780). Z. Säugetierk. 33:129-149.

Miller, W.W. 1908, *Hepatozoon perniciosum* n.g. n.sp., a hemogregarine pathogenic for the white rats; with a brief description of the sexual cycle in the intermediate host, a mite (*Laelaps echidninus* Berlese). Hyg. Lab. Bull. Wash. No. 46.

Miyairi, K. 1934, Zur Entwicklungsgeschichte einer Haemogregarine (richtiger Karyolysus) aus der Feldmaus *Arvicola hatanezumi* Sasaki. Proc. Imp. Acad. Tokyo 8:328-329.

Pessoa, S.B., L.Sacchetta & J.Cavalheiro 1971, Notas sobre hemogregarinas brasileiras. XV. Sobre uma nove especie do genero *Haemogregarina* S.S. Parasita da *Thamnodynastes pallidus nattereri* (Thunberg) e sua evolução em mosquitos. Rev. Latinoamer. Microbiol. 13: 29-32.

Pessoa, S.B., P.de Biasi & L.Sacchetta 1974, Nota sobre o *Hepatozoon tupinambis* (Laveran et Salibeni, 1909) (Protozoa, Apicomplexa) parasita do teju (*Tupinambis teguixin* Linne, 1758) (Sauria, Teiidae). Mem. Inst. Butantan 38:123-130.

Peus, F. 1970, Zur Kenntnis der Flöhe Deutschlands (Insecta, Siphonaptera). III. Faunistik und Ökologie der Säugetierflöhe, Insectivora, Lagomorpha, Rodentia. Zool. Jb. Syst. 97: 1-54.

Pramer, D. & S.Al Rabiai 1973, Regulation of insect populations by protozoa and nematodes. Ann. N.Y. Acad. Sci. 217:85-92.

Redington, B.C. & L.A.Jachowski, Jr. 1972, Role of *Haemogamasus reidi* (Acari: Mesostigmata) in the life cycle of the grey squirrel protozoan *Hepatozoon griseisciuri* (Protozoa: Haemogregarinidae). J. Parasitol. 58:401-403.

Runzheimer, J. 1969, Quantitative Untersuchungen an der 5. Gefangenschaftsgeneration von *Clethrionomys glareolus* (Schreber, 1780). Ein Beitrag zum Studium des Domestikationsgeschehens. Z. Säugetierk. 34:9-37.

Sachtleben, P. & G.Albrecht 1967, Eine einfache Anreicherung von Blutleukozyten für die Bestimmung des Kerngeschlechts. Deutsch. Med. Wschr. 92:2270-2272.

Šebek, Z. 1975, Parasitische Gewebsprotozoen der Wildlebenden Kleinsäugetiere in der Tschechoslowakei. Folia Parasit. 22:111-124.

Šebek, Z. 1977, Parasitische Blut- und Gewebsprotozoen der Kleinsäuger mit besonderer Berucksichtigung der für den Menschen pathogenen Arten. 2. Internat. Arbeitskoll. Naturh. Inf. Krh. Zentraleur. Graz 1976, p. 57-74.

Splendore, A. 1920, Sui parassiti delle arvicole. Ann. Ig. Sper. 30:560-582.

Figures 1, 2. Sections through two males of *Megabothris turbidus* with sporocysts of *H.erhardovae* in the marginal parts of their abdominal cavities (Gomori stain). 60x.
Figure 3 (left). Mature oocysts with numerous sporocysts in the flea's coelom. ca. 700x.
Figure 4 (right). Smear of the same development stage, stained with Giemsa. ca. 800x.
Figure 5 (left). Tissue section of three different sporogonic developmental stages of *H.erhardovae* in the flea (Gomori stain). ca. 120x.
Figure 6 (right). Mature oocyst, in toto, isolated from the flea's abdomen. 25x.

Figure 7, 8 (left and middle). Immature oocysts in a Giemsa stained smear.
Figure 9 (right). The same stages in a section (note the scanty oocyst membrane). 1200x.
Figure 10 (left). Mature sporocysts. (Phase contrast, unstained).
Figure 11 (middle). The same sporocyst (Giemsa stained).
Figure 12 (right). Mature sporocyst of *Hepatozoon* isolated from a gamasid mite. ca. 1600x.
Figure 13 (left). Elongate stage from a flea's gut content. First step of the sporogonic development (taken 24 hours after the blood meal).
Figure 14 (middle). Mature infective sporozoite (last stage of the sexual development). (Phase contrast, unstained).
Figure 15 (right). The same sporozoite (Giemsa stained). ca. 1800x.

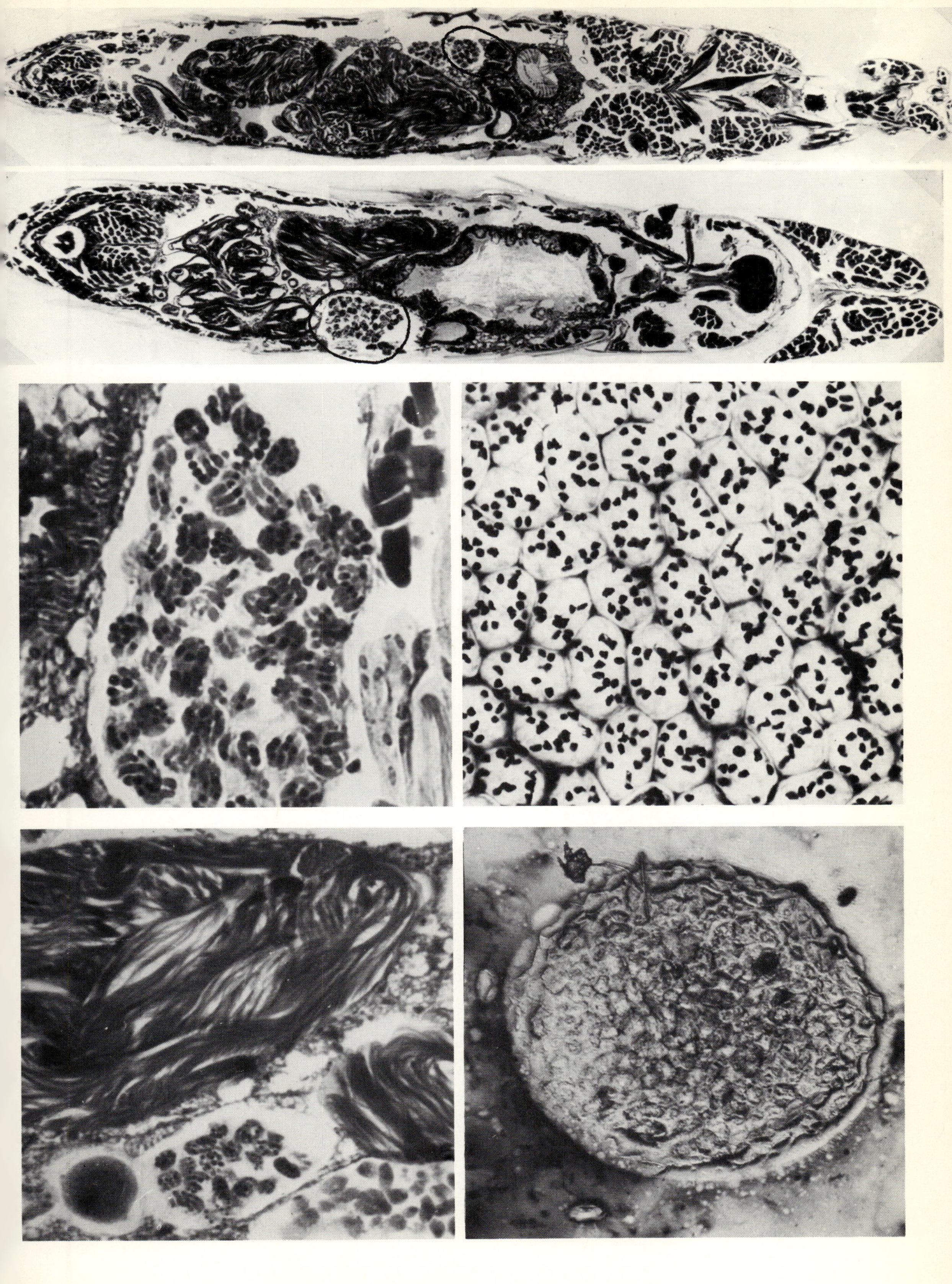

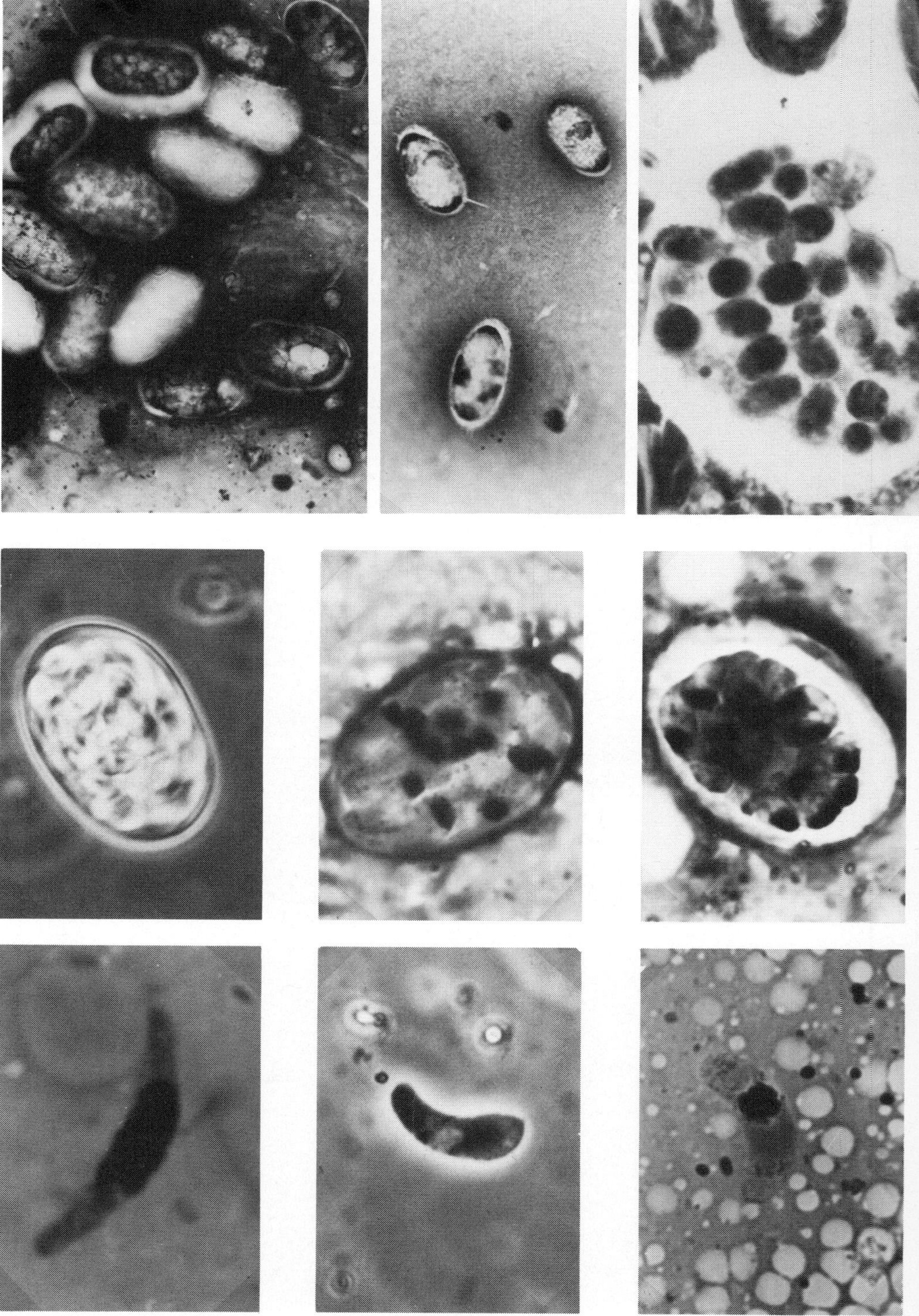

YOSEF SCHLEIN
Hadassah Medical School, Jerusalem, Israel

MORPHOLOGICAL SIMILARITIES BETWEEN THE SKELETAL STRUCTURES OF SIPHONAPTERA AND MECOPTERA

ABSTRACT

In a morphological comparison of some Siphonaptera and Mecoptera it was found that the first abdominal spiracle is located in the thorax, the eyes lie in a skeletal socket in *Boreus* and fleas, and a coxo-propleural articulation similar to that of fleas is present in *Panorpa.* Sclerites in the thorax of *Merope* are suggested as homologues to the flea link plates, and the upper part of the pleural arch of fleas is compared to the first axillary sclerite in the wing base of *Boreus.* The result supports the concept that the fleas developed from a *Boreus*-like insect.

INTRODUCTION

The origin of fleas from a *Mecoptera*-like ancestor was first suggested by Tillyard (1935) and this assumption was supported by Hinton (1958) on the basis of a comparison of larval features. There is further evidence for relationship between the two Orders. Thus the spermatozoa of the fleas and Mecoptera so far examined have two unusual features in common: the flagella spiral round the mitochondria and the inner circle of nine fibrils is lacking (Baccetti, 1972; Rothschild, 1975); the process of resilin secretion in the flea (pleural arch) and *Boreus* (wing base) is similar and is different from that in the locust and dragonfly (Rothschild & Schlein, 1975); the unusual proventricular spines of *Boreus* and fleas are morphologically remarkably similar (Richards & Richards, 1969); both Orders share a) multiple sex chromosomes (Bayreuther & Bräuning, 1971) and b) sexual dimorphism in the nerve cord (the males possessing an additional abdominal ganglion). The Siphonaptera and Mecoptera are the only two Orders in which both panoistic and polytrophic ovaries are recorded (Kunitskaya, 1960; King & Teasley, present volume; Rothschild, 1975).

The present study attempts to investigate further the possibly close relationship between the two Orders by looking for structural similarities and homologies in the exoskeleton. Some of these similarities throw light on the characteristics of fleas which do not conform with the usual plan found in other Orders.

The following points are considered: the link plates; the coxo-propleural articulation; the skeletal frame of the eyes; the first abdominal spiracle and the pleural arch.

TERMINOLOGY

The specific morphological terms applied to fleas are those used by Rothschild & Traub (1971) and the terminology used for the identification of homologies is the one used by Snodgrass (1935, 1946).

The link plates

The thoracic segments of fleas are closely united with each other, and with the head and abdomen, with very little individual movement. They are separated by arthrodial membrane, and the only sclerotised connection between them is the two pairs of link plates.

The first pair of link plates is attached to the pronotum anteriorly, and to the internal protrusion at the upper part of the mesopleura on the other side. The second pair of link plates connects the posterior part of the mesopleura to the upper internal protrusion of the lateral metanotal area (fig. 1).

This description does not agree with that of Snodgrass (1946) who describes both pairs as being connected anteriorly with the pleuron and posteriorly with the notum.

In the protomecopteran *Merope* (fig. 2) there are two lateral posteriorly directed hook-like protruberances in the first and second segments of the thorax. The first pair arise from the pronotum as half crescent shapes, and taper to their posterior attachment with the mesopleura. This plate incorporates the spiracle as does the first link plate of the flea. The second sclerite extends backwards from the mesopleura and ends in the intersegmental membrane. A pronotal-mesopleura-connecting-sclerite, found in *Merope,* is also present in *Panorpa* and *Boreus* (fig. 4) in which the second sclerite is absent. Considering the resemblance in shape, the similarity of the connections and the location of the first thoracic spiracle, it is suggested that the link plates of the flea be considered as homologues of the two sclerites of *Merope.*

The coxo-propleural articulation

The articulation of the first coxa to the propleuron consists, in fleas (fig. 1), of a coxal pivot and a receptive acetabulum in the pleura. This arrangement according to Snodgrass (1946) is 'just the reverse of that which ordinarily hinges the leg to the body'. In Mecoptera, in both *Bittacus* (fig. 3a) and *Boreus,* this hinge has a double articulation: one pleural process descends to the coxa and a second coxal process which connects to the pleura. In *Panorpa* (fig. 3b) there are also two processes but only the coxal process has a socket, whereas the propleural process ends free in the membrane as it does in fleas. A double coxal articulation as in *Bittacus* and *Boreus* should be regarded as a possible intermediate stage in the development of the coxal hinge of *Panorpa* and the Siphonaptera and that of other insects.

The skeletal frame of the eyes

Adult fleas have a pair of small simple eyes, which on account of the origins of the optic nerve at the anterior ventral corner of the forebrain, must be regarded as reduced compound eyes (Hopkins & Rothschild, 1966, plate 12). They are situated in deep cup-shaped inflections of the cuticle, penetrated at their inner ends by the optic nerves and situated anterior to the antennal fossae. Due to this circumocular inner socket and the external cornea, the eyes are almost completely surrounded by a cuticular layer.

The eyes of *Boreus,* which otherwise are normal compound eyes, also lie within a cuticular socket (fig. 4) formed by a deep internal flange of the circumocular sulcus. *Panorpa* also has an internal flange of the circumocular sulcus but it is less deep than that of *Boreus.*

The similarity of the eye socket of *Boreus* and Siphonaptera is not necessarily an indication of close phylogenetic relationship as, according to Snodgrass (1960) there are several cases of a deep skeletal partition around the eyes among insects.

The position of the first abdominal spiracle

The location of the first abdominal spiracle of insects is usually in the membrane close to the thorax. In the fleas (fig. 1), it is incorporated into the third segment of the thorax and 'always lies within the epimeral flange' (Snodgrass, 1946). In *Bittacus, Panorpa* and *Merope* this spiracle is situated in its usual place – in the membrane behind the thorax, whereas in *Boreus* (fig. 4) its position is in the third thoracic epimeron as in the fleas.

The pleural arch

The pleural arch of fleas (fig. 1) is situated at the apex of the inflexion which forms the metapleural ridge. It consists of a cup-shaped sclerotization which fuses with the ventral end of the notal ridge into which enters the roughly conical extension of the upper end of the pleural ridge. The resilin pad is sandwiched between the two structures. In the pleural arch, the presence of the resilin identifies this structure as the site of the wing base of the winged flea-ancestors (Neville & Rothschild, 1976).

A comparison of the base of the wing of *Panorpa,* which has full-sized wings, and *Boreus hiemalis,* where the wings are reduced, show that the pattern of the reduction is apparently similar to the one that took place in the evolution of fleas. The structure of the wing base with the large first axillary sclerite (the only sclerite present), closely resembles the pleural arch area of some fleas. The first axillary sclerite is already relatively large in the wing base of *Panorpa* (fig. 5) where all the usual sclerites are present. This sclerite, which is shaped like a short 'Y', looks like a socket for the head of the pleural wing process, when the wing is turned down. In *Boreus,* this sclerite is the only axillary sclerite left at the base of the deformed wings (fig. 6a). Its normal position is that of a socket for the apex of the pleural wing process. Both the sclerite and the pleural wing process of *Boreus* are, as in the pleural arch of fleas, almost completely invaginated in the thorax. The displaced scutoscutellar suture extends upwards from the base of the axillary sclerite. In *Ceratophyllus* (fig. 6b) and in other fleas (fig. 7), the outer sclerotization above the pleural arch is in fact a sclerite which is incompletely separated from the notum by delicate external sutures. This sclerite has a shape and position similar to the first axillary sclerite of *Panorpa* and *Boreus* and should be considered its homologue. The external superficial suture in fleas, that begins above the sclerite, roughly parallel to the internal notal ridge (figs. 6b, 7a) may be defined as the scutoscutellar suture similar to this ridge in *Boreus.* According to these landmarks the triangular sclerite posterior to the ridge is the postnotum. The internal sclerotizations which emerge from the first axillary sclerite of fleas (fig. 7) generally correspond to the external sutures of the lateral area of the metanotum and the notal ridge (Rothschild & Traub, 1971; Rothschild, 1975). They may have evolved as internal inflections of the latter, but the true sclerite borders are the external sutures of the skeleton (fig. 7a, b) and they seem to retain more of the sclerite pattern common to insects. The internal sclerotizations are constituents of the frame of struts of the jump system of fleas (Rothschild & Schlein, 1975), and, as previously pointed out (ibid., p. 483), there are two main types of the 'line of force' Broadly speaking the less specialized jumpers which spend considerable periods away from the body of the host have a less vertical notal ridge and, with the lateral internal sclerotizations (= transverse ridge of the metanotal area, Rothschild & Schlein, 1975), form a V-shaped joint with the pleural arch (fig. 7a, b). In such fleas, the arch is situated nearer the dorsum but in the Pulicoid fleas, in which its situation is more ventral (Rothschild, 1969), there is a corresponding lengthening of the stem of the notal ridge and a shortening of the pleural ridge; the line of force then becomes vertical. In those fleas which spend most of their adult lives fixed to the host, there is an associated loss of the lateral internal sclerotizations which reaches a maximum in *Echidnophaga* but is also noticeable in *Pulex* (figs. 7c, d). A close study of these ridges, which display a bewildering array of variation on the main themes, will reveal the evolution of the subtly different types of jumping activity.

DISCUSSION

The two aspects considered in this work are the phylogenetic relations of Siphonaptera and Mecoptera and the identification of morphological structures that were changed and reduced by the loss of wings and the adaptation to parasitic life. The difficulty in tracing the origin of a certain trait from another group of insects lies in the possibility of a parallel and independent development or convergence occurring in two distantly related insect orders. A characteristic that had been considered specific to fleas, e.g. the pivot of the first coxa (Snodgrass, 1946) has here been shown also to exist in Mecoptera. It is therefore not possible to state categorically that any one of the anatomical features described here is specific to Mecoptera and Siphonaptera. An attempted explanation of the development of certain structures, based on a comparison of different insect groups as in the present study, may be valid even when based on convergent evolution.

The evidence (with the reservation that given characteristics may not be confined to the two orders discussed), supports the concept that the Siphonaptera developed from some Mecoptera-like insect. No structures such as the 'link plates' of *Merope* and the link plates of fleas are present in other insects according to the figures in Matsuda's (1970) extensive study of the insect thorax. The special coxo-propleural hinge with a coxal pivot found in fleas is also present in Mecoptera in which there are intermediate stages of its development.

The first axillary sclerite which apparently constitutes the upper part of the pleural arch of fleas is relatively large in *Panorpa.* It is large and is the only one that remained of the wing base sclerites of *Boreus.* The displacement of the first abdominal spiracle to the thoracic pleura as seen in fleas and *Boreus* has occurred within the order Mecoptera as this spiracle is situated in the abdominal membrane in *Bittacus, Merope* and *Panorpa.* Hinton (1958) came to the conclusion that 'fleas could be derived from a *Boreus*-like ancestor'. According to the present study, the *Boreus*-like hypothetical flea-ancestor may have had a *Panorpa*-like first leg-hinge and a *Merope*-like second pair of link plates.

ACKNOWLEDGEMENTS

I would like to thank Dr Miriam Rothschild for suggesting this project and for many interesting discussions in the course of the investigation.

REFERENCES

Baccetti, B. 1972, Insect sperm cells. In: J.E. Treherne, M.J.Berridge & V.B.Wigglesworth (eds.), Advances in Insect Physiology. 9: 315-397. New York, Academic Press.

Bayreuther, K. & Bräuning, S. 1971, Die Cytogenetik der Flöhe II. *Xenopsylla cheopis* Rothschild 1903, und *Leptopsylla segnis* Schönherr, 1811. Chromosoma 33: 19-29.

Hinton, H.E. 1958, The phylogeny of the Panorpoid orders. Ann. Rev. Ent. 3: 181-206.

Hopkins, G.H.E. & Rothschild, M. 1966, An Illustrated Catalogue of the Rothschild Collection of Fleas (Siphonaptera) in the British Museum (Natural History). Vol. IV. London, British Museum (Natural History).

King, R.C. & Teasley, M. 1980, Insect oogenesis: some generalities and their bearing on the ovarian development of fleas. This volume.

Kunitskaya, N.T. 1960, On the reproductive organs in female fleas and determination of their physiological age. Med. Parazitol. Parazit. Bolez. 29(6): 688-701.

Matsuda, R. 1970, Morphology and evolution of the insect thorax. Mem. ent. Soc. Canada No. 76.

Mickoleit, G. 1967, Das Thoraxskelet von *Merope tuber* Newman. Zool. Y.B. Anat. 84: 313-342.

Neville, C. & Rothschild, M. 1967, Fleas – insects which fly with their legs. Proc. R. ent. Soc. Lond. Ser. C 32(3): 9-10.

Richards, P.A. & Richards, A.G. 1969, Acanthae: a new type of cuticular process in the proventriculus of Mecoptera and Siphonaptera. Zool. Jb. (Anat.) 86:158-176.

Rothschild, M. 1969, Notes on fleas: with the first record of a Mermithid Nematode from the Order. Proc. Br. ent. Nat. Hist. Soc. 2: 9-16.

Rothschild, M. 1975, Recent advances in our knowledge of the order Siphonaptera. Ann. Rev. Ent. 20: 241-259.

Rothschild, M. 1976, Fleas (Part II). The Internal Organs: can they throw any light on relationships within the Order? Proc. Br. ent. Nat. Hist. Soc. 9: 97-110.

Rothschild, M. & Schlein, Y. 1975, The jumping mechanism of *Xenopsylla cheopis.* I. Exoskeletal structures and musculature. Phil. Trans. R. Soc. Lond. Ser. B. 271: 499-515.

Rothschild, M. & Traub, R. 1971, A revised glossary of terms used in the taxonomy and morphology of fleas. In: G.H.E.Hopkins & M. Rothschild, An Illustrated Catalogue of the Rothschild Collection of Fleas (Siphonaptera) in the British Museum (Natural History). Vol. V. London, British Museum (Natural History).

Snodgrass, R.E. 1935, Principles of Insect Morphology. New York, Toronto, London, McGraw-Hill.

Snodgrass, R.E. 1946, The skeletal anatomy of fleas (Siphonaptera). Smithsn. misc. Coll. 104. No. 18.

Snodgrass, R.E. 1960, Facts and theories concerning the insect head. Smithsn. misc. Coll. 142. No. 1.

Tillyard, R.Y. 1935, The evolution of the Scorpion-flies and their derivatives (order Mecoptera). Ann. Rev. ent. Soc. Amer. 28(1): 37-45.

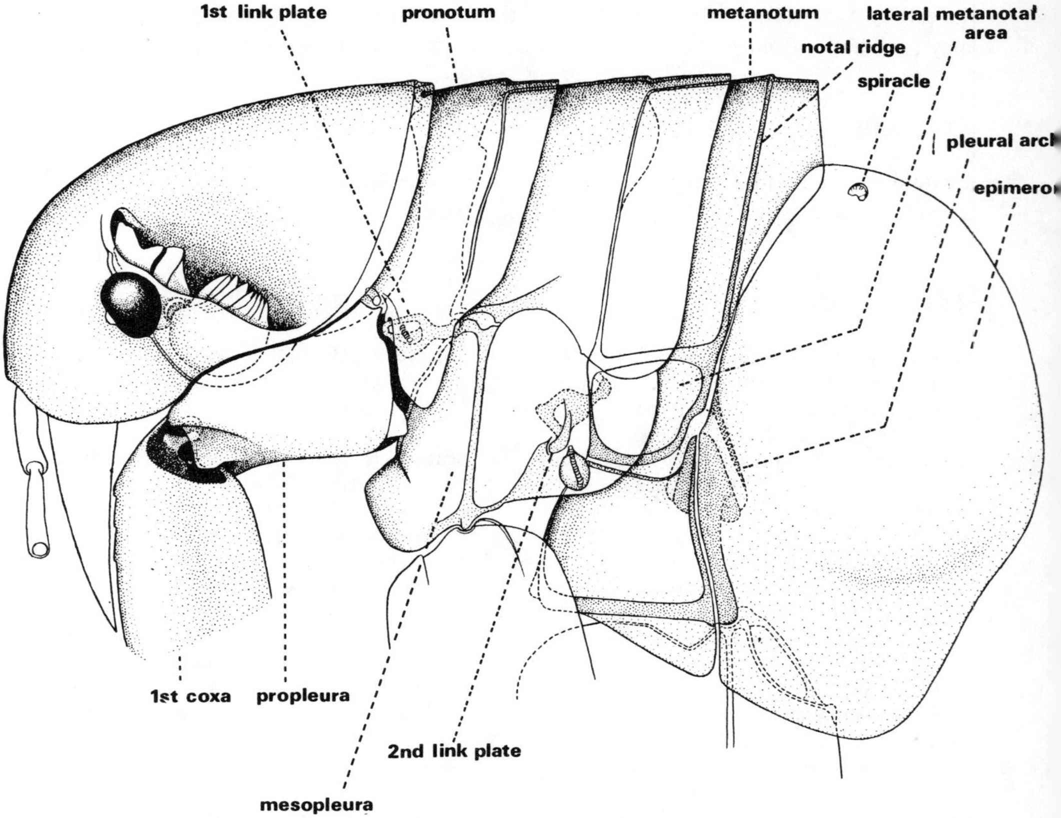

Figure 1. The thorax and the head of *Xenopsylla cheopis,* lateral view.

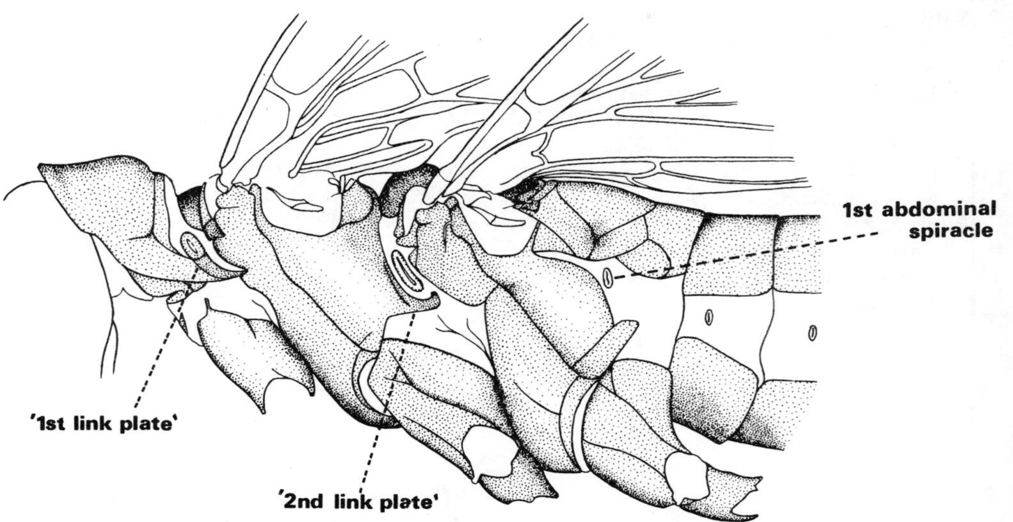

Figure 2. The thorax of *Merope tuber,* lateral view (after Mikoleit, 1967).

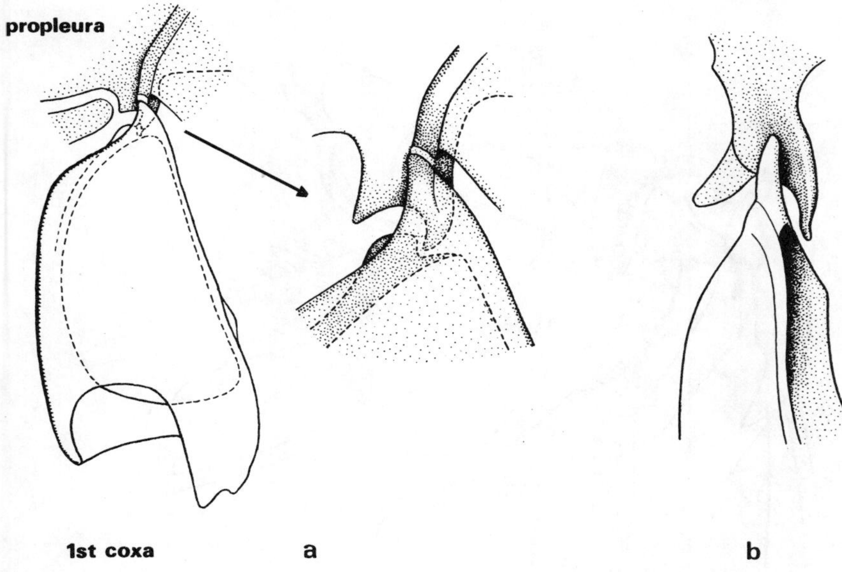

Figure 3. The coxo-propleura articulation of: a) *Bittacus;* b) *Panorpa.*

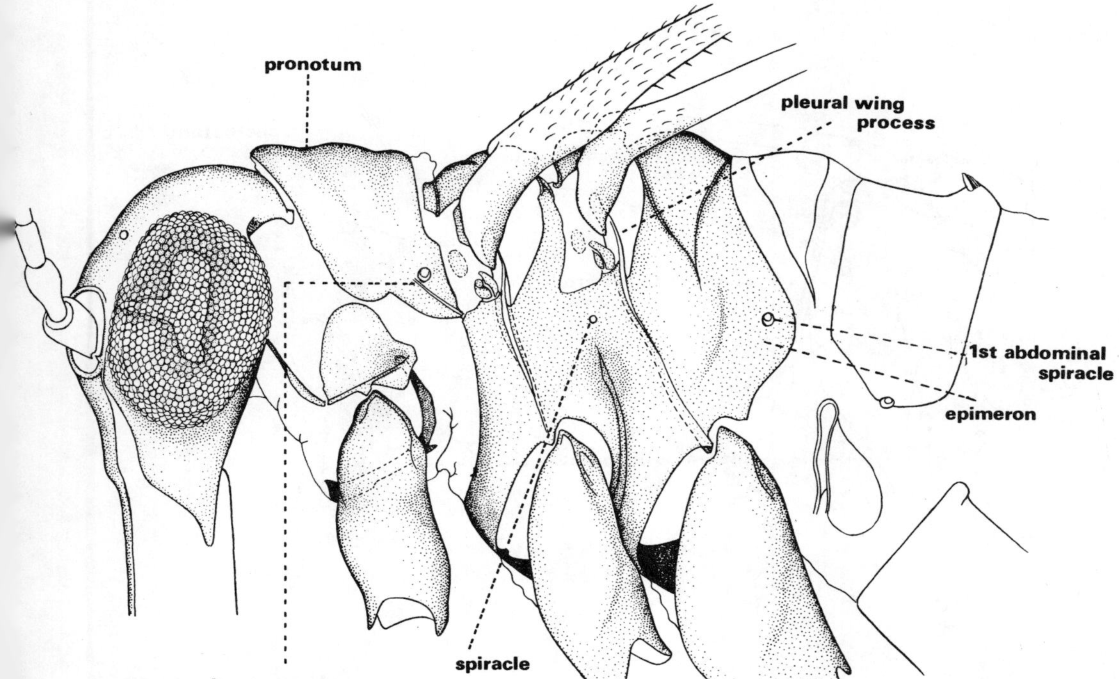

Figure 4. Lateral view of the head and thorax of *Boreus.*

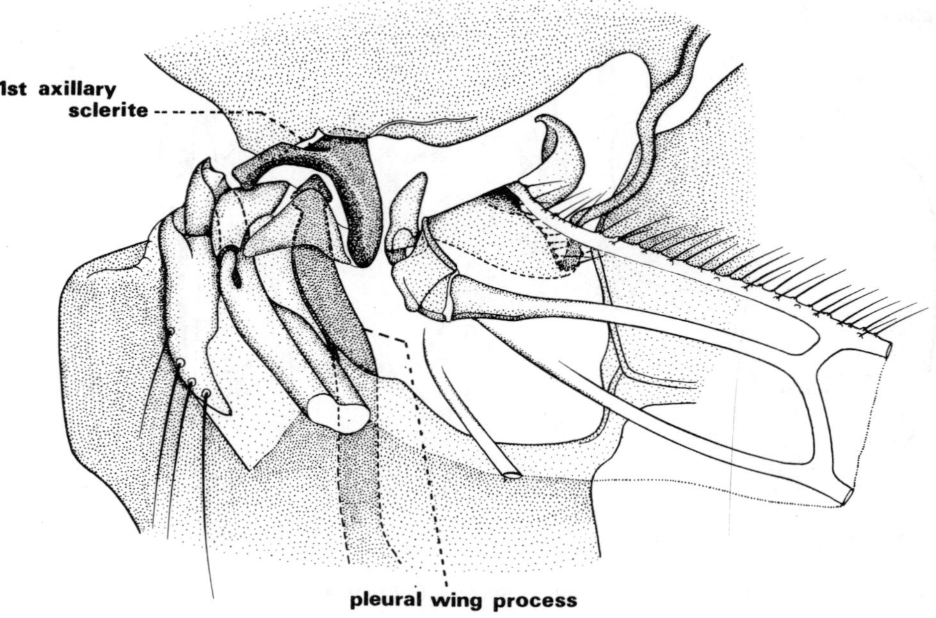

Figure 5. The wing base sclerites of *Panorpa*.

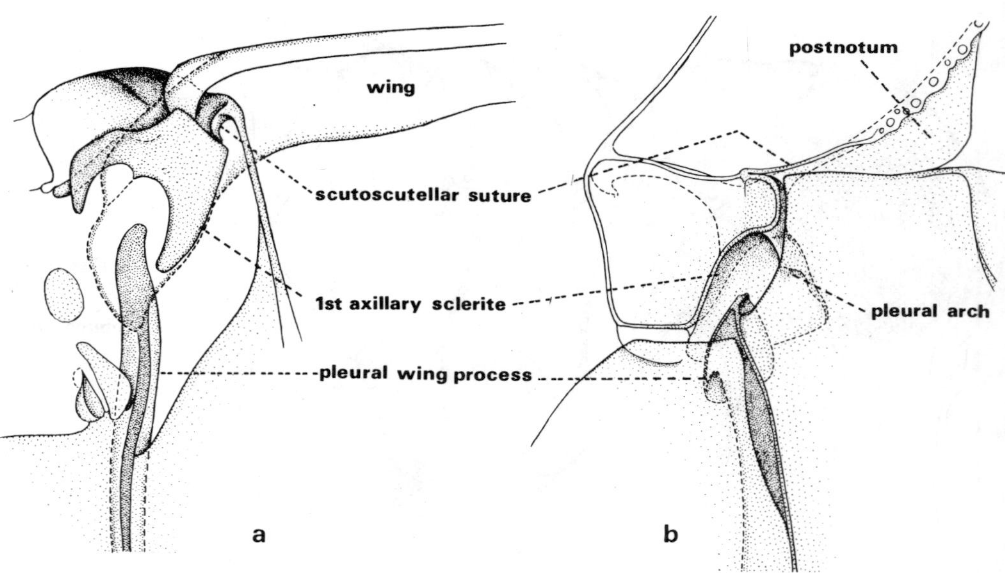

Figure 6. a) The first axillary sclerite and the pleural wing process of *Boreus.* b) The pleural arch of *Ceratophyllus styx.*

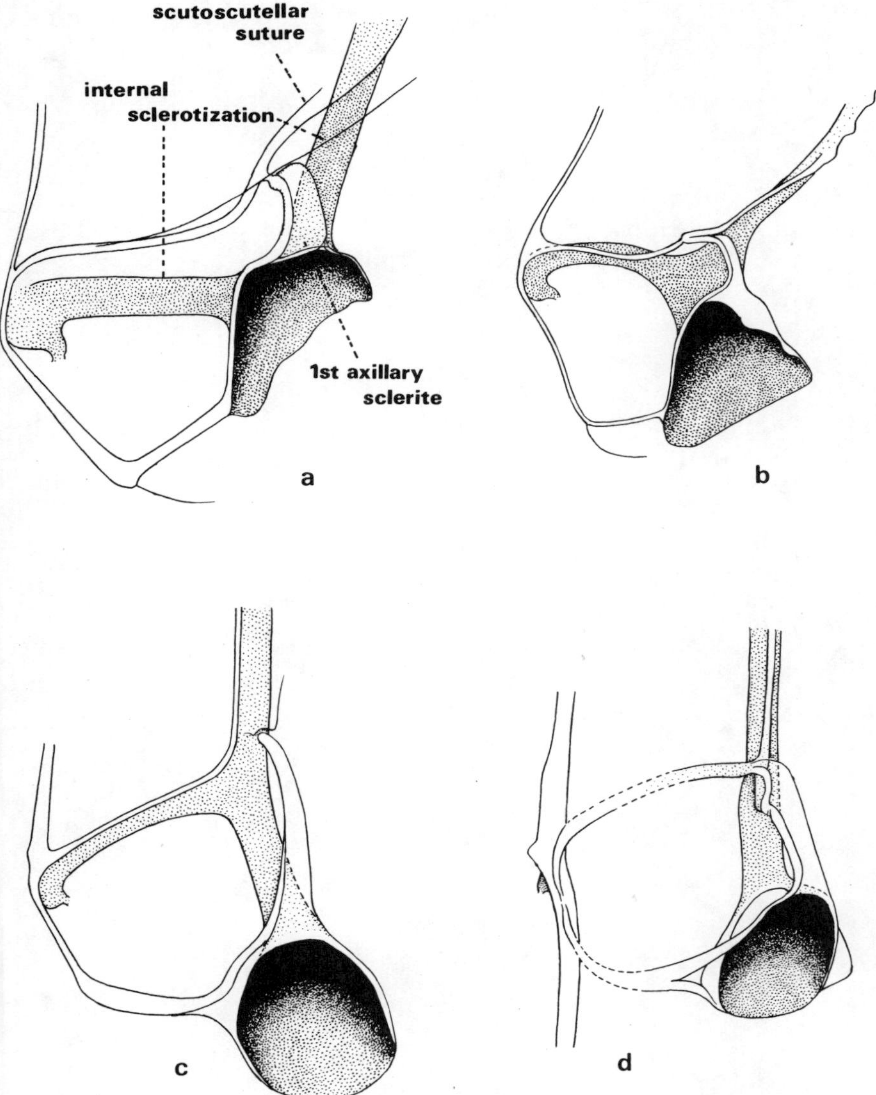

Figure 7. The external sutures and the internal sclerotization of the 'first axillary sclerite' (lateral metanotal area and pleural arch) of: a) *Hystrichopsylla* sp.; b) *Ceratophyllus styx;* c) *Xenopsylla cheopis;* d) *Pulex irritans.*

ECOLOGY AND FAUNISTICS

HEINZ E. KRAMPITZ
Universität München, Germany

HOST PREFERENCE, SESSILITY AND MATING BEHAVIOUR OF *LEPTOPSYLLA SEGNIS* REARED IN CAPTIVITY

ABSTRACT

Nine species of rodents (including two commensal and two introduced species) were compared regarding their suitability for rearing the house-mouse flea, *Leptopsylla segnis* under controlled conditions, and for observing its behaviour on the host. Murine rodents were found to be superior to voles or hamsters (Cricetidae) as hosts for this purpose. The degree of successs was correlated with host phylogeny – if one species was suitable (or unfavourable) for rearing this species, so were its relatives. Ambient room temperature and humidity provided the best conditions for the culture of *L.segnis.* This species regularly acted as a semi-sessile flea regardless of whether or not it was on a suitable host – provided the rodent had hairs to which *L.segnis* could attach. The rate of survival, and of reproduction, was as high in *L.segnis,* when on a favourable host, as in the other species, indicating that the phenomenon of 'sessility' did not render the flea more susceptible to the grooming activities of the host. Copulation and ovulation occurred while *L.segnis* was attached to the host, but the eggs are forcibly expelled so that they do not adhere to the hairs of the rodent, and, instead, fall onto the nest litter and remain attached there. A defaecation ritual, whereby the fleas detached from a site on the head and moved to a common locus on the back, was occasionally noted on laboratory mice. On nude hosts, *L.segnis* fared poorly even though copulation occurred, whereas *Nosopsyllus fasciatus* and *Xenopsylla cheopis,* which are active, not semi-sedentary, fur fleas, thrived.

1 INTRODUCTION

Despite the rapid and sustained growth in the numbers of species of fleas known to Science, the bionomics and habits of only a few have been intensively studied. Holland (1964) commented upon this dearth of information on the life history and ecology of fleas, and pointed out that such detailed knowledge could contribute to our understanding of the systematics of the Order. In an effort to obtain critical data on the behaviour of fleas on the host, on their reproduction and regarding host-specificity, we have been rearing and observing several species of rodent fleas under standardized laboratory conditions. This article presents a summary of our findings, primarily with *Leptopsylla segnis* (a parasite of the house mouse, *Mus musculus,* or commensal *Rattus*), which has been introduced into many parts of the world, and which is common in the Munich area. Some of the results are compared with those observed for certain other species of fleas. In an earlier paper (Krampitz, 1965), it was observed that: 1) All tested strains of *Mus musculus* proved to be suitable for rearing *L. segnis.* 2) The flea is semi-sedentary in habit, remaining fixed in position for long periods instead of leaving the mouse, when engorged, in order to rest or copulate or oviposit in the nest material (thereby differing from many species of fleas previously observed in the laboratory). 3) The numbers of *L. segnis* bred in the colony depended to a great extent upon the reproductive phase of the host, e.g. a larger population was obtained when the parasite fed upon families of *Mus* rather than upon selected male, female or castrated mice of the same size.

The intimate association of *L. segnis* with *M. musculus* raised the question as to whether this species could thrive in the absence of what appeared to be its natural host, particularly since field data suggested that such might be the case. Accordingly, detailed investigations were undertaken, under varying but controlled environmental conditions, on such factors as experimental host range and the degree and stability of the phenomenon of attachment to the host (the semi-sedentary condition of 'sessility').

2 MATERIALS AND METHODS

The colonies of *Nosopsyllus fasciatus, Megabothris turbidus* and *Ctenophthalmus agyrtes,* used in these tests for comparison with *L. segnis,* were also derived from fleas collected in the vicinity of Munich. The *X. cheopis* were from a long-established culture, maintained on laboratory rats and golden hamsters. Experience has shown that for best results in the rearing of fleas, the hosts must be changed regularly, e.g. the pairs of *Mus* used for *L. segnis* were replaced monthly.

The containers employed were made of glass, with dimensions 23 x 23 x 28 cm. The substrate for rearing the larvae consisted of a mixture of 1 000 g of sand, 100 g sawdust, 300 g pulverized dried blood, 30 g of finely ground laboratory food-pellets, and 1.5 g yeast. For bedding of the host, only wads of soft, crepe-like paper was provided. In a series of tests, pairs of laboratory-born and reared rodents, representing nine species as listed in table 1, were used as hosts. The cultures were maintained in rooms at constant temperature and humidity, at levels indicated in the tables presented.

Each pair of rodents was exposed to 50 newly emerged, hungry fleas for each species under test. The cultures were examined four weeks later to determine the number of fleas present on the hosts and in the litter. The former was done by direct observation and collection after the mice were killed – and the fleas then placed on new hosts. The litter was sifted and examined for adults, rechecked after a ten day lapse, and then discarded. Larvae and pupae were not counted. The experiments were repeated thrice, and for each the data were expressed in terms of the total number of fleas observed minus the 50 originally placed on the hosts (Krampitz, 1974). When studying attachment-behaviour of *L. segnis,* individual hosts were maintained and examined at regular intervals.

3 RESULTS

The results obtained after exposing pairs of rodents of nine species under three sets of conditions (as indicated by A, B and C) are shown in table 1.

Table 1. Numbers of *Leptopsylla segnis* found on pairs of nine species of hosts under indicated three different sets of environmental conditions for a period of 28 to 38 days after exposure to 50 fleas

Host species	Reproduction indices			
	A 16±2°C/RH60±10%	B 24±2°C/RH50±10%	C 28±1°C/RH70±10%	Σ
Rattus norvegicus (dom.)	− 45	+ 11	− 48	− 82
Mus musculus (dom.)	+349	+303	+ 58	+710
Apodemus sylvaticus	+385	+327	+ 87	+799
Apodemus agrarius	− 46	+392	+212	+558
Mastomys natalensis	− 19	− 12	+459	+428
Σ	+624	+1 021	+768	+2 413
Microtus arvalis	− 50	+ 27	− 42	− 65
Microtus agrestis	− 50	− 41	− 49	−140
Clethrionomys glareolus	− 46	− 49	− 50	−145
Mesocricetus auratus	+ 44	+138	− 47	+135
Σ	−102	+ 75	−188	−215

(dom.) = domestic strains

It is immediately apparent that the various murids (the upper five rodents listed) are far better hosts for rearing *L. segnis* than are the cricetids tested (the cricetine *Mesocricetus* and the microtine voles *Microtus* and *Clethrionomys*). Thus, no reproduction seems to have occurred in the fleas fed on voles, and limited success was attained with hamsters under conditions A and B. Of the murids, *R. norvegicus* was clearly inferior to *Mus, Apodemus* and *Mastomys,* although good results were attained only at 28°C and a relative humidity of 70 per cent. The greatest success was achieved with *A. sylvaticus.* The degree of acceptability of the various hosts, as indicated by this method, was associated with the affinities of the rodent; if one species served well, or poorly, so did its close relatives. It would appear that for sympatric species, one acceptable host could readily replace the other.

The data indicate that room temperature, i.e. about 21-24°C, is the most suitable of those tested regarding reproduction of *L. segnis* but the species is remarkably tolerant of extremes, provided a favourable host is available. Since *L. segnis* is frequently common on commensal *Rattus* in countries like Ethiopia (R. Traub), the failure of *L. segnis* to thrive on this host in our tests may be due to the experimental conditions employed (e.g. a small container). This criticism may also apply to the data on the hamster, but off-hand it seems that *Mesocricetus* is not suitable for propagation of *L. segnis.*

It is noteworthy that on all of these hosts, *L. segnis* retained its typical semi-sedentary habit, even when it suffered a shortened life span under clearly unfavourable conditions. In order to express the sessility phenomenon in a quantitative manner, and to evaluate its biological significance, the behaviour of five species of fleas on *A. sylvaticus* at room temperature was observed and compared. That host was selected, on the basis of previous experience, as the best available for the spectrum of fleas listed in table 2. The numbers of fleas found in the ground-litter and on the coat of the host on 1, 3 and 7 days after exposure to the rodent, are shown in table 2.

We have never observed an adult *L. segnis* elect a resting place in the litter after feeding,

Table 2. Percentage of fleas (five species) found *on* the body of the host (solid bar) as compared to that for fleas *off* the host (striped bar), 1, 3 and 7 days after exposure.
Host = *Apodemus sylvaticus.* I = Day 1; II = Day 3 and III = Day 7.

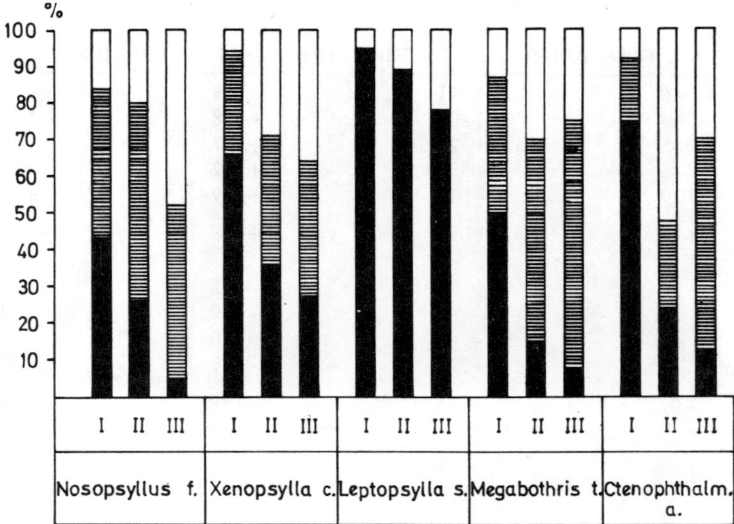

all remaining on the pelt of the host, and even copulating while in the sessile position. A species of flea in which both sexes constantly remain on the host, forgoes the apparent advantages of the mobile fur flea which is exposed, only for relatively short periods of time, to the danger of being killed by the host, and which copulates and lays its eggs in the security of the nest of the rodent. In the case of *L.segnis,* the eggs are forcibly expelled and thus fall free of the hairs, and since they are covered with an adhesive substance, the eggs readily adhere to the debris in the nest, where the larvae easily find appropriate conditions for development.

L.segnis may attach to any part of the host's body for feeding or resting, but in general they were found on the head (fig. 1), particularly when the population is high. A curious observation made at times on heavily infested white mice was that the fleas detached from the head and moved to a temporary site which is not normally selected for feeding, and there they defaecated in common, producing a conspicuous patch of specks of faeces on the hairs. The fleas then returned to their usual site of attachment. The host did not seem disturbed by the procedure, although it may be significant that the phenomenon was never observed twice on the same mouse.

The other species of fleas under observation lack specific resting places and shifted about on the host, without any discernible pattern. The rate of attrition due to the grooming action of the host was materially greater in these active fur fleas than in *L.segnis,* whose semi-sedentary habit undoubtedly has survival value for the flea.

In order to further explore the question of the biological significance of such long-term attachment, experiments were carried out utilizing hairless laboratory mice (heterozygotic strain). Such hosts are satisfactory for *L.segnis* nutritionally, but their nude condition prohibits attachment by the flea, and survival was severely affected in *L.segnis,* as can be seen by comparing tables 3 and 4. The results for each of the three designated

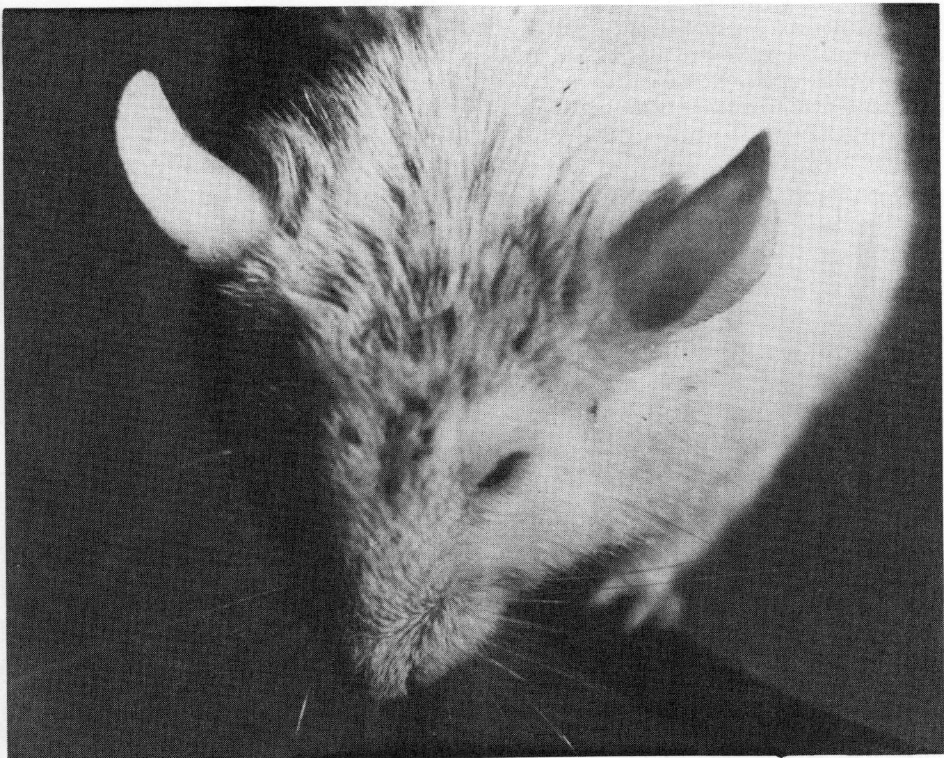

Figure 1. Patch of *L.segnis* fleas attached to the head of a laboratory mouse.

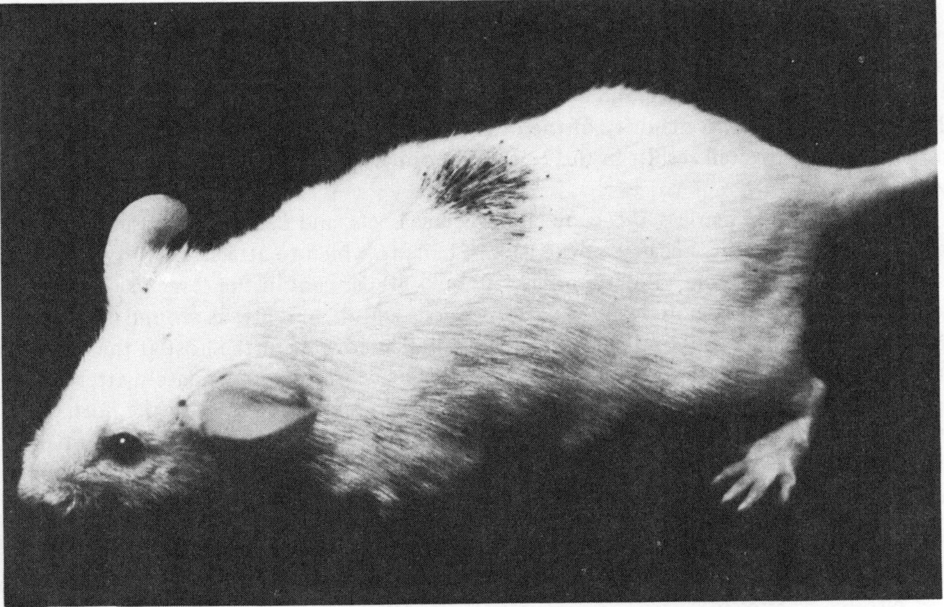

Figure 2. Faeces of *L.segnis* fleas in special defaecation site on back of laboratory mouse.

Table 3. Distribution and survival of *L.segnis* (A), *X.cheopis* (B) and *N.fasciatus* (C) exposed to normal hairy Swiss mice for varying periods of time, up to 14 days.
Solid bar = percentage of fleas found on the fur.
Striped bar = adult fleas found in the nest litter.

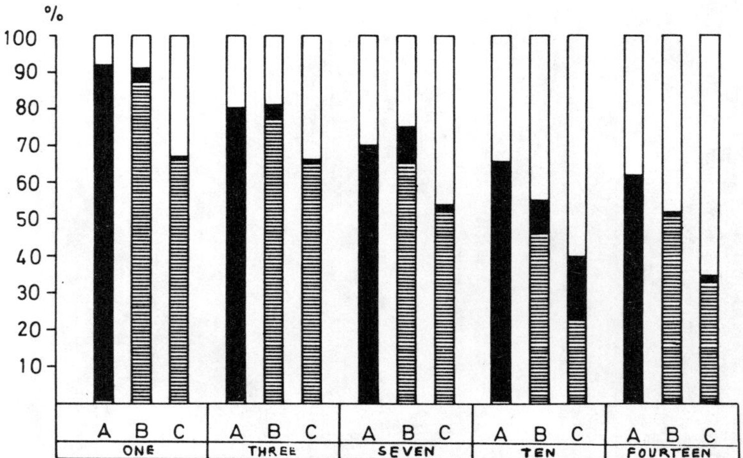

species of fleas recovered from normal (hairy) adult laboratory mice after varying periods of exposure, are shown in table 3. Inasmuch as the maximum interval was 14 days, and the life cycle (from egg to adult) generally takes that long, even in these proven, eminently suitable larval media and ambient conditions, the figures probably represent only the survivors of the 50 fleas originally placed on the mice. In each instance, the number of fleas recovered dropped appreciably between each check in the two-week period. In virtually every test, the recovery-rate (and hence, presumed survival-rate) in *L.segnis* surpassed that for *X.cheopis* and *N.fasciatus,* and essentially all the *L.segnis* were attached to the host, when collected. The vast majority of the other species of fleas were found in the litter, not on the mice. The overall results in this test with normal *Mus* resembled those for *Apodemus* (table 2).

In contrast, when hairless *Mus* were employed as hosts, and *L.segnis* could not assume its semi-sedentary habits because of the lack of hairs to which to attach, the survival rate markedly decreased (table 4). Within one day, only 50 per cent of the *L.segnis* could be recovered, and the rate soon dropped to 25 per cent and then as low as around 10 per cent, while at the maximum only 10 per cent of those collected were at the host at the time. No larvae were found, even though pairing of the *L.segnis* was noted immediately after feeding. The fleas in copula moved actively to escape the grooming activities of the host. The other two species of fleas survived as well, or better, on the nude *Mus* as on the normal one, although they definitely spent less time on the body of the mice, departing as soon as engorged.

4 DISCUSSION

Rothschild (1964) pointed out that in the superfamily Pulicoidea there is a tendency to

Table 4. Distribution and survival of *L.segnis* (A), *X.cheopis* (B) and *N.fasciatus* (C) exposed to hairless *Mus* for varying periods of up to 14 days.
Solid bar = percentage of fleas found on the body of the host.
Striped bar = the percentage collected in the nest litter.

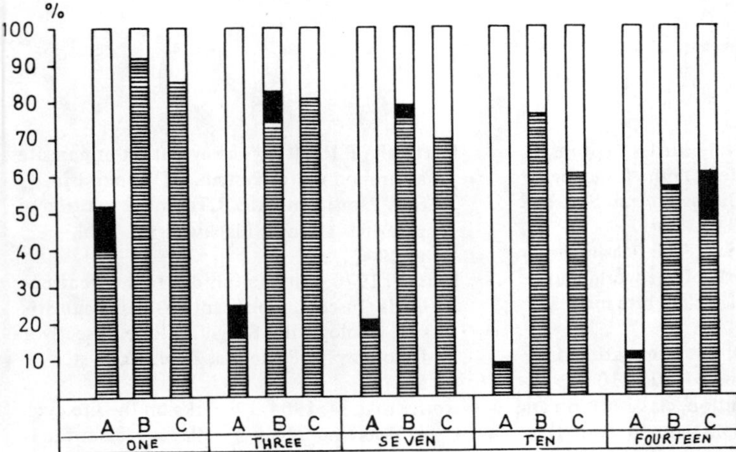

become attached to the body of the host by means of anchoring mouthparts. In most of the other higher taxa, the fleas are mobile. However, fleas like *L.segnis* (Ceratophylloidea) exhibit an intermediate stage between those two life styles, remaining, for long periods, in a semi-sedentary position by means of its modified bristles, but readily able to detach therefrom. The type and caliber of observations so well made by Geigy & Suter (1964) on the techniques of copulation and ovulation are needed for fleas like *Leptopsylla,* for no doubt there are other species which are semi-sessile. Ioff (1942) differentiated between nest fleas and body fleas, but believed that the relative distinctions were difficult to express in quantitative terms. Peus (1970) also emphasized the 'approximate accuracy of such differentiation'. Mattingly (1965) regarded the 'relative sessility' of fleas as a secondary adaptation. Traub (1972) treated various categories of nest fleas, as well as other adaptive modifications of Siphonaptera. Elsewhere in this present volume, Traub also discusses the semi-sedentary behaviour of *L.segnis* and other fleas which possess what he terms a 'crown of thorns', i.e. a frontal row of spiniform bristles or a tiara of spines on the head, which presumably serve as a temporary means of attachment.

Our research on *L.segnis* is an example of the experimental approach to the question of behavioural patterns which are host-related. Peus (1970) suggested that the sessility of *L. segnis* reported by Krampitz (1965) might prove to be more apparent than real if more nests of *Mus* were examined for the presence of this species. The new data, herein reported, makes this seem unlikely, particularly since *L.segnis* has now been shown to behave similarly on a variety of murid rodents, and also since field data support the contention that this flea naturally parasitizes sundry murines (Chick & Martin, 1911). As indicated by Lewis (1974), *L.segnis* infests only synanthropic murines, and then only in restricted areas. Like its most important host, the house mouse, *L.segnis* is well adapted for residence in the drier parts of the world. Laboratory colonies do best at relatively low levels of humidity (table 1), and fare poorly if kept under mesic conditions. Nevertheless, the flea probably is not truly xerophilic, as Beaucournu (1973) has pointed out. Its known distribution and populations probably reflects man's activities, through the concomitant effects on the commensal host indoors, rather than external natural conditions.

Acknowledgement is made to R.Traub for revising this article because of linguistic difficulties.

REFERENCES

Beaucournu, J.C. 1973, Note sur les Leptopsyllidae (Siphonaptera) de la faune française (Ire partie): Répartition, Biologie. Ann. Soc. ent. Franc. N.S. 9:483-499.

Chick, M. & C.J.Martin 1911, The fleas common on rats in different parts of the world and the readiness with which they bite man. J. Hyg. 11:122-123.

Geigy, R. & P.Suter 1960, Zur Kopulation der Flöhe. Rev. Suisse Zool. 67:206-210.

Holland, G.P. 1964, Evolution, classification and host relationship of Siphonaptera. Ann. Rev. Ent. 9:123-146.

Ioff, I.G. 1942, Problems of the ecology of fleas in relation to their epidemiological importance. (In Russian) Pyatigorsk p. 1-116.

Krampitz, H.E. 1965, Beobachtungen an einer Laboratoriumszucht von *Leptopsylla segnis* Schönherr 1811 (Insecta: Siphonaptera). Z. Parasitenk. 26:197-214.

Krampitz, H.E. 1974, Reproduction studies on some common European fleas reared in captivity. Paper read on the III. Internat. Congr. Parasitol. München, Aug. 28th. (Proc. p. 880).

Lewis, R.E. 1974, Notes on the geographical distribution and host preferences in the order Siphonaptera. Part 5. Chimaeropsyllidae, Ischnopsyllidae, Leptopsyllidae and Macropsyllidae. J. med. Ent. 11:525-540.

Mattingly, P.F. 1965, The evolution of parasite-arthropod vector system. III. Symp. Brit. Soc. Parasit. In: A.E.R.Taylor, Evolution of parasites. Oxford, Blackwell Sci. Publ. p. 29-45.

Peus, F. 1970, Zur Kentnis der Flöhe Deutschlands (Insecta, Siphonaptera). III. Faunistik und Ökologie der Säugetierflöhe, Insectivora, Lagomorpha, Rodentia. Zool. Jb. Syst. p. 1-54.

Rothschild, M. 1964, Remarks on the life cycles of fleas (Siphonaptera). Paper read at the I. Internat. Congr. Parasitol. Rome (Proc. I, 29/1966).

Rothschild, M. 1975, Recent advances in our knowledge of the Order Siphonaptera. Ann. Rev. Ent. 20:241-259.

Sikes, E.K. 1931, Notes on breeding fleas, with reference to humidity and feeding. Parasitology 23:243-423.

Traub, R. 1972, Notes on zoogeography, convergent evolution and taxonomy of fleas (Siphonaptera), based on collections from Gunong Benom and elsewhere in South-east Asia. II. Convergent evolution. Bull. Br. Mus. nat. Hist. (Zool.) 23(10):307-387.

Traub, R. 1980, Some adaptive modifications in fleas. Proc. Int. Conf. Fleas, Ashton, England, June 1977. p. 33-68.

ISTVAN SZABÓ
Hungarian Natural History Museum, Budapest, Hungary

THE BORDERS OF THE DISTRIBUTION OF CERTAIN SUBSPECIES OF FLEAS IN HUNGARY

ABSTRACT

Certain European subspecies of fleas have their confines in Hungary, and where two or more subspecies meet, intermediate forms may be observed. The geographical distribution of the following subspecies in the Hungarian part of the Carpathian Basin is discussed on the basis of 20 years' intensive collecting: *Palaeopsylla soricis rosickyi* Smit, and *P. s. starki* Wagner; *Doratopsylla d. dasycnema* (Rothschild) and *D. d. cuspis* Rothschild; *Ctenophthalmus a. agyrtes* (Heller), *C. a. bosnicus* Wagner, *C. a. peusianus* Rosicky, *C. a. eurous* Jordan & Rothschild and *C. a. kleinschmidtianus* Peus; *Ctenophthalmus as. assimilis* (Taschenberg) and *C. as. erectus* Smit & Rosicky; *Hystrichopsylla t. talpae* (Curtis) and *H. t. orientalis* Smit.

In 1960, when I decided to study the Siphonapteran fauna of Hungary, there were only about 20 species of fleas known to occur in that country, and precise data on hosts and localities were lacking for some of them. The paucity of species and inadequacies in the records clearly suggested that knowledge of the Hungarian fauna was half a century behind that pertaining to most European countries. On the basis of what was known regarding the Siphonaptera in the neighbouring areas, I anticipated that the Hungarian fauna would total about 75 species and subspecies, and my faunal opus (Szabó, 1975) lists 58 such taxa. More recent observations add an additional two species to the list for Hungary (Szabó, unpublished).

The distributional limits of subspecies, particularly where two such forms meet, are of zoogeographical significance and scientific interest, as indicated by Jordan (1938), who pointed out that two subspecies may be expected to hybridize and form hybrid populations in districts of contact. Jordan cited some subspecies of *Ctenophthalmus agyrtes* (Heller) as examples where such intermediate forms occurred, but also stressed that in other instances two subspecies of *C. agyrtes* may be found in the same geographical area and yet be highly restricted therein to specific biotopes, without the formation of intergrade populations. Johnson & Traub (1954) also cited examples of intermediate forms in areas where two subspecies met, e.g. in *Peromyscopsylla hesperomys* (Baker). New data on the range of subspecies contribute to our understanding of the Siphonapteran faunistics of the Carpathian Basin and are treated in this paper.

The useful map on the distribution of *Palaeopsylla soricis* (Dale) published by Smit (1960b) indicates at a glance that no records had been reported for Hungary, as shown in fig. 1. Today, hundreds of records are available, and these show that in our country the most abundant soricid flea is *P. soricis rosickyi* Smit. The occurrence there of the subspecies *P. soricis starki* Wagner, in the northeastern region, was first demonstrated in 1974 (Szabó, 1975). At the interface of the distribution of these two subspecies (fig. 2), intergrades or presumed hybrids have been collected. The available data are adequate to demonstrate that the distribution of these subspecies in Hungary is limited to the northeastern region.

The range of the subspecies of *Doratopsylla dasycnema* (Rothschild) is more difficult to determine. Smit (1960a), in the absence of Hungarian records, employed data from

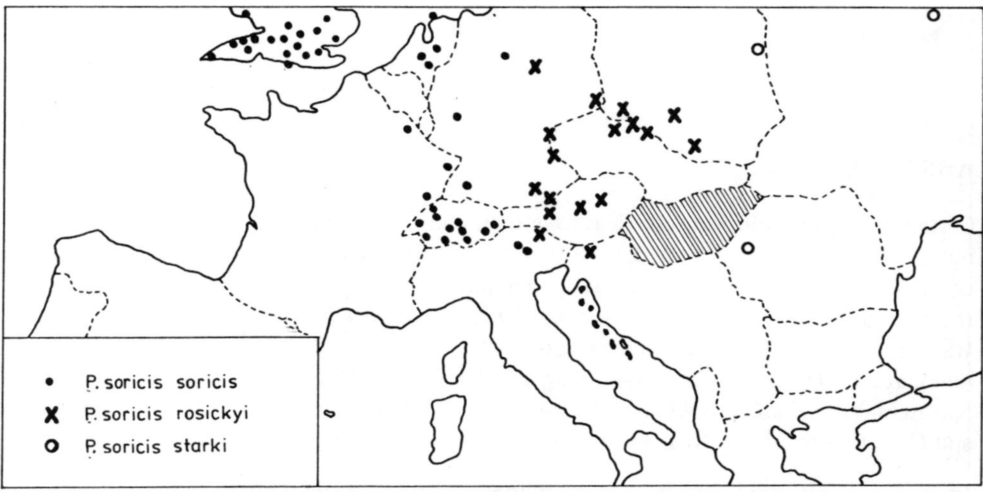

Figure 1. The distribution of the subspecies of *Palaeopsylla soricis* as shown by Smit (1960b).

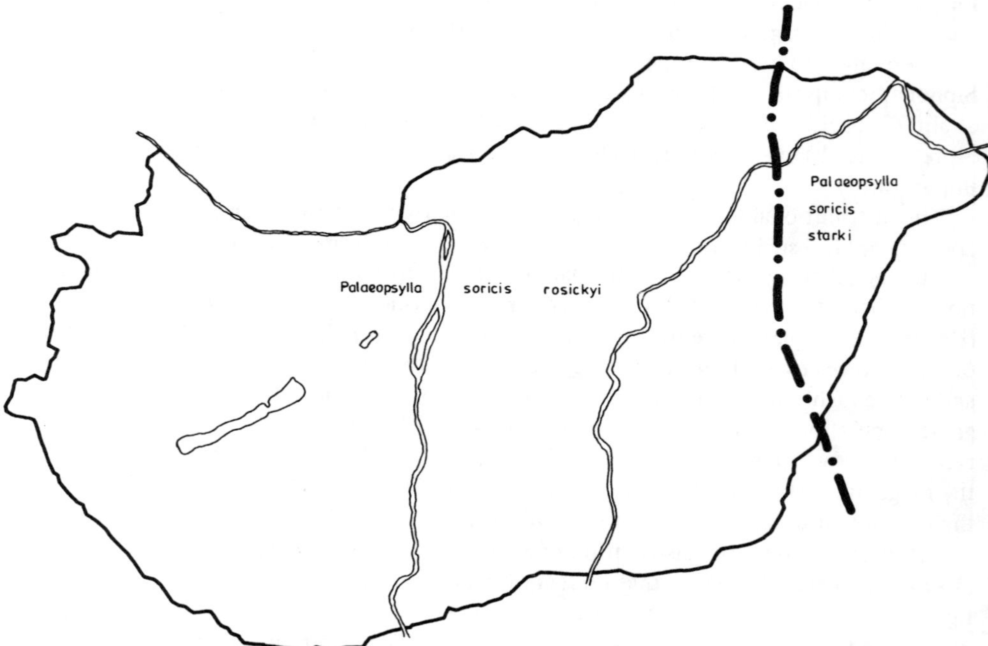

Figure 2. Distribution of *Palaeopsylla soricis rosickyi* and *P. soricis starki* in Hungary.

neighbouring countries to indicate where *D.d.dasycnema* (Rothschild) and *D.d.cuspis* Rothschild might meet (fig. 3). The validity of this assumption has been substantiated by my collections of *D.d.dasycnema* in western and northern Hungary. The taxon is restricted to hilly and mountainous regions and hence its eastern limits cannot be determined in view of the low, flat nature of southeastern Hungary. The only existing records for *D.d.cuspis* are all to the east of our country or southwest thereof, and it is possible that this subspecies exists in the southernmost parts of Hungary or in the eastern regions of the North Hungarian mountain range, which are areas where Siphonapteran data from soricids are scanty. Until such definitive clarification is obtained, it seems advisable to exclude *D.d. cuspis* from our faunal lists.

The most common fleas found on small mammals in Hungary are the various subspecies of *Ctenophthalmus agyrtes.* No host specificity has been noted, despite my access to thousands of specimens from hundreds of such localities. The distribution of the subspecies of *C.agyrtes* in Hungary was reported by Smit & Szabó (1967), who listed the occurrence there of no fewer than five subspecies. Of these, the smallest range was noted for *C.a. agyrtes* (Heller) which is found in the northwestern part of this country, as shown in fig. 4. In marked contrast is *C.a.bosnicus* Wagner, which, in Hungary, occurs throughout the whole of Transdanubia and in part of the Danube-Tisza Mid-Region. *C.a.peusianus* Rosicky inhabits the greater portion of the North Hungarian Mountain Range. Only *C.a. kleinschmidtianus* Peus is found east of this region, along the upper reaches of the Tisza River and on the flatlands of Bereg. *C.a.eurous* Jordan & Rothschild is represented by a few specimens from the southeastern region of Hungary.

At the borders of the distribution of the various subspecies, intermediate forms have been collected which display characteristics of two, or even three, of the forms concerned.

The limits of the subspecies of *Ctenophthalmus assimilis* (Taschenberg) have not yet been determined for Hungary. Both *C.as.assimilis* (Taschenberg) and *C.as.erectus* Smit & Rosicky are fairly specific fleas of *Microtus arvalis* (Smit & Rosicky, 1965) and have been collected together in various localities in our country. It does seem clear that *C.as.erectus* is known only from southeastern Europe, and that it occurs in northern Hungary.

Although *Hystrichopsylla t.talpae* (Curtis) has been collected in the Bratislava, Czechoslovakia area and in the environs of Vienna and Moosbrunn, Austria (Peus & Smit, 1957),

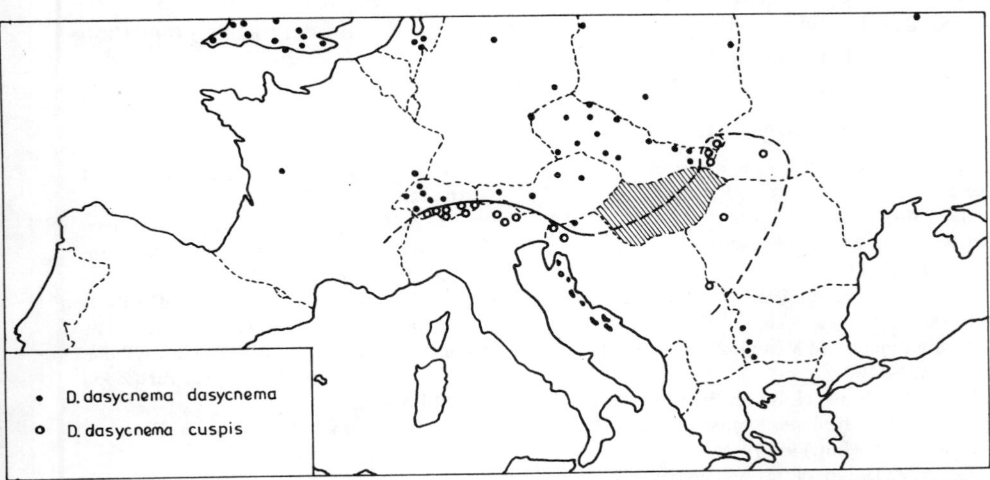

Figure 3. The distribution of the subspecies of *Doratopsylla dasycnema* as shown by Smit (1960a).

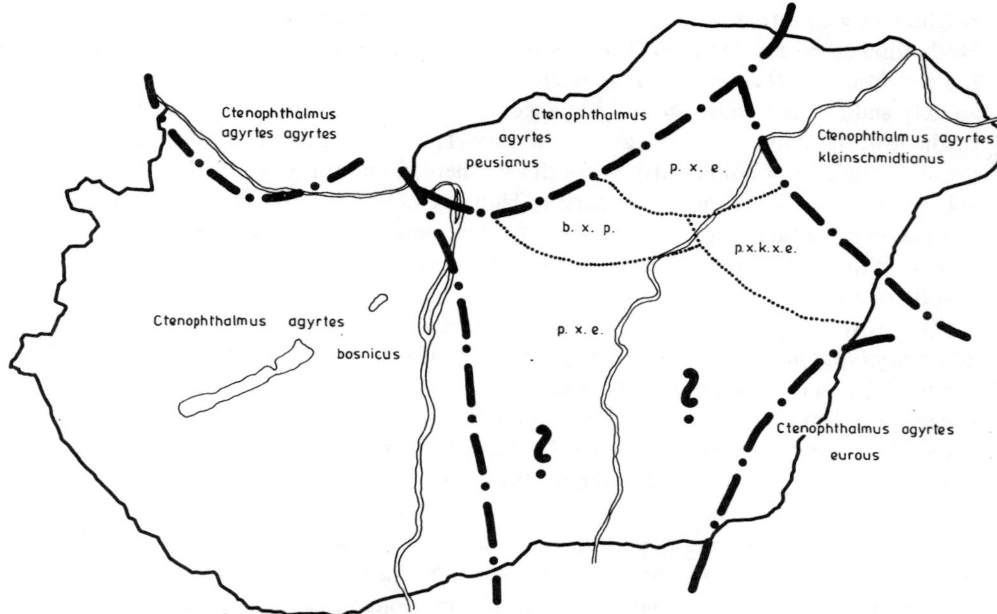

Figure 4. The distribution of the subspecies of *Ctenophthalmus agyrtes* in Hungary.
Intermediates: b.x.p. = *C.a.bosnicus* and *C.a.peusianus.*
p.x.e. = *C.a.peusianus* and *C.a.eurous.*
p.x.k.x.e. = *C.a.peusianus; C.a.kleinschmidtianus* and *C.a.eurous.*
? = additional data required for this area.

and thus has been reported merely a few kilometers from the borders of western Hungary, this subspecies has not yet been found in our country. Even in the Sopron area, very close to the borders, only another subspecies, *H.t.orientalis* Smit, was taken, and intermediate forms were not observed. Recently those two forms were promoted to species status by various authors.

These findings, then, add to our knowledge of the ranges of subspecies of fleas in Central Europe and provide additional examples of the formation of intergrading populations at the districts of contact by two or more subspecies.

REFERENCES

Johnson, P.T. & R.Traub 1954, Revision of the flea genus *Peromyscopsylla.* Smithson. misc. Coll. 123(4):1-68.

Jordan, K. 1938, Where subspecies meet. Novit. Zool. 41:103-111.

Peus, F. & F.G.A.M.Smit 1957, Über die Beiden Subspecies von *Hystrichopsylla talpae* Curtis (Ins., Siphonaptera). Mitteil. Zool. Mus. Berlin, 33:391-410.

Smit, F.G.A.M. 1960a, Notes on the shrew-flea *Doratopsylla dasycnema* Rothschild. Bull. Br. Mus. nat. Hist. (Ent.) 9:359-367.

Smit, F.G.A.M. 1960b, Notes on *Palaeopsylla,* a genus of Siphonaptera. Bull. Br. Mus. nat. Hist. (Ent.) 9:369-386.

Smit, F.G.A.M. & B.Rosicky 1965, Three new Siphonaptera from the Balkans. Acta Faunist. Ent. Mus. Nat. Prague 10:177-186.

Smit, F.G.A.M. & I.Szabó 1967, The distribution of subspecies of *Ctenophthalmus agyrtes* in Hungary (Siphonaptera: Hystrichopsyllidae). Ann. Hist.-Nat. Mus. Natl. Hung. Pars Zool. 59:345-351.

Szabó, I. 1975, Bolhák-Siphonaptera. Fauna Hungariae, 15, 18, p. 97.

J.-C. BEAUCOURNU
Faculté mixte de Médecine et de Pharmacie, Rennes, France

LES PUCES DU LAPIN DE GARENNE, *ORYCTOLAGUS CUNICULUS* (L.)

ABSTRACT

The Siphonaptera of the European rabbit, *Oryctolagus cuniculus* (L.).

The European rabbit *Oryctolagus cuniculus* is, since the last ice age, endemic in the Mediterranean occidental region (i.e. Iberian peninsula and southeastern France). Nevertheless, its origins are unknown. Palaeontologists believe in an Asiatic origin for lagomorphs, assuming that leporids differentiated principally in the Nearctic region. A study of the rabbit fleas in France confirms its dual origins, but the Asiatic 'cradle' is most evident.

Spilopsyllus cuniculi is clearly more nearly related to the Nearctic genus *Cediopsylla,* a parasite of *Sylvilagus.*

Xenopsylla cunicularis is related to fleas of gerbils from central Asia to north Africa. It is a phenomenon of host change which occurred before the glaciation (end of the Miocene to the early Pleistocene); this species is known from France, Spain and Morocco.

Odontopsyllus quirosi (with two subspecies) is known from France and Spain. There are only two other known described species of this genus, both Nearctic parasites of *Sylvilagus.*

Caenopsylla laptevi is a parasite of carnivores in central Asia, Asia Minor and Egypt, but parasitises the rabbit in France and Spain (subspecies *relicta*). Two other *Caenopsylla* are known from the north of Africa, and are morphologically further placed from the ancestral type of *Caenopsylla.* They parasitise primitive rodents, Ctenodactylidae. *Odontopsyllus* and *Caenopsylla* are allied to *Ctenophyllus,* essentially distributed in central Asia; this last genus parasitises ochotonid lagomorphs.

Les mammifères endémiques, surtout lorsqu'ils sont isolés sur le plan zoologique, sont toujours d'un intérêt extrême pour le parasitologiste.

Le banal lapin européen *Oryctolagus cuniculus* (L.), est l'un des éléments les plus intéressants, à ce titre, de la faune d'Europe occidentale et plus particulièrement de la France méridionale et de l'Europe du Sud-Ouest. Si, actuellement, ce caractère d'endémicité ne parait plus évident cela est dû aux multiples introductions qui, du fait de l'homme, ont transformé le lapin méditerranéen en animal cosmopolite.

Son aire primitive est inconnue. Pour Lavocat (in litt., 13.5.75) notre lapin semble s'être différencié en Eurasie à partir d'un stock de Leporidae primitifs nords-américains. Toutefois si de nombreux gisements, quelquefois mal datés, l'ont révélé en Europe, aucun ne nous le montre ni en Asie, ni en Afrique du Nord, les références le concernant pour ce

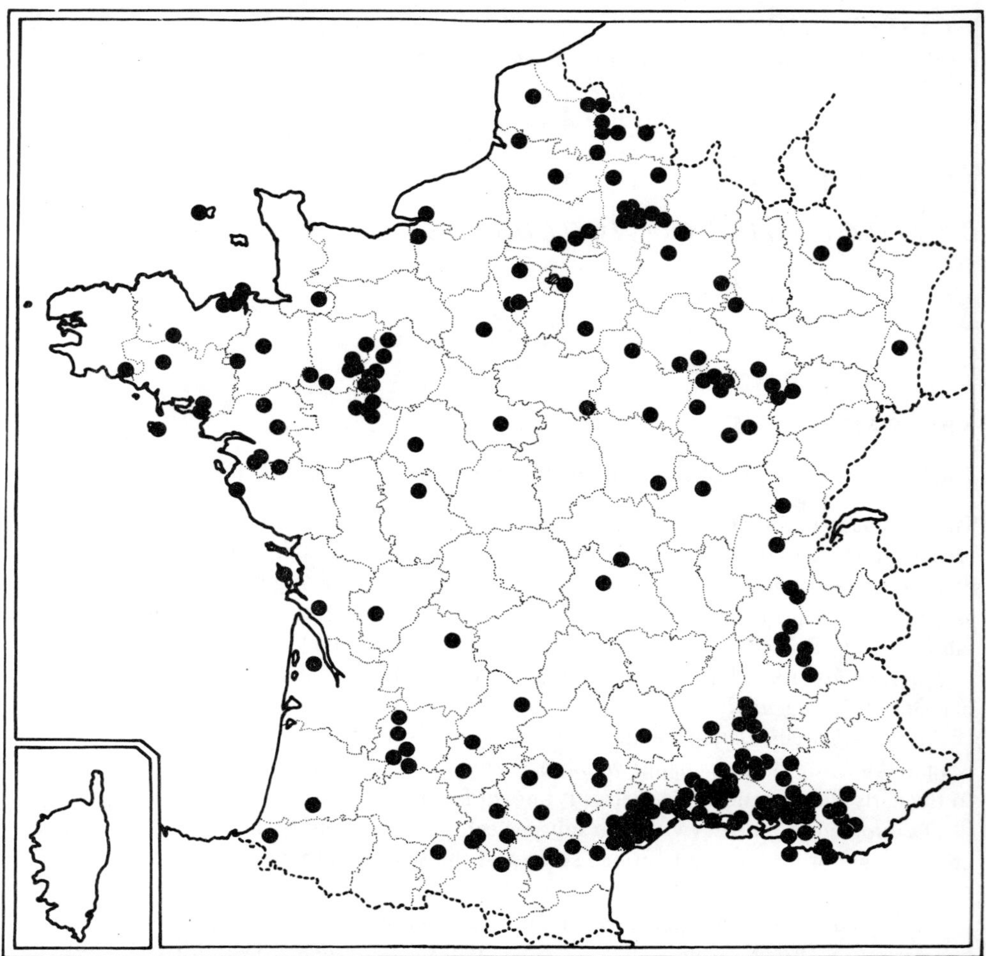

Carte 1. Captures de *Spilopsyllus cuniculi,* en France.

pays (Egypte, Maroc) se rapportant apparemment à *Lepus crawshayi. O.cuniculus* semble apparaître en Europe au Villafranchien (Pleistocène inférieur) et y avoir eu une assez vaste répartition. Très vite les glaciations vont le repousser vers le sud.

Son aire, pendant et immédiatement après la période glaciaire, est limitée à l'extrêmité Ouest du Bassin méditerranéen: Péninsule ibérique à coup sûr, pour certains l'Afrique du Nord (ce qui ne s'appuie à l'heure actuelle sur aucune base, paléontologique ou autre); pour nous, une partie de la Provence, c'est-à-dire le Sud Est méditerranéen français situé entre Rhône et Alpes, fait partie de cette aire relictuelle: ceci nous semble attesté par l'existence dans cette seule région de deux Amphipsyllinae parasites du lapin, l'une correspondant à une sous-éspèce endémique. Ce n'est qu'avec le début de la période historique (avec les Phéniciens en particulier) mais surtout à partir du XVIIIe siècle que le lapin a recommencé à accroître son aire presque toujours avec l'aide de l'Homme.

De tous les Leporidae, *O.cuniculus* est celui qui présente la faune la plus riche en

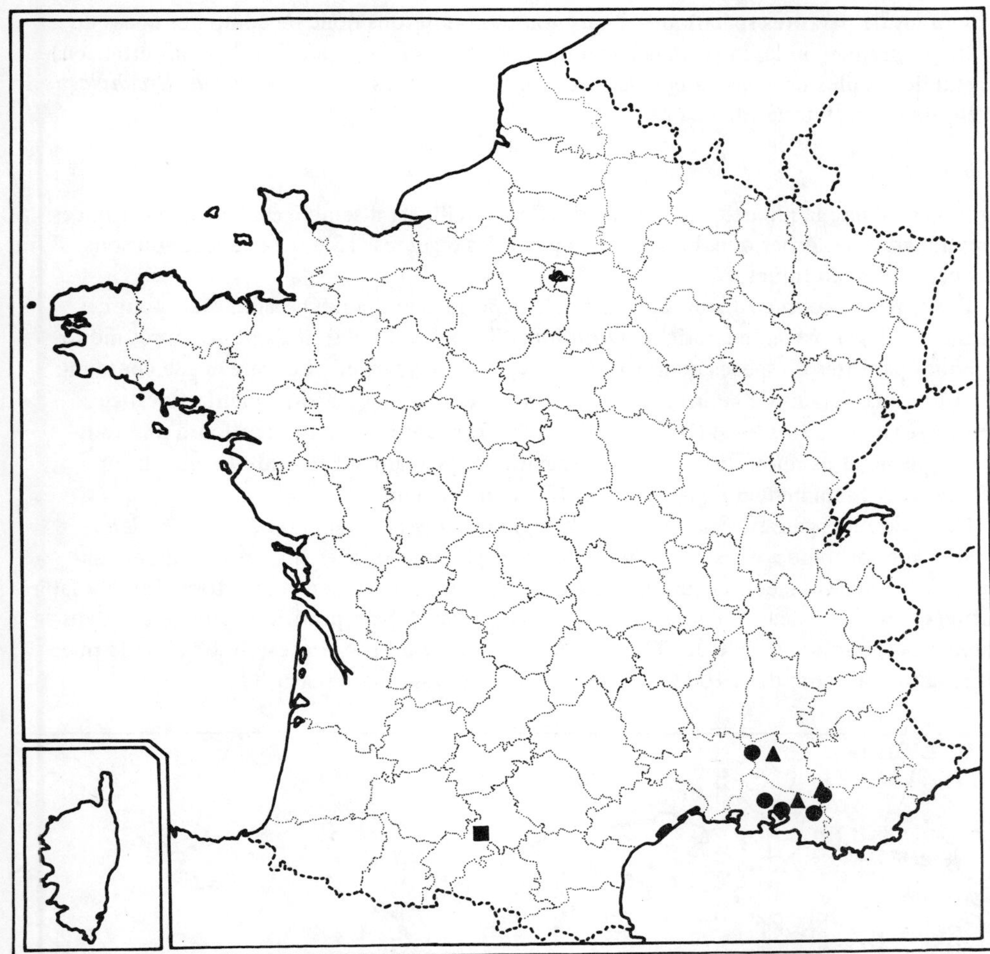

Carte 2. Captures, en France, de *Xenopsylla cunicularis* (■), *Odontopsyllus quirosi episcopalis* (●) et *Caenopsylla laptevi relicta* (▲).

Siphonaptères spécifiques, tous se rencontrant dans le Sud de la France et en Espagne, c'est-à-dire dans les régions où l'endémisme du lapin est le plus ancien (cartes 1 et 2): 2 Pulicidae: *Spilopsyllus cuniculi* (Dale, 1878), *Xenopsylla cunicularis* Smit, 1957; 3 *Leptopsyllidae: Odontopsyllus q.quirosi* (Gil Collado, 1934), *O.quirosi episcopalis* Beaucournu & Gilot, 1974, *Caenopsylla laptevi relicta* Beaucournu, Gil Collado & Gilot, 1975.

Nous pensons que l'étude de ces parasites peut aider à comprendre l'origine biogéographique de leur hôte.

Il serait intéressant, dans cette optique, de pouvoir comparer les puces d'*Oryctolagus* avec celles des autres 'lapins'. Il nous faut négliger, faute de documents suffisants, le cas des Archaeopsyllinae, parasites sténoxènes (*Nesolagobius callosus* Jordan & Rothschild, 1922) ou euryxènes (*Ctenocephalides damarensis* Jordan, 1936, *C.felis strongylus* (Jordan, 1925)) de leporidés en Asie (Sumatra) et en Afrique. *Caprolagus* et *Pentalagus,* 'lapins' connus respectivement de la sous-région indienne et des îles Ryou-Kyou, semblent eux dépourvus de parasites spécifiques. Nous nous contenterons donc de comparer le cas du lapin de garenne, ou lapin européen (qu'il serait plus exact de qualifier de méditerranéen), à celui de ses plus proches parents les 'Cotton-tails' (soit les genres *Sylvilagus, Brachylagus* et *Romerolagus*), tous américains.

Spilopsyllinae

Si l'on considère la répartition actuelle des Spilopsyllinae, il semble évident que ces puces sont apparues en Amérique du Nord puisque 4 des 6 genres, 15 des 18 espèces ou sous-espèces sont néarctiques.

Des 6 genres connus, 4 sont inféodés aux Leporidae et 2 aux Oiseaux, mais pour ces derniers (*Actenopsylla,* néarctique; *Ornithopsylla,* paléarctique), il s'agit de genres monotypiques parasites d'espèces migratrices nichant dans les terriers. Les oiseaux se contaminèrent-ils aux dépens de lapins fouisseurs comme on l'admet classiquement, ou est-ce l'inverse? Quant à *Hoplopsyllus* et *Euhoplopsyllus* néarctiques à l'exception d'une sous-espèce, ils ne nous retiendront pas, leur spécificité les axant sur le genre *Lepus* ou sur *Romerolagus,* lapin endémique de la Cordillière mexicaine.

Des 2 derniers genres, *Cediopsylla* polytypique est exclusivement néarctique; *Spilopsyllus,* monotypique, est sténoxène du lapin de garenne et, avec lui, normalement cantonné à la partie occidentale du bassin méditerranéen. L'origine de cet ectoparasite du lapin européen est donc néarctique selon toute vraisemblance. Sa répartition actuelle est, dans la région paléarctique, celle de l'*O.cuniculus* Sauvage. En France, c'est de très loin la puce dominante, présente dans 100 % de nos prélèvements sur lapin (carte 1).

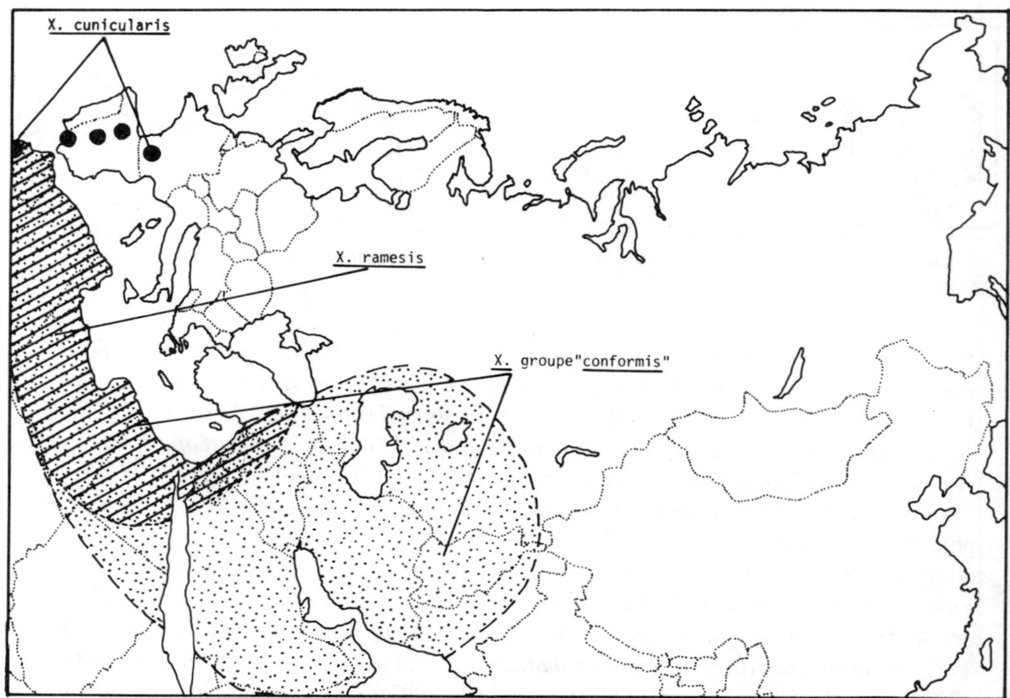

Carte 3. Répartition connue des *Xenopsylla* du groupe '*conformis*'.

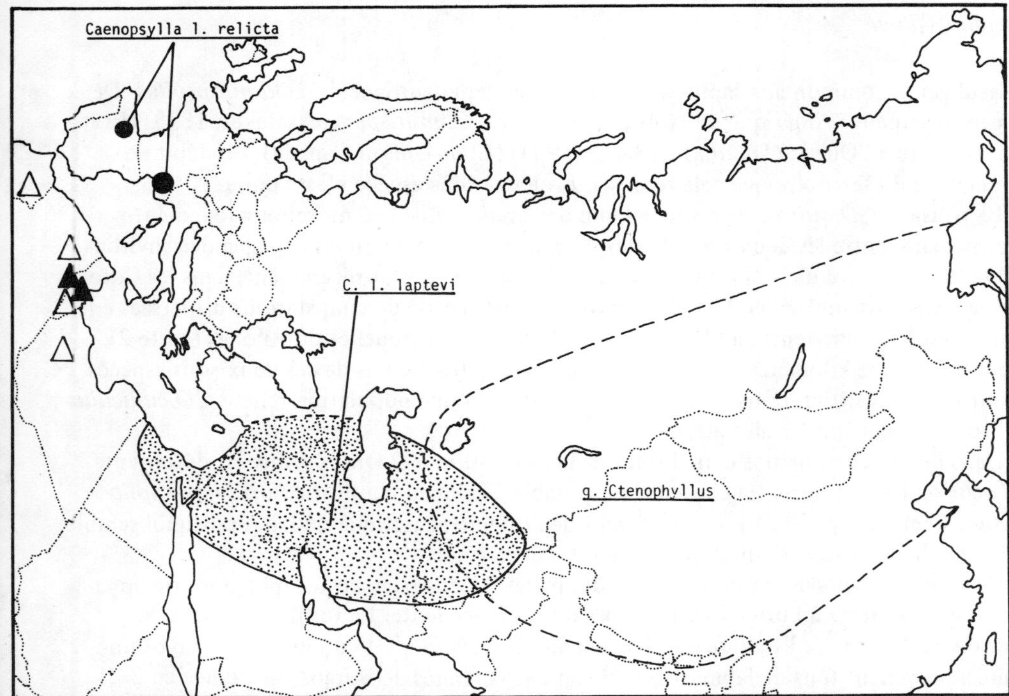

Carte 4. Répartition connue des genres *Caenopsylla* (△: *C.mira;* ▲: *C.assimulata*) et *Ctenophyllus.*

Actuellement aucune donnée ne permet de savoir si l'implantation de cette puce en Europe s'est faite, à l'Eocène par le 'Pont islandais' ou plus tard par le détroit de Behring, et avec quel hôte.

Xenopsyllinae

X.cunicularis fait partie d'un groupe d'espèces paléarctique (group *'conformis'*) répandu d'Asie Centrale au Maroc (carte 3). Elle appartient à une genre inféodé aux rongeurs et est particulièrement proche de *X.ramesis* (Rothschild, 1904), parasite de Gerbillinae, dont elle est sympatrique au Maroc et dont elle semble constituer un mutant écologique.

Dès 1934, Cartana et Gil Collado l'avait très vraisemblablement rencontré sur le continent européen (en Espagne près de Madrid), mais ce n'est qu'en 1966 qu'elle y est signalée avec certitude par Lane. Par la suite, nous l'identifions près de Burgos et, tout récemment, dans le Sud-Est de la France (carte 2).

Il est évident que le problème biogéographique posé par la présence de cette puce en Europe se pose différemment suivant que l'on admet, ou non, la présence du lapin au Maroc au Pléistocène. Dans la négative, qui semble hautement probable, il faut considérer ce parasite comme une relique villafranchienne.

Le lapin étant venue d'Asie en Europe a amené avec lui cette *Xenopsylla* de l'extrémité orientale du Bassin méditerranéen où le groupe *'conformis'*, et en particulier *X.ramesis* est présent. Cette opinion est confortée par le fait que les *X.cunicularis* d'Europe sont plus proches, morphologiquement, de *X.ramesis* que les *X.cunicularis* du Maroc.

Amphipsyllinae

Le seul genre commun aux lapins américains et au lapin européen est *Odontopsyllus.* Deux espèces occupent l'Amérique du Nord, l'une, à l'Est, *O.multispinosus* (Baker, 1898) (Etats-Unis), l'autre à l'Ouest, *O.dentatus* (Baker, 1904) (Etats-Unis et Canada). En dépit de l'immensité du territoire que cela représente aucune sous-espèce n'est connue.

La troisième, *O.quirosi,* parasite le lapin de garenne. Elle est, morphologiquement, intermédiaire entre les deux précédentes. *O.q.quirosi* à été décrit en Espagne des environs de Madrid, retrouvé dans la province voisine de Tolède et, plus récemment, dans les Coto Doana (M.Rothschild, in verbis); *O.q.episcopalis* est signalé de cinq stations dispersées en France, en Basse-Provence, à l'Est du Rhône (Vaucluse et Bouches-du-Rhône) (carte 2). Sept à huit cents kilomètres seulement séparent les gîtes connus de ces deux sous-espèces ce qui semble justifier, nous l'avons dit, l'hypothèse d'une population relicte d'*O.cuniculus* en Provence pendant les glaciations.

Dans l'état actuel de nos connaissances, la répartition d'*Odontopsyllus* est déroutante. On peut seulement noter qu'elle est superposable à celle du couple *Cediopsylla – Spilopsyllus.* Ceci n'est pas, toutefois, en faveur de l'origine néarctique du lapin puisqu'il semble avéré que le berceau des Amphipsyllinae est paléarctique.

Caenopsylla laptevi, dernière en date des puces découvertes sur le lapin, apporte apparemment davantage à l'origine de son hôte que la précédente. La forme nominale est parasite de *Vulpes,* de l'Egypte à la Turkménie; *C.l.relicta* n'est connue que par 6 exemplaires, provenant tous de lapins, capturés en Espagne près de Burgos, en France en 3 communes des départements du Vaucluse et des Bouches-du-Rhône (cartes 2 et 4). Près de 3.000 km séparent les aires connues des deux sous-espèces de *C.laptevi.* La discontinuité de cette aire s'accompagne de curieuses divergences morphologiques. Si les genitalia mâles sont très proches, dents génales, genitalia femelles et chétotaxie tarsale, en particulier, sont différents.

En Europe occidentale, *C.laptevi* n'est pas un parasite du renard. Nous avons étudié en France plus de 300 de ces animaux dont plusieurs dizaines dans la région où a été récolté *C.l.relicta:* aucun ne l'a livré. En contrepartie, *C.l.laptevi* n'est manifestement pas inféodé aux lagomorphes: si le lapin n'existe pas dans son aire, des lièvres y vivent: une seule capture provient de cet hôte. Aucun *Ochotona* n'a été trouvé porteur de cette puce.

Nous pouvons donc estimer acquis qu'à chaque sous-espèce correspond un tropisme parasitaire différent. Ceci ne peut s'expliquer que par une inféodation primitive de *C.laptevi* à *O.cuniculus* ou à son ancêtre direct, originaire, nous l'avons vu, d'Eurasie. Au moment des glaciations, l'aire de l'hôte, fouisseur et lié à la terre meuble, se restreint à la partie occidentale de l'Europe et finit par former deux ilôts résiduels, l'un en Provence, l'autre en Péninsule ibérique, séparés par la barrière pyrénéenne (nous rappelons que les exemplaires français et ibériques ne sont pas rigoureusement identiques). En Europe, elle se maintient en ces deux foyers où le climat est compatible avec l'écologie du lapin, foyers qui permettent parallèlement la subspéciation d'*O.quirosi;* en Asie, la puce ne survit qu'en s'adaptant au prédateur classique du lapin, le renard. Il faut, par ailleurs, souligner que *C. l.laptevi* est non seulement la plus proche du foyer originel des Amphipsyllinae, mais aussi, morphologiquement, la moins différenciée des 3 espèces du genre *Caenopsylla* (carte 4). On peut donc concevoir que si *Caenopsylla mira* Rothschild, 1909 et *C.assimulata* (Weiss, 1913), d'Afrique du Nord sont actuellement inféodées aux rongeurs primitifs Ctenodactylidae, ce n'est que par un phénomène de capture. Attribuer *Caenopsylla* aux Leporidae n'a rien d'arbitraire puisque le genre apparenté *Ctenophyllus* essentiellement distribué en Asie Centrale (carte 4) est inféodé aux Ochotonidae, tandis qu'*Odontopsyllus,* également proche de ces 2 genres, ne parasite que des 'lapins'. Des Ochotonidae viennent, d'ailleurs, d'être signalés dans le Miocène du Maroc.

On peut noter qu'une tique spécifique du lapin de garenne, *Haemaphysalis hispanica* Gil Collado, semble avoir la même origine géographique que *C.laptevi.* Elle n'est actuellement connue que d'Espagne et de Provence. Les espèces les plus voisines sont les *Haemaphysalis* du sous-groupe *asiaticus,* parasites de carnivores (en particulier *H.adleri* connu d'Iraq, du Liban, d'Israël et d'Oman), à l'exception de *H.caucasica* Olenev inféodé au genre *Lepus* en Ukraine et en Asie Centrale.

En conclusion l'étude des puces du lapin de garenne, montre la dualité d'origine supposée pour ce Lagomorphe par les Paléontologistes:

– néarctique, et sans doute ancienne, avec *Spilopsyllus cuniculi,*

– paléarctique, et manifestement plus récente, avec *Caenopsylla, Odontopsyllus* et *Xenopsylla cunicularis.*

Les trois premiers genres sont primitivement liés aux Lagomorphes. Le dernier représente un phénomène de capture, paléontologiquement très proche dans le temps, aux dépens de rongeurs Gerbillidae.

MARIA HÁLIA ABREU
Museu e Laboratório Zoológico e Antropológico, Lisboa, Portugal

QUELQUES ASPECTS PARTICULIERS DU CYCLE ANNUEL D'INFESTATION DU LAPIN DE GARENNE PAR DEUX ESPECES DE PUCES

ABSTRACT – SPECIAL ASPECTS OF THE NATURAL INFESTATION OF THE RABBIT BY TWO SPECIES OF FLEAS

The two species of fleas, *Spilopsyllus cuniculi* and *Xenopsylla cunicularis,* parasitic on rabbits in Portugal have specific temporal distributions. There are cyclical variations throughout the year, reaching a peak for *S.cuniculi* in February and *X.cunicularis* in September.

As anticipated, the population curve of *Spilopsyllus* therefore fits the hormonal cycle of the host (which peaks between January and March). In contrast, if *X.cunicularis* were bound to the hormone cycle of the rabbit, then a population peak in September would not be expected, since there is no second hormonal cycle of the rabbit between August and October. We, therefore, conclude that if there is hormonal dependence of *X.cunicularis* on its host, it is of a different type from that exhibited by *S.cuniculi.*

The geographical and seasonal distribution of myxomatosis in Portugal (i.e., mainly central and southern, and in the summer) raises some questions about hormone dependence which will be the subject of future investigations.

Une des lacunes de la bibliographie entomologique portugaise est, justement, celle que l'on trouve dans des travaux de caractère éco-physiologique sur Siphonaptera et leur hôtes. Ce fait nous a motivés à l'initiation d'une étude sur l'infestation ectoparasitaire du Lapin de Garenne, *Oryctolagus cuniculus huxleyi* Haeckel.

Au Portugal, cette étude est d'une importance capitale par diverses raisons. D'abord, le Lapin de Garenne constitue une espèce cynégétique indigène des plus recherchées, sinon la plus recherchée de toutes, ce fait étant dû à différents facteurs que nous n'irons pas discuter ici, mais parmi lesquels sont considérables son pouvoir dispersif élevé et les quantatifs cynégétiques disponibles. Deuxièmement, les organismes du Gouvernement liés aux problèmes de cynégétique ont constaté depuis longtemps l'apparition régulière d'essors de myxomatose sans que, cependant, ces mêmes services aient, jusqu'à présent, envisagé le problème de leur possible enraiement ou même éventuelle extinction.

Appuyés surtout aux conditions citées, nous avons commencé les captures du Lapin de Garenne en 1974, celles-ci ayant été totalement réalisées dans la région de Campo Maior (Herdade do Baldio – Santa Eulália) Alto Alentejo.

Pendant la première année de travail, et pour obvier aux innombrables difficultés de toute sorte qui ont surgi, nous avons fait des récoltes mensuelles de dix individus, cinq de chaque sexe, de l'espèce déjà mentionnée. En procédant à l'analyse des résultats de ces récoltes nous avons jugé nécessaire d'effectuer de nouvelles captures, pendant les mois de Septembre et d'Octobre, dans le but de vérifier les résultats obtenus. En chacun de ces deux mois nous avons capturé près de quarante animaux, avec un rapport inter-sexes d'environ 1:1.

Encore à la campagne, et immédiatement après chaque capture, nous avons isolé tous les exemplaires dans des sacs en matière pastique et, en mettant à découvert seulement la partie postéro-ventrale du corps pour éviter au maximum l'éventuelle fuite de parasites, nous avons ensuite procédé à l'extraction des viscères, et à la fixation en San Felice d'ovaires et de testicules pour une étude histologique postérieure. L'extraction des parasites n'était effectuée qu'après environ dix-huit à vingt-quatre heures de séjour des hôtes, à l'intérieur d'une chambre frigorifique. Cette dernière opération était destinée à faciliter la récolte des parasites, rendus inactifs ou même morts par l'action du froid.

Avant d'entrer dans la discussion des résultats obtenus, il sera peut-être nécessaire de fournir quelques éclaircissements par quel moyen le matériel recueilli à été identifié.

Tableau 1. Distribution des deux espèces *S.cuniculi* et *X.cunicularis* au cours de l'année 1974-1975

Mois	*S.cuniculi*		*X.cunicularis*	
	♂	♀	♂	♀
Décembre	126	134	1	1
Janvier	219	229	–	1
Février	245	343	–	1
Mars	142	238	–	–
Avril	47	68	–	–
Mai	4	7	–	–
Juin	17	19	1	2
Juillet	1	1	1	1
Août	1	1	3	4
Septembre	1	5	24	20
Octobre	–	–	5	1
Novembre	3	45	10	11

Tableau 2.

Mois	Nombre total de parasites	Nombre de parasites ♂	Nombre de parasites ♀	Moyenne générale P/H	Moyenne P/H ♂	Moyenne P/H ♀
Décembre	262	127	135	26,20	8,60	43,80
Janvier	449	219	230	44,90	9,00	80,80
Fébrier	589	245	344	58,90	14,60	103,20
Mars	380	142	238	38,00	6,40	69,60
Avril	115	47	68	11,50	2,80	20,20
Mai	11	4	7	1,10	0,40	1,80
Juin	39	18	21	3,90	2,40	5,40
Juillet	4	2	2	0,40	0,00	0,80
Août	9	4	5	0,90	1,00	0,80
Septembre	50	25	25	5,00	6,00	4,00
Octobre	6	5	1	0,60	0,60	0,60
Novembre	99	43	56	9,90	8,20	11,60

Comme nous ne disposions pas de personnel auxiliaire pouvant nous donner l'appui technique indispensable à l'éxécution d'un travail de cette nature, il ne nous a pas été possible d'identifier mensuellement et au moment dû, le matériel que nous étions en train de récolter. Ainsi, et ignorant à priori les éspèces en présence, le premier fait que nous avons noté concernait le nombre total de parasites par hôte: pendant les mois de Décembre à Mars, ce nombre était très élevé, et on remarquait une particulière incidence au moment où les femelles étaient grosses. A partir de Mai, non seulement nous avons constaté une diminution progressive du niveau d'infestation, mais aussi que les parasites infestaient sans choix, les hôtes des deux sexes. Cette situation s'est maintenue sensiblement jusqu'à la fin d'Octobre, moment où nous avons remarqué un nouvel accroissement du niveau d'infestation des hôtes, tout en respectant le caractère d'indistinction sexuelle déjà mentionné.

Au fur et à mesure que nous procédions à l'identification du matériel nous avons remarqué la présence de deux espèces différentes de parasites, *Spilopsyllus cuniculi* (Dale, 1878) et *Xenopsylla cunicularis* (Smit, 1957) distribuées comme indiqué dans le tableau 1.

En partant des valeurs indiquées ci-dessus nous avons procédé au calcul des moyennes P/H, P/H ♂ et P/H ♀ (moyenne de parasites par hôte, moyenne de parasites par hôte male et moyenne de parasites par hôte femelle) que nous avons disposés sur le tableau 2, avec les totalités mensuelles de parasites et les totalités mensuelles de parasites des deux sexes.

Nous avons cru que ces données pourraient nous aider à interpréter les résultats obtenus.

Les données les plus significatives concernent les P/H ♂ et P/H ♀. Pendant certains mois, de Décembre à Mars, par exemple, les valeurs de P/H ♀ atteignent des niveaux de 5 à 10 fois supérieurs à celles de P/H ♂ correspondants; d'ailleurs, le P/H se présente assez élevé. L'infestation est due presque exclusivement à des exemplaires de l'espèce *S.cuniculi.*

A partir de Mai, les valeurs de P/H ♂ et P/H ♀ sont semblables et on peut même remarquer une légère augmentation de P/H ♂ pendant le mois de Septembre, mais il ne nous semble pas qu'un tel fait aît une signification spéciale.

P/H ♀ ne subit un nouvel acroissement qu'à partir de Novembre, quand les exemplaires de l'espèce *S.cuniculi* réapparaissent en nombre relativement élevé, sans que, cependant, leur nombre n'atteigne les niveaux des valeurs les plus élevées que nous avons détectées.

Quant à l'espèce *X.cunicularis,* sa présence pendant les mois froids est absolument

Tableau 3.

Mois	*S.cuniculi* ♂	♀	*X.cunicularis* ♂	♀
Septembre	1	3	62	49
Octobre	33	45	27	15

Tableau 4.

Mois	Nombre total de parasites	Nombre de parasites ♂	Nombre de parasites ♀	Moyenne générale P/H	Moyenne P/H ♂	Moyenne P/H ♀
Septembre	115	63	52	3,03	1,78	4,93
Octobre	120	60	60	3,00	2,86	3,16

sporadique. Elle n'apparaît en pourcentages significatifs que de Juin à Octobre, le mois de Septembre étant celui pendant lequel elle présente un taux d'infestation plus élevé.

C'est justement la constatation de cette incidence plus ou moins marquée, qui nous à incités à faire de nouvelles récoltes, aux mois de Septembre et d'Octobre de l'année suivante, dont les résultats sont présentés dans le tableau 3.

Le tableau 4 comprend les valeurs qui concernent ces deux mois de Septembre et d'Octobre, des moyennes de parasites par hôte – P/H, de parasites par hôte mâle – P/H ♂ et de parasites par hôte femelle – P/H ♀, conjointement avec les totaux mensuels de parasites et les totaux mensuels de parasites de chacun des sexes.

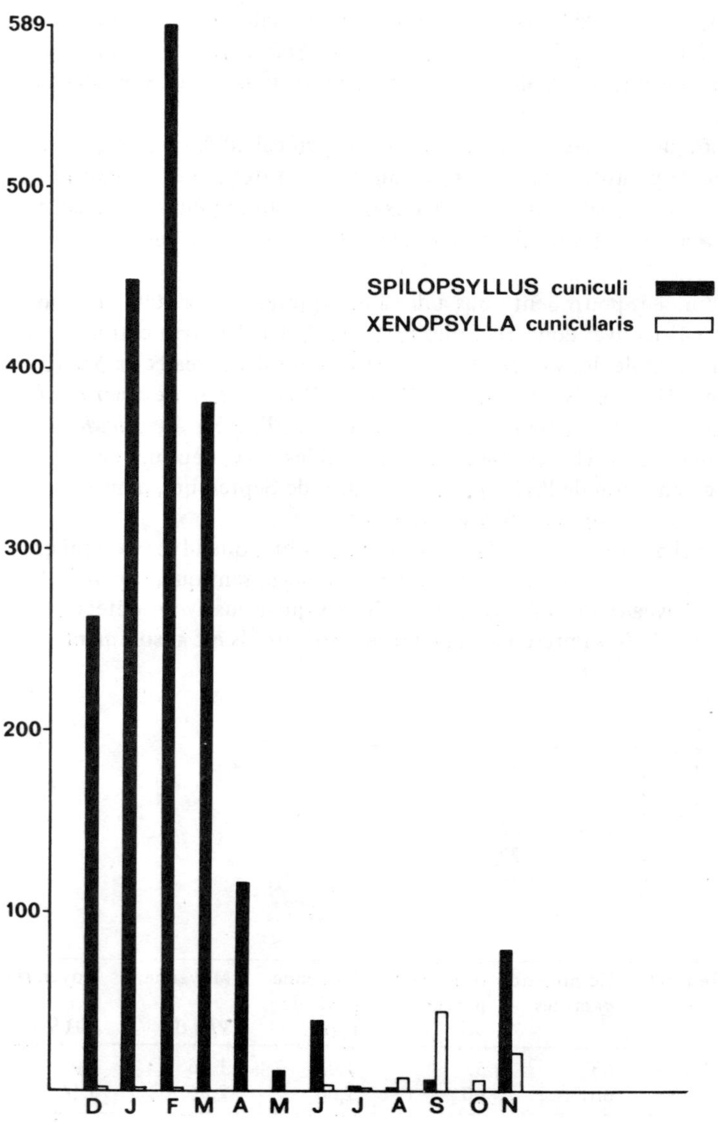

Après une première analyse des résultats présentés, nous remarquerons tout de suite l'existence d'une certaine variation entre les valeurs concernants les mois de Septembre et d'Octobre 1975 et celles que nous avons exposées postérieurement, concernant les mêmes mois de 1976.

Plusieurs facteurs ont contribué, bien sûr, à ces différences: variation du nombre des composants de l'échantillon utilisé par nous (dix dans le premier cas et près de quarante dans le deuxième); altération des endroits de capture dans une aire légèrement supérieure à 2 000 hectares; tout un ensemble de facteurs ambiants tels que la température, l'humidité, la plus ou moins grande abondance de nourriture; la propre altération du régime cynégétique auquel était soumise l'aire citée: lors des premières captures c'était une réserve de chasse particulière, en passant à la suite, et avant la deuxième période de captures, à figurer sous le régime de terrain libre, situation dans laquelle la pratique de la chasse est complètement libre, protégée à peine par la législation en vigueur.

De toute façon, et indépendement des facteurs auxquels les populations de Lapin de Garenne qui existent dans cette aire puissent être soumises, on remarque la présence d'un facteur constant, dont l'indiscutable intérêt mérite d'être mis en relief: la présence de *X. cunicularis* en nombres particulièrement élevés pendant les mois de Septembre et d'Octobre.

Une observation du graphique annexe nous conduit immédiatement à la conclusion suivante: toutes les espèces de parasites en présence varient cycliquement le long de l'année, et leurs distributions respectives présentent un pic au mois de Février pour *S. cuniculi* et au mois de Septembre pour *X.cunicularis.* La variation des valeurs du niveau d'infestation autour de cette valeur maximale se déroule de façon analogue quelle que soit l'espèce considérée.

Après notre travail sommaire nous avons le sentiment que l'espèce *X.cunicularis* doit présenter au Portugal un comportement assez semblable à celui qu'elle manifeste au Maroc (Smit, 1957), malgré notre impossibilité de constater si, en l'absence de parasites de cette espèce sur l'hôte, ils se trouvent dans son nid ou dans son terrier.

Analysant attentivement la distribution des espèces et le nombre de leurs intégrants au long de l'année, nous sommes menés à admettre que les rapports parasite – hôte sont sûrement différents quand ils sont évalués relativement à *S.cuniculi* et à *X.cunicularis.*

Etant définitivement admis que la première est en intime rapport avec les cycle hormonal de l'hôte, (Mead-Briggs & Rudge, 1960; Rothschild & Ford, 1964), ce fait ne semble pas se produire en ce qui concerne la deuxième. Le graphique nous montre que les pics se trouvent séparés par une période de sensiblement six mois. Evidement le cycle hormonal du Lapin de Garenne ne se trouve pas en des points équivalents entre les mois de Janvier à Mars et entre ceux d'Août à Octobre. Nous sommes donc menés à admettre que si *X.cunicularis* possède en effet une hormono – dépendence de l'hôte (Lapin de Garenne), cette dépendance est d'un type différent de celle qui existe dans *S.cuniculi.*

En outre, au Portugal la myxomatose se manifeste plus fortement dans les régions du Centre et du Sud du pays – la région où nous avons fait notre étude se situe dans la partie inférieure de la zone centrale – et les essors se produisent surtout pendant les mois d'été. A cette inégalité de la distribution de l'incidence pourront contribuer, et sûrement contribuent plusiers facteurs que nous ignorons, car notre expérience de campagne ne dépasse pas de beaucoup celle que nous mentionnons dans ce travail. Cependant, nous sommes convaincus qu'elle ne sera pas exclusivement due à la variation quantitative des contingents cuniculaires, plus importants au centre et au sud du pays.

Etant donné le manque de bibliographie portugaise et étrangère sur ce sujet, auquel il faut ajouter notre impossibilité d'avoir déjà complété l'étude histologique des quelques centaines de préparations d'ovaires et de testicules que nous avons fait, et avec toutes les limitations que nous avons déjà rapportées, notre inexpérience dans cette matière ne nous permet que de poser quelques questions que nous essaierons d'étudier plus tard:

1. Taux d'infestation élevés pour *S.cuniculi* en rapport avec la période de reproduction de l'hôte – dépendance hormonale du parasite, sans doute, mais jusqu'à quel niveau en essais de campagne?

2. Pendant les mois d'été, *S.cuniculi* diminue son activité. Reste-t-elle dans les terriers ou bien cherche-t-elle un nouvel hôte? Nous rappelons que dans la région où nous avons travaillé, les terriers des lapins sont souvent cohabités par d'autres mammifères.

3. *X.cunicularis* manifeste un niveau d'infestation semblable pour les mâles et pour les femelles; ce fait sera-t-il une conséquence de son éventuelle indépendance du système hormonal de l'hôte, comme on constate pour *X.astia* (Prasad, 1973, 1976) et pour *X.cheopis* (Rothschild, 1964; Prasad, 1973, 1976)?

4. Chercher une explication plausible pour l'absence de *X.cunicularis* sur son hôte, pendant l'époque de reproduction de ce dernier.

5. Rapport possible de *X.cunicularis* avec la température ambiante, fait possible de constituer, par lui seul, une explication suffisante pour les deux phases distinctes qui se présentent à nous dans l'analyse de son cycle annuel de vie.

6. Rapports entre les cycles de vie des deux espèces parasites. Sont-ils analogues, différents, diamétralement opposés ou simplement différés par une période de temps déterminée?

7. Investigation sur la possibilité de *X.cunicularis* de constituer également un bon vecteur pour le myxome. En cas affirmatif, étude comparée des évolutions de l'affections quand elle est transmise par chacun de ces deux espèces de parasites.

Nous pensons que les hypothèses soulevées constituent les fondements du programme de nos futurs travaux de recherche. Malgré toutes les difficultés que nous avons déjà signalées, nous persévérons dans notre effort d'étudier 'in loco' le problème des interrelations puce – Lapin de Garenne, en toute leur étendue et dans la mesure de nos possibilités effectives.

REFERENCES

Mead-Briggs, A.R. & J.B.Rudge 1960, Breeding of the rabbit flea, *Spilopsyllus cuniculi* (Dale): Requirement of a 'factor' from a pregnant rabbit for ovarian maturation. Nature 187:1136-1137.

Prasad, R.S. 1973, Studies on host-flea relationship. II. Sex hormones of the host and fecundity of rat fleas *X.astia* and *X.cheopis.* Indian J. med. Res. 61:38-44.

Prasad, R.S. 1976, Studies on host-flea relationship. IV. Progesterone and cortisone do not influence the reproductive potentials of rat fleas *Xenopsylla cheopis* (Rothschild) and *X.astia* (Rothschild). Z. Parasitenk. 50:81-86.

Rothschild, M. 1964, (Note of an exhibit), The rat flea (*Xenopsylla cheopis* Rothschild) breeding on adrenalectomised and castrated rats. Proc. Roy. ent. Soc. Lond. (C) 29(9):35.

Rothschild, M. & B.Ford 1964, Breeding of the rabbit flea (*Spilopsyllus cuniculi* Dale) controlled by the reproductive hormones of the host. Nature 201:103-104.

Smit, F.G.A.M. 1957, A new species of *Xenopsylla* parasitic on rabbits in Morocco. Arch. Inst. Pasteur Maroc 5(6):201-205.

KULDIP RAI
Zoological Survey of India, Calcutta, India

ECOLOGY, GEOGRAPHICAL DISTRIBUTION AND HOST PREFERENCES OF THE INDIAN FLEA FAUNA

ABSTRACT

Despite spectacular advances in science and technology over the past several decades – the harnessing of nuclear energy, the mastery of space flight, the breaking of the genetic code – mankind has still to go a long way in its age-old gigantic crusade against harmful insects. In the present day of an exploding human population, and with the consequent multitude of inherent problems, there is a great interest in the field of ecology. Man may acquire certain arthropod-borne pathogens by ranging into haunts of vectors, or he may encourage the vectors to enter domestic situations and thereby, set the stage for epidemics. In the former situation, man may acquire a pathogen which is ordinarily found in other mammals, while in the latter instance, man more frequently is both donor and recipient of the agent of disease. These ecological bonds must be understood before we can control diseases borne by arthropods. The environment of any animal includes four major components – climatic conditions, food, other organisms (of the same or different species), and a place in which to live. Theoretically, an animal's chance to survive and multiply may depend on its whole environment, but in practice, one or several components usually turn out to be important enough to account for nearly all the variability that has been observed in the density of the population (or populations). The usual transmission routes for zoonotic agents are from wild or domestic animals to man, but some organisms may pass from human beings to other mammals. Recently, in the USSR and Germany, quite a large number of faunistic surveys and ecological studies have been conducted to provide a background for epidemiological investigations relating to fleas. Some preliminary work on the effects of constant temperature and humidity on the development of the larvae and pupae of the three Indian species of *Xenopsylla* (Siphonaptera: Pulicidae) was done by Sharif long ago (1949). It is intended to make detailed observations on the following aspects of flea ecology in India: 1) breeding during estivation of host; 2) changes in host specificity with altitude and temperature limitations; 3) sojourn of fleas in burrows; 4) competition for feeding sites on the body of the host, and 5) many other aspects such as feeding habits, seasonal variation of populations and their distribution.

As indicated by the Hon. Miriam Rothschild (1975): 'there is an urgent need for analysis and co-ordination of this widely scattered information, and a comprehensive volume on the ecology and biology of fleas is the primary requirement of the day'. In India, research work is going to be conducted on these neglected aspects of flea fauna in the light of these highly valuable suggestions of Dr Rothschild, who is one of the most distinguished students of Siphonaptera active today.

The distribution of animals is determined not only by a number of geomorphological and meteorological factors, but also by biotic factors which include plants and other animals. Just as plants and their habitats are closely related, so are animals and the microclimates they inhabit. A study of the microclimates and their influences on diurnal and seasonal rhythms of activity is of paramount importance in the understanding of animal ecology. Fleas are basically confined to animals that customarily have a den or recurrent resting place. Only the more advanced species have escaped this restriction. Insectivores and rodents are almost always infested because of their tendency to live in holes and burrows.

A resurgence of interest in Wegener's theory of continental drift has stimulated research into geographical distribution of fleas and it also shows how the knowledge of evolution and distribution of mammals and their ectoparasites can contribute to our understanding of infectious diseases (like plague and murine typhus) associated with fleas. The host selection by insect parasites is regulated by a combination of factors, the most important of which appears to be chemical. Pheromones contribute a primary means of communication. Various species of animals use them to attract mating partners, to direct others to suitable food or resting sites, to cause others to stay away when appropriate, and for a variety of other behavioural functions. Successful parasitism requires a sequence of the following steps – a) host habitat location; b) host location; c) host acceptance; d) host suitability; and e) host regulation.

REFERENCES

Rothschild, M. 1975, Recent advances in our knowledge of the order Siphonaptera. Ann. Rev. Ent. 20:241-259.

Sharif, M. 1949, Effects of constant temperature and humidity on the development of the larvae and the pupae of the three Indian species of *Xenopsylla* (Insecta: Siphonaptera). Phil. Trans. R. Soc. Lond. (B) 233:587-635.

PER BRINCK
University of Lund, Sweden

ECTOPARASITE INFESTATIONS OF RODENTS AND SHREWS IN NORTHERNMOST SCANDINAVIA

ABSTRACT

In the years of 1965-70, the parasitological unit of the Ecology Building in Lund carried out an investigation of small mammals and their external parasites in five areas in northernmost Scandinavia – well north of the Arctic Circle. The small mammals (9 000 specimens) represented in the material are six species of shrews (five *Sorex* and one *Neomys*) and five species of voles (three *Clethrionomys* and two *Microtus*).

The populations of the small rodents of North Scandinavia fluctuate in cycles. Shrews likewise have cycles, and those of both rodents and shrews are synchronous. A complete cycle usually lasts four years. Sometimes, the shrews are out of phase with the rodents, and their maximum comes before that of the rodents.

According to present models of host parasite systems, there is a delay of the parasite maximum in population oscillations. It is generally accepted that the main reason is that there is a form of discontinuity between parasite reproduction and the establishment of reproductive individuals in new hosts. The magnitude of the delay depends on a complex of factors which refer to two different phenomena: 1) the reproduction rate of the parasite and its timing in relation to the reproduction of the host and 2) the dispersal strategies and transmission rates of parasites. In the non-permanent groups of parasites, reproduction rates are lower than in the permanent groups and the same applies to transmission rates. Therefore, there is a delay of one year in the flea cycle and two years in the cycle of the northern tick species. The ticks take two years for the development from egg to invasive larva. In the fleas under discussion, there might be a few generations a year. In the permanent species of mites, as in lice, the complete generation from egg to sexually mature adult is well timed to that of the host. Therefore, these parasites should be more in phase with their hosts, but such is not the case. The mean number of parasites per host increases due to the early increase in the number of parasites at the height of the rodent cycle and to the following decrease in the number of hosts.

For ectoparasites, a clumped or aggregate occurrence in the hosts is normal. The present material shows a good fit to a negative binomial distribution. Of the fleas, 60 per cent of the population live on host animals with one parasite only. For the ticks, 50 per cent live on hosts with five parasites or less, and among the resident mites 50 per cent of the population occur on hosts with 13 specimens or less.

The overdispersion of the parasite populations is a characteristic of most parasites: the number of animals without parasites or with very few parasites are very high. As a matter of fact, the parasite load is so low in northernmost Scandinavia that the medium animal of shrews has no parasites at all and the median rodent has three parasites only. On the other hand, during the low in the rodent cycle, the pressure of parasites is much heavier than during the high. It develops during the social and physical deterioration of the host populations which follows the highs.

The exposure of the host to infestation of possible species is random. The simultaneous arrival of a second or third species is a rare event. The host's unspecific predation on invading parasites keeps the total level down, particularly so for species that do not multiply on the host. When reproducing species multiply – lice and certain mites – an ecological release in a broad ecological niche may be the first result, followed by a compression of the niche if another parasites starts a successful reproduction on the host. Species-packing is not an obvious feature among northern ectoparasites.

ALLEN H. BENTON & JONATHAN F. DAY
State University College, Fredonia, N.Y., USA

SEASONAL CHANGES IN THE FLEA FAUNA OF NESTS OF THE SOUTHERN FLYING SQUIRREL

ABSTRACT

Four flea species regularly infest the nests of *Glaucomys volans* in the northeastern United States: *Opisodasys pseudarctomys, Orchopeas howardii, Epitedia faceta* and *Conorhinopsylla stanfordi.* From November 1976 to May 1977, more than 1,000 fleas were collected from these nests and the seasonal changes studied. Preliminary data indicate that the life cycles of the dominant species are adjusted to maximize success and minimize competition. The position of the nest also appears to influence distribution of the parasites.

When several species of fleas infest the same host, some modifications in behaviour, life cycle or spatial distribution must be made in order to maximize the success of each species and minimize competition for whatever resources may limit the number of ectoparasites. There are at least three ways in which different species may exploit the same host with minimum impact on each other:

1. Spatial – one species may tend to live and feed on a particular part of the body of its host, while another species uses a different body area. This has been shown to be true for some rabbit fleas.
2. Temporal – life cycles of different species may vary in such a way that adults of each species are dominant at different times.
3. Ecological – one species may dominate under certain ecological conditions, while another may dominate if circumstances are different. This has been demonstrated for certain species of bird fleas.

The southern flying squirrel, *Glaucomys volans,* is an abundant small mammal in deciduous forests of eastern North America. It lives in holes in trees, often adapting old woodpecker holes to its needs, but it will also utilize appropriate nest boxes placed in its habitat. In October 1976, we erected nest boxes in two areas known to be inhabited by these squirrels. One area was in the vicinity of Fredonia, Chautauqua County, New York, in second-growth deciduous forest along both sides of a ravine. The second area was near West Pawlet, Vermont, about 400 miles (640 km) to the east of the first area.

After the nests had been in place for three weeks, we began to collect nest material and extract the fleas. Nests were collected in plastic bags and taken to our laboratory, where the fleas were removed by hand collection over a period of several days. Collections were made at approximately monthly intervals, depending upon schedules and

Table 1. Collections of fleas from nests of *Glaucomys* in Chautauqua County, N.Y., November 1976 - June 1977.

Percent of total collection.

	Nov. 2 1976	Nov. 16, 26 1976	Dec. 21, 31 1976	Feb. 6 1977	Feb. 28 1977	Apr. 2 1977	May 26 1977	June 23 1977
Opisodasys pseudarctomys	1.9	17.4	25.5	8.2	7.2	88.0	9.2	26.5
Orchopeas howardii	15.3	28.2	32.1	64.7	85.5	9.3	90.8	73.5
Epitedia faceta	67.5	38.0	45.6	27.1	7.2	2.7	–	–
Conorhinopsylla stanfordi	7.1	7.6	4.1	–	–	–	–	–
n =	268	92	364	85	55	291	54	283

weather conditions. The results of our first eight months of collections are shown in table 1.

During the late autumn and early winter, *Epitedia faceta* is the most abundant flea species. During the winter, *Orchopeas howardii* dominates, as adults of the other species die out and adults of this species continue to survive. *O.howardii* overwinters in the adult stage, while *Opisodasys pseudarctomys,* and presumably the other two species as well, reproduce in early winter and overwinter as larvae or pupae. In the late December collections, numerous larvae were found. When reared to adulthood, all of them proved to be *O.pseudarctomys.* On the basis of this fact, we expected that there would be a gradually decreasing number of adults through the winter months, with *O.howardii* predominating. We further anticipated that the early spring collections would show a resurgence of *O.pseudarctomys,* as the overwintering larvae and pupae emerged, under the influence of warmer spring weather. As the figures in table 1 indicate, this is precisely what happened. We then predicted that figures for early summer would show a declining population of *O.pseudarctomys* and an increasing population of *O.howardii,* as the offspring of overwintering adults began to emerge. As the figures for May and June indicate, our prediction was borne out by the collections.

It appears, then, that *Epitedia faceta* is a cold-weather species, which is most abundant during fall and early winter. The largest collections of this species were taken from nests on the cooler north-facing slopes of our study area. Adults decline rapidly in numbers during the winter, and apparently the species must spend the spring and summer months in immature stages.

Conorhinopsylla stanfordi is also a fall and early winter flea, although we have not yet found it to be dominant in any nest. This may be because we have not yet made any collections in September or October, or it may be because we have not yet encountered the particular ecological conditions under which the species might become dominant. It is also possible that the species may always occur in relatively small numbers, and never achieve dominance.

Opisodasys pseudarctomys is dominant during the early spring, when pupae which have overwintered reach maturity. We anticipate that there may be another peak later in the summer, when the offspring of the spring generation reach maturity, but we do not yet have data for those months.

Orchopeas howardii dominates during the winter, when it approaches 100 per cent of adults in the nests. The offspring of this generation then mature in early summer, to dominate collections made at that time.

The data seem to indicate that *Orchopeas howardii* and *Opisodasys pseudarctomys* have evolved seasonally synchronized life cycles which permit alternate domination of the two species in the nests of the southern flying squirrel through most of the year. A cold-adapted species, *Epitedia faceta,* dominates in late fall and early winter. Complete studies of the life cycles of these species, involving laboratory rearing of larvae, will be conducted during the next year.

This research has been supported by a grant from the Edmund Niles Huyck Foundation, Rensselaerville, New York, and the authors wish to express their gratitude for this support.

GUNVOR BRINCK-LINDROTH
University of Lund, Sweden

A COMPARISON BETWEEN THE FLEA FAUNA OF TERRESTRIAL SMALL RODENTS AND SHREWS IN SCANDINAVIAN MOUNTAINS AND THE PYRENEES

ABSTRACT

From small rodents and shrews are recorded 20 species and subspecies of fleas from Scandinavian mountain districts and 24 from the Pyrenees. Six species occur in both areas and may be grouped according to their latitudinal and altitudinal distribution, while some genera and species are strictly confined to north or south Europe. The association of fleas and host animals differs greatly between Andorra and a Swedish Lapland locality, although the general composition of the flea fauna in the areas is fairly stable.

Members of the parasitology unit of the Ecology Building in Lund have for many years studied ectoparasites of small mammals. Vast collections have been brought together, mainly from mountain areas in Scandinavia. In the same way, we have in each of the last three years collected in the Pyrenees for about a week.

In the following I shall present a brief comparison between the flea fauna of terrestrial small rodents and shrews of the Scandinavian mountains and that of the Pyrenees (mainly Andorra) (fig. 1).

Mountain faunas are interesting everywhere because of the importance of the insularity factors, that is the splitting and subsequent isolation of the populations and the ecological and taxonomic consequences (Brinck, 1974). In this case, both regions have been influenced by the Pleistocene ice age (fig. 1), which have greatly contributed to form their present faunas.

Because of the ice ages the biota in Scandinavia are young, compared with flora and fauna in areas south of the maximum glaciation limit. The northern ice-sheets started to melt away some 12-15 000 years ago, which is a comparatively short period. In the mountain chains of south Europe vast glaciers acted as barriers for the dispersal of biota, and probably played an important role in the displacement of the small mammal populations and subsequently for the flea distribution.

Host animals

The small mammal fauna comprises species of five families (table 1). There are 28 species, of which 19 are represented in Scandinavia and 20 in the Pyrenees. Ten species – including the cosmopolitan *Rattus* and *Mus* species – are met with in both areas. *Talpa europaea* in Scandinavia is in brackets because it is a typical lowland form in this area. The family

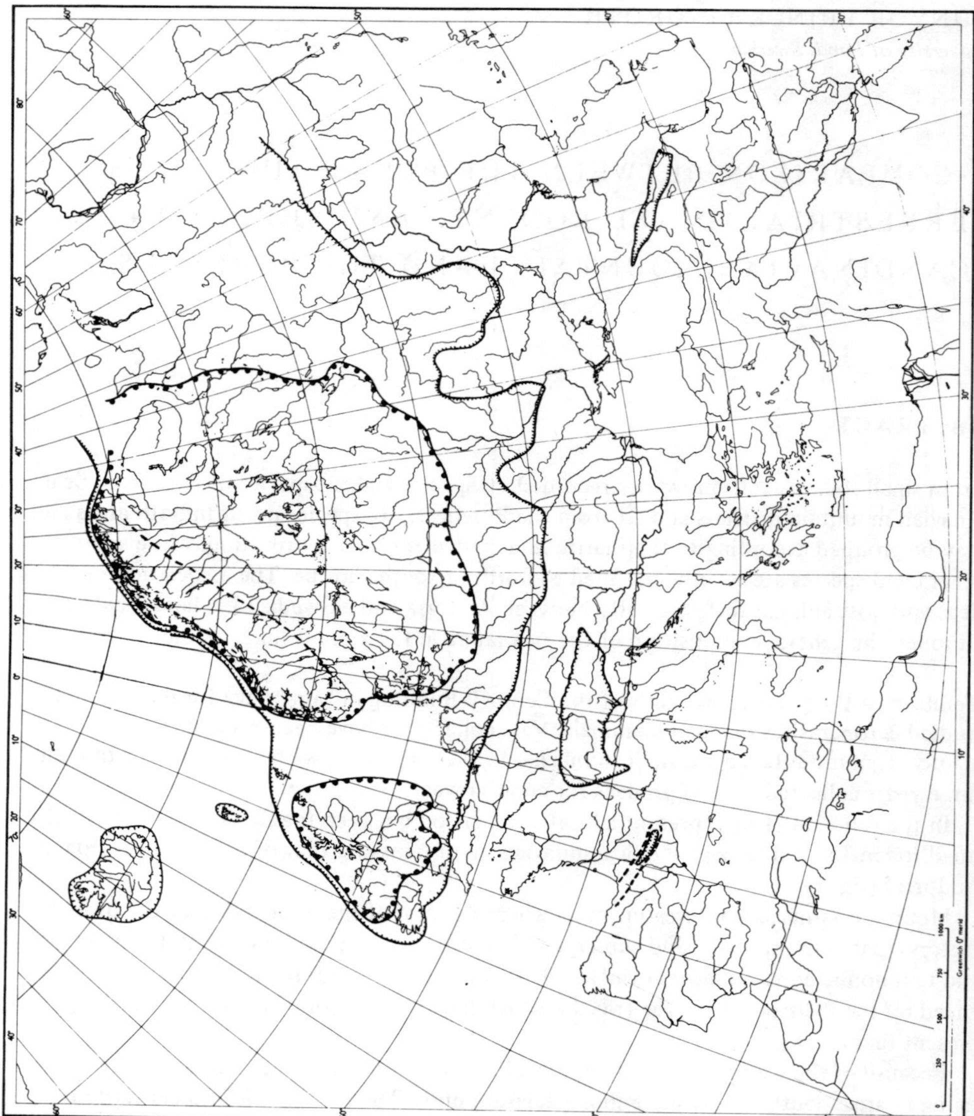

Figure 1. The Scandinavian and Pyrenean mountain areas (broken lines) are at approximately 59-70° and at 43° lat. N respectively. The ice limits of the maximum glaciation (unbroken lines) as well as those of the last glaciation (dotted lines) in north and west Europe have been indicated.

Table 1. The occurrence of terrestrial small rodents and shrews in Scandinavia and the Pyrenees. *T.europaea* is absent in north Scandinavia

Host taxa	Scandinavia		The Pyrenees Approximate limit		
Talpidae	(1)				2
Galemys pyrenaicus			60-2 000 m	x	
Talpa europaea		(x)	2 000 m	x	
Soricidae	6				4
Sorex araneus s.l.		x	2 000 m	x	
S.isodon		x			
S.caecutiens		x			
S.minutus		x	1 800 m	x	
S.minutissimus		x			
Neomys fodiens		x	2 000 m	x	
N.anomalus			2 000 m	x	
Gliridae					2
Glis glis			2 000 m	x	
Eliomys quercinus			2 000 m	x	
Cricetidae	8				8
Lemmus lemmus		x			
Myopus schisticolor		x			
Clethrionomys glareolus		x	2 000 m	x	
C.rutilus		x			
C.rufocanus		x			
Arvicola sapidus			2 000 m	x	
A.terrestris		x	2 000 m	x	
Pitymys pyrenaicus			2 000 m	x	
P.mariae			?	x	
Microtus arvalis			2 250 m	x	
M.agrestis		x	1 800 m	x	
M.oeconomus		x			
M.nivalis			2 000 m	x	
Muridae	4				4
Apodemus sylvaticus		x	2 400 m	x	
A.flavicollis		x	?	x	
Rattus sp.		x	?	x	
Mus musculus		x	?	x	
Total number of species 28	18 (19)				20

Gliridae occurs in the Pyrenees only. The same applies to the genera *Galemys* and *Pitymys*, while *Lemmus* and *Myopus* are restricted to north Europe. For *Sorex* and *Clethrionomys*, the number of species is higher in Scandinavia, while it is the reverse for *Arvicola* and *Microtus*. Only the Muridae family has a similar number of genera and species.

The flea fauna

Forty-three species and subspecies have been found on the small mammal species in question (table 2). Twenty taxa are met with in Scandinavia (Brinck-Lindroth & Smit, in litt.) – or 23 if the 'lowland-forms' are included – 24 were collected in the Pyrenees (Beaucournu, 1976).

Table 2. The occurrence of flea species in Scandinavia and the Pyrenees. Species in brackets are lowland forms in Scandinavia

Flea taxa	Scandinavia		The Pyrenees			
	Norway	Sweden	Atlantiques	Hautes	Andorra	Orientalis
Hystrichopsyllidae						
Doratopsylla d.dasycnema	○	○	○	○	○	○
Corrodopsylla birulai	○	○				
Hystrichopsylla talpae	(○)		○	○	○	○
Hystrichopsylla orientalis	○	○				
Typhloceras poppei	(○)	(○)				○
Atyphloceras nuperum			○			○
Catallagia dacenkoi		○				
Palaeopsylla s.soricis	○					
Palaeopsylla s.rosickyi	○	○				
Palaeopsylla s.starki	○	○				
Palaeopsylla s.vesperis			○	○	○	○
Palaeopsylla minor			○	○	○	○
Ctenophthalmus a.agyrtes	○	○				
Ctenophthalmus nobilis dobyi						○
Ctenophthalmus a.andorrensis			○		○	○
Ctenophthalmus a.catalaniensis						○
Ctenophthalmus a.apertus						○
Ctenophthalmus baeticus arvernus			○	○	○	○
Ctenophthalmus nivalis ianlinni				○		
Ctenophthalmus uncinatus	○	○				
Ctenophthalmus obtusus	○					
Rhadinopsylla integella	○	○	○	○	○	○
Rhadinopsylla pentacantha		(○)	○	○		
Rhadinopsylla pitymydis						○
Rhadinopsylla mesoides				○		
Leptopsyllidae						
(Leptopsylla segnis – cosmopolitan*)*						
Leptopsylla taschenbergi amitina			○	○	○	○
Peromyscopsylla b.bidentata	○	○				
Peromyscopsylla b.geryasii			○	○	○	○
Peromyscopsylla s.silvatica	○	○				
Peromyscopsylla s.spectabilis			○	○	○	○
Amphipsylla s.sibirica	○	○				
Amphipsylla sp.					○	
Ceratophyllidae						
Myoxopsylla l.laverani			○	○	○	○
(Nosopsyllus fasciatus – cosmopolitan*)*						
Amalaraeus penicilliger mustelae	○	○				
Amalaraeus p.pedias	○	○				
Amalaraeus p.pyrenaicus			○	○	○	○
Callopsylla saxatilis		○	○	○	○	
Megabothris calcarifer	?	○				
Megabothris walkeri	○	○				
Megabothris turbidus	○	○		○	○	○
Megabothris rectangulatus	○	○				

Table 2 (continued)

No. of species	Total	Scandinavia	Pyrenees
Hystrichopsyllidae	25	11(14)	16
Leptopsyllidae	8	3	4
Ceratophyllidae	10	6	4

Table 3. Flea species occurring in Scandinavia and the Pyrenees. Distribution according to latitude and altitude

Flea species	Approximate distribution limit Scandinavia	The Pyrenees
T.poppei	Southernmost Sweden (57°N) and Norway (58°N) lowland	1 000 m
R.pentacantha	Southernmost Sweden (56°N) lowland	1 800 m
H.talpae	Southermost Norway (58°N) lowland	2 200 m
1.		
D.d.dasycnema	Central Sweden (63°N) 400 m (along the coast 64-65°N) South Norway (62°N) 600 m	2 200 m
M.turbidus	Central Sweden and South Norway (61°N) 300 m	2 200 m
P.soricis rosickyi	Latitude as above – 800 m	2 300 m *(s.vesperis)*
2.		
R.integella	Entire Scandinavia, also mountain areas	1 000-1 900 m
P.b.bidentata	As above	500-2 000 m *(b.gervasii)*
3.		
A.penicilliger pedias	From ca. 66°N northwards, all elevations	800-2 500 m *(p.pyrenaicus)*
P.s.starki	As above	See above *(s.vesperis)*

With very restricted distribution:
On the west coast of Norway to ca. 68°N: *P.s.soricis*
In central Sweden and south Norway: *A.p.mustelae.*

Most species belong to the comprehensive Hystrichopsyllidae family. Two genera, *Corrodopsylla* and *Catallagia,* are confined to the north, while *Atyphloceras* is absent there. *Ctenophthalmus* is represented by a number of species in central and south Europe. In the Leptopsyllidae family, the genus *Leptopsylla* is represented by a cosmopolitan species in Scandinavia and in Ceratophyllidae the genus *Myoxopsylla* is absent in the north. Although mammal fleas of *Callopsylla* are not recorded from Scandinavia the genus is represented by a bird-flea there.

Only six species occur in both areas (table 3). Besides these, three species are included in the table, as they live in the lowland in Scandinavia, although they go far up in the Pyrenees. These species are more or less restricted to western Europe, although *Typhloceras poppei* is also present in Greece and Algeria.

The montane species may be placed in three groups according to their Scandinavian distribution. This grouping corresponds to the altitudinal distribution of the species in the

Pyrenees. They all have a very wide distribution in Europe and only one – *Peromyscopsylla bidentata* – is absent from the British Isles (George, 1974).

The three species in the first group (with *Doratopsylla dasycnema* in its nominate form and *P.soricis* in the central European subspecies) extend northwards to central Scandinavia, although their exact distribution and altitudinal limit is not known. *D.d.dasycnema* and *Megabothris turbidus* are successively substituted by the northern *Corrodopsylla birulai* and *Megabothris rectangulatus.* In the Pyrenees the species in this group ascend to more than 2 000 m – with *Palaeopsylla soricis* represented by different allopatric subspecies. In the second group, *Rhadinopsylla integella* and the nominate form of *P.bidentata* occur all over Scandinavia during the cold season. Apparently, they were found in the Pyrenees only at high elevations. *R.integella* occurs only in mountain areas in central and southern parts of Europe – for instance, it is restricted to northern parts of the British Isles (George, 1974). *P.bidentata* is absent from most parts of western Europe; in the Massif Central and the Pyrenees, it is represented by ssp. *gervasii.* The third group comprises forms of *Amalaraeus penicilliger* and *P.soricis. A.p.mustelae* and *P.s.soricis* occur in western Europe; in the north they are known from central Scandinavia and the west coast of Norway, respectively. In northernmost Scandinavia, *A.p.pedias* and *P.s.starki* are met with, while *A.p.pyrenaicus* is restricted to montane areas in southern France and northernmost Spain. *P.s.vesperi* occurs in western parts of Great Britain and France as well as in north Spain.

When we try to analyse these distribution patterns, a first question will be: do host specificity and host preference restrict the distribution of the flea species in question? As far as I can see, there are no such limits set by the host animals. Although *T.poppei* is specific on *A.sylvaticus,* the range of *T.poppei* is within that of the host species. The other flea species all parasitize widely distributed shrews, voles and mice.

The number, distribution and derivation of subspecies always have some interest for a study of the composition and origin of a local fauna. In the present case, two subspecies are confined to the Massif Central and the Pyrenees, namely *A.p.pyrenaicus* and *P.bidentata gervasii.* They are also present in the Iberian peninsula. *P.s.vesperi* has a wider distribution: it is Lusitanian. It seems possible that these taxa have derived from isolated populations during the Pleistocene, in the Iberian peninsula or southern France. As regards the subspecies occurring in Scandinavia, these may all have spread to this region from surrounding areas.

A few genera and/or species are strictly restricted to northern or southern parts of Europe (table 4). *C.birulai* is a common parasite of shrews in the northernmost Palaearctic. It has a very wide distribution from Norway to Japan. *Catallagia dacenkoi* is a rare species

Table 4. Northern and southern distribution of some flea genera and species

Scandinavia		The Pyrenees
Corrodopsylla birulai (Soricidae) *Catallagia dacenkoi (C.glareolus)*		*Atyphloceras nuperum (C.glareolus)* *Leptopsylla taschenbergi amitina (A.sylvaticus)* *Myoxopsylla l.laverani (E.quercinus)* *Callopsylla saxatilis (M.nivalis)*
Genera	no. sp. and ssp.	no. sp. and ssp.
Ctenophthalmus	3	6
Megabothris	4	1

in the far north: there are so far, only three localities known in Scandinavia and Finland. Like *C. birulai* it has a wide distribution eastwards.

Most species found in the Pyrenees are restricted to southern Europe, although *Atyphloceras nuperum* occurs as far north as Poland and Germany. *Leptopsylla taschenbergi* in spp. *amitina* was also found in North Africa. The nominate form of *Myoxopsylla laverani* has a more or less west Mediterranean distribution, although there are records from central Europe. *Callopsylla saxatilis* seems to be a 'rare' species: few localities are known (Beaucournu, 1976; Peus, 1976), which may be due to its occurrence at high elevations in mountains.

Apart from *C. birulai* which was collected on various species of shrews, there is a certain host preference among fleas in this group. A case of pronounced host specificity is that of *C. saxatilis* on *Microtus nivalis.* The other flea species occur in part of the range of their host animals.

Although several flea genera occur in both the Pyrenees and Scandinavia (as *Rhadinopsylla* and *Peromyscopsylla*), *Ctenophthalmus* in its narrow sense, and *Megabothris* are of particular interest – they are represented by inverse numbers of species (table 4). There are six taxa of *Ctenophthalmus* recorded from the Pyrenees: some have a very restricted distribution (i.e. *C. nobilis dobyi, C. nivalis ianlinni*), while others were found in south France and in more or less restricted areas of the Iberian peninsula (i.e. *C. baeticus arvernus, C. a. apertus*). They occur on various vole and mouse species and probably they all originate from the Iberian peninsula. In Scandinavia there are few species of *Ctenophthalmus:* in the present material there are only three species. Of these, *Ctenophthalmus obtusus* has an unusual boreo-montane distribution: outside Norway it is recorded from Czechoslovakia and Rumania. *C. a. agyrtes* is widely spread in south Scandinavia, while *C. uncinatus* is found all over Scandinavia, although less common. They are all parasites on voles and mice.

Among the *Megabothris*-species, *turbidus* is present in both areas (cf. table 2). *M. walkeri* has a similar distribution in Scandinavia, but goes further to the east. A rather uncommon species in the North is *M. calcarifer,* while the more abundant *M. rectangulatus* has a clear boreo-alpine or boreo-montane distribution: it occurs in Scotland and in central European mountains. The *Megabothris*-species listed are mainly parasites of voles (the genera *Clethrionomys* and *Microtus*), although *M. turbidus* in Scandinavia is often met with on species of *Apodemus.*

As expected, the differences between the two flea faunas are considerable. The Scandinavian species are mainly northeastern and southern immigrants, in a few cases with a British connection, while the Pyrenean fauna has a more composite provenance.

In Scandinavia, the northeastern invaders are particularly interesting, because they show nearctic connections. Among species of the genus *Megabothris, calcarifer* is present in Alaska, and so is *C. dacenkoi.* Although *Amphipsylla* is mainly a palaearctic genus, the species *A. sibirica* is also present in the new world with a particular subspecies. On the other hand, *C. birulai,* which belongs to the mainly nearctic genus *Corrodopsylla* is a palaearctic species, not represented in North America.

Many host species have had opportunities to disperse these flea species through the northern Palaearctic Region – there are several species of shrews and voles with wide distribution. Some host species occur also in the Nearctic as *Clethrionomys rutilus* and *Microtus oeconomus.*

The flea fauna of the Pyrenees comprises elements with Iberian, Mediterranean and

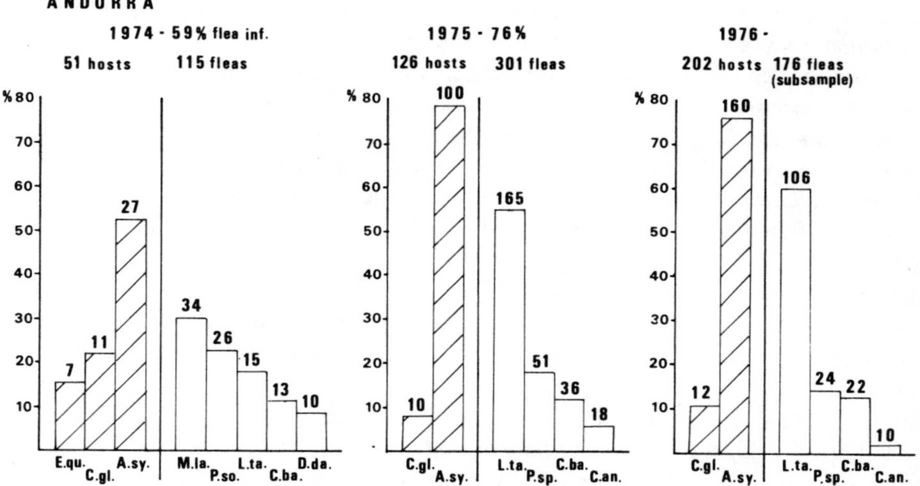

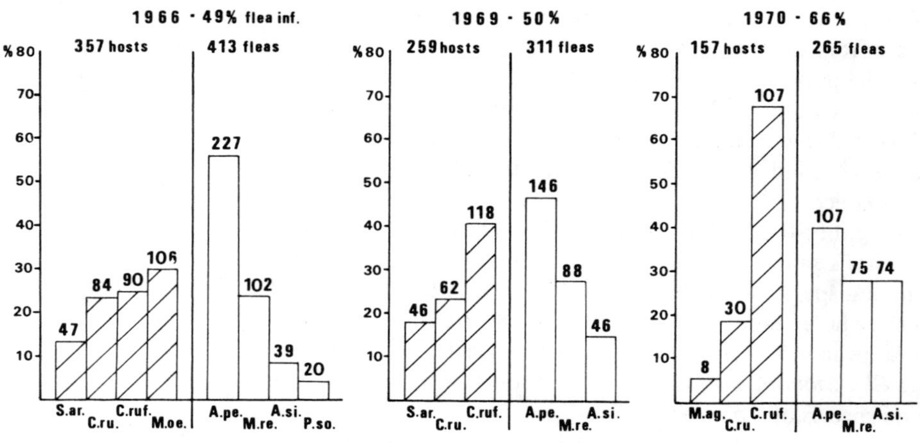

Abbreviations:

Hosts –
- S.ar. *Sorex araneus*
- E.qu. *Eliomys quercinus*
- C.gl. *Clethrionomys glareolus*
- C.ru. *C.rutilus*
- C.ruf. *C.rufocanus*
- M.ag. *Microtus agrestis*
- M.oe. *M.oeconomus*
- A.sy. *Apodemus sylvaticus*

Fleas –
- D.da. *Doratopsylla dasycnema*
- P.so. *Palaeopsylla soricis*
- C.an. *Ctenophthalmus andorrensis*
- C.ba. *C.baeticus*
- L.ta. *Leptopsylla taschenbergi*
- P.sp. *Peromyscopsylla spectabilis*
- A.si. *Amphipsylla sibirica*
- M.la. *Myoxopsylla laverani*
- A.pe. *Amalaraeus penicilliger*
- M.re. *Megabothris rectangulatus*

Figure 2. Flea infestation of terrestrial small rodents and shrews in south and north Europe. Only dominating species are included.

south European origin, apart from others with west, central and north European provenance. According to Beaucournu (1976), there is only one taxon which does not occur outside the Pyrenees, namely *C.nivalis ianlinni.* Whether mountain forms as *A.penicilliger pyrenaicus* and *P.bidentata gervasii* have originated in the Pyrenees or in mountain areas south or north of this area is an open question so far. The great variety of host species in the south adds a number of flea species to the fauna and at the same time the possibilities for evolution of new subspecies increase.

The flea fauna in Andorra and Abisko, Sweden

Finally, I should like to report briefly on our Andorra material (fig. 2). Only the dominant host and flea species have been included in the histogram. For comparison, I have inserted species of fleas and their hosts at a classical Swedish Lapland locality: Abisko, about 200 km north of the Arctic Circle. The Lapland material is part of our comprehensive North Scandinavian collections during six years. Included are only years with small mammals, as there are years in the cycle when no or few small rodents and shrews are found (Hansson et al., 1978). Thus, 1967 was a low population year, which influenced the abundance of host animals and fleas in 1968. The composition of host and flea species differs very much between Andorra and Abisko. On the other hand, the conditions in the areas are fairly stable. The first year in Andorra gave fewer *Apodemus* than the others, yet it is the dominating host species. Consequently, the host specific *L.taschenbergi* (spp. *amitina*) is abundant. I shall also mention that collecting was carried out at different altitudes, which is not analyzed here. *Apodemus sylvaticus* was, for instance, collected as high up as to 2 400 m.

In north Scandinavia the *Clethrionomys* species dominate among voles and *A.penicilliger* (spp. *pedias*) and *M.rectangulatus* among the flea species. There are also a few *A. sibirica* (the nominate form), which seems to prefer *C.rufocanus* – or perhaps more likely the habitat where this vole lives.

The flea infestation is somewhat higher in Andorra which may be due to the abundance of the host specific *L.taschenbergi* on *Apodemus* (1975-76).

Among interesting flea species collected in Andorra but not mentioned previously is *A.sibirica.* As far as I know, this species was not recorded from the area. Unfortunately, the material consists of two female specimens, which cannot be identified to subspecies. One of them is filled up with nematodes, which is often the case in Scandinavian material of the species (Brinck-Lindroth & Smit, 1973).

ACKNOWLEDGEMENT

I am much obliged to Dr J.-C.Beaucournu, Rennes, who kindly checked the lists of Pyrenean flea and host species.

REFERENCES

Beaucournu, J.-C. 1976, Contribution à l'étude des Siphonaptères de Mammifères du Nord-Ouest de la région méditérranéenne (France, Italie, Péninsule Ibérique). Thèse, 284 pp.

Brinck, P. 1974, Strategy and dynamics of high altitude faunas. Arctic and Alpine Research 6(2):107-116.

Brinck-Lindroth, G. 1968, Host spectra and distribution of fleas of small mammals in Swedish Lapland. Opusc. Ent. 33:327-358.

Brinck-Lindroth, G. 1972, Subspecific differentiation and distribution of the fleas *Palaeopsylla soricis* (Dale) and *Malaraeus penicilliger* (Grube) in Fennoscandia. Ent. Scand. 3:297-312.

Brinck-Lindroth, G. 1974, Differentiation and distribution of the flea *Amalaraeus penicilliger* (Grube, 1851) in Western and Central Europe, with a description of a new subspecies, *A.p.pyrenaicus.* Ent. Scand. 5:265-276.

Brinck-Lindroth, G. & F.G.A.M.Smit 1973, Parasitic nematodes in fleas in northern Scandinavia and notes on intersexuality and castration in *Amphipsylla sibirica* Wagn. Ent. Scand. 4: 302-322.

George, R.S. 1974, Provisional atlas of the insects of the British Isles, 4, Siphonaptera. European invertebrate survey. 12 pp.

Hansson, L., J.Löfqvist & A.Nilsson 1978, Population fluctuations in insectivores and small rodents in northernmost Fennoscandia. Zt. f. Säugetierk. 43:75-92.

Magnusson, N.H., G.Lundqvist & G.Regnell 1963, Sveriges Geologi. 698 pp.

Peus, F. 1976, Flöhe aus Anatolien und anderen Ländern des Nahen Ostens. Abhandl. Zool. Bot. Gesell. Wien, 20, 111 pp.

Saint Girons, M.-C. 1973, Les Mammifères de France et du Benelux. 481 pp.

Siivonen, L. 1968, Nordeuropas däggdjur. 183 pp.

Smit, F.G.A.M. 1966, Insecta Helvetica Catalogus, 1, Siphonaptera. 106 pp.

Smit, F.G.A.M. 1969, A catalogue of the Siphonaptera of Finland with distribution maps of all Fennoscandian species. Ann. Zool. Fenn. 6:47-86.

G. HOSIE
Department of Agriculture and Fisheries for Scotland, Glasgow, UK

OBSERVATIONS ON THE OCCURRENCE OF *CERATOPHYLLUS GALLINAE* AROUND NEW HOUSING ESTATES IN THE WEST OF SCOTLAND

ABSTRACT

Ceratophyllus gallinae (Schrank) dispersing from the nests of sparrows (*Passar domesticus*) and starlings (*Sturnus vulgaris*) in housing estates less than five years old have attacked people out of doors and proved difficult for local authorities to control though no difficulty was found in controlling indoor infestations of *Pulex irritans* L., *Ctenocephalides felis* (Bouché) and *Ctenocephalides canis* (Curtis). Affected areas outside the houses presented few physical barriers, unlike older property with nests, but no complaints about fleas. Such established nesting sites may have provided a source of fleas and contributed to a more rapid build-up of infestation in nearby new property. Fleas sometimes emerged from nests, built during construction of the houses, that were no longer accessible to the birds.

Complaints were not always received from areas known to be infested, because some people were reluctant to admit that they had fleas around their house and others were not sufficiently badly affected by the bites to complain. Effects of bites varied considerably between individuals from mild to severe irritation. Out of 15 complaints involving *C.gallinae,* eight were made during May, the rest occurred from March to July.

C.gallinae active in March and reported active in November were from nests near sources of heat.

Control measures carried out without proper identification to species failed because the wrong hosts – cat, dog, man were suspected and the true source – bird nests – went untreated. Once the true source was located and treated, control was effective.

1 INTRODUCTION

These observations were made following requests for advice, by Local Authorities, when they experienced difficulty controlling fleas attacking people out of doors.

Control measures using insecticidal dusts and sprays, though normally effective against fleas indoors, were not proving effective against these outdoor infestations. Reports were appearing in newspapers and something had to be done.

2 COMPLAINTS AND INFESTATIONS

Discussions with the Local Authority Environmental Health Department staff indicated

Table 1. Flea infestations notified to DAFS Infestation Control

Notif. no.	Date of notif.	Identification	In-doors	Out-doors	Remarks
		Period 1/1/66 to 31/8/70			
1	March 1966	*Ceratophyllus gallinae* (Schrank)	o		Old house – dog suspected by local authority
2	May 1966	*C.gallinae*		o	Attacking children playing in garden – house four years old – dog suspected
3	May 1966	*C.gallinae*		o	Infesting recently cleared area of ground – 40,000 ft^2
4	June 1966	*C.gallinae*	o		Office infested from starlings' nest
5	April 1967	*C.gallinae*	o		Old tenement flat
6 *	May 1967	*C.gallinae*		o	New house – owner attacked several times when working in garden. First complaint April, still there in June
7 *	May 1967	*C.gallinae*		o	Infesting gardens and paths outside new houses
8 *	May 1967	*C.gallinae*		o	Infesting gardens and paths outside new houses
9	July 1967	*C.gallinae*		o	Children attacked in grass paddock of agricultural small holding
10	May 1968	*C.gallinae*	o		Private house – specimen from Servicing Company – no further details
11 *	April 1969	*C.gallinae*		o	Infesting garden of new house built two years, but only occupied ten months
12 *	May 1969	*C.gallinae*	o	o	Infesting houses and gardens on a new council estate, poultry farm nearby
		Period 1/9/70 to 30/4/72			
13	Oct. 1970	*Pulex irritans* L.	o		Factory infested
14 *	April 1971	*C.gallinae*	o	o	Attacking children in garden and bedroom of new house
15	May 1971	*C.gallinae*	o		Specimen from old tenement house
16	May 1971	*P.irritans*	o		House – persisting on dog
17 *	March 1972	*C.gallinae*		o	Attacking people in garden of new house. First noticed in November 1971 and started again February 1972. Poultry farm nearby

that they received complaints of fleas regularly, but they presented no special problems. The complaints were normally about fleas indoors. The staff rarely bothered to get them identified to species, assuming that they were human, cat or dog fleas, depending on which animals were available. It was only when fleas occurred out of doors that they failed to achieve lasting control and so sought specialist advice.

Records of flea problems notified to us confirmed this – see table 1. Out of 17 infestations of fleas notified to us over a period of six years involving four Local Authorities in the west of Scotland, 15 were *Ceratophyllus gallinae* (Schrank) – The Hen Flea. In ten of these cases, the fleas were attacking people out of doors. The other two infestations were of *Pulex irritans* L. and both were indoor infestations.

To check if *Ceratophyllus gallinae* was always more difficult to control than other species, fleas from all the complaints by one Local Authority were identified to species. This was continued for a period of 18 months – see table 2. They received 21 complaints during this time involving human, cat or dog fleas but no *Ceratophyllus gallinae.* All the fleas had occurred indoors and all had been easily controlled.

Because this Local Authority was exceptional in that it consisted almost entirely of built-up areas, the records of another involving more rural areas were also checked for roughly the same period – see table 3. This Local Authority had received complaints involving *Ceratophyllus gallinae* in the past and asked for our advice to control them. Twenty-four complaints had been received during this period, none had been identified to species, but none had occurred outside and none proved difficult to control. It seemed probable therefore that none had involved *Ceratophyllus gallinae.*

Table 2. Flea complaints received by one local authority between 1/9/70 and 30/4/72

Complaint no.	Month	Identification	In-doors	Out-doors	Remarks
1	Oct.	*Pulex irritans* L.	o		House – flat
2	Oct.	*P.irritans*	o		House – flat
3	Nov.	*P.irritans*	o		Old house
4	Dec.	*Ctenocephalides felis* (Bouché)	o		Old property
5	Dec.	*P.irritans*	o		House – flat
6	Jan.	*P.irritans*	o		House
7	Feb.	*P.irritans*	o		House – flat
8	April	*C.felis*	o		House flat – new multi-storey – on cat and dog
9	April	*Ctenocephalides canis* (Curtis)	o		House – flat
10	May	*C.canis*	o		House – flat
11	May	*C.canis*	o		House – flat
12	June	*C.canis*	o		House – flat – old
13	July	*C.felis*	o		House – flat
14	July	*C.canis*	o		House – flat – old
15	Aug.	*P.irritans*	o		House – flat
16	Aug.	*P.irritans*	o		House – flat
17	Sept.	*C.felis*	o		House – flat
18	Sept.	*P.irritans*	o		House – flat
19	Nov.	*P.irritans*	o		Office
20	March	*P.irritans*	o		House – flat
21	April	*P.irritans*	o		House – flat

Table 3. Complaints recorded as 'Fleas' by local authorities in one county between 26/7/70 and 4/2/72

Complaint no.	Month	Age of house in years	Complaint no.	Month	Age of house in years
1	August	4	13	June/July	10
2	July	over 20	14	August	Not known
3	September	10	15	June	10
4	September	over 20	16	July	10
5	November	Not known	17	September	20
6	December	10	18	October	10
7	January	10	19	September	20
8	November	20	20	September	5
9	November	20	21	Sept./Oct.	20
10	April	10	22	Oct./Nov.	10
11	May/June	Not known	23	Oct./Nov.	Not known
12	June	10	24	Nov./Dec.	10

Complaints of fleas received by these Local Authorities involved human, cat, dog and hen fleas, but it was only the hen fleas that they were finding difficult to control effectively.

2.1 Outdoor infestations

Of the 15 complaints concerning *Ceratophyllus gallinae* notified to this Department and listed in table 1, seven marked * were investigated in the field. In all of these cases, the fleas were dispersing from sparrow (*Passar domesticus*) or starling (*Sturnus vulgaris*) nests and were occurring outside the houses in gardens and on footpaths where people, often children, were being bitten. All the houses were new, i.e. built within the past five years and gardens and surrounding vegetation were still not well established and provided few physical barriers. This was in contrast to older property sometimes nearby where there were long-standing nesting populations of sparrows and starlings and presumably fleas. The gardens here were well-established with thick hedges and dense vegetation and so far as was known no trouble with fleas had been experienced. However, it was thought that these established sparrow and starling nesting sites and possibly rookeries in some cases may be contributing to the trouble in the infested area in two ways.

First – by direct migration of fleas. The fleas will be more obvious and move more easily over ground with few physical barriers, as was found around the new infested houses. For example, at Nos. 7 and 8 there were old established nesting populations of sparrows and starlings nearby and few physical barriers between them and the infested gardens. No nests could be found on the houses themselves and it seemed probable that the fleas were migrating from the old nesting sites.

Second – by providing a readily available source of fleas for the infestation of new nests in the new property and so possibly resulting in a rather larger population developing more quickly than usual in the new nests. For example, at Nos. 12 and 17 there were obviously infested nests in the houses themselves but nearby poultry houses may have been a source of these infestations.

In some cases, for example, Nos. 8 and 17 the fleas started to give trouble within a few weeks of the houses being occupied. In each case, the birds' nest had been built during construction of the house and the fleas had emerged since its completion. In such cases, it can be difficult to find the source, especially where, as at No. 17, the birds can no longer reach the nest which had been sealed off when the house had been completed. Such a case would not present a recurring problem, but such possibilities had to be borne in mind when looking for possible sources of fleas.

It was sometimes obvious that several houses in the neighbourhood of those from which complaints had been received must have been similarly infested but the occupants had not complained. There appeared to be two main reasons for this.

First – the natural reticence of some people to admit they had fleas about their house.

Second – the degree to which some individuals were affected by the bites. Some hardly noticed them, others developed red spots and varying degrees of transient irritation, and yet others were affected by swelling of the bitten areas and intense irritation.

2.2 Period of complaints

All the complaints of fleas involving *Ceratophyllus gallinae* occurred from March to July – see table 4. On the one occasion they occurred out of doors in March (No. 17, table 1), the nest was situated in the wall cavity adjacent to an electric storage heater with the only outlet at the eaves. The wall was warm and this may have been the reason for the fleas emerging earlier than usual. This house owner said he had also seen fleas in November the previous year when the storage heater had been operating for a few weeks but there had not been many and they had not been particularly troublesome. Notifications involving fleas out of doors were received by us mainly during May, but fleas had probably been active for some little time before that. In most cases, the activity of the fleas probably did not precede notification by more than one or two weeks and the fleas were certainly still active when we were notified. We do not have any information on the total period of activity in most cases, but they were found to be active for two to three months on at least two occasions.

2.3 Control measures

The main reason for these observations was to enable us to give effective advice on how to control infestations of fleas out of doors. The obvious mistakes made had been that the fleas had not been properly identified to species and the wrong hosts had been suspected and as a result treatment had been directed to the wrong places. Once a proper identification

Table 4. Total number of complaints per month for each species of flea between 1/1/66 and 11/4/72

Species	Jan.	Feb.	Mar.	Apr.	May	June	July	Aug.	Sept.	Oct.	Nov.	Dec.
Ceratophyllus gallinae	–	–	2	3	8	1	1	–	–	–	–	–
Pulex irritans	1	1	1	1	1	–	–	2	1	3	2	1
Ctenocephalides felis	–	–	–	1	–	–	1	–	1	–	–	1
Ctenocephalides canis	–	–	–	1	2	1	1	–	–	–	–	–
C.gallinae – indoors	–	–	1	2	4	1	–	–	–	–	–	–
C.gallinae – outdoors	–	–	1	2	6	–	1	–	–	–	–	–

had been made and the true source of the fleas located, control presented no special problems. Removal of the nest so far as possible, treatment of infested areas with insecticide and finally proofing of the nesting site to prevent re-entry of the birds was usually effective.

REFERENCES

Rothschild, M. & T.Clay 1952, Fleas, Flukes and Cuckoos – A Study of Bird Parasites. London.

Smit, F.G.A.M. 1957, Siphonaptera – Handbook for the Identification of British Insects. Roy. ent. Soc. London 1(16):1-94.

Thompson, G.B. 1958, The Pests of British Birds and Mammals XXXIII – The Insect Ectoparasites of the House Sparrow. Ent. Mon. Mag. 94:1-5.

Waterston, J. 1910, On Some Habits and Hosts of Bird Ceratophylli Taken in Scotland in 1909. Proc. Roy. Phys. Soc. Edinburgh 188: 73-91.